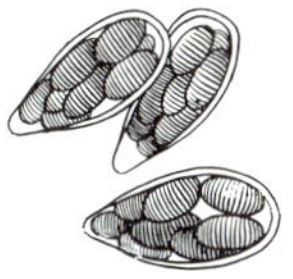

An introduction to plant biology

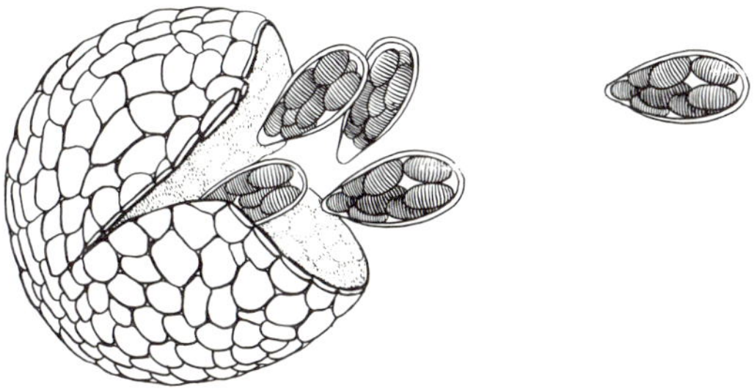

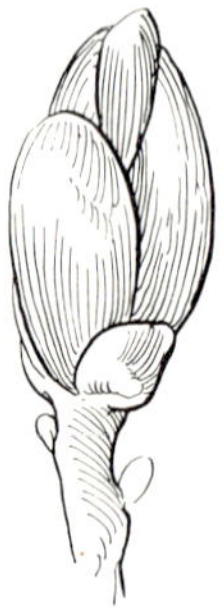

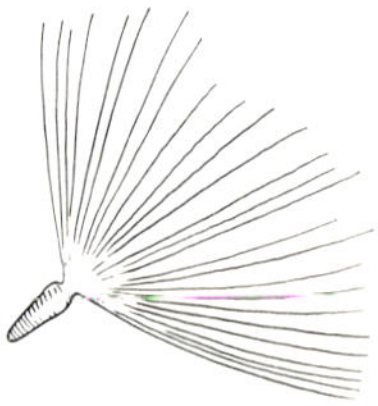

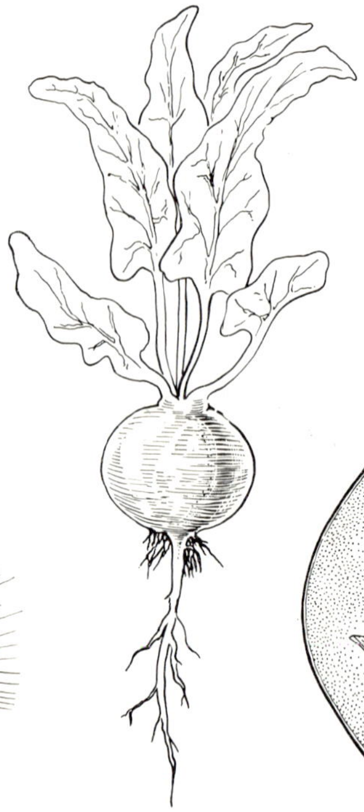

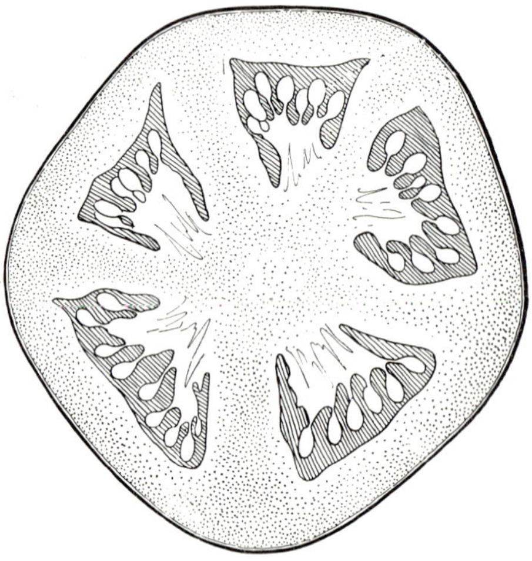

An introduction to PLANT BIOLOGY

ROSS H. ARNETT, Jr., Ph.D.
Professor in the School of Agriculture,
Purdue University, Lafayette, Indiana

DALE C. BRAUNGART, Ph.D.
Associate Professor of Biology,
The Catholic University of America,
Washington, D. C.

Third edition
with 429 illustrations

Saint Louis
THE C. V. MOSBY COMPANY
1970

Third edition

Printed in the United States of America

Standard Book Number 8016-0315-3

Library of Congress Catalog Card Number 75-108299

Distributed in Great Britain by Henry Kimpton, London

To Mary E. Arnett

and

Mary E. Braungart

For demo where plants are Copy

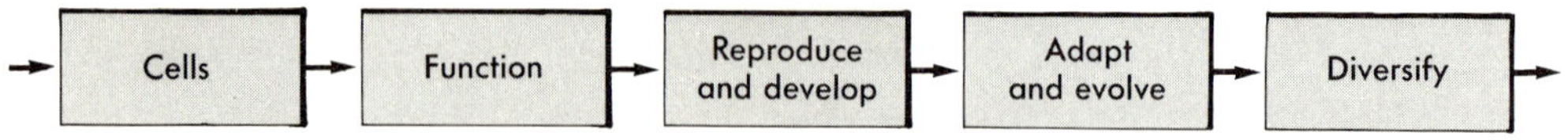

All living organisms are the result of a web of interactions. These life systems, with simple beginnings, evolve to greater complexities of relationships, each with its own life line.

Plant life systems are seen as:

- 1. **chemical** combinations with specific physical properties in the form of **cells,** which, with energy input
- 2. **function** in an **environment,**
- 3. **reproduce** and **develop,**
- 4. **adapt** and **evolve** in time, **distributed** over the earth to form
- 5. the great **diversity** of the past and the present.

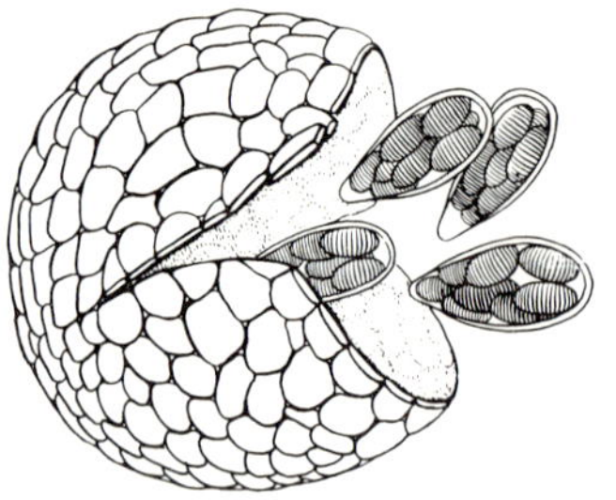

Preface for the student

The world in which we live is sometimes called "Man's World," but we cannot call it ours—yet, not until we know everything on it and, most of all, how to treat it properly. As knowledge conquers fear, so does it lead us to an appreciation of beauty and through beauty to relaxation and enjoyment.

As you read this book we hope you will learn to see more of the beauty about you—beauty in organization, in detail, in function, and in development. The selfish destroy all beauty around them. The lazy become overwhelmed by detail. The truly educated usually find true happiness in beauty and in detail.

President Kennedy once said that learning is like sin—the more you do it, the easier it gets. We have tried every means we know to make this book interesting and easy to use. The learning is up to you.

So many new developments have taken place during the last decade that a complete rewriting of the book was necessary to keep it within its present scope and to maintain a proper balance of subject matter. In fact, there has been a revolution in biology as in other aspects of life. Molecular studies related to living processes have so changed botanical sciences that earlier books can hardly be recognized as covering the same subjects. We debated for some time whether we should actually change the title of the book, but finally agreed that since our purpose and the major subjects remained the same, we would keep the present title.

Once again we are grateful to the many teachers who took the time to write to our publishers with corrections and suggestions. Very careful consideration and study have been given to each. We have gone to the monographic literature, rechecked each statement in the book, and made any modification necessary to modernize this text.

Great efforts were directed toward improved learning aids by the addition of flow charts, new diagrams, and new illustrations. These we believe will be very useful. To help the learning process, we have completely reorganized the text to follow the theme on page vi of the book. This is followed through by a summary statement at the beginning of each of the six parts of the text. Usually the first paragraph of each chapter contains a summary of its contents, and each section likewise has a summary. Scattered throughout the text are "study boxes" designed to review each topic discussed in detail. All these features are intended for quick reading, quick review, and rapid learning. The user may go very rapidly through the entire book reading highlights, he may move more slowly reading summaries, or he may spend time reading the details. In this way we have presented a "world's fair" of plant science, where one may spend a day or a semester or, upon completion of this study, decide to spend a lifetime.

When you have finished using the book, we hope you will be better able to call earth "Man's World" because you will have learned more about it and will want to treat it as if it were your own.

Ross H. Arnett, Jr.
Dale C. Braungart

Contents

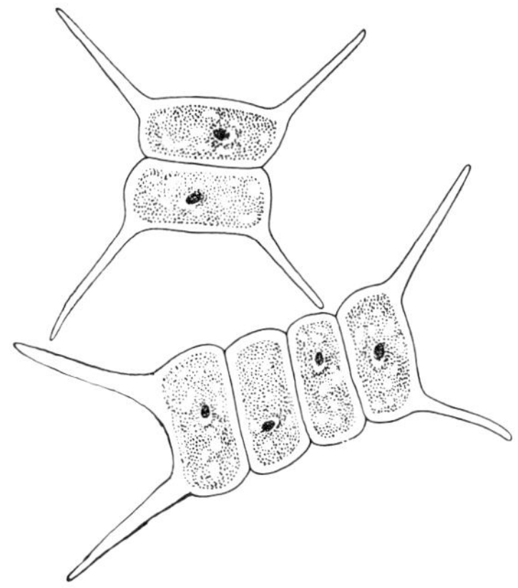

Part I Organization of data

Part II Chemicals and cells

Part III Function and environment

Part IV Reproduction and development

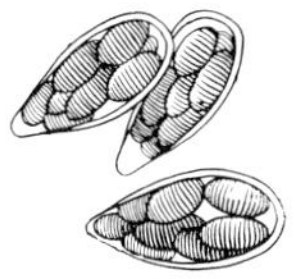

An introduction to plant biology

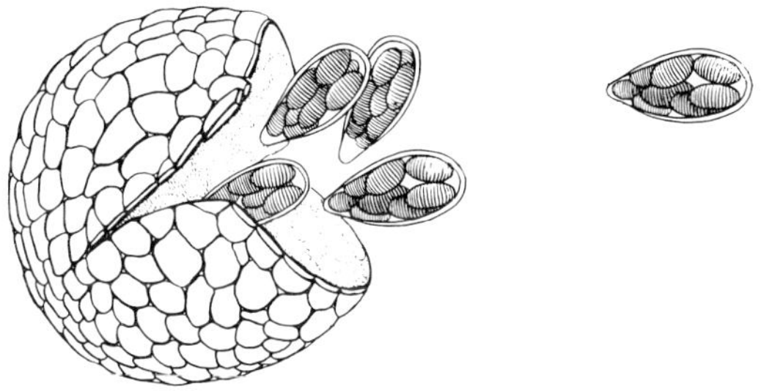

Part I

Organization of data

- Man and many other animals observe and react. Only man asks "why?"
- The search for "why" has led from random to organized observation, then to experimentation, and finally to a vast system of data processing.
- Data are factual statements. Ideas are tentative associations or classifications of factual statements—the hypotheses of the scientist.
- Tests verify data by repeated observation and measurement, using associated ideas to arrive at a concept.
- Concepts, the relationship of facts, as theory, show the significance of data in the life of man, in his environment, and in his other ideas.
- This is the scientific method.

Chapter 1

Introduction

No man-made structures can be seen on the earth in the above picture taken from a satellite. At this distance our thousands of miles of highways and our great cities are not visible, nor is the plant cover we know to be the dominant feature of the earth's land surface. But we know the plants of the hills, the valleys, the mountains, the plains, and the deserts are there, spread out before us in this magnificent, detached view of our world, all necessary for the survival of man and animals in a delicately balanced biological economy. Conquering space is one frontier; survival in a crowded world is another.

The need for more knowledge is obvious, and equally obvious is the great amount of complex **information** gathered and recorded by humans during the approximately 5,000 years' accumulation of records. So vast are these records that they are rapidly burying us in a sea of paper—ironically, a plant product. To efficiently gather more knowledge, without repeating what is already known, we must know what we now have. As a result, our systems of organization of **data** are changing from simple indexing systems of days gone by to modern methods of **data processing.**

Primitive humans, forced to act either positively or negatively to their observations of environmental changes and the conclusions they drew, had little time to ask "why?" Once they learned to shelter themselves, to plant, and to store, they began to contemplate the many aspects of their environment. Thus, in some long forgotten cave, there began on an unrecorded day the rough **scientific method,** the basic technique we still use. At the same time the animal became man, the rational being.

SOURCES OF SCIENTIFIC INFORMATION

The natural scientist depends on observations as the source for all the information he uses. From these observations made directly upon the subject, or indirectly with the aid of instruments, he is able to form concepts and discover the organization of the matter he is investigating. Science, however, does not consist of research alone, and scientists are not just researchers. Rather, science is a way of thinking based on logic, and scientists, like other people, must think as well as observe. These apparently simple statements will be developed in detail.

INFORMATION SOURCES

Observation and experimentation
Specimens and samples
Literature and related stored records

The modern scientist enters gradually into his particular field. As a student he begins to restrict his professional interest and delve deeper into a branch that attracts his attention. Through advanced studies and association with teachers and research workers he begins to master a subject. This requires extensive reading as well as practical experience in the laboratory and in the field. At the proper time he becomes a specialist in a particular subject, supported by a broad knowledge of science in general. So equipped, he may enter into a life of research, development of process or product, teaching, or a combination of these activities.

The ultimate and therefore the most important source of information for a scientist is observation. The matter of observation and **experimentation** is too complex to be treated at once; we shall return to this later. Still it must be kept in mind that observations are recorded from specimens and samples and that the published records are the literature and related documents that form the storehouse of human knowedge.

The biological sciences utilize as natural study materials, specimens, either as whole organisms living or preserved or as samples of materials taken from these organisms. The botanical sciences use, as specimens, plants collected in natural growth areas, grown in the greenhouse, or preserved in herbarium collections. Biochemistry, molecular biology, and **physiology** are examples of investigations requiring samples extracted from parts of living plants. Other studies, particularly **cytology** and **histology,** use microscope preparations of plant parts. All these materials are gathered and carefully selected for the accurate and detailed information they will reveal.

Printed information distributed to libraries and individuals is referred to as literature. Not all available information is stored as **literature.** Photographs, movies, slides, and charts, some of which may be available in duplicate copies and some only as originals, are also information sources. Modern data-processing equipment uses punch cards, tapes, and iron cores to store data for easy retrieval. The latter method of information storage has developed very rapidly and is now widely used. Because the term literature has restricted meaning, the term **record** is now used to refer to any durable packet of information used in the communication of information. Of great concern is an efficient system for the storage and retrieval of records. Records suitable for storage that utilize recently developed automatic data-processing equipment are referred to as **source documents.** These are converted to punch cards, electronic tape, or iron core for storage and later retrieval. Such a system is known as an **information retrieval system** (IRS).

KINDS OF DATA USED IN SCIENCE

Scientific progress, practically speaking, is the result of (1) the organization of the available information and (2) the active search for supplementary data to complete the information necessary to demonstrate a fact. It is not by chance that many of the great discoveries of the past have been made by several persons almost simultaneously. Once a certain amount of information becomes available, it is natural that the educated minds of several scientists will piece together the separate facts and test these ideas, often in different ways but at about the same time, by gathering additional facts to support their ideas until enough information is available to write a report of their discovery. Such was the pattern of discovery of the evolutionary processes. This, the scientific method, is considered in further detail in the section on the technique of data organization.

KINDS OF DATA USED IN PLANT SCIENCE
Observational
Mensural
Experimental

Observation data are records of information gathered by human senses, generally unaided or with the help of optical instruments. Light microscopes and cameras are generally the aids used for gathering these data. For the plant scientist, information about the shape, size, structure, and color of plants, where they live, and what other kinds of plants and animals live with them constitutes the recorded data.

Measurements of weight, lengths, and proportion add to the information about plants. Sometimes complex measurements are necessary when making physiological studies. Oscilloscopes and other electronic instruments record mensural data. Computers may be used for the rapid processing of the great amount of mensural information that may be gathered.

Once enough observational and mensural information about a plant or a process is available, it may be possible to conduct experiments leading to the discovery and proof of many facts. Experimental data are gathered to demonstrate growth processes, biochemical actions, reproduction, and genetic change. The validity of observational and experimental facts is demonstrated if, after repetition under the same conditions, the results are the same.

OBSERVATION AND EXPERIMENTATION

Even though basic studies usually begin with observational data, there is no real distinction between these and data derived through experimentation. Observations may meet all the requirements of an experiment. This may at first seem to be an obscure statement, but as generally understood, an experiment is essentially the measurement of phenomena resulting after the stabilization of most of the factors involved and after allowing the variation of one or a few of the others for comparison. Direct observation is a tabulation or recording of phenomena as they occur unchanged in the field or in the laboratory. Refined experimentation is possible first by observations made with or without instrumentation and second upon comparison with other similar observations. The degree of sensitivity of the instruments does not define the difference between the two procedures. Laboratory and most field experimentation involves a **closed system** and a means of obtaining comparative data by the use of an unchanging control. As the system functions, it is compared with the control. It follows that if a scientist makes comparative observations in the field after careful advance planning, he is experimenting. By the selection of two closely related phenomena he is able to make comparisons, and he is therefore an experimenter. Viewed in this light, the distinction between observation and experimentation fades. The essentials for scientific work are careful planning and detailed testing. It is clear that the only difference between ordinary observation and the usual experiment is the lack of a comparative controlled system in the first and its presence in the latter.

THE SCIENCE OF BOTANY

Most of this introductory chapter pertains to science in general, but only selected topics from such a vast study can be covered in any depth. Science may be fractured by considering particular processes or by selecting particular objects for examination. Either way, there is considerable overlap of subject matter between the two, and the edges of any of these artificial sections are fuzzy by nature. The student of quantum physics has an interest in the nature of life, and the physiologist is concerned with the laws of physics.

The layman can easily distinguish plants from animals because he concerns himself with the outer appearance of the higher plants and the higher animals, but, as will be discussed in other chapters of this textbook, the scientific dichotomy of living organisms is impossible. Although it is applied to the study of those organisms ordinarily called plants, the term **botany** has no real meaning, which is the reason the title of our book does not use that term. The long tradition of separating organisms into plants and animals is so firmly established in our colleges, universities, museums, and scientific societies, as well as in our scientific and everyday writing, that it is doubtful if we will discard soon the words botany and **zoology.**

The plant sciences investigate members of the **Monera,** certain **Protista,** the **Plantae,** and the **Fungi.** Thus botany includes all or part of four of the five kingdoms of organisms now recognized, omitting only part of the Protista and all of Animalia. This includes study of the structure, classification, and history of these organisms. Each of these subjects is considered in detail in the appropriate chapters.

Other significant areas of investigation of plant life include genetics, **ecology,** physiology, and life cycle or life history studies. These topics are by themselves large areas of specialization.

> **SUBJECTS INCLUDED IN PLANT SCIENCE**
> Plants and plantlike organisms
> Genetics and populations
> Physiology and ecology
> Reproduction, development, and evolution

Research, using plants as specimens, has both theoretical and practical application —theoretical as the life processes are explored and practical as discoveries are made that relate to agriculture and industry. Universities, government agencies, and private organizations, each with somewhat different objectives, provide the place and equipment for these investigations.

Many coordinated conditions must be present before research of any kind may be conducted. The proper education of the research staff is first. An adequate library with current information easily and promptly available is required for the beginning of any research project. Some research can be done with very simple equipment and by a single individual. Other research areas require sophisticated equipment that is often expensive. The more complex the research, the more likely a team of research scientists will be necessary, and since

World War II, this has been the trend in scientific research.

REQUIREMENTS FOR RESEARCH PROJECTS
Educated personnel
Adequate libraries
Proper equipment
Worthy subject matter
Financial backing

Most scientists will agree that one of the greatest difficulties in doing research is the selection of a worthy problem. All aspects of the approaches to the investigation must be carefully considered and harmonized. The problem should be one that can be handled by the investigator with the facilities available and within a reasonable length of time. Most important of all, the project should be concerned with a topic that occupies a real place in the scheme of nature. Or, to state this in another way, the phenomenon being investigated should be a real one, unknown, and when completed, logically fill a gap in our knowledge of the universe. (See Chapter 3 for the branches of the botanical sciences.)

TECHNIQUE OF DATA ORGANIZATION

The classical description of the scientific method does not adequately describe the process of organization of factual knowledge as it is practiced in the modern, computer-benefited world of science. As usually described, the scientific method is restricted to the philosophical process of gathering observations as data, forming a **hypothesis,** testing the latter, and announcing a new **theory.** Although this has all the elements of today's technique, it lacks working detail, and most important it neglects the idea of **feedback** controls, now a part of the continuing research practiced in today's rapid pace of living.

The **flow chart** (Fig. 1-1) of the modern method of data organization shows how the various phases of research and study are related. The boxes in the chart represent source records, and the arrows con-

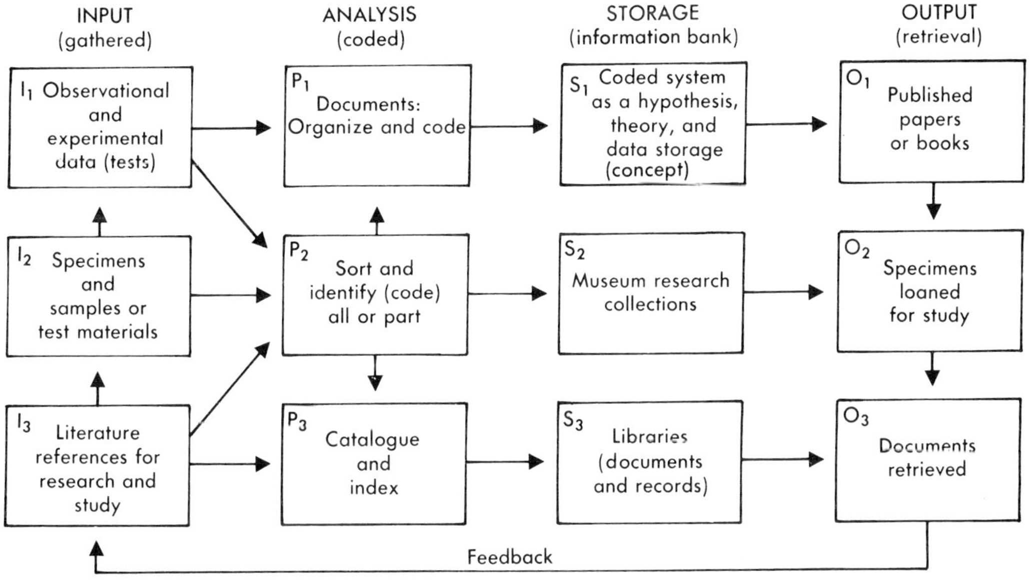

Fig. 1-1
Flow chart showing the stages and pathways of information and data processing.

nect related records or the same records being processed. The term record is used broadly and indicates any information source; it is not restricted to written documents. The simplest statement made to describe this system is "gather, sort, store, and retrieve." Notice the similarity of this to the classical scientific method. The gathering of data is the **input.** The association of the data permits the formation of a hypothesis or coding. The repetition of these first two steps is the testing of the hypothesis. Finally, enough information is associated, stored, and synthesized to formulate a theory, which is made available to all through retrieval as the **output** in some form of a publication. These steps of scientific research are repeated each time new facts are discovered.

A detailed examination of the flow chart (Fig. 1-1) shows, in somewhat simplified form, all aspects of research and data processing treated as a part of research. It should be understood that data processing is used in many aspects of modern life other than in research and that this discussion is restricted to the area of academic study and research.

All input materials are grouped into three categories. Data derived from observation and experimentation are the original or primary source of all information (source document I_1). These are the records in notebooks, on cards, charts, photographs, and tapes that are to be organized. In certain kinds of research, information is stored as specimens or as samples of material (source document I_2). Through the use of preserved specimens, information is stored in near original condition so that additional data may be gathered at a later time. This is noted on the flow chart by the arrow that connects with source document I_1. Finally, information is gathered by a study of the literature, indicated by source document I_3, called literature references. These are the information inputs in a computerized system. Here all the output, if used again, is fed back into the system and used in other stages of the process.

As the input materials are gathered, the processing begins. Each type of material is treated according to its nature, but each is organized following a carefully thought-out code description system (P_1). Key words are used to describe these documents, as, for example, growth, reproduction, genetics, or classification. When dealing with specimens and samples (I_2), these materials must be identified (P_2) as chemicals or organisms. Both the input data (I_1) and the literature (I_3) are used to make these identifications. Identification, therefore, is by code names. These are added to the documents (arrow to P_1) or immediately catalogued and indexed as a literature document (P_3). If these documents are published, they will pass through as source documents S_1 or O_1 on the flow chart. They may be stored as a part of the literature storage without publication, in libraries (S_3) or stored with the collections (S_2). Note that literature reference requests made for input I_3 must be coded for eventual retrieval in libraries. This coding is, of course, the library call number system.

Information now processed is stored for use. Source records are stored either as hypotheses, theories, or coded data (S_1). Prepared specimens (S_2) are stored as collections arranged according to the identification code and according to the theories proposed. Other samples may be stored or discarded. Libraries (S_3) are arranged according to the coding system they have adopted.

Information retrieval is one of the most pressing problems of the modern world. Vast amounts of information have become available through the increased pace of research since World War II. Stored information, unless it can be quickly and thoroughly retrieved, is useless. If the coding system is not adequate, source documents are lost. Therefore output is dependent

Fig. 1-2
Information processing using electronic data-processing equipment at BIOSIS. (Courtesy Biological Abstracts, Philadelphia, Pa.)

upon an adequate retrieval system, of which there are many.

Published papers or books (O_1) are retrieved through the cataloguing system (P_3) used by libraries. Their contents are coded also by means of an index or by key words in the title. Specimens and samples (O_2) needed for further study or reference are retrieved by looking for information under the names of the organisms or chemicals (see also naming plants in the next chapter). Items filed in libraries as books, papers, or typescripts are termed information documents (O_3). These documents, once retrieved by any of the several means available, may be studied, copied, or loaned. Many large libraries have converted to a punch card, tape, or iron core system of information retrieval, using a **computer** (Fig. 1-2) to search for the documents. Abstracting journals, such as the one published for the biological sciences, *Biological Abstracts,* are the result of such a system. Once the information is coded, the computer, properly programmed, will search for the documents. By requesting information using the proper code words, a complete bibliography of the current information on the subject becomes available very rapidly.

GATHERING BOTANICAL DATA

The general aspects of data organization or data processing explained in the previous section have specific application to the study of plants. Specimens mounted on standard-size sheets of paper and stored in cabinets as **herbarium** collections are an important source of botanical information. The classification, distribution, genetics, and evolution of plants are studied by using herbarium collections and the data associated with the specimens. As might be expected, the major collections of plants are to be found in the large cities of the western world. Some of the most important collections are maintained by the Royal Botanical Gardens, Kew; British Museum

(Natural History), London; Museum of Natural History, Paris; United States National Herbarium, Washington; New York Botanical Garden; Gray Herbarium and other herbaria of Harvard University, Cambridge; Missouri Botanical Garden, St. Louis; Field Museum of Natural History, Chicago; and the Academy of Natural Science, Philadelphia. These institutions maintain large botanical libraries associated with the collections, and several taxonomic botanists are employed to care for and do research on the collections.

Other kinds of research, for example, physiological, genetics, ultrastructure, and horticultural, are conducted in national and state laboratories and in the laboratories of many universities. The state agricultural experiment stations are frequently a part of a state university and either physically located on the campus or usually not a great distance away. The staffs of these stations often hold joint appointments at the university, and graduate students may conduct research under the direction of their research staff.

USING BOTANICAL DATA

Data gathered through research are used for the preparation of scientific papers for publication in journals and similar research outlets. This forms the literature resources available in our libraries and laboratories and is used, in turn, as an aid for the identification of plants through descriptions and identification keys. When a new project is undertaken, the literature must be thoroughly reviewed for pertinent background information; thus the reason for the information retrieval procedures.

Literature is available, as previously mentioned, in forms other than journal articles. By understanding the nature of these publications, one can better judge the reliability of a document. There is no guarantee that published information is accurate. In fact, as a rule, it seems that the further the information is removed from the original publication, the less likely it is to be accurate. One may call a report of original research a primary publication. All other publications bear upon these reports; therefore they are secondary publications. Examples of these are manuals for identification, monographs, textbooks, and popularized accounts in magazines, newspapers, popular books, or radio and television programs.

Because of the great amount of published information, abstracts of research have become an important means of keeping up-to-date with research progress. In the United States the most readily available abstracting journal for the botanical sciences is *Biological Abstracts,* published by BIOSIS (Biological Science Information Service) (Fig. 1-2). This publication and the various aids for information retrieval associated with the publication are available in every university and most college libraries.

Most recently, information processing, using the computer machinery described earlier is being developed as a means of searching for data to use in research projects. It seems obvious that the procedure described as the technique for the organization of data is a cycle that yields more information at each revolution. Ultimately this information, in one way or another, affects modern society.

SCIENCE AND MODERN SOCIETY

The impact of **science** on man's welfare through technological advance is obvious no matter where we look about us. Evidence of this may be found even in the most remote areas of the earth, if in no other way than as a discarded can, an unsightly reminder of our triumph over bacterial spoilage and molding.

To detail any of the many advances in recent years in the plant sciences would be beyond the scope of this text, but an understanding of the broad areas of progress and of the way the plant sciences fit into this should be discussed.

Progress in transportation affects botany

in several ways, both practically and in research. Plant scientists are able to travel widely and to collect specimens in areas previously difficult to reach. This has greatly expanded knowledge of distribution and floristics. As a result, additional information has been brought into the classroom so that the student of today is not as provincial as in the past.

The study of **genetics** and **biochemistry** has led to remarkable benefits in crop production all over the world, through greater yields of wheat, corn (Fig. 1-3), and other grains. Biochemical research revealed the type of proteins that give the greatest food value to the grain, and research in the genetics of the plant showed how, through selection and cross-breeding, the crop could be improved.

The details of photosynthesis are explained in Chapter 7, but the importance of plant life is emphasized by pointing out that until recently all sources of power came from plants or from water. Atomic power is now being developed and probably will replace all other power sources. Petroleum and coal represent the greatest source of energy in our modern world. These materials were formed by the process of photosynthesis using energy from the sun, fossilized, and stored as oil and coal deposits beneath the surface of the earth. Scientists estimate that the annual oil formation still occurring today exceeds the amount being used. It is doubtful if this eases any of the current alarm over excessive use of these reserves. The atomic power sources eventually may change our current pessimistic view of the danger of running out of power.

Advances in medicine and the virtual elimination of bacterial diseases through

Fig. 1-3
A new high-lysine corn, discovered by Purdue University scientists, known as opaque-2. This corn has twice as much lysine, a protein building block, as normal corn and therefore greater food value than other corn hybrids. (Courtesy Purdue University Agriculture Experiment Station Information Service.)

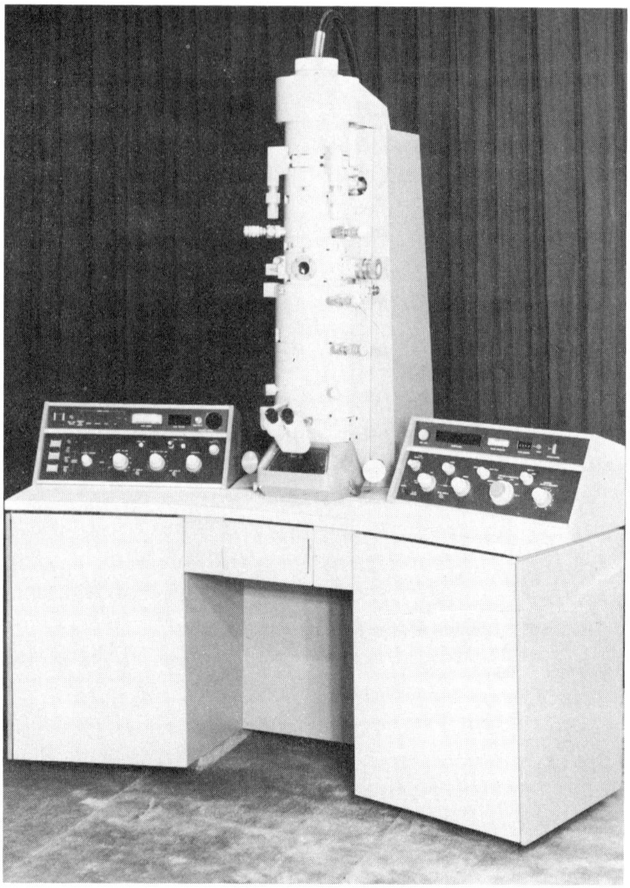

Fig. 1-4
A high-resolution electron microscope capable of viewing macromolecules to 2 Å. (Courtesy Jeolco, Inc., Medford, Mass.)

the use of plant antibiotics increase longevity. Similar advances in technology increase time for research as well as for leisure. The use of computers and automatic machinery may even eliminate any need for manual work. The sociological problems resulting from this progress are well beyond the areas covered by this text.

Not all problems have been solved by science. In fact, it seems that as soon as one problem is solved, another is created. The role of plant sciences is apparent in the following brief account of our current ills.

The population explosion continues to keep pace or exceed increases in crop production so that poverty and undernourishment affect at least one-third of the world's people. The rapid depletion of the soil and other natural resources, mainly through ignorance of the principles of ecology, increases the threat of even wider starvation. Air and water pollution problems follow as a result of this same ignorance and of the added indifference and greed on the part of some who refuse to accept the responsibility for these transgressions of natural balance. Multicolored smoke pouring from factory buildings spells death, signs that should be associated with the trademark of the firms so advertising.

Disease, although now usually not of plant origin, is far from eliminated, and viruses responsible for diseases, for example, the common cold, remain to be controlled. One might even add to the list such social diseases as war and riot, as well as psychological unbalance, which are at least partly due to technological advances and are indirectly related to plant science through the lack of food in the tension areas.

We may learn through the conquest of space that plant life is not restricted to our planet. But for botanists, one remaining frontier is the ocean. Knowledge of marine plant life and of its balance with animal life promises additional food sources to help ease population pressures.

The great problems needing research attention in the plant sciences and related areas are both practical and theoretical. The invention of the electron microscope (Fig. 1-4), which has a magnification power sufficient to bring into view objects as small as 2 Ångström (Å) units, resulted in a great interest in **molecular biology** and ultrastructure. Although progress in these studies has been rapid, much remains to be accomplished. Development and growth of plant life have not received the attention given to parallel studies of animals, so that this remains a potential area of study. Finally, studies that will lead to a better understanding of how life originated, its evolution, how speciation takes place, and eventually even the creation of higher life forms are important modern research areas.

In the minds of some, science is the magic solution to all problems. Even with the remarkable advances possible through science, there are still dangers to man and his world. So far in this discussion we have not attempted to define science. Indeed, such is hardly necessary because our description is itself a definition. But perhaps some distinctions are necessary for a better realization of the possible dangers of science. In a restricted sense, science deals only with facts and their arrangement into a system. So defined, it differs little from technology, which is restricted to the practical use of scientific systems for the development of a machine, product, or process. More broadly defined, science deals with the orderly arrangement of any body of knowledge. Thus the treatment of history, philosophy, and religion may be included in science. Probably the middle ground is to be preferred, so that science is restricted to the knowledge of natural objects, associated phenomena, and theories or concepts about these objects. The systematic arrangement of this knowledge depends upon logic. Modern thinkers have developed a system of symbolic logic that is actually a mathematical treatment of thought processes. This has greatly improved the accuracy of scientific thought and the lucid exposition of scientific knowledge.

Perhaps the first danger of science as it is often presented to the public is the large-scale oversimplification of complex matters. People are led to believe that the discovery of a theory will immediately eliminate one or another kind of problem. For example, the biological control of plant pests is theoretically possible, and the many dangers of insecticides can be prevented. Pest population management is exceedingly complex, and no simple controls can be expected. Even after scientific discoveries are made, as in the use of atomic energy, their practical application may be long in developing.

A second danger is the acceptance with little question of statements made from scientific laboratories. Acting with insufficient data has led to disappointing results and sometimes disaster. One example is the overpopulation of deer in certain areas through the elimination of natural enemies and the overrestriction of hunting. Failure to properly test some drugs before release to the public has resulted in deformities and even death, although these are certainly rare exceptions. The greatest danger of public acceptance of some of the scientific announcements is the false security that may result.

Returning to the dangers of pesticides, probably the most important scientific dangers directly related to plant science are shown by the public indignation that recently erupted mainly through the publication of Rachel Carson's *Silent Spring.* Laboratory developments in pest control are far advanced over public education in their use. The scientist cannot be blamed entirely for the failure of the public to understand that scientific discovery may not be handled properly by the irresponsible. It is obvious that the use of dangerous insecticides, like drugs, should be restricted to those who are qualified to use them. Fortunately, the overreaction to these problems has resulted in further study, correction, and advances in biological control.

The conflict of interests between conservationists and industralists is hardly fitting in today's world of scientific achievement. Yet, like war, the illogical situation still exists, and correction comes only at a snail's pace. Unless water and air pollution and soil and forest depletion cease almost immediately, the prospect for future generations seems grim. Without intending to proselytize, we cannot fail to point out that only through wide public and professional understanding of the principles of **ecosystems** can balances be reestablished at a more rapid rate. One way to bring about this understanding is through the study of the plant sciences at the college level, where large numbers of young people can discuss in some depth the facts behind the problem.

Widespread alarm and indignation have been expressed over the increased use of computers in daily life. Where at first they were used to solve lengthy and complex mathematical problems with speed far beyond any previous ability, they now enter the daily lives of each of us, not as a research tool, but as a machine seriously robbing us of our human freedom. This newest danger of science does not differ in principle from any of the others. The solution rests in the acceptance of responsibility. Manual labor is not replaced only by skilled labor. Responsible people who know the need for accuracy are required as well. No computer yet invented will assume the responsibility of accurate input, and none are capable of making decisions requiring moral judgment. Only the operator can be held responsible for the output.

Science of the future will differ from that of today only in detail, refinement, and perhaps speed of discovery. People are inclined to complain when science seems too slow in solving every problem they pose. So many discoveries have been made so rapidly that they have flooded the market, and we accept this with the weary nonchalance of overindulged children. But this is only in its technological advances. The computerized world of the future, with its space travel, controlled environment, disease-free populations, and almost complete leisure time, will be boring indeed to minds. Overindulgence, too, will bring its problems. Hopefully the mind will turn more to pure intellectual pursuits. The heroes of battle will be forgotten, and in their place will be heroes of theoretical discovery.

Scientists differ from people with other interests and professions only in the zeal for organizing knowledge into systems. Certain qualities seem to dominate in men of science as in many with other professional interests. A high intelligence is noted in all academic pursuits, coupled with a passion for accuracy. Most scientists seem to enjoy what they do, and early retirement and change of work does not seem to be their hope. Instead they prefer to work long hours to complete a project and find their reward in the acceptance of their discoveries and in the sympathetic understanding of their colleagues. This has resulted in the public view of the scientist as a man aloof to social contact and unaware of today's social problems. Lack of communication between scientists and the public has sometimes caused great problems that can be corrected only through education.

SPECIAL TERMS USED IN CHAPTER 1

Ångström unit (Å)	information
biochemistry	information retrieval system (IRS)
botany	input
closed system	literature
computer	molecular biology
cytology	Monera
data	output
data processing	physiology
ecology	Plantae
ecosystem	Protista
experimentation	record
feedback	science
flow chart	scientific method
Fungi	source document
genetics	theory
herbarium	zoology
histology	
hypothesis	

QUESTIONS TO CONSIDER

1. Name some of the uses of computers that you know about in your immediate community. What kind of input do they have? What is their output? What, if anything, is changed by this process?
2. Do these computer processes meet the requirements of the scientific method? If not, how do they differ?
3. Some of the scientific experiments, such as those performed in chemistry, have predictable results. Is this likely to be true of experiments in the life sciences? Discuss.
4. In what laboratories with which you are familiar is research being conducted in the plant sciences? What kinds? Is the output practical, theoretical, or both?
5. In the context of the discussion in this chapter, what is pure science?
6. What are he aims of science? Of scientists?
7. What are the limits of science? The dangers?
8. What are the two essential features of an experiment?
9. How would you design an experiment to show the following:
   a. Whether plants use water at night
   b. The proper amount of fertilizer to use for optimum growth of a common house plant
   c. The effect of the length of day on the flowering of a plant
   d. Whether a plant is self-pollinated or cross-pollinated
   e. Whether the absence of a particular kind of plant in your area is due to environmental causes or simply to not having been distributed to your area from its natural range

DISCUSSION FOR LEARNING

1. The system shown in the chart (Fig. 1-1) can be applied to many kinds of investigations. Discuss several such applications; for example, whether the juice of a certain plant is beneficial as a drug; the use of the system to solve a murder mystery; the use of the system as a way to study for an examination.
2. List and discuss some of the problems you feel have been caused by technological advancement through scientific discovery.
3. Considering the rapid discovery of the past thirty years, list and discuss some of the advances you feel will be made by the year 2000.

Chapter 2

Development of plant sciences

The triad shown in the picture above was millions of years in the making. Extremely slow and gradual changes took place in the evolution of man. About 35,000 years ago the Cro-Magnon man of France settled down to planting crops and painting magic images on walls of caves. Thus began human culture, that mass of behavior human beings in any society learn from their elders and pass on to the young generation. As man progressed he transformed from a rare animal into one of the most numerous. It is difficult to say just how agriculture started, but it seems certain that it was the women who first fenced off a good patch of greens to protect them from herbivores. Progress toward cultivation varied from place to place. In one place, plants were cultivated before animals were domesticated; in another, the reverse. The domestication of plants and animals developed in uncertain steps until recent times. Nevertheless, with each step came more freedom from work. With time to contemplate, plans could be made and crude experiments attempted, and eventually science developed.

The need to recognize plants as sources of food, drugs, building materials, and fabrics led to the classification of plants as one of the early developments of botany and the establishment of plant taxonomy as a science. With the thousands of plant species that are now under cultivation on farms and about our houses it is natural that they should be named and classified. Hardly a house is without an ornamental plant of some kind, making horticulture the most popular recreation activity in today's busy world. Through science and industry, farming as shown in our opening picture is no longer known. Today it is a business akin to manufacturing; tomorrow the actual manufacturing of food from the raw material of the sea will ensure man's future.

EARLY HISTORY

Many early writings show that men have long speculated on their own origin and on life around them in addition to learning how to plant. The search for useful plants, although the beginning of biological science, was not entirely utilitarian. The early development of magic followed by logical reasoning led to a confusing history of human thought. Woven into these thoughts was always the theme of change and development. Contrary to popular belief, the theory of evolution has a history as long as that of all biology. The ancient Chinese, the Babylonians, and the Egyptians speculated about the origin of life on earth. The Chinese in their ancient writings said that five elements were on the earth at the beginning and were the foundation of the universe—water, wood, fire, soil, and gold. Their anthropocentric thinking is obvious. The universe had the elements that man needed. Not recognizing wood as a complex living material but knowing its value to their lives, these writers believed that this substance was so basic that it must have always been present. With these elements they were able to postulate that life came from positive and negative interactions of the primary elements.

The act of water putting out fire is a negative interaction, but the act of fire burning wood is a positive interaction. Water added to the soil produces life, and so on. Thus Confucians and Taoists had some idea of change, descent, complexity, and progress, from the least complex to the more complex.

In their legends of a Creator the ancient Egyptians based their ideas of the origin of life on a first cause. These concepts centered about an abyss of water out of which came life from a state of inertness. These same ideas were proposed in Genesis. The transition from nonliving to living substance was always associated with water.

In their writings the Babylonians allowed mystical powers to permeate the universe. With this as their theme they went on to develop a classification of organisms that was well advanced. Unfortunately, even with the many thousands of examples of Babylonian writing preserved, we know little about the thoughts of these people.

The Israelites recognized plants for various reasons—as sources of food, as being harmful, or some as even sacred. Many of these observations were recorded on a plant product, papyrus, obtained from a tall sedge native to Egypt.

During the golden age of ideas man began to use his intellect to toy with abstract ideas and found this much to his liking. Thales (ca. 640-546 B.C.) proposed laws about the origin of the universe, including various mathematical studies on the subject. He recognized the basic nature of water. The whole universe, he said, was composed of water, and he thought of it as but another substance that permits life, believing that the earth floated on water.

Anaximander (611-547 B.C.) proposed the limitlessness of space, which now opposes Einstein's theory of finite space. Anaximander believed that living things arose from mud and recognized an order in them—first plants, then animals, and finally man.

Anaximenes (588-524 B.C.) thought only in terms of spontaneous generation. He agreed with Anaximander on the concept of the limitlessness of space and believed that air was the universal medium. Heraclitus (ca. 540-475 B.C.) proposed the theory that fire produced the changes in the world, and as a result, all things were in constant motion or in a state of ceaseless change. So many of the concepts that we take for granted today were important and new ideas in early times.

Parmenides (born ca. 540 B.C.) thought of the world as immutable and uniform: as it always was, so it will be always. This concept led later thinkers such as Democritus (ca. 460-400 B.C.) to inquire into the nature of change and to proposed the term atom as the stable element, but he postulated an infinite number in the universe.

Empedocles (495-435 B.C.) is often called the father of the idea of evolution. His elements were reduced to four—fire, air, water, and earth (Fig. 2-1). Change was brought about by two forces, attraction and repulsion. These forces acted in unison upon the four elements and in this way effected change. He also had a concept of adaptation for survival, which was basic to Darwin's theory—once the sexes were separate, species arose. This concept shows that the roots of the idea of evolution were well established, even in those early times. Empedocles also believed in the concept of order in nature—plants, animals, and man.

Anaxagoras (488-429 B.C.), during the same period as Socrates and Democritus, thought of matter as originating from an infinite number of "seeds." Living things sprang from atmospheric germs, not from terrestrial earth. Hippocrates (460-370 B.C.) is remembered as an ancient who practiced the art of healing by use of plant drugs. Today Hippocrates is considered the Father of Medicine, and young physicians take what is called the Hippocratic oath upon their graduation from medical school. Hippocrates also taught that there were four elements: earth, air, water, and fire. He also believed in four humors: blood, yellow bile, black bile, and phlegm (Fig. 2-1). Hippocrates taught that health depended upon the proper balance of these humors. From him we get the term "good humor" and other expressions.

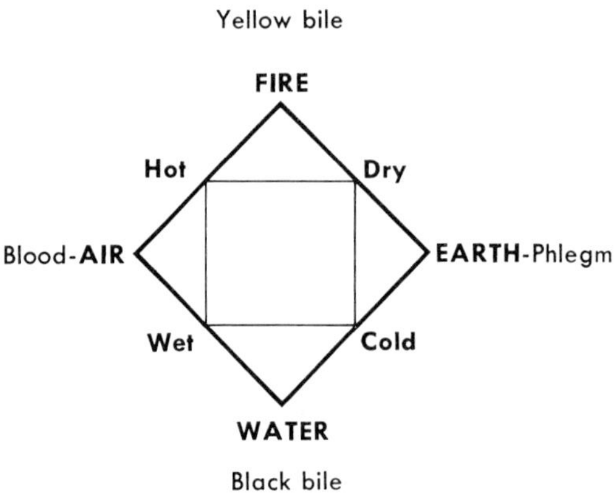

Fig. 2-1
Relationships of the elements according to Empedocles and the humors according to Hippocrates, ancient philosophical and scientific ideas.

Aristotle (384-322 B.C.) was a natural philosopher whose careful observations formed the basis of his many biological writings, most of which, however, were devoted to animals. It was he who first noted that sunlight was necessary for the formation of the green coloring matter (chorophyll) in plants. He recognized a hierarchy of being. One reason the theory of evolution took so long to be recognized and accepted was that the basic ideas previously proposed were not absorbed by Aristotle. If they had been, it is likely that the idea of evolution would have developed much sooner because Aristotle's texts were used exclusively for many hundreds of years. His classification of plants is essentially very close to that used today, but he did not directly classify animals and plants. Later compilers arranged his ideas into a classification. He knew that anatomy was necessary for classification or that classification was based on anatomy. He recognized the individual, the species, and the genus as used in the modern logical sense. This concept was, philosophically, part of his four predictables. He considered genera as a category, the members of which have many parts in common, and that it was not necessary to repeat all of these in a description. This was, in effect, the binomial system, but it was not again used consistently until 1753 was Linnaeus restated it. Linnaeus was a great reader of both Aristotle and the Bible and quoted from both extensively. The genus that Aristotle recognized is equivalent to the modern category of "class," but his categories could change, as do those used in logic. Now, under the Linnaean system, the various categories have been stabilized. Aristotle also recognized that species exist and that they have reality in nature, whether this is recognized or not.

Aristotle was a believer in spontaneous generation. His writings may give the idea that he was an evolutionist, but only when they are taken out of context. Certain statements he made might indicate that he had a real concept of evolution, but he did not. He arranged organisms in a seemingly evolutionary order, from the least complex to the more complex, which in modern terms is evolutionary. His relationship was mathematical. Even though he believed in the immutability of the species, he also believed that motion is imparted to the universe by an immaterial first cause. First causality is responsible for all change, he said, thus absolving himself from all credit for evolutionary thought.

Formal botany had its origin in the efforts of Theophrastus (370-287 B.C.), who was an apt pupil of Aristotle. Much of the scientific data collected during the campaigns of Alexander the Great were used by Theophrastus to learn about plant functions. He also described the plants brought back by Alexander and attempted to classify them. Theophrastus distinguished bulbs, tubers, and rhizomes from true roots and, in addition, had an understanding of the sexual reproduction of the higher plants. All the information Theophrastus accumulated was lost to the world until it was recovered by Andrea Cesalpino (1519-1603) during the Renaissance. Around 30 B.C. the art of agriculture was described in Vergil's Georgics. Another book, by Varro (117-27 B.C.), appeared, in which he described the growth of plants and also suggested that contagious diseases were due to microscopic plants. Finally, about the middle of the first century A.D., Dioscorides (ca. 50-100), who was both a botanist and a military physician, wrote a treatise on plants and pharmacy that lists some 600 plants and their medical properties. Botany therefore had its beginnings in the observations and curiosities of the ancient philosophers.

During the second half of the first century learning was revised mainly through the efforts of Pliny the Elder (A.D. 23-79), a Roman. His thirty-seven books on natural history were the encyclopedias of science of the period and included the knowledge and beliefs of the forgotten writers of Greece and Rome. This encyclopedia included a discussion of the general theory of the universe, of the stars in space, and of the earth and its contents. He preserved for all times the superstitions of his era and recounted in good faith the practices and uses of all forms of magic.

Unfortunately, Pliny, a poor compiler, not an original researcher, passed on much misinformation, some of which persists even today. For instance, it is he who is responsible for the idea that every plant and animal must have some use directly or indirectly for man. This idea is frequently expressed by the nonbiologist when confronted by some particularly strange and objectionable organism.

During the Dark Ages, from A.D. 400 to 1000, few new ideas were added. This was, biologically, the age of the herbalists. The Arabs reintroduced the work of Aristotle to the western world about A.D. 1000, and his work remained unquestioned through the Middle Ages. Otherwise little remained except the herbals. These were massive books illustrating plants of particular interest as drugs and were the source of information for the alchemists and doctors of the time.

The Age of Scholasticism gives us only two really outstanding men of science who contributed to the field of biology. Roger Bacon (1214-1294) was an English Franciscan monk who was frequently in trouble for his advanced thinking. His outstanding contribution to biology was the invention of the magnifying glass, the most valuable tool of the biologist. Albertus Magnus (1206-1280), a Dominican from Germany, was also far ahead of his time. He wrote many accurate descriptions of plant anatomy.

The use of plants in the treatment of disease led to an awakening of interest in the study of botany, a science that was traditionally practiced in the monastery and convent gardens. The shape of a leaf or the color of a flower was regarded as a symbol of its use as designed for the plant by the Creator.

PLANT SCIENCES AFTER THE INVENTION OF THE PRINTING PRESS

The invention of the printing press in the 1450's permitted the beginning of modern biology, just as it permitted the development of all fields of knowledge. The works of Aristotle were among the first books to be published.

The detailed history of biology after the era of printing began is too involved for anything more here than an account of a few highlights. We have already reviewed the ideas that led up to the concept of evolution. Mainly during these later periods, tools such as the microscope were brought into use so that great masses of data were gathered which formed for the first time the factual basis of biology.

William Harvey's (1578-1657) proposal of a biological maxim *"ex ova omnia"* was forgotten for years. It remained for Redi, Spallanzani, and Pasteur to prove empirically that which seemed evident to the logical mind of Harvey and which seems so obvious to us today: the proof that all living things come from other living organisms. Yet this represents one of the longest, most involved, and most interesting research projects in the history of science. The theory of spontaneous generation held by men during the Dark Ages was finally disproved in the world of the educated. Still, in the dark corners of uneducated minds, there lurks this old misconception.

THE FOUR MAIN ADVANCEMENTS IN THE HISTORY OF PLANT SCIENCES

1. Discovery of the nature of cells
2. Development of a classification system for plants
3. Experiments that led to the knowledge of heredity, genetics, and the nature of DNA
4. Proposal and refinement of the idea of the evolutionary process

In the seventeenth century Jan Baptista van Helmont (1577-1644) showed how plants used water. By planting a willow twig in 200 pounds of soil, he found that it grew; it finally weighed 164 pounds, although the soil lost only several ounces. He concluded that its weight was due to the water he had added and not to the nutrients in the soil. Later, other experiments showed that certain minerals from the soil were also necessary for plant growth.

Francesco Redi (1621-1697) tried to disprove spontaneous generation by a series of interesting experiments showing that substances wouldn't spoil at high altitudes if they had been previously protected from contamination. The final proof came when Louis Pasteur (1822-1895) showed that spoilage was due to microorganisms. He established a new law upon which biologists could build a unified conceptual scheme for biology, and it led to the process we now know as pasteurization. No longer do intelligent people believe that decaying bodies generate flies or that swamps generate disease. Persons who now understand the nature of microorganisms, the production of spores, and the process of decay find it hard to believe in the generation of new lower forms of life from dust and decay. Because of the former lack of understanding of these simple processes, we can see why the theory of evolution took so long to be accepted and why some people still refuse to accept it. Those things that they cannot understand or that they are warned against by persons in whom they trust are not accepted. Once the concepts of biology are elaborated and unless the minds are completely rigid, the logic of evolution should at least be apparent.

The development of the microscope is credited by some to Zacharias Janssen (1590) and by others to Galilei Galileo (1564-1642), who invented an instrument in 1610. More than likely both can be given the credit. Anton van Leeuwenhoek (1632-1723) refined the microscope and described for the first time microscopic plants and animals. In 1665 Robert Hooke (1635-1703) named the empty spaces in cork tissue, calling them cells. Marcello Malpighi (1628-1694) described the details of plant vessels. Nehemiah Grew (1641-1712) in 1682 made drawings of plants in which he depicted cells, although unaware of their structure and significance. It took almost two centuries for someone to realize that cells were not the empty spaces described by Hooke. In 1835 Felix Dujardin (1801-1860) recognized that each cell contained some living material, and he labeled it sarcode. Today we call it protoplasm, a term coined in the early part of the nineteenth century. Both Johannes Purkinje (1787-1869) in 1839 and Hugo von Mohl (1805-1872) in 1846 are given credit for using the term.

Matthias Schleiden (1804-1881) in 1838 and Theodor Schwann (1810-1882) in 1839 independently proposed the cell theory, a theory that maintains that all life comes from cells and is composed of one or more cells as the operational unit. This was one of the first unifying concepts of biology.

After the Renaissance, the growth of wealth and the development of artistic tastes resulted in planned parks and gardens, and there was more general cultivation of various trees, flowers, and vegetables. With this increased love for beauty and the use of herbs as remedies for the sick, as well as a natural curiosity about life in general, the sixteenth century saw a rise in the development of botanical knowledge. Familiarity with many new kinds of plants awakened the desire to catalog and compare them with more familiar ones.

Louis Pasteur first demonstrated conclusively the cause of many infectious diseases as a direct result of his studies on spontaneous generation, which we discussed previously. As a result of this find, Joseph Lister (1827-1912) was able to devise aseptic methods of surgery. Robert Koch (1843-1910) was the first to isolate the bacilli of tuberculosis and establish through his knowledge of these minute microorganisms the basic theory of disease caused by living organisms.

Otto Brunfels (1489-1534) was one of the first workers in modern botany. It was he who introduced the modern concept of the genus as used in botanical nomenclature. Early classifications of plants were based upon utilitarian ideas or on such obvious external characteristics as growth, and these classifications led to the division of plants into herbs, shrubs, and trees. In 1660 John Ray (1627-1705), who was an all-around naturalist, published the first of a series of works on systematic botany. He was able to distinguish between monocots and dicots, using the seeds of these plants in addition to using the flowers, fruits, and leaves, all of which led to a more natural classification of the plants. It was through the efforts of this man that the terminology used by earlier workers passed down to the time of Carl von Linné (Latin, Linnaeus; 1707-1778). It was he who first consistently used the now famous system of nomenclature, the binominal system. This has provided us with a means of cataloging plants (and animals) so that accumulated data may be easily indexed. Over the years it has proved to be a workable scheme of nomenclature that, although not invented by him, used all the best features of preceding schemes, was refined, and resulted in a highly practical system. It is so flexible that it is capable of expansion and adjustment, making it possible to include all known organisms as well as those yet to be discovered. For example, only 5,950 species of plants were known to Linnaeus. Today over 300,000 different species of plants are recognized, all of which can be and have been named and included in the Linnaean system.

By the end of the eighteenth century the process of photosynthesis had been discovered, and the sexual process in plants was also beginning to be understood.

DARWIN TO THE TWENTIETH CENTURY

The publication of Darwin's *Origin of Species* in 1859 marked the culmination of twenty years of thought on his part and centuries of slow accumulation of evidence. Charles Darwin (1809-1882) was responsible for changing the field of biology from an organized mass of unrelated facts into a unified science. Although he did not actually propose the theory of evolution, his theories were responsible for what is perhaps now the most unifying concept of biology—the application of the knowledge of the processes of evolution to understanding the significance of biological processes and organisms. This was a turning point for biology, from which issued modern biology based on the principle that all life on the earth has a common origin and operates under the same physical and chemical laws.

The nineteenth century saw the study of the geographical distribution of plants and the development of a knowledge of the relations between climatic and other environmental factors. The presence of a nucleus and chromosomes in a cell was established, and the method of cell division was discovered. The function of plant organs was determined, thereby establishing plant physiology as a recognized science.

Darwin's theory. In his early life Darwin unknowingly collected the facts he needed, as well as gained the broad experience necessary to see at a conceptual level the processes of life that he was to explain. His voyage around the world from December 1831 to October 1836 as naturalist and geologist on the *Beagle* was the primary source for this experience. During the voyage he took notes on everything that he saw, and these notes later formed the basis for several books published long before his most famous one. Also, he read the works of Reverend Thomas Robert Malthus (1766-1834) on population increase, which suggested to both Darwin and Alfred Russell Wallace (1822-1913) the idea of a struggle to survive. Darwin contributed to the scientific literature enough original research to give him a firm position in the annals of biology even if he had never written the *Origin of Species*. His studies of the formation of coral reefs, of insectivorous plants, and of earthworms are still basic works, although they are seldom remembered because his later work so overshadows them.

At the same time Darwin proposed his theory, Wallace proposed a similar theory. Once again the time was ripe for discovery. One of the great tributes to scientific objectiveness was Wallace's act of withdrawing his paper in preference of Darwin's, and thus giving up a first place in history. The two men collaborated and pooled their ideas to their great credit. Wallace is now best known for his studies of plant and animal distribution.

When his famous book was published, the effect was immediate and widespread. The fashion of the time became the reading of Darwin, a daring act because to many it denied the existence of God. Surely if monthly book clubs had been in existence at that time, social pressure would have demanded that this special selection be read by all, and it probably would have outsold even such popular novels as our modern classic *Gone With the Wind*. Unfortunately, few understood the book, even as today. Although it still remains one of the great books to be read by generally educated people, one should not assume that it represents current thinking on the subject.

Darwin lived to see his work scientifically accepted by most but the brunt of great public controversy at the same time. Nevertheless, Darwin and Newton are buried side by side in Westminster Abbey in London as recognized great men, great scientists, and great contributors to human knowledge. Doubtlessly the theories of both men number among the greatest of scientific achievements.

At the time Darwin's theory was first published, it was known as the "survival of the fittest." It should be noted that he never used the term "evolution." His proposal will be discussed in more detail in later chapters. The following are his main points.

SUMMARY OF THE DARWINIAN THEORY OF EVOLUTION

1. Species change or mutate.
2. There is descent with modification.
3. Variation in individuals permits change by selection.
4. Therefore the best-fitted individuals in a given environment survive; the least fitted perish; "the survival of the fittest."
5. Change is always toward more complex forms, even when parts are reduced or lost through parasitism.
6. Therefore, once a change is made, the new form cannot change back to its original state.

Although each of these concepts had been vaguely presented by others, no one had previously brought them together in one book documented with the detail and logic of Darwin.

In summary, Darwin laid the foundation for all future evolutionary thought. Since then, biologists have been elaborating and refining his theory by adding facts as well as concepts.

TWENTIETH CENTURY BOTANY

Even though the theory of heredity, frequently termed mendelism, was actually published many years before 1900, it remained unused, buried in the literature. Actually our knowledge of the sex of plants grew out of the work of Rudolf Jakob Camerarius (1665-1721) published in 1694. The first intentionally artificial plant hybrid was produced by Thomas Fairchild in 1717. As early as 1779, color blindness in man was reported to the Royal Society of London as being inherited. Other rather unorganized and poorly demonstrated examples of thoughts on heredity could be cited, but none can be considered as contributing to genetic theory.

Mendelism. Gregor Mendel (1822-1884) (Fig. 2-2), an ordained priest of the Au-

Fig. 2-2
Gregor Mendel (1822-1884), the first person to publish on the laws of heredity.

gustinian order, lived in their monastery at Brunn, Austria. In the gardens of this monastery he carried on his epoch-making experiments, which practically ceased with his election to position of abbot in 1868. He is best known for his experiments with the garden pea plant *Pisum sativum*, although he experimented with several other kinds of plants as well as with bees. Mendel was fortunate that he studied characteristics that were conspicuous and easily distinguishable. He was successful because he studied single differences, one at a time as well as in combination with others, because he kept accurate records of all the progeny, and because he chose a plant that was normally self-pollinated and therefore self-fertilized but that also could be cross-pollinated. Self-pollination leads to genetic purity so that each of the plant varieties he used for his experiments initially were genetically pure for the characteristics he chose.

The results of Mendel's pea experiments were published in 1866 in the *Proceedings of the Brunn Society for the Study of Natural Science* but were completely overlooked for thirty-four years. At the time his work was published there was no theory of the germ plasm, no description of chromosomes, and no theory of the gene, and, furthermore, little was known about cell structure and behavior. Mendel's ignorance of what is today common knowledge makes his conclusions about heredity much more significant. Mendel died in 1884, unrecognized and disappointed that he had been unable to continue his investigations. His great contribution to the field of heredity was rediscovered independently in 1900 by three botanists whose research had led to similar discoveries. Hugo DeVries (1848-1935) in Holland, Erich von Tschermak (1871-1962) in Austria, and Karl Correns (1864-1933) in Germany recognized the significance of Mendel's work. Many important research papers followed the disclosure of these discoveries, and what came to be known as the mendelian principles were widely studied in many organisms. Thus Mendel's work led to the modern study of genetics.

SYSTEMATICS

Darwin's theory had more immediate effect on other branches of biology than it did on plant taxonomy, where one might suppose it would have its greatest application. In fact, the most immediate reaction took place in social thought instead of in biology. Not all biologists of the period accepted the theory, and many of those who did were reluctant to change from a pigeonhole type of classification to the natural one made possible by Darwin's work. For one reason, the mechanism was not clear. Gregor Mendel's principles were not available to Darwin. For another reason, the idea accepted by Jean Baptiste Lamarck (1744-1829), that the inheritance of characters acquired by parents was possible, was still prevalent. Lamarck did not actually propose the theory of the inheritance of acquired characters; he simply did not question the concept that was generally believed during his time. Inasmuch as Darwin did not explain the mechanism of evolution, it took time for the old ideas to disappear from the work of the early post-darwinian period.

Much of the development of modern evolutionary thought was the result of German zoologists. It was not until after Mendel's work was rediscovered that botanists took an active part in establishing the theory of evolution. The discovery of the gene theory and work with plants as well as with animals rapidly developed the field of genetics, and Wilhelm Ludwig Johannsen (1857-1929), DeVries, and other botanists took leading roles. By the 1920's the theory of evolution, while generally accepted, was considered in scientific circles to be proved and was no longer of great concern to biologists.

The past thirty or more years have seen

a greatly renewed interest in the subject. New ideas of speciation, new definitions of species, and the development of systematic biology as an important fundamental branch of biology, all basic to the theory of evolution, have arisen so that at the present time books on evolution are appearing at a rapid rate. It may be safely said that since 1937, when Theodosius Dobzhansky published *Genetics and the Origin of Species,* more scientific information has been published on the theory than in all the previous books and papers on the subject combined. Both zoologists and botanists have taken an active part. Among the leading botanical works are E. B. Babcock's study of the genus *Crepis,* Edgar S. Anderson's (1897-1969) work on hybridization, and G. L. Stebbin's work with grasses. A series of books by Clausen, Keck, and Hiesey deals with experimental studies on the behavior of plants when grown in widely different habitats. These experiments opened a whole new field—experimental evolution.

MOLECULAR BIOLOGY

Post World War II biology shows little resemblance to that of earlier years. The invention of the electron microscope by V. Zwarykin in 1939 proved to be a powerful tool for the study of biochemical substances such as viruses, proteins, and other macromolecules. Ever since the laws of heredity were discovered and the genetic agents seen in the nucleus, the greatest biological mystery needing solution was the way these materials duplicated. James D. Watson and Francis Crick sorted through the clues and finally in April of 1953 announced their discovery of the double helix chain of amino acids in the DNA molecule. This firmly established the physical-chemical basis of life and closed the final chapter of the ancient vitalistic theory of life. For this discovery they were presented the Nobel prize in 1962. Their work has completely changed biology from primarily a descriptive science to one that is primarily analytical.

QUESTIONS TO CONSIDER

1. Name four advances in the botanical sciences.
2. What important observation made repeatedly throughout recorded history eventually led to the discovery of the evolutionary process?
3. Which scientists referred to in this chapter made discoveries that relate only to the botanical sciences? Can the history of botany be separated from the history of zoology?
4. What do each of the major discoveries reported in this chapter have in common?
5. Give two reasons for the slow development of science until after A.D. 1000.

DISCUSSION FOR LEARNING

1. How might the history of biology differ if Darwin had not published his book on the origin of species and if Mendel's work had been widely read?
2. Make a chronological list of the major discoveries in biology and discuss how one discovery led to another and another.

Chapter 3

Modern botanical sciences

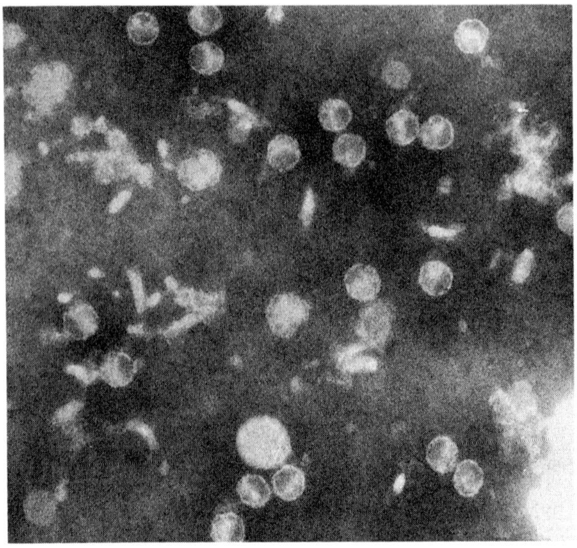

Jeolco, Inc., Medford, Mass.

Plant sciences, except for certain applied branches, merge with other life sciences because they deal with living organisms and share the same principles. The only logical reason for separate study is the different study techniques used by the botanists. The dichotomy of the biological science is most evident in studies relating to **taxonomy** and classification and least evident in investigations of the life systems. The modern botanical scientist is most concerned with molecular processes within the plant cell, with the **ultrastructure** of the cell, with the process of genetic change in plant species, with **viruses,** and with the developmental mechanisms in plants. He makes use of the electron microscope and is able to photograph the structure of complex molecules such as those that form the ribonucleic acid (RNA) phages in the above photograph. The plant taxonomist makes use of the results of this work as a means of demonstrating the unity of the plant world, and he is concerned with the evolution of these organisms as well. It is the taxonomist who must organize and process the available information about plants.

Applied botany includes such topics as plant disease (pathology), bacteriology (the study of bacterial diseases), forestry, horticulture, floriculture, landscape gardening, agronomy, plant breeding, and certain aspects of pharmacy. These subjects are beyond the scope of this text.

PLANT CLASSIFICATION

The great variety of plant life and the fact that plants have an evolutionary history have led to a system of classification

SPECIALIZED DIVISIONS OF BOTANY

Plant physiology—the study of functional processes in plants (Chapter 7)

Plant morphology—the study of the structure of plants, including **cytology** and **histology** (Chapters 6, 11, and 12)

Plant ecology—the study of the relationships of plants to their environment (Chapters 8 and 15)

Plant taxonomy—the identification and classification of plants, which with other sciences forms the study of **systematic botany** (Chapters 17 to 28)

Plant geography—the study of the distribution of plants (Chapter 13)

Plant genetics and evolution—the study of the genes in the cells of plants and their effect (**heredity**) on changes of species (Chapter 10)

Paleobotany—the study of fossil plants and their evolutionary history (Chapter 16)

Mycology—the study of the Fungi and related plants (Chapter 28)

CLASSIFICATION OF PLANTLIKE ORGANISMS

Monera—Single-celled plants without distinctive nuclei
- Blue-green algae
- Bacteria
- Viruses and viruslike organisms

Protista—Single-celled or colonial organisms; cells with distinct nuclei
- Euglenoid algae
- Golden algae
- Dinoflagellata and allies*

Plantae—The true plants
- Red algae
- Brown algae
- Green algae
- Stoneworts
- Mosses
- Liverworts and hornworts
- Psilophyta, the whisk fern
- Ferns
- Club mosses and allies
- Horsetails and allies
- Cycads
- Ginkgos
- Conifers
- Gnetales
- Flowering plants

Fungi

*The word allies is used to indicate other members of the taxa, but with a different or without a common name.

designed to group similar and probably related plants together. In this way the description of plants can go by steps, from the more generalized statements to the more specific. Thus it is possible to say that most members of the rose family Rosaceae have flower parts arranged in multiples of five, and so on until, through hierarchical steps, the **species** (plural also **species**) is reached. In some cases even the species rank is divided, as is seen in Table 3-1. The distinction between **category** and **taxon** is apparent in the table. When the plant or group of plants is referred to, the name is a taxon, but when its position in the hierarchical scale is considered, the rank is a category.

The last part of this text gives details of the classification of the various plant groups. Before these details can be considered, it is necessary to refer to many kinds of plants, especially when considering the history, evolution, and ecology of the plants. At the left is a simplified classification of the common names used in these preliminary chapters.

NAMING OF PLANTS

All kinds of plants, like other living organisms, have formal names, either in Latin or Latinized Greek words, and use the

Table 3-1
Categories of plants and corresponding examples of taxa

Category*	Taxa†
Kingdom	Monera, Plantae
Subkingdom	Euchlorophyta
Division	Anthophyta
Class	Angiospermae
Subclass	Dicotyledoneae
Order	Rosales, Cornales
Suborder	Caesalpinieae
Family	Rosaceae, Cornaceae
Subfamily	Caesalpinioideae
Tribe	Sophoreae
Genus	*Crataegus*
Series	Pavifoliae
Species	*C. intricata*
Variety	*C. i.* var. *straminea*

*There are other divisions of these categories, but they are seldom used.
†The examples chosen are taken at random and do not show the classification of any single group of plants.

binominal system. The system follows the rules and procedures outlined and described in the "International Code of Botanical Nomenclature." This code has been developed and refined over a long period of time and is continually studied and modified to serve the taxonomic botanists. The current code was adopted by the Tenth International Botanical Congress in 1964. The purpose of these rules is to help assure that each taxonomic group will bear only one correct name and that the name will be the earliest name applied, according to the rules, to the group, except in particular cases where particular names are conserved in a special list found in the published edition of the code.

In addition to the Latin name, many well-known plants also have a common name in the language of the country or region where they occur. There are distinct advantages in knowing the **scientific name** as well as the common one. First, the scientific name is recognized as the name of the organism in every part of the world, even though the common name may be more easily remembered. Second, the common name may be different in every language and could even vary from one locality to another within the same country. The adder's tongue *(Erythronium americanum)* is sometimes known as the dogtooth violet. Thus it has two common names, and unless both are given, there is no way to tell that they refer to the same plant. Naming a plant by the use of the botanical code provides a single name for a single kind of organism and results in less confusion, which is one of the aims of the scientist. The scientific name of corn is *Zea mays.* (Note how all generic and **specific names** are placed in italics, but other scientific names are not. This is because they are foreign words and it has always been the custom to indicate them in this manner.) The first name indicates the genus to which the species is assigned and is the **generic name.** This word is always capitalized, and it is always a noun. The second name, the specific epithet, is not capitalized, and lower case is recommended even for the name of a person or a place, although in some books this kind of specific name is capitalized. The word is an adjective modifying the generic name, a noun, and agrees with it in gender. The third name, when used, is the author of the original species description, which in the example given here is Linnaeus. The author's name is not placed in italics; it serves as a very abbreviated bibliographical reference. Sometimes this name is placed in parentheses and is followed by another author's name to indicate that the species has been moved, for botanical reasons, to another genus other than the one in which it was originally placed when the species was described. Names in Latin are used because of their universality. Latin, now a dead language, was formerly the language of educated persons and the language of textbooks and scholarly writings. About all that remains today is the rule that the description of the new taxa (singular, taxon) must be written in Latin. These descriptions may be followed by less formal and usually more detailed descriptions in the language of the author.

By using the first validly described name of a plant any ambiguity is almost entirely overcome. All other Latin names applied to the same species of plant are synonyms and are not used except in catalogs that show the history of the names. Synonymy occurs because the older literature does not clearly define the plant, because older names have been overlooked, or because two or more workers from different geographical areas described the same plant species. As soon as these facts are discovered, the more recent names are placed in synonymy. Sometimes it develops that a very well-known plant has two names and that the older name seldom has been used. To avoid confusion, the more recent name is conserved by placing it on the official list of conserved family and generic names. Priority is established by accepting the first edition of *Species Plantarum* of Carolus Linnaeus, published in 1753, the starting point for most plant nomenclature.

New plants are being described all the time. These descriptions and keys for their identification are published in botanical journals and monographs. As new plants are discovered and added to the flora list, the manuals for the identification of plants have to be revised and brought up-to-date.

IDENTIFICATION

Plant scientists must identify wild plants before they can report certain kinds of research data, particularly studies that involve distribution, variation, certain kinds of genetic studies, and investigations on the processes of speciation and evolution. The process of identification involves several steps, each of which lead to greater accuracy.

1. Spot identifications—often botanists are able to recognize common local plants on sight. They learn these plants by comparing them with pictures and herbarium specimens and, through practice, are able to recall the names, often with a high percentage of accuracy. This method has the disadvantage that plants from other regions are unrecognized or incorrectly identified, and even some local plants may not be separated as distinctive due to lack of critical appraisal of characters.

2. Botanists, fortunately, have a wealth of regional manuals, termed floras, to use for accurate identification of the flowering plants. These contain keys for the identification of the taxa, descriptions of the plants, their distribution, and, frequently, illustrations of the species. The lower plants are not well covered.

3. Specimens collected for research purposes may need to be identified with the use of technical monographs. To find these, a search of the available literature must be made using the various library tools.

4. Specimens identified with the aid of manuals should be verified by comparison with correctly identified herbarium specimens. Research material is often sent to specialists for vouchering, that is, verification of the identification.

PREPARATION OF PLANT SPECIMENS

Because a detailed study of plants cannot be conveniently made in the field, many techniques for laboratory study have been developed. These processes include the proper method of collecting, preparation, and storing of specimens. Preparation for the study of microscopic anatomy demands that special reagents be used in the field so that these parts can be preserved in as near their lifelike state as possible. Special treatment must be given to particular groups of plants so that their parts may be properly preserved for later identification. Details of such methods can be readily demonstrated in the laboratory, and reference to them is made in several manuals that describe these techniques.

Students of taxonomic botany usually

begin by making a collection of plants. The plants are gathered in the field and are placed in cellophane bags within a **vasculum,** a metal can that is easily carried. After the day's collection, the specimens must be pressed for preservation. This is done by means of a plant press consisting of two frames of standard size, 12 by 18 inches. Between these frames are placed a packet of felt blotters and corrugated cardboard ventilators cut to the size of the frames. The plants are carefully arranged between sheets of folded newsprint or newspaper, which are next arranged between two blotters, one plant per pair of blotters. These are alternated with cardboard ventilators until a package about 2 feet thick is obtained. Then the frames are laid over and under the package and drawn tightly together with straps. These presses are placed over low heat so that the water absorbed by the blotters from the plants is removed through the channels in the corrugated cardboard. After about 24 hours the presses are opened and the now limp plants are arranged for the best display of parts. They are then relocked in the press and dried for another 24 hours. Then the dried and flattened plants are removed and are ready to be glued to the herbarium sheets made of high-quality paper. Attached to each of these sheets is a label identifying the plant and, most important, field notes telling where each of the plants was growing and any other information about the plant and its immediate environment.

If chromosome counts are to be made, it is necessary that growing tips or developing parts of the flower be preserved in the field in some specific fixing fluid, such as Newcomer's,* which will preserve the cells in various stages of mitosis or meiosis so that the chromosomes may be studied later in the laboratory.

Sometimes in the study of various kinds of plants it is desirable to observe the microscopic structures of the stem, roots, or other parts. The preparation of these parts requires a long and involved **microtechnique.** The cross section of a stem, for example, must first be fixed in a solution like alcohol or formalin, dehydrated through a series of graduated steps of increased alcohol and decreased water, and, finally, after complete dehydration, placed in xylol, which is miscible with paraffin. Once the specimen is embedded in paraffin it can be cut in sections only a few microns thick by means of a microtome. The paraffin sections are then fixed to a glass slide, the paraffin removed by xylol, and the xylol replaced by a gradual series of alcohols until the specimen is again in water. The section is stained, the whole process of dehydration repeated, and a cover glass fixed to a slide with a material like Canada balsam. Because of this procedure, the specimen may now be easily and safely examined under the high magnification of a compound microscope without danger of breaking the specimen.

EXPERIMENTAL DESIGN

Both the method and the history of science, as shown in the previous chapters, begin with the recognition of problems and finish with a scientific report. These usually contain a review of the literature already available on the subject, followed by a statement of the problem to be solved, the method used to discover new facts about the subject, the data gathered, and the results. A discussion points out the significance of the discovery made in relation to previous knowledge. The theory of this method and examples of broad problems needing study have been discussed in Chapter 1. The actual selection of a specific research problem, as pointed out there,

*Science **118**:161, 1953. The formula is 6 parts isopropyl alcohol, 3 parts propionic acid, 1 part ether (petroleum), 1 part acetone, and 1 part dioxane.

is possible only after considerable experience.

Experimentation usually involves rather sophisticated equipment. **Chromatography,** a widely used laboratory technique for separating similar materials in a mixture, is used in amino acid analysis for many types of research projects. **Isotope tracers** are used by biologists to follow the movement of ions from place to place within a plant or to determine the chemical pathway of a reaction taking place in a living system. Common tracers are carbon 14 or cobalt 60. For use in enzyme and growth studies, it is possible to obtain many different kinds of biochemical compounds labeled with any of several tracer isotopes. Even the particular atom in the compound may be tagged so that the progress of the particular radical may be traced. The following are some instruments used in botanical research:

MEASUREMENT

Wheatstone bridge—a device using matched resistors for the measurement of electrical quantities. Any change that can be expressed electrically, including temperature, pH, light, and energy, can be measured with this device.

Oscilloscope—an electronic measuring instrument used to show and measure wave patterns.

ANALYSIS

Centrifuge—a machine used to separate materials of different densities by using centrifugal force.

Spectrophotometer—any material that absorbs radiant energy within the range of light may be identified and the rate of its reaction measured with this instrument.

Manometer—a U-shaped tube used to measure changes in gas pressure, which is indirectly a measurement of the rate of respiration or photosynthesis within living systems.

Scintillation counter—an instrument used to count radioactivity in studies involving isotope tracers.

Experimental design is the application of the scientific method by planning in advance the procedure to be used to investigate a scientific problem. Such experiments are likely to yield more significant data than those otherwise performed because they can be made to meet the requirements of statistically significant data gathering. Not only are controls important in experimental investigations, but also randomization is necessary to ensure that the data are not biased by the experimenter. The careful selection of the instruments used, the test organisms, and the control of the environmental conditions during the test all contribute to the successful gathering of significant data for analysis.

Modern botanical science, like all science, uses three sources of information—the previous literature, samples as stored data, and observed or sensed information. This is the principle described in Chapter 1. The following chapters report the findings of botanists over the past several hundred years, but most of the information discussed has been discovered only in the past several decades.

SPECIAL TERMS USED IN CHAPTER 3

binominal system	scientific name
category	species
chromatography	specific name
experimental design	systematic botany
species	taxon
genetic name	taxonomy
heredity	ultrastructure
isotope tracer	vasculum
microtechnique	virus
morphology	

QUESTIONS TO CONSIDER

1. What features of modern botany are the same as those of other branches of biology? In what way does it differ?
2. What are two advantages of using Latin names for plants?
3. What is the greatest disadvantage of using common names for plants?
4. What are the three principal ways one might learn the name of a plant?
5. What are the uses of a plant name once an identification has been made?
6. When someone has a question about a botanical phenomena, how is it determined that experimentation is necessary?
7. What features of an experiment are absolutely

necessary if the results are to be scientifically significant?

DISCUSSION FOR LEARNING

1. Chapter 1 discussed in some detail the scientific method, and this chapter shows some of the branches of plant science. Discuss some of the reasons for dealing with broad problems.
2. A distinction is often made between "pure" and "applied" science. Discuss these differences and consider how one may influence the other.

Chapter 4

Economic botany

USDA photograph.

A large number of plants such as the wine grape shown above are directly or indirectly of some use to man. The food we eat, our shelter, and clothing come to us from these organisms. The animals that feed on grasses, the fish that live on algae, and the whole variety of cereals are sources of food for man. Thus the basic food of all animals is produced by green plants during photosynthesis, and oxygen is liberated into the atmosphere for reuse by both plants and animals. In many cases the available nitrogen in the soil is due to plants. Wood provides housing, and the cotton plant, clothing. In addition, coal, furniture, paper, medicines, including the antibiotics, many chemicals, and some dyes, all plant products, are by no means of minor importance. Forests prevent erosion and regulate humidity. Thus plants play a major role in the world's economy. The green plants are primarily important as an original source of food in anmial food chains.

KINDS OF ECONOMIC PLANTS

1. Food plants
2. Fiber plants
3. Lumber trees and forest products
4. Medicinal plants
5. Plants for conservation and land development
6. Poisonous palnts
7. Disease-producing plants
8. Weeds

FOOD PLANTS

Food plants not only provide us with energy in the form of carbohydrates (Fig. 4-1), proteins, and fats, but also furnish us with minerals and vitamins. All animals are dependent directly or indirectly on the food stored in plant cells as the result of photosynthesis. The cereals (grasses such as corn, rice, and wheat) are the principal source of carbohydrates for breadlike foods. Corn and soybeans provide us with a major

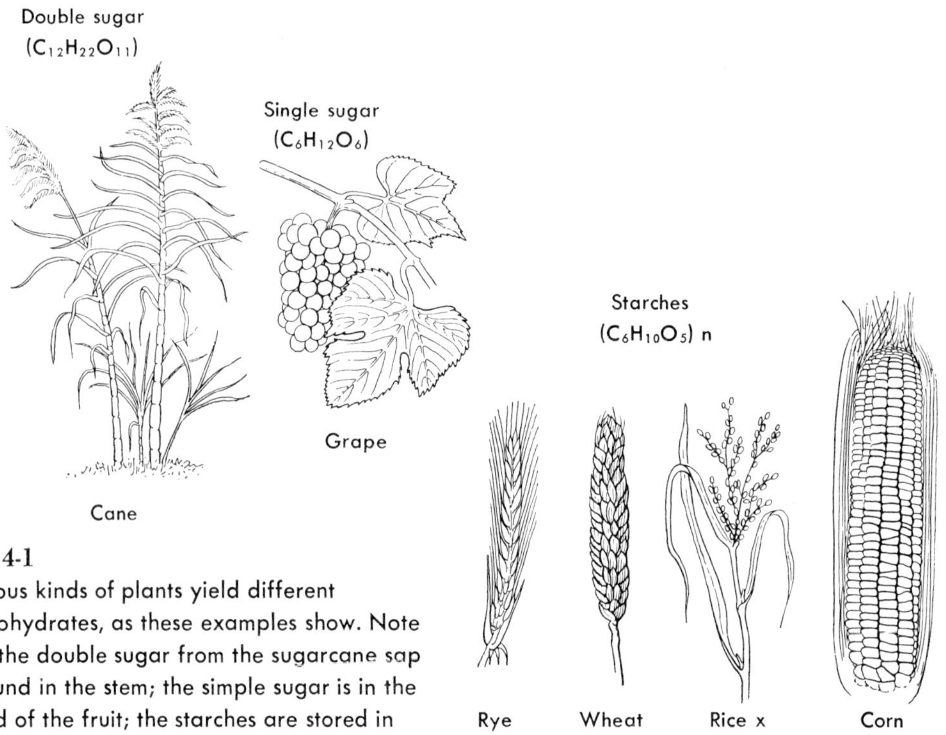

Fig. 4-1
Various kinds of plants yield different carbohydrates, as these examples show. Note that the double sugar from the sugarcane sap is found in the stem; the simple sugar is in the liquid of the fruit; the starches are stored in the dry seeds.

source of fat-producing foods. Legumes such as beans, peas, and peanuts are an important source of protein, along with nuts such as almonds and cashews. Forage plants, for example, clover and grasses, indirectly provide us with food by feeding livestock. Many plants provide us with food seasonings or spices, the common names of which are familiar to all (sage, garlic, bay leaves, cinnamon, thyme, rosemary, savory, cloves, anise, and wintergreen to name a few). Most of our vegetable plants are primary sources of essential minerals in our diet, provided that they are properly prepared. Corn, cauliflower, carrots, spinach, and potatoes are high in food value. Others, such as eggplant and squash, help to round out our diet. Plants by means of their fruits (for example, apples, cherries, and peaches) provide us with vitamins, essential elements of our diet that we cannot manufacture ourselves. Citrus fruits and tomatoes are important sources of these vitamins. Some plants provide us with the essential ingredients of beverages, such as coffee, tea, and cola, and with the hops, malt, and wine grapes, the raw materials for alcoholic drinks. In addition, distilled spirits are prepared from rye, corn, wheat, juniper berries, and sugar cane. A few of the lower groups, fungi, molds, and bacteria, are essential to cheese production, and some can be eaten directly, such as mushrooms, puffballs, truffles, and others.

FIBER PLANTS

Fiber plants are essential to many manufacturing processes. Cotton (Fig. 4-2), linen (made from flax), jute, Indian hemp, Manila hemp, sisal hemp, kapok, and other plants produce long cells or closely packed groups of cells with tough cell walls to form fibers used in the making of cotton cloth, rope, burlap, thread, string, and

Fig. 4-2
Fibers of the cotton boll, a form of fruit, supply much of the clothing and other fabrics used throughout the world. (USDA photograph.)

packing materials. After suitable chemical treatment these plants are also sources of rayon, paper, cellulose, lacquers, cellophane, and many other products used in modern industry.

LUMBER TREES AND FOREST PRODUCTS

Lumber trees (Fig. 4-3) are among the most useful plants known to man. The use of wood for building materials is as old as civilized man. Some of the best lumber is cut from oak, mahogany, ash, and maple (hardwoods), and from redwood, cedar, and pine (soft woods). Oak, hickory, and pine are used for fuel, pine being used for the manufacturing of charcoal. Posts, mine timbers, poles, and marine pilings are cut from oak, ash, and osage orange. Veneers and plywood are made from pines, maple, and mahogany. Railroads use red oak for ties. White oak is used in the manufacture of barrels, but the cooperage trade has become almost entirely extinct, since cardboard containers made from pine and redwood scraps are now in general use. Excelsior for packing material is made from aspen and cottonwood. The few wooden shingles still used come principally from red cedar. Differences in the cell structure and in the amount of gum and resins in the wood determine the quality of wood and the use to which it is put. Wood is also used for the manufacture of tars, resins, alcohol, acetone, wood gas, and turpentine. Chicle for chewing gum, rubber, cork, and some spices are additional wood products. A few secondary forest products deserve mention—cork from cork oak, Christmas trees from pine, fir, spruce, and hemlock, and woods for inlay and other ornamentation from hardwoods such as locust, persimmon, ebony, and mahogany.

MEDICINAL PLANTS

Medicinal plants have contributed greatly to the welfare of man. Among the

Fig. 4-3
Young spruce forest high in the western mountains. Note the absence of undergrowth, due mainly to the overgrazing of cattle.

Fig. 4-4
Digitalis used in the treatment of some heart ailments is manufactured from foxglove plants such as these shown under cultivation. (Courtesy Merck Sharp & Dohme, Inc., Division of Merck & Co., Inc., Rahway, N. J.)

Fig. 4-5
The green mold shown here is *Penicillium chrysogenum,* a mutant form from which almost all the world's supply of penicillin is obtained. (Courtesy Chas. Pfizer & Co., Inc., New York, N. Y.)

drug-producing plants and their products are belladonna, a dilator extracted from the belladonna plant; foxglove (Fig. 4-4), the source of digitalis for the treatment of heart disease; cascara and senna, yielding cathartics; and morphine and opium, powerful drugs from the poppy. Many drugs formerly came from plants, but they are now made synthetically from coal tar. Since World War II a new series of plants has come into use, the fungus plants such as *Penicillium* sp. (Fig. 4-5), and *Streptomyces* sp. (Fig. 4-6), the organisms producing terramycin and others. These plants are grown by drug companies, using special culture methods (Fig. 4-7).

PLANTS FOR CONSERVATION AND LAND DEVELOPMENT

More subtle uses of plants include water conservation and the prevention of land erosion. Conservation may be defined most simply as the prevention of the disturbances of natural balances or the restoration of these balances. These disturbances are many; some are natural, but most are man created.

Natural communities of plants form a cover over the soil, which holds back large quantities of water and prevents rapid runoff of rain water. This not only prevents erosion but also maintains a water reserve, which in turn prevents drought conditions. When the natural vegetation cover is disturbed through cultivation and landscaping or highway, airport, and community development, this water-retention effect is destroyed. Floods and erosion result.

Erosion of soil is one of the greatest

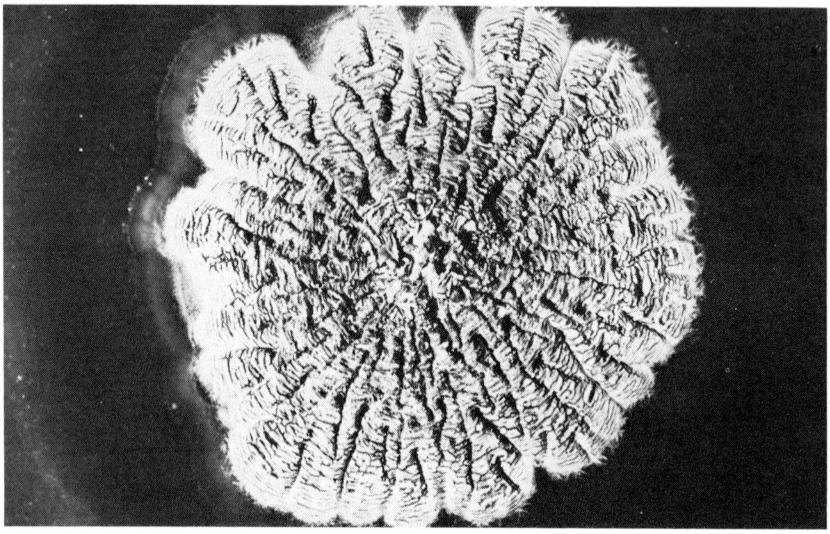

Fig. 4-6
A colony of *Streptomyces rimosum* such as this is the source of Terramycin, discovered and developed for treatment of disease in 1949. (Courtesy Chas. Pfizer & Co., Inc., New York, N. Y.)

Fig. 4-7
Masked technicians in sterile clothing operate freshly sterilized machinery to fill bottles with liquid antibiotics. The room in which they are working has specially filtered air and special lighting to maintain sterile conditions. (Courtesy Chas. Pfizer & Co., Inc., New York, N. Y.)

conservation problems. Water and wind move vast quantities of soil from place to place, much of which ends up in the sea where it is of little use. Wind erosion is responsible for much of the configuration of land in arid regions. Gorges and mountains are the result of water erosion. Both types may be prevented by stopping the useless removal of trees that hold back both wind and water. The current practice of flood prevention, rather than control, has cost this nation dearly. The practice of cutting down trees and digging straight channels simply removes water in a hurry, along with millions of tons of topsoil, and leaves behind a water shortage of ever-increasing magnitude. Planting trees and providing for backwaters as a reservoir for flood waters are solutions to the problem in all but montainous areas where the damage is already too great (see also Fig. 13-3). There, only expensive dams can keep back the water. Wind erosion can be prevented by planting windbreaking trees (Fig. 4-8). In irrigated areas these trees should be grown along the edges of all fields, and a plant cover should then be maintained at all times. The practice in the Midwest of early harvest and immediate plowing to prevent the growth of weeds, followed by late planting, is a continuing folly that has turned vast areas into dust bowls in the past.

Leaching of the soil by the removal of humus and plant cover causes a deficiency in mineral content, which soon renders such areas barren of all possible plant growth. The solution for the prevention of this loss is obvious. Such areas may be reclaimed only by extensive fertilization, as well as by water control and planting of erosion resisting grass (Fig. 4-9).

Trees affect local weather conditions in several ways. They increase the relative humidity by transpiration and, at the same time, produce a cooling effect in the shade by evaporation of the water transpired. This in turn contributes to cloud formation

Fig. 4-8
Windbreaks along the edge of irrigation ditches surrounding cultivated fields of young plants help to prevent wind erosion of the soil. (U. S. Forest Service.)

Fig. 4-9
Holly grass is planted by soil conservationists in barren areas like this to anchor the soil. This grass grows naturally in windswept alpine areas. Here it holds the soil until other plants can become established. (USDA photograph.)

from the rise of water vapor, thus making more moisture available for rain (Fig. 8-3). Trees, herbs, and grasses all provide food for wildlife, which in turn carry on complex food chains, and some of these animals, especially the game animals, are of direct interest to man. Many food chains, however, only remotely affect man's economy and unless he is interested in and appreciates the esthetic value of understanding these complexities of nature, he does not concern himself with them.

In agricultural areas a highly organized soil conservation program is sponsored by governmental agencies. Farmers participating in this program are required to grow windbreaks, terrace their land, practice contour plowing, that is, plow across slope lines, add required fertilizers, and practice proper crop rotation. Following the latter practice, a natural building of the soil can be effected by first planting soil-building plants such as legumes, after which may be planted crops that tend to deplete the mineral content of the soil. Some plants, for instance tobacco, rapidly deplete the mineral content of soil. This may be easily seen by looking into any ash tray. The soil of the tobacco land of the nation rests there in the form of the high ash content of the burnt tobacco leaves. The ash is nearly pure mineral salts. Eventually the conservation practices just described will bring back the land, but a long program of proper education lies ahead before the lands of the nation are finally stabilized.

The development of new agricultural land is becoming more and more important. As urban areas squeeze out more farmland, arid areas are developed by means of irrigation. All too frequently, however, this results in the exploitation of both land and water, and after a short period the land is abandoned in worse condition than before. There is great need for research in the proper development and management of new lands. Unfortunately, politics often enters into this problem, and strong lobbies prevent the proper control of such areas. As an example of this, the cattle interests have prevented proper legislation that would control grazing in western arid areas (Fig. 4-10) and in our national forests. Unless these selfish interests are soon controlled, irreparable damage will be done. As it is now, the once extensive western grasslands will take many years to restore.

Fig. 4-10
Right and in background, the effect of overgrazing. Left, luxuriant growth in an area in which no grazing has occurred. Much of our western land was once grassland, but through improperly controlled grazing it has been turned into unproductive desert. (U. S. Forest Service.)

Equally important to future generations is the conservation of natural areas. In these days of rapid transportation and relative abundance of leisure time, millions of people flee the urban areas in search of recreation in natural areas of this country. The natural areas are rapidly diminishing. Man already has exterminated many animals such as some kinds of quail and the passenger pigeons and nearly has exterminated many others. These he knows about, but many other interesting animals as well as plants have been exterminated unknowingly. National parks have been set aside, plus many natural areas, as wildlife refuges to preserve as much of the natural country as possible. Much of the interest in these areas centers around the preservation of game animals, with little regard, however, for the conservation of all wildlife, including plants, insects, and other forms of life that are of interest for themselves, not as something to shoot or fish so that a man can assert his superiority. A policy that will fully preserve these areas is urgently needed. Visitors to natural areas should be educated to appreciate fully the delicate ecological balances and the widespread effect of disruption of even the most subtle of these balances. This we owe to future generations.

POISONOUS PLANTS

Many organisms are harmful in one way or another to some other organisms. The harm may be simply in the form of competition for space or food, or both. Other harmful effects may be more direct, such as parasites living at the expense of other organisms or in the case of plants, producing substances that are directly poisonous when eaten; for example, water hemlock (Fig. 4-11) and arrow grass (Fig.

Fig. 4-11
Water hemlock, *Cicuta douglasii* shown here, is probably the most poisonous plant in the United States; the poison is found principally in the roots, which are often mistaken for parsnips. (USDA photograph.)

4-12). Molds such as the genera *Alternaria, Hormodendrum, Penicillium, Aspergillus, Mucor*, and *Rhizopus* are major wind-borne spore allergens, second only in importance to the pollens in causing asthma and rhinitis. Spores originating in the midwestern states can be wind borne to eastern cities in 24 hours. Western states have the least incidence of the disease. July, September, and October are the important months for this allergy. A few plants are poisonous to touch, for example, poison ivy (Fig. 4-13) and the greatly feared manchineel (Fig. 4-14), cause of much discomfort in tropical regions.

DISEASE-PRODUCING PLANTS

Many plants cause diseases of animals and man. The bacteria and fungi are the greatest offenders of this kind. Bacterial diseases will be discussed in greater detail in Chapter 18. Fungal diseases, such as those of the skin and lungs, are briefly mentioned in Chapter 28. Some bacteria and fungi help to limit animal populations and thus are useful to man in the control of certain pest animals. For example, the fungus causing milky disease in the Japanese beetle is a practical control of this pest. It is now being spread artificially to control the beetle, thereby preventing serious damage to fruit crops and ornamental plants.

WEEDS

Weeds are not a particular kind of plant, as many people suppose. Any plant growing in a lawn, garden, or cultivated field that is not desirable to the grower is considered to be a weed. There are a great many such plants, most of which are more tenacious of life than the cultivated plants protected by man. For this reason there is a struggle between the two kinds of plants, and the

Fig. 4-12
Arrow grass grows in wet areas, as shown in this view, where it is sometimes eaten by sheep and cattle with fatal results because the poisonous substance is hydrocyanic acid.

Fig. 4-13
Poison ivy causes irritation to the skin of most people. The shiny leaves in groups of three make it easily recognizable.

Fig. 4-14
Manchineel, a small tree native to the beaches of Central America, the West Indies, and Florida. The milky sap of the fruit and leaves causes severe irritation and even blindness.

cultivated plant soon loses if the grower falters in the battle. Much time and expense are spent on weed control. Removing the weeds bodily, applying weed-killing chemicals, and searching for insects to control the weeds, as well as diseases specific to the weed, are some of the means used in this struggle. In many cases it would be better to join the weeds than to fight them. Traditionally, many of the so-called weeds have been excluded from our list of desirable plants, even though many are attractive and useful. For instance, the common dandelion (Fig. 4-15) has all the beauty of a chrysanthemum and is a delicious green with the food value of similar cultivated plants. Among the many kinds of weeds of cultivated crops, all of which are obnoxious because they reduce the yield per acre, are wild garlic, locoweed, crabgrass, mustard, pigweed, dock, and plantain.

Fig. 4-15
The dandelion is an ubiquitous weed of lawns, yet has the beauty of a garden flower.

ORIGIN OF CULTIVATED PLANTS

Plants have been under cultivation by man for so long that most of them are taken as a matter of course. The following lists show that the isolation of the Old World and the New World resulted in separate cultivation of food plants. A surprising mixture of plants after voyages of Columbus has promoted a more varied diet, which is evident by the New World plants that have taken Old World names (Jerusalem artichoke) or that are now grown more extensively in the Old World (pineapple), or by the Old World plants (orange) that are now grown more extensively in the New World.

New World plants not grown in the Old World before the time of Columbus:

corn	squash
lima bean	pumpkin
potato	peanut
sweet potato	papaya
kidney and other beans	avocado
tomato	pineapple
green pepper	guava
Jerusalem artichoke	chocolate
sunflower	cashew

Old World plants not grown in the New World before the time of Columbus:

wheat	carrot	plum
rye	breadfruit	cherry
barley	mangosteen	alfalfa
oats	onion	wine grape
millet	garlic	apricot
sorghum	spinach	peach
rice	eggplant	olive
buckwheat	lettuce	fig
turnip	endive	almond
cabbage	celery	quince
rutabaga	asparagus	pomegranate
chard	peas	watermelon
mustard	soybean	cucumber
radish	yam	mango
beet	apple	clover
parsnip	pear	

The first cultivated plants were selected from wild stock because they were the best tasting and the easiest to grow. This natural selection process brought about man-made changes in many of our plants so that they bear little resemblance to their ancestors. It is obvious that this selection is possible through the genetic mechanisms inherent in the plants. Only relatively recently in the history of agriculture have systematic attempts been made to develop new crops through crossbreeding. However, almost all our modern crop plants are hybrids (Fig. 4-16) developed for higher yields, greater food value, or as more attractive ornamentals.

Fig. 4-16
Close-up of Wichita wheat, an example of increased food production from hybrids. These grains of wheat are much larger than those of wild wheat or earlier cultivated varieties. (USDA photograph.)

Most of the original wild forms of our modern cultivated plants are known, and these forms can be traced back easily to a certain wild species. In fact, many of these plants will revert back to the wild form if allowed to grow unattended, the very existence in their present form depends upon the continued attention of man.

Organization of data

Cultivated plants are usually larger than their wild relatives. They have broader and thicker leaves, sturdier sprouts, stems, and stalks, fleshier roots, and larger flowers (Fig. 10-12). Their fruit and seeds show these same characteristics. This gigantism is due to **polyploidy,** the multiplication of the **chromosome** compliment in the cells (see Chapter 10 for an explanation of this). The study of these phenomena in plants has led to the science of plant breeding. Further consideration of this topic is beyond the scope of this text.

SPECIAL TERMS USED IN CHAPTER 4

chromosome   polyploidy   weed

QUESTIONS TO CONSIDER

1. Considering the many synthetic products manufactured from petroleum, what are still the most important cultivated plants producing useful products?
2. Name one or more cultivated plants that have similar wild relatives.
3. Name one or more cultivated plants that are very different from their wild relatives.
4. There are many plants in the word that are grown locally that never are shipped to markets in other sections of the country or the world; what are some that may be found in your area? Can you name others?
5. From what you have read so far, what kind of knowledge is necessary for plant-breeding experiments?

DISCUSSION FOR LEARNING

1. Discuss the possibilities of continuing to feed the world's expanding population through the development of greater yields of plant products. What types of plants might be most useful?
2. Many drug manufacturing companies are sending "plant explorers" (see photograph at the beginning of Chapter 18) throughout the world in search for new drug plants. Discuss the reason for this interest and why it continues in spite of the rapid development of synthetic products.
3. Conservation reforms are critically needed. The popular press is full of pleas to prevent stream and land erosion, air, water, and soil pollution, and for proper land management. Discuss some of the things each private individual might do to hasten these reforms.

Part II

Chemicals and cells

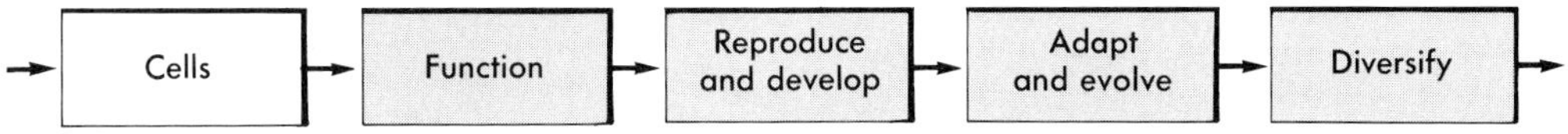

- Cells are the units of life activities.
- All life systems involve cellular intake, cellular work, flow, feedback, and replication.
- All life's activities take place in, or at the command of, the cell.
- These complex chemical units of molecules are programmed in details that almost defy the imagination as well as man's ingenuity to devise ways of probing into cellular secrets.
- The basis of all biological theory is to be found in the study of the cell.
- The whole organism is the result of variation and activity of cells.

Chapter 5

Chemical and biological systems

Atoms and **molecules** have a unique feature, the members of each species are exactly alike. Living organisms are composed of molecules, yet each member of a species of organism is different from another. Since all biological phenomena have a physicochemical basis, it is important to understand how one individual can differ from another when each is composed of the same materials. The reason behind this lies in the processes that take place through the transfer of energy from one substance to another during molecular activity. Energy in biological systems is, however, a controlled exchange, unlike the violent reactions seen during storms, as in the photograph above.

ATOMS AND MOLECULES

According to cosmic theory the universe is composed either of matter or energy, and matter, or material substances, represents stored or potential energy. Potential energy, once released, becomes motion or kinetic energy. Matter itself is composed of atoms, the smallest piece of a particular element. Two or more atoms joined by energy bonds are molecules if they form a substance. Thus a molecule is the smallest piece of a particular substance that can exist independently and still possess the properties of a substance.

The number of known elements (103) is far more than is involved in any biological substance or process, but all have the same general properties. For example, they are composed of electrons, neutrons, and protrons, which gives each element an atomic weight (Table 5-1). The atomic weight is a relative measure based on normal carbon, C^{12}, with an atomic weight slightly over 12. Atoms of an element do not all have the same weight. These variants are called **isotopes.** Among the first known were the isotopes of oxygen that comprise heavy water, O^{17} and O^{18}, or the widely used isotope of carbon, C^{14}. The discovery of radioactive isotopes induced further study that eventually led to our knowledge of atomic fission.

Atoms have properties known as the laws of chemical combination. For instance,

Table 5-1
Atomic weights of common elements found in biological substances

Element	Symbol	Atomic weight*
Carbon	C	12.0111
Hydrogen	H	1.0079
Oxygen	O	15.9994
Nitrogen	N	14.0067
Sulfur	S	32.0640
Phosphorus	P	30.9738
Potassium	K	39.1020
Iron	Fe	55.8470
Copper	Cu	63.5400
Zinc	Zn	65.3700
Iodine	I	126.9044
Manganese	Mn	54.9380
Chlorine	Cl	35.4530
Magnesium	Mg	24.3120

*When using atomic weights it is customary to round off these numbers to the nearest whole.

when two or more elements combine to form a particular compound, they unite in a constant and definite way, depending upon the **valence** of the element. This is always in a simple ratio, and this ratio remains the same throughout any subsequent combination with other elements. The simplest example of this is water, H_2O. Two atoms of hydrogen always combine with one of oxygen to form water in a ratio of 2:1. Carbon dioxide is formed from one atom of carbon to two of oxygen, CO_2, but methane is formed from one atom of carbon and four of hydrogen, CH_4. Therefore the ratio in carbon dioxide is 1:2, and in methane, 1:4.

Molecules formed from these atoms have certain properties as well. First, they have a molecular weight that is easily computed by finding the total weight of each atom in the molecule, using the approximate weights given in Table 5-1. Water has a molecular weight of 18, carbon dioxide, 44, and methane, 16. Based on the assumption that equal amounts of chemicals must contain equal numbers of molecules, the gram-molecular weight, or gram equivalent of the molecular weight, that is, the molecular weight expressed in grams, is frequently used in chemical experiments and is termed a **mole** (sometimes mol). This is reflected to some extent also in the actual size of the molecule once the density is known.

Arrangement of atoms and molecules

Molecules also have a particular structure, determined by the arrangement of the atoms of which they are composed. Two factors determine these arrangements. The first is the valence, or the number of possible links for each atom, and the second is the possible number of positions that an atom may occupy in relation to the other atoms. This is called **isomerism.** It was learned that two different chemicals could have the same atomic formula but have different structural formulas. For example, both acetic acid and methyl formate are represented by the atomic formula, $C_2H_4O_2$, but by the following structural formulas:

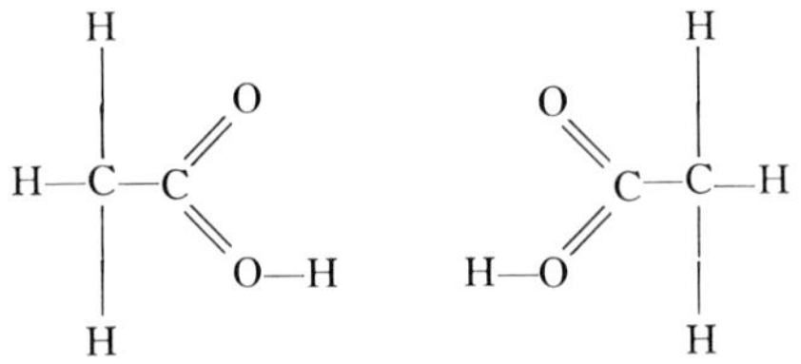

Acetic acid Methyl formate

Note that one is a mirror image of the other. They are similar substances but have somewhat different properties.

Energy and molecules

Various well-known chemical reactions take place to release energy for work. A simple example is the burning of methane to release energy for cooking food:

$$CH_4 + 2O_2 \rightarrow CO_2 + 2H_2O$$

Reactions vary according to the type of materials involved. It is the storing and releasing of energy in chemical reactions that is of particular interest to the botanist because the two main physiological functions, **photosynthesis** and **respiration**, are processes of storing (photosynthesis) and releasing (respiration or burning) of energy. The energy is stored in molecules and is represented as an energy bond holding the molecules together. Before the atoms of a molecule can be separated, a certain amount of energy must be absorbed, and when the atoms are joined, energy is released. The rest of the process in living systems involves the substructure of cells and a special group of proteins called **enzymes.** This will be discussed in detail in Chapter 7.

Chemical reactions

A chemical compound, in one sense, represents an energy storage unit, since if the proper energy is applied to the compound to separate the component molecules, the parts absorb this energy, which is later released when the atoms again unite. Chemical reactions will take place with varying amounts of activity, depending upon their environment. H_2 molecules react violently with O_2 molecules to form water. Under certain conditions the reaction is an explosion. Disastrous explosions of hydrogen have destroyed property and taken lives so that the gas is known as a dangerous material. At the same time, billions of hydrogen atoms are reacting with billions of oxygen atoms in living cells for ordinary life functions. The energy for these reactions is taken from heat, light, electricity, X rays, or pressure energy.

Reactions of a biological nature require rather high activation energy sources. If there were no other means of obtaining these reactions than by the direct application of energy to the chemical systems within the cell, life would be much different from the way we now know it. Perhaps somewhere in the universe such life forms do exist unrecognized. In our system the reactions occur with the help of special chemicals termed **catalysts.** In biochemical systems the catalysts are termed enzymes. Most biological reactions have specific enzymes for each reaction. The exact reaction mechanism is unknown, but it is known that each enzyme is a huge molecule with a specific geometric structure. Only recently has an enzyme, ribonuclease, been synthesized. When it functions, it forms a temporary bond with the compounds that are reacting. This releases energy that enables the system to function. The specific

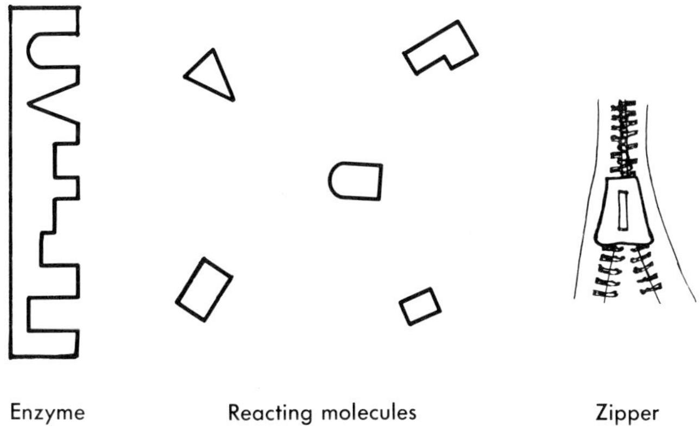

Fig. 5-1
Diagram showing how an enzyme is able to fit into a reacting molecule to guide its reaction and then be released to take part in additional reactions, much the same as the head of a zipper joins and separates the two halves.

action of an enzyme, simply stated, is due to its particular shape and to the shape of the molecules that are reacting (Fig. 5-1). The enzyme forms a template into which fits the reacting compound. Once the action is completed, the new compound is released and the enzyme remains intact. Conversely, the same enzyme may aid in the separation of molecules. The enzymes merely assist in the speed of the reaction and do not determine its direction. There is no permanent change in the enzyme. The reacting material is referred to as the **substrate.** The reaction takes place faster and at a lower temperature simply because the surface template of the enzyme assures the close contact of the reactants, whereas, without the enzyme, the chance of the two coming into close contact through random movement is much less. Therefore, for the reaction to take place without the enzyme, the process would have to be speeded by the addition of heat or some other energy source.

Enzymes are named after the substrate that they activate by adding the suffix, -ase. Thus an enzyme taking part in a reaction of the sugar sucrose is termed sucrase, and so on. There are six general groups of enzymes, classified according to their type of catalyzing action (Table 5-2).

Enzyme activation often requires an **activator,** a small molecule, frequently an inorganic ion. The activator does not take part in the reaction, but the process will not take place unless the activator is present.

A final group of substances concerned in the understanding of chemical reactions are the **coenzymes**—nonprotein chemicals that take part in the process along with the enzyme. One of the best known coenzymes is **adenosine triphosphate (ATP).** Its role in energy transfer will be discussed in Chapter 7.

BIOLOGICAL ORGANIZATION

Throughout the previous discussion little mention was made of the very important fact that biological structure and biochemical processes exist only in living cells (except, of course, under experimental conditions in the laboratory, where certain of the processes may be duplicated outside of a cell). Before the structure of the cell is considered (Chapter 6), consideration must be given to the living substance itself, **protoplasm.** All cells contain protoplasm,

Table 5-2
Classification of enzymes

No.	Systematic name	Example	Reaction
1.	Oxidoreductases	Glucose oxidase	Oxidation-reduction
2.	Transferases	Glucokinase Ribokinase Triokinase	Group transfer
3.	Hydrolases	α-Amylase Pepsin Trypsin	Hydrolytic
4.	Lyases	Isocitrate lyase Citrate synthase	Addition of groups to double bonds
5.	Isomerases	Lysine racemase Ribulose phosphate epimerase	Isomerizations
6.	Ligases (synthetases)	GMP synthetase CTP synthetase	Condensation of two molecules in ATP or similar triphosphate pathways

which is a very general term that applies to the various materials found in a living cell. It will be noted later that not all cells found in mature plants are living.

Certain generalizations may be stated about protoplasm that will be true for all kinds of living organisms.

Chemical characteristics

Approximately 95% by weight of protoplasm is composed of four elements—oxygen (62%), carbon (20%), hydrogen (10%), and nitrogen (3%). It so happens that all of them are present in some form or other in the atmosphere, the oceans, and the fresh waters. The remaining 5% by weight of protoplasm is made up of approximately thirty elements, some in relatively large amounts, such as calcium, and others only in trace amounts.

Most of these elements are present in a combined form, so that it is possible to distinguish both classes of compounds—inorganic and organic. **Inorganic compounds** are combinations of mineral substances, water being one of the most prevalent. The protoplasm of organisms rarely contains less than 60% water, and it may be present in amounts as high as 99%. Other inorganic compounds are the phosphates, carbonates, chlorides, and sulfates of calcium, sodium, potassium, and magnesium. Although most of these may be in solution in water, some make up the more solid parts of the organism. In addition, such gases as oxygen and carbon dioxide may be present.

Organic compounds are complex combinations that always include the element carbon. Carbon dioxide and carbonates are not usually considered organic compounds because of their simple nature and mineral origin. The carbosydrates, the proteins, and the fats are the chief kinds of organic compounds found in protoplasm. Generally, they are dissolved or suspended in water, and in some cases they may combine to form fat-protein complexes.

The simplest organic compounds are found in the **carbohydrates,** which are composed of carbon, hydrogen, and oxygen. The carbons are usually in a chain —C—C—C— in which the hydrogen and oxygen are present in the same 2:1 ratio, as in water. The simplest sugars, such as glucose and fructose, are made up of 6-carbon chains and have the formula $C_6H_{12}O_6$ (**monosaccharide**).

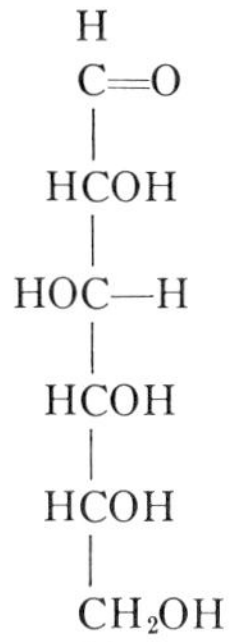

D-Glucose

Two 6-carbon sugars may combine to form a 12-carbon sugar (**disaccharide**), with the loss of water, as follows:

$$\text{Glucose} + \text{Fructose} \rightarrow \text{Sucrose} + \text{Water}$$
$$(C_{12}H_{22}O_{11} + H_2O)$$

Several disaccharides may combine to form longer chains called **polysaccharides,** such as starch and cellulose, which are chains of about 2,000 glucose units.

Carbohydrates are capable of decomposing with the release of energy into water and carbon dioxide or fragments of several linked carbon atoms, as illustrated below:

$$C_6H_{12}O_6 \rightarrow 6\ CO_2 + 6\ H_2O + \text{Energy}$$
$$C_6H_{12}O_6 \rightarrow 2\ CO_2 + 2\ C_2H_6O + \text{Energy}$$
$$(\text{Alcohol})$$

The main effect of the carbohydrates in protoplasm is to release stored energy by this reduction.

The **fats,** like the carbohydrates, are composed of carbon, hydrogen, and oxygen. However, the ratio of oxygen and hydrogen is not the same as in the carbohydrates

(2 H's to 1 O). The fats are composed of a 3-carbon compound called **glycerin** and longer carbon chains, **fatty acids,** as illustrated below:

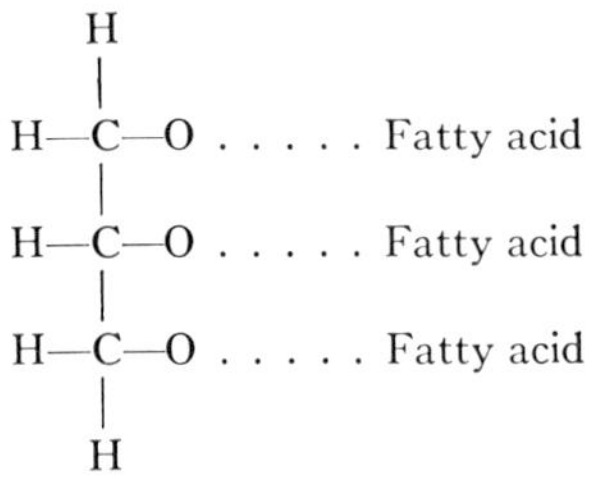

Glycerin

The fatty acids usually consist of even-numbered carbon chains and may be saturated or unsaturated. In other words, they may or may not combine readily with some other atom. Long and saturated fatty acids are usually the hard fats, such as tallow. The unsaturated fatty acids have shorter carbon chains and are generally known as oils. A wax is a long fatty acid that has combined with some substance other than glycerin. Complex ring compounds known as **sterols,** which form the basic structure of many hormones, are closely related to the fats.

The fats, like the carbohydrates, are important in that they release energy through their decomposition. As the carbohydrates can be converted to fats, so also can the fats be converted to carbohydrates. All the organic compounds (fats, carbohydrates, and proteins) are utilized in the synthesis of protoplasm. However, fats are the fundamental building blocks of protoplasm; they contribute to the framework of the membranes and thereby have an important role in diffusion and osmosis.

The most complex of the organic compounds found in protoplasm are the **proteins.** They are macromolecules with a very high molecular weight that may vary from 5,000 to many millions. Those substances with a lower weight are termed **polypeptides.** In addition to carbon, hydrogen, and oxygen, nitrogen and sulfur are usually present. Proteins are composed of **amino acids,** represented by the general formula NH_2-CHR-COOH. The NH_2 radical is the amino group; the R is a ring or chain of carbon atoms having side linkages of H or OH or even of other groups; and finally, the COOH is the acid or carboxyl group that can release H^+ ions under certain conditions.

The amino acids are the structural units of the proteins, and protein specificity is determined by the kind, number, and arrangement of these substances. The almost unlimited number of arrangements accounts for the great variation of life forms and processes found on the earth today and in the past. Herein lies the basis not only of evolution, but also of the origin of life itself.

So far about sixty amino acids have been isolated or identified in plant tissues, but only about twenty are concerned in protein formation and five in nucleic acids. The nature of the protein is determined by the arrangement of the amino acids. The series of amino acids are linked together by peptide bonds. The structures resulting from the formation of peptide bonds are termed **peptides,** and the individual amino acids of peptides are termed **residues.** Thus a dipeptide (one with two amino acid residues) looks like this:

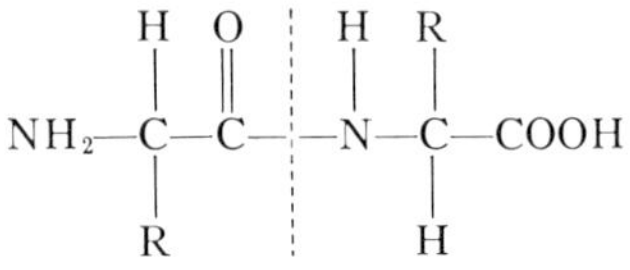

or an actual polypeptide (in this case a tetrapeptide) is shown at the top of opposite page.

A long series of these peptide chains comprise a protein. The complexities of these are beyond the scope of this text, except that a brief consideration of the DNA molecule will demonstrate one type of protein structure (Fig. 6-5). The various amino acids that form a protein chain will

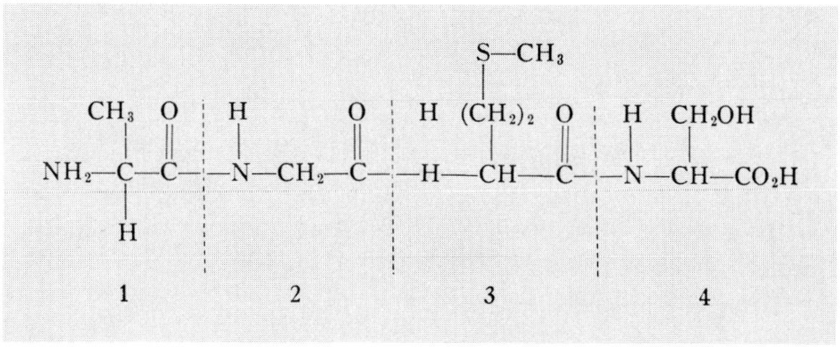

follow a certain sequence, and these are represented according to their abbreviations given in Table 5-3.

Plant scientists are particularly interested in a special group of proteins, the **nucleoproteins,** because of their role in the transmission of hereditary characters and in the biosynthesis of proteins. They are found in both the nucleus and the cytoplasm of the cell (Chapter 6) as protein-rich in the basic amino acids, and combined with a nonprotein material termed a nucleic acid, threadlike particles of high molecular weight. The following diagram shows the highly complex structure of these chemicals:

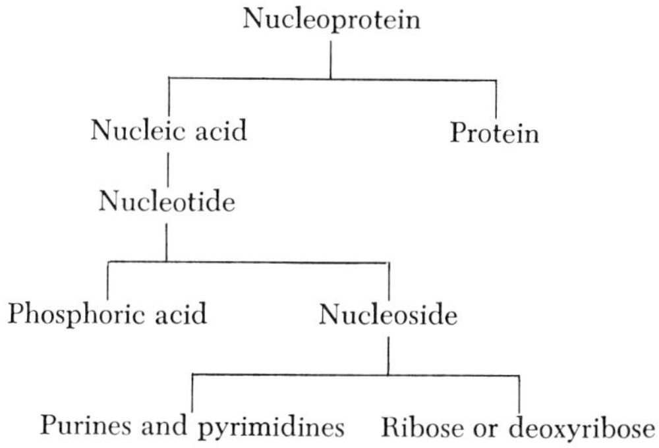

Space does not permit further consideration of the biochemistry of each of the substances, but it is considered in detail in any of several references listed for this chapter. It should be understood, however, that there are two main groups of nucleic acids distinguished by their sugars, ribose or deoxyribose. Ribonucleic acid (RNA) is similar to deoxyribonucleic acid (DNA)

Table 5-3
Common amino acids and bases

Name	Abbreviation	Where found
Adenine	Ade	DNA, RNA
Guanine	Gua	DNA, RNA
Thymine	Thy	DNA
Cytosine	Cyt	DNA, RNA
Uracil	Ura	RNA
Glycine	Gly	Protein
Alanine	Ala	Protein
Valine	Val	Protein
Leucine	Leu	Protein
Isoleucine	Ileu	Protein
Serine	Ser	Protein
Threonine	Thr	Protein
Cysteine	Csh	Protein
Methionine	Met	Protein
Glutamic acid	Glu	Protein
Aspartic acid	Asp	Protein
Lysine	Lys	Protein
Arginine	Arg	Protein
Tyrosine	Tyr	Protein
Histidine	His	Protein
Proline	Pro	Protein
Hydroxyproline	Hyp	Protein
Cystine	Cys	Protein
Tryptophan	Try	Protein
Phenylalanine	Phe	Protein

but differs somewhat in addition to the different sugars they contain. RNA contains the base uracil, which is not found in DNA, and, conversely, the latter has thymine, not found in RNA. The two have a distinctly different role in life processes as will become apparent in later discussions in this text. DNA is found exclusively in the cell nucleus where it occurs in an even more complex condition as a part of the **chromatin,** the genetic material of the cell. RNA, on the other hand, occurs mainly in the cytoplasm and plays an important part in the biosynthesis of protein (Chapter 7).

LIFE SYSTEMS

At each level of organization, molecular and chemical activities effect the functions we recognize as life activity. Therefore, in simple terms, the life systems are a series of rather clearly defined steps that increase in complexity as they progress:

Cell → Organism → Population → Species

As the organization of plant life becomes more complex, these systems, too, become complex, but they still may be reduced to the same basic processes, that is, all life systems involve processes that take place in four stages:

Intake → Flow → Feedback → Replication

The remainder of this text deals with the details of each of these phases of life activity at each level of organization.

SPECIAL TERMS USED IN CHAPTER 5

activator
adenosine triphosphate (ATP)
amino acid
atom
carbohydrate
catalyst
chromatin
coenzyme
disaccharide
enzyme
fat
fatty acid
glycerin
inorganic compound
isomerism
isotope
mole
molecule
monosaccharide
nucleoprotein
organic compound
peptide
photosynthesis
polypeptide
polysaccharide
protein
protoplasm
residue
respiration
sterol
substrate
valence

QUESTIONS TO CONSIDER

1. As a review of some basic chemistry, describe the process known as "burning." What would happen if an enzyme could be incorporated in the process of burning wood or paper?
2. What is the molecular weight of the simple sugar, $C_6H_{12}O_6$?
3. What is a chemical bond?
4. Why are isotopes important in the atomic age?
5. Protoplasm formerly was referred to as a vital substance. Why is this unimportant in today's consideration of the nature of life?

DISCUSSION FOR LEARNING

1. The limits of chemistry, biochemistry, and physiology are difficult to draw. Discuss why this is so.
2. Some biochemists are interested in constructing protoplasm from simple inorganic chemicals. Discuss some of the problems in attempting to do this.
3. Reference is often made to the DNA molecule as being coded and to the amino acid alphabet. Discuss how the possible combinations of amino acid radicals in these protein molecules might be considered a language.

Chapter 6

Cells

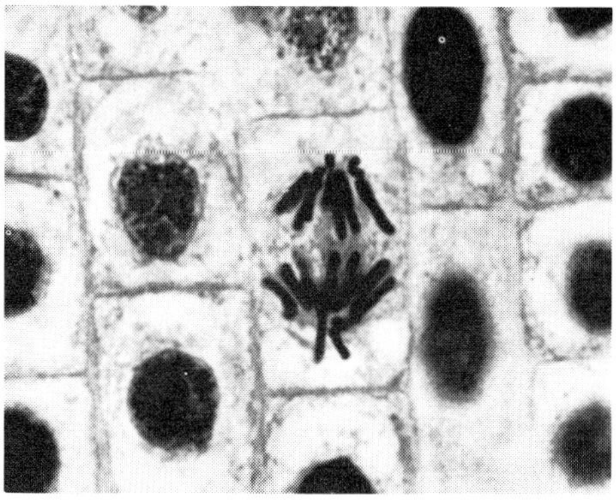

Carolina Biological Supply Co., Burlington, N. C.

The world of atoms, molecules, chemical bonds, and energy exchange is transformed into the world of living substances when we focus on the **cell.** The content of a cell is often referred to as protoplasm, but the term now has relatively little meaning when used to describe the living substance. Cells are never at rest, as can be seen in the photograph above. Even when they are not dividing, they are otherwise active chemically.

Cells have a general structure and size (Fig. 6-1), but their physical appearance varies according to their specific physiological functions. Their chemical composition is much the same in general terms, but it, too, varies from moment to moment. This variation accounts for the great diversity of living organisms, yet at the same time it is the cell that is the unit of life.

Cell shape ranges from spherical, in the freely floating cell, to flattened, irregular, cylindrical, cuboidal, or rectangular, or spindle-shaped. Cell size in plants ranges from that of certain kinds of bacteria with a diameter of about 1 μ, in which each cell is a single protist, to the largest fiber cell in certain plants, with a single cell as long as 22 inches. Although the cell size ranges as much as a million to one, the range in the size of plants is much greater. The smallest flowering plant is approximately 1 mm. long (*Wolffia* sp., a floating aquatic plant), whereas the largest (the giant redwood tree in the west) measures 360 feet and weighs at least 1,000 tons.

A cell may be described as a small mass of protoplasm, the **cytoplasm** and a **nucleus,** surrounded by a **plasma membrane.** Each cell, at least at the beginning of its life, is a complete unit, carrying on all life functions, but some cells of higher organisms, for example, cells in the stem of woody plants, are incomplete and do not live long after they are formed.

In cooperation with the cytoplasm the nucleus controls and regulates growth, development, and continued existence of the cell. Even though a cell can function without a nucleus for a short time, the enucleated cell will soon cease to function.

The nucleus cannot extend its control

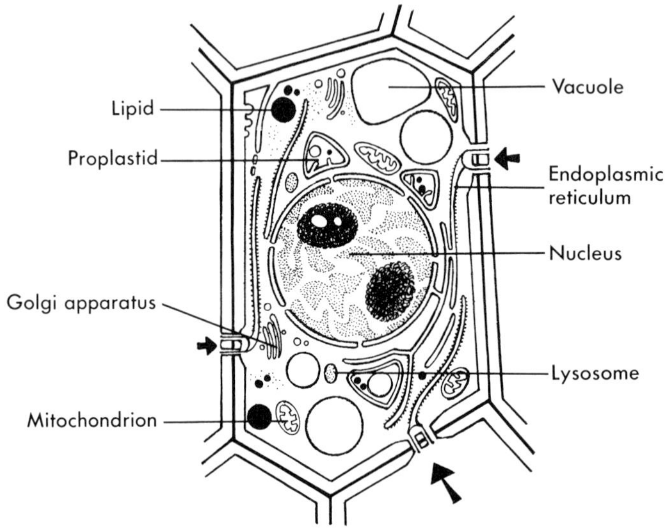

Fig. 6-1
Diagram of a generalized plant cell showing the relative size and position of the parts. The protoplasmic connecting pores are shown at the arrows.

over an indefinite amount of cytoplasm, since as a cell enlarges, its volume increases as the cube of the radius of the cell, whereas that of the surface area of the nucleus increases merely as the square. Such a disproportionate increase in cytoplasm would soon upset the metabolism of the cell. However, a nucleus can increase its surface area by changing its shape or by doubling its chromosome number, thereby doubling its volume. Mature cells tend to maintain a constant nuclear-cytoplasmic ratio, and growing cells will divide, thus keeping that ratio below the maximum value.

Surface area also limits the cell's size. The substances required for metabolism can only pass in and out of a cell through its surface membrane. For example, nearly all cells require oxygen, and if sufficient oxygen is to reach the center of a cell, the concentration outside the cell must be at or above a critical value. The amount of any material in a cell depends upon the ease with which it can enter the cell, its rate of utilization within the cell, its diffusibility after it has entered, and the dimensions of the cell. Since the surface area increases by the square of the radius and the volume increases as its cube, any increase in volume is governed by the ability of the cell surface to provide for the entrance or exit of the various gases, water, and foods necessary for the demands of metabolism. These limitations can be overcome by cells when they abandon their spherical shape by flattening, elongating, developing large vacuoles filled with liquids, or otherwise increasing their surface area.

The rate of metabolic activity within the cell is a third factor affecting cell size. Small cells and small organisms metabolize more rapidly than do larger ones. Thus surface exchanges in smaller organisms must be accomplished more rapidly, and the cells must be smaller so that the amount of surface area compared to volume can be kept at a maximum. The cell contents are confined by the cell membrane that is 0.1 μ or less thick, forming an elastic but firm support. Small cells do not make efficient use of this support, whereas large cells could burst the cell membrane as a result of internal pressure. Therefore a balance must be maintained between the need for support and the need for adequate surface-volume relationship.

In summary, three factors govern size: (1) the nuclear-cytoplasmic ratio, (2) the ratio of the cell surface area to the cell volume, and (3) the rate of cellular activity or metabolism.

METHODS FOR STUDYING CELLS

Two approaches are used in cytological studies: (1) direct study using microscopes and photographs and (2) chemical testing of the composition of the cell. Although the discoveries of recent decades have greatly expanded our knowledge of cell structure and function, cytological investigation is still an extremely active area of research.

It is almost impossible at present to study structure and function simultaneously. By using special techniques we can study the functions of a cell while the component parts are in motion and chemical reactions are taking place. At the same time the composition and the position of a cell's parts will be undergoing alterations. Hence, it is not possible to study the details of structure at the same time function is studied. When a cell is stained so that its structure can be studied, the cell usually ceases to function. Consequently, at the time it is possible to study the structure, it may not be the same as it was while it still functioned. In other words, structure is not static but is as dynamic as function. Therefore there can be no such thing as the study of the structure of a living cell. Nevertheless, much can be learned of structure by investigating a dead cell at any one instant.

The observation of a living cell under the microscope reveals that protoplasm is a relatively transparent material, the components of which, when unstained, refract light to some and the same degree. One part cannot be distinguished from another in the living state. To overcome this problem, cells are stained with dyes, either in the living state or after the cell has been preserved in as lifelike a state as possible. Because the chemical composition of cellular components varies, differential dyes can be used to demonstarte the cellular components. In addition, a number of other techniques are available today, such as microsurgery, electron microscopy, and phase microscopy—all of which contribute to our knowledge of the cell and its makeup.

The ordinary light microscope, similar to those used for classroom study, is sufficient to identify cell types and gross cellular structure. Their limit of resolution, the minimum distance between two points that allows them to be discriminated with the light microscope is about 0.25 μ. Obtaining increased constrast with a phase microscope, a device that through the interference of light rays by special condensers causes a transparent object to appear in shades of gray depending on its thickness and its refractive index, is a system particularly useful in studying living cells and tissues because they need not be harmed by contrasting dyes. A similar instrument, the interference microscope, has the advantage of giving quantitative data. In the dark-field microscope the ordinary condenser is replaced by one that illuminates the object obliquely so that no direct light enters the objective. The object appears bright because of the scattered light, and the background remains dark. With this type of microscope, objects can be seen that are smaller than those detected with the ordinary microscope. A polarizing microscope is still another instrument that increases resolution by increasing contrast in an object through the polarization of the light passed through it. The electron microscope, already described in Chapter 3, permits the study of the cell ultrastructure. The even newer scanning electron microscope, while it lacks the resolving power of the electron microscope, has the advantage of high contrast so that shape and dimension of extremely minute structures can be photographed.

Remarkable advances have been made in our knowledge of the molecular biology of cells through x-ray diffraction, a technique making use of a beam of X rays brought into parallel lines by means of a defining slit. The beam passes through the specimen to be analyzed, and the rays are recorded on a photographic plate. The diffraction pattern shown is then interpreted in terms of the molecular size and structure. It was this technique that led to the discovery of the double helix structure of the DNA molecule.

The chemistry of cells is determined by selective staining. The careful selection of these stains and their use on fixed specimens, killed at various stages or during a living process, will allow the assay of small quantities of material. The details of these techniques may be found in manuals dealing with the subject. Closely allied to this technique, however, is a method using substances tagged (marked) with radioisotopes. A living plant cell is treated with an enzyme, such as tritiated thymidine, which has a radioactive hydrogen, tritium (H^3). The tagged thymidine is taken up during DNA replication. Cells so treated are then brought into contact with a photographic emulsion, and the position or change of position of the radioisotope within the cell is recorded on the photograph. By comparing this radioautograph with the cells as seen under a microscope, the radioisotope is localized.

The items on the checklist of cell structure shown in the preceding study box are considered one by one in the following paragraphs. The student should not forget that the structure and content of the living cell is constantly changing. These minute factories are sites of great activity and energy exchange difficult to describe in mere words. The best way to understand what is taking place is to imagine oneself as an observer taking a tour of one of these factories. As one passes from place to place, he should remember that activity continues throughout the factory at all times.

THE CELL

- Nucleus
  - Nucleolus
  - Chromatin (chromosomes)
  - Nuclear membrane
- Cytoplasm
  - Matrix
  - Vacuolar system
  - Ribosomes
  - Microsomes
  - Mitochondria
  - Plastids
- Cell membrane and cell wall

CELL MEMBRANE AND CELL WALL

The cell, to remain a discrete unit, must have a boundary, but not solely to outline the cell. The **cell**, or plasma, **membrane** is a functioning structure, composed of lipoid and protein molecules of about 40 Å in thickness. Through its particular structure the cell permeability is determined. Thus the entrance and exit of such substances as water, oxygen, and çarbon dioxide, as simple compounds or as dissolved substances, are regulated. Differences in cell function determine which substances can enter the cell through osmosis or by active transport involving energy expenditure (Chapter 7).

Most of our knowledge about the plasma membrane is based on the function of unicellular organisms or on animal cells bathed by body fluids. In the case of plant cells the outer living limit is the cell membrane, but it is not necessarily the outer limit of the cell. On the outer surface of the cell is a permanent cell structure in the form of a **cell wall.** This fairly rigid surface is characteristic of most plant cells and one that makes them different from animal cells. The cell wall is secreted by the cell. It is formed in various stages during the cell's development, first as a jellylike pectin compound, the middle lamella, which acts as a cell cement. It later becomes infiltrated with **cellulose**, which is usually a polysaccharide but in some cases is a woody lignin. Because of its rigidity, the cell wall gives form to the plant cell and serves as a skeletal structure. Plant cells in particular acquire walls that give permanent shape and support to the plant. In addition, in some plants the surface cells accumulate a layer of cutin (waxlike material) that forms a **cuticle** on the exposed cell surface; for example, the shiny surface of a poison ivy leaflet is a cuticle.

CYTOPLASM AND CYTOPLASMIC ORGANOIDS

Once inside the cell membrane we see that the cytoplasm is the metabolic center of the cell, where its functions are executed. It should be emphasized that parts of the cytoplasm also replicate. Therefore both the nucleus and the cytoplasm execute the twofold purpose of carrying out metabolic functions and self-perpetuation. Thus they are both interdependent; one cannot usually survive very long without the other.

In the cytoplasm may be found various organoids such as the round or rod-shaped mitochondria, microsomes, and plastids. All of these are involved in the energy-liberating processes of the cell, as well as in other chemical transformation. This maze of activity must be studied in units divided according to the gross appearance of the parts.

Matrix. Various descriptions of the nature of cytoplasm have been given in the past. It was once thought of as a mixture suspended in the cell membrane, either with a changing structure or an unknown structure. Our present idea, due to the great resolution of the electron microscope, is that the cytoplasmic matrix is the basic molecular fabric of the cell and is composed of ribosomes, ribonucleic acid (RNA) molecules, globular and fibrous proteins (the enzymes), small molecules of salts, and water. This condition obtains in the procaryotic cells and some of the embryonic plant cells. In the eucaryotic cells there are in addition many intracellular membranes (the vacuolar system) and cytoplasmic organoids such as mitochondria, plastids, and microsomes. The matrix is colloidal in nature (see Chapter 7), and it is these macromolecules that carry out the biosynthetic functions and energy transfers of the cell. The movements of the cytoplasm during photosynthesis, cell cleavage, and the formation of fibers are all the result of matrix activity.

Vacuolar system. The more advanced plant cells become vacuolated; that is, open, interconnected spaces form in the matrix so that there is a complex system of sacs, tubules, and vesicles lined by membrane. Thus we conceive of the cytoplasm as composed of two main parts—and the cytoplasmic matrix, which is outside of the vacuolar membrane, and the vacuole compartments surrounded by the vacuolar membrane. Because of the fine divisions of this system, there appear to be an endoplasmic reticulum both granular and agranular, depending on whether ribosomes are present. Part of the vacuolar membrane system is the double nuclear membrane that isolates the nucleus from the rest of the cell. It appears that in reality a flattened vacuole surrounds the nucleus, giving the appearance of a double membrane, the outer one with numerous plugged pores. Through these pass substances concerned with cell metabolism, protein synthesis, and gene activity—the normal function of the nucleus.

The **Golgi complex** is the term applied to the membranes of the vacuolar system lacking ribosomes; it is similar to the agranular endoplasmic reticulum. The distinctions of the Golgi complex are the flattened sacs composed of parallel membranes, often with small terminal vesicles. These are scattered throughout the cytoplasm of plant cells, and the special term, **dictyosome,** is applied to them. These structures are usually associated with the storage of secretions.

Ribosomes. Definite bodies composed of ribonucleic acid and protein are abundant in the matrix of cells. These organoids are usually referred to as the "engines" of the cell because it is in these particles that protein synthesis takes place. Here, with the aid of various RNA molecules produced by the nucleus, the various amino acids are assembled into a definite sequence to produce the polypeptide chains. Ribosomes of many of the eucaryotic cells are associated with the intracellular membranes of the vacuolar system, but in bacterial cells and other single cells, particularly developing cells, they are free in the matrix.

Lyosomes. These organoids, only recently discovered in animal cells, are known to occur among some fungi also. They are membrane-covered storage vesicles slightly smaller than mitochondria. Enzymes used for digestion are kept from acting on the cytoplasm of the cell by their tough membrane.

Microsomes. Among the least known of the cellular contents are the microsomes, small bodies located on or near the vacuolar membrane. These may be seen as small particles in electron micrographs. But when they are recovered from cell **subfractionations** using the ultracentrifuge, it is found that they are not actually specific organoids of the cell, but rather a heterogenous group of cellular substances, including lipids and enzymes, that comprises both the granular and agranular reticulum.

Mitochondria. The organoids known as mitochondria are perhaps both the most distinctive and most interesting of those found in the cytoplasm. In several respects they are comparable to the nucleus: they contain a distinctive DNA (mitochondria DNA or MDNA); they are capable of replication, although this is not yet understood; and finally, they control the chemistry of the cell. In fact, it is in these structures that great chemical activity takes place, since this is the site of the Krebs tricarboxylic acid cycle (Chapter 7) and of respiration. Over seventy enzymes, coenzymes, cofactors, vitamins, and metallic ions take part in the activities in these organoids. Mitochondria (Fig. 6-2) are usually rather uniform throughout, but they may at times swell apically, especially during times of great activity. They are composed of a double external wall of protein, the outer and inner membranes, between which is an inner space. The inner matrix is divided by mitochondrial **cristae,** which appear like baffles or septa formed from invaginations of the inner membrane. Much of the chemical activity seems to take place on or near these cristae, and this appears as small granules. In summary, these organoids are best described by the term cell "engines" or cell "machines."

Plastids. Plastids are nucleoprotein-containing globular bodies present in the cytoplasm of plant cells. The most important of these plastids (**chloroplasts**) contain **chlorophyll** and are the seat of photosynthesis in the green cell (Fig. 6-3). Other plastids, which lack the layered structure and the photosynthetic apparatus of the chloroplast, may be present. Colorless plastids are called **leukoplasts,** whereas those containing concentrations of yellow, orange, or red carotenoid pigment are known as **chromoplasts.** Chloroplasts are vital to the manufacture of food in those plants in which they are present.

The ultrastructure of chloroplasts consists of stacks of small granules, **grana,** embedded within the stroma or matrix of the organoid. The grana are interconnected by membranous tubules. This gives the plastid the appearance of being lamellate (see Chapter 7 for the function of chloroplasts).

Nucleus. Function and replication are under the control of the DNA molecules of

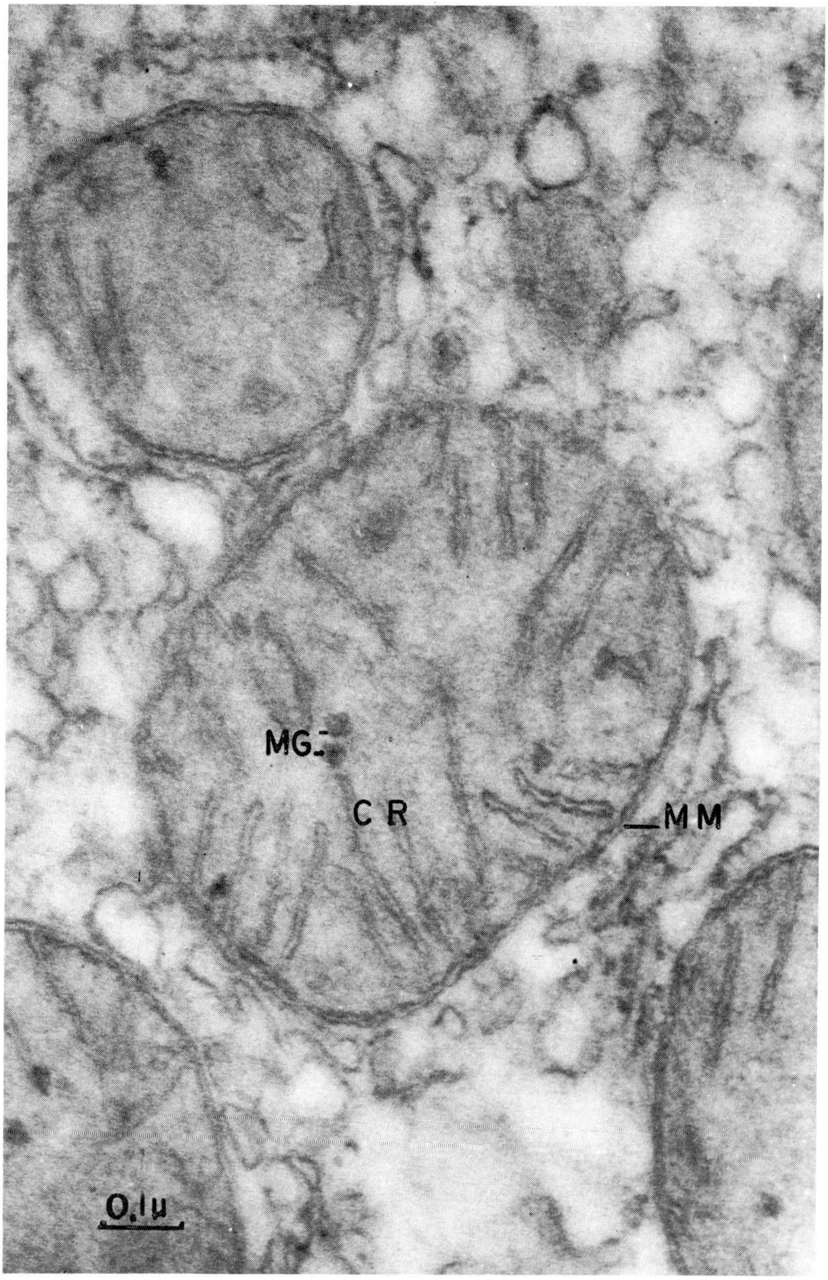

Fig. 6-2
Mitochondria showing cristae, **CR**; mitochondrial granules, **MG**; and mitochondrial membrane, **MM**. (×250,000.)

Fig. 6-3
Cell of a filament of *Spirogyra* sp. showing the extent of a chloroplast in an active vegetative cell.

the nucleus or of nuclear material. Procaryotic cells have their nuclearplasm distributed throughout the cell, but eucaryotic cells have the nucleus as a discrete organoid surrounded by a membrane (previously described). The most conspicuous nuclear components are the slender threads of chromatin (DNA, RNA, and nucleoproteins). During cell division these thicken, and each has a distinct shape. They are then known as chromosomes. The number of chromosomes for each cell nucleus is constant for a given plant species. This is known as the chromosome number (Chapter 10).

Within the nucleus itself is a differentiated body, the **nucleolus,** which seems to act independently of the nucleus both during cell division and between cell divisions. This complex structure produces RNA, which then acts as the messenger from the nucleus to the point of growth and enzyme synthesis in the cytoplasm. There may be only one or there may be several nucleoli within the nucleus. Both the nucleus and the nucleoli are suspended in an aqueous **nuclear sap.**

CHROMOSOMES

Chromosomes consist mainly of three kinds of chemical substances: protein, **deoxyribonucleic acid (DNA),** and **ribonucleic acid (RNA).** Both the nucleic acids and the proteins are giant molecules having a main stem and many side branches, as discussed previously. There are approximately twenty different kinds of side branches in the proteins, whereas in the nucleic acids there are only four, and they are of a different type. All proteins and nucleic acids occur in no more than one optical form; therefore there are no mirror-image molecules or isomers. It appears that these are basic components of living matter, having a simplicity that is beyond expectation.

The most important role of the proteins is in the formation of enzymes, which, as we have described previously, are highly specific for the reaction they catalyze. The structure of each enzyme, moreover, is rigidly determined, and its side branches are arranged in a fixed order. Enzymes are assembled in the correct order through a coding system in the DNA molecule that is transmitted to the messenger RNA molecule. The amount of DNA in each set of chromosomes is usually fixed for any species. Its constancy led to the belief that it is this material that determines the hereditary pattern. DNA alone, free of protein, may carry genetic information when added to a bacterial cell, where it will produce an inherited change. Other evidence gained through a study of the infection of a bacterial cell by a bacteriophage (a virus that parasitizes bacteria) has shown that the DNA of the bacteriophage enters the cell and there replicates.

DNA is a chemical of high molecular weight (over 1 million) that is made of a number of smaller molecules linked together. These molecules include sugar, deoxyribose, phosphoric acid, and four bases, of which two are pyrimidines (thymine and cytosine) and two are purines (adenine and guanine). The alternate arrangement of sugar and phosphoric acid forms the outside boundary of DNA, and the base pairs link the two sides together. The bases are always paired, adenine and thymine together and guanine and cytosine together. Bases are linked together by hydrogen bonds. From x-ray analysis it has been determined that DNA has its molecules arranged like a double helix or loose spiral. It is similar to a spiral staircase, the handrails of which are alternating sugar and phosphate and the steps of which are base pairs. Fig. 6-4 shows a model of DNA as determined by Watson and Crick.

There may be thousands of folds in the spiral configuration of a DNA molecule, and the base pairs may be arranged in any order. Such a molecule may have a multitude of possible variations, thereby confirming that the sequence of base pairs is the key to the heredity-determining qualities of DNA and is responsible for the formation of proteins and RNA. DNA is the key molecule, as its structure confers hereditary uniqueness on cells as well as on organisms. The base pairs of the DNA molecule act as a pattern or template to initiate formation of other complex molecules that compose a living cell, in other words, the genetic code.

The genetic code is entirely a matter of the arrangement of the amino acids in the proper sequence on the DNA or RNA molecules, which in turn determines the

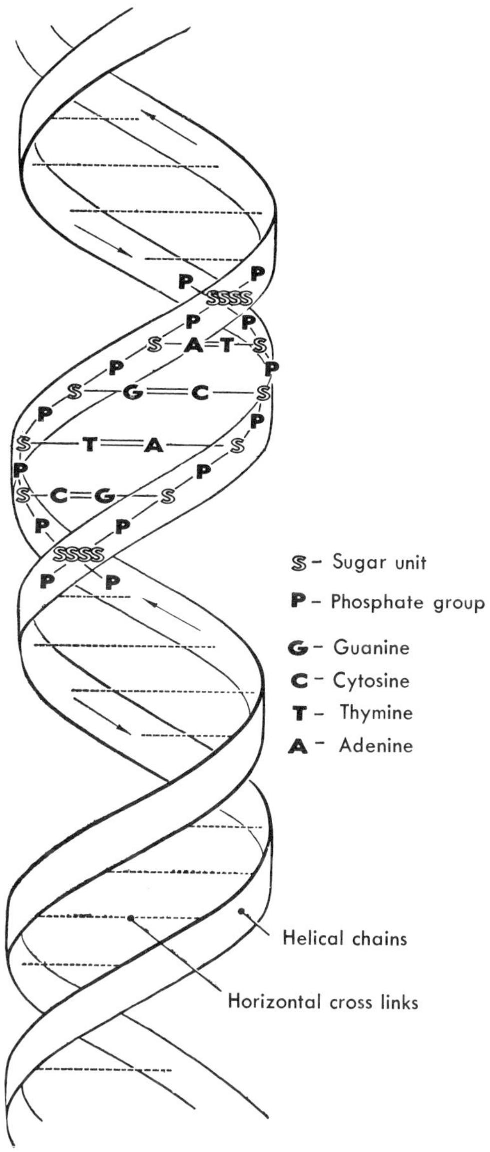

Fig. 6-4
Model of the double helix DNA molecule as described by Watson and Crick.

sequence of the amino acids that form the proteins manufactured in the ribosomes.

In Fig. 6-5 the DNA model consists of two parts, each of which is the complement of the other. Therefore either chain can act as a mold on which a complementary chain

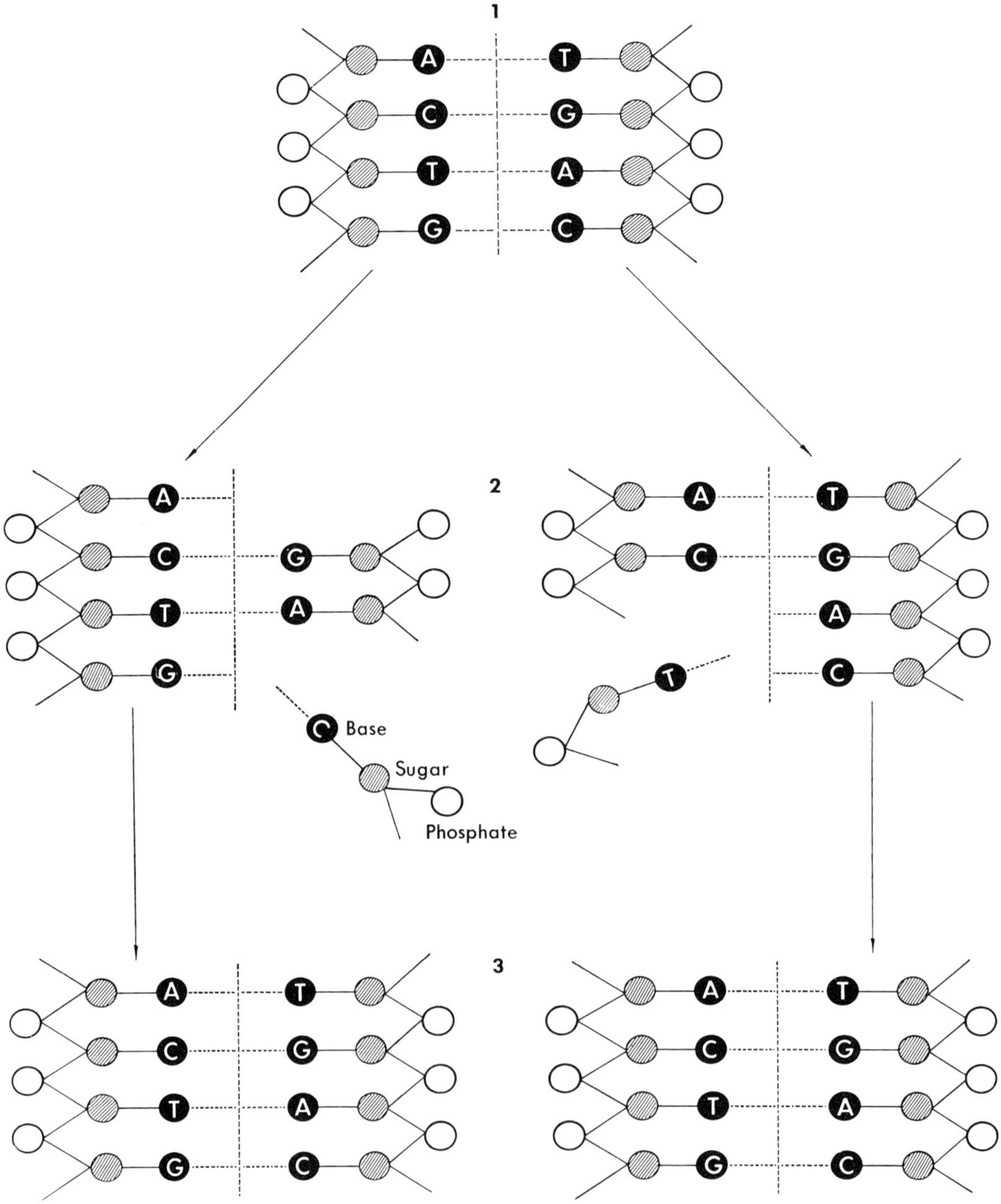

Fig. 6-5
Diagram of DNA replication. **1**, One "loop" of molecule. **2**, The molecule splits and bonding of nucleotides occurs, through enzyme action, to form new "loops." **3**, Two identical helices formed by each half of the original helix, completing replication.

can be synthesized. As the two chains unwind and separate, each builds a new complement to itself so that when the process is complete, two pairs of chains exist where there was only one before. Since the pairing of the bases is specific, the sequence of the pairs of bases is duplicated exactly. The mold has not only assembled the amino acid and sugar molecules but also has placed them together in exactly the same order.

If we assume a single helical chain of DNA and, in the cell, an available supply of the four kinds of building blocks necessary to make a new chain, they may then duplicate themselves, as shown in Fig. 6-5. The building blocks are adenine, thymine, guanine, or cytosine attached to a sugar and a phosphate, forming four varieties of nucleotides. They may join together in an unlimited number of sequences, one of which is illustrated. From time to time one of these free nucleotides will attach itself by its base to one of the bases of the single chain and another will attach itself to an adjoining base. If they are not the correct mates, that is, if they are not mated properly, the nucleotides drift apart, to be replaced by the correct partners so that they can be in the proper position to complete their linkages and form a new chain. Therefore only those nucleotides that are compatible with their adjacent partners can move into a permanent position.

If it requires water or any other inorganic substance, the cell accumulates this substance through diffusion and osmosis. On the other hand, when a cell requires a carbohydrate or fat or their derivatives, the cell must synthesize them from simpler raw materials, with the aid of appropriate enzymes. These organic molecules automatically will be exact duplicates of those present, provided that the enzymes remain the same. Enzymes also are required to link amino acids together in the duplication of proteins. Evidently the genes of the chromosomes must provide specific blueprints or templates for this linking. Finally, the gene must duplicate itself, acting as its own template and controlling its own replication.

MITOSIS

All cells at one time or another undergo **cell division** or **mitosis** (Fig. 6-6). They either divide regularly, as in unicellular organisms and growing tissue, or they divide only when stimulated under specific accidental, abnormal, or experimental conditions, as in x-irradiation. It is probable that all cells posses the potentialities for mitosis, but whether and with what frequency they do divide depends, under normal conditions, upon whether they are in contact with other cells and upon their growth state, that is, volume in relation to normal size. In those cells having a definite nucleus, mitosis requires duplication of the nucleus and division of the cytoplasm into two parts.

Duplication of the nucleus involves duplication of the chromosomes. The chromosomal changes and the events associated with them are grouped under the term mitosis. However, in those cells, such as in *Oscillatoria formosa* (Monera), in which a distinct nucleus and chromosomes are not evident, cell division includes gene replication without the characteristic mitotic changes. Sometimes division of the cell may take place without these visible, characteristic mitotic events.

The immediate stimulus necessary to start cell division is probably enzymatic. Some cells divide when they have grown to double their size; other cells grow only after mitosis has occurred. Some chemicals are known to promote cell division; others inhibit the process. All that can be said is that certain environmental factors stimulate cells to divide. Acknowledging that by one means or another the process of mitosis begins, we can arbitrarily distinguish four stages that gradually merge into one another; it is difficult to determine where one

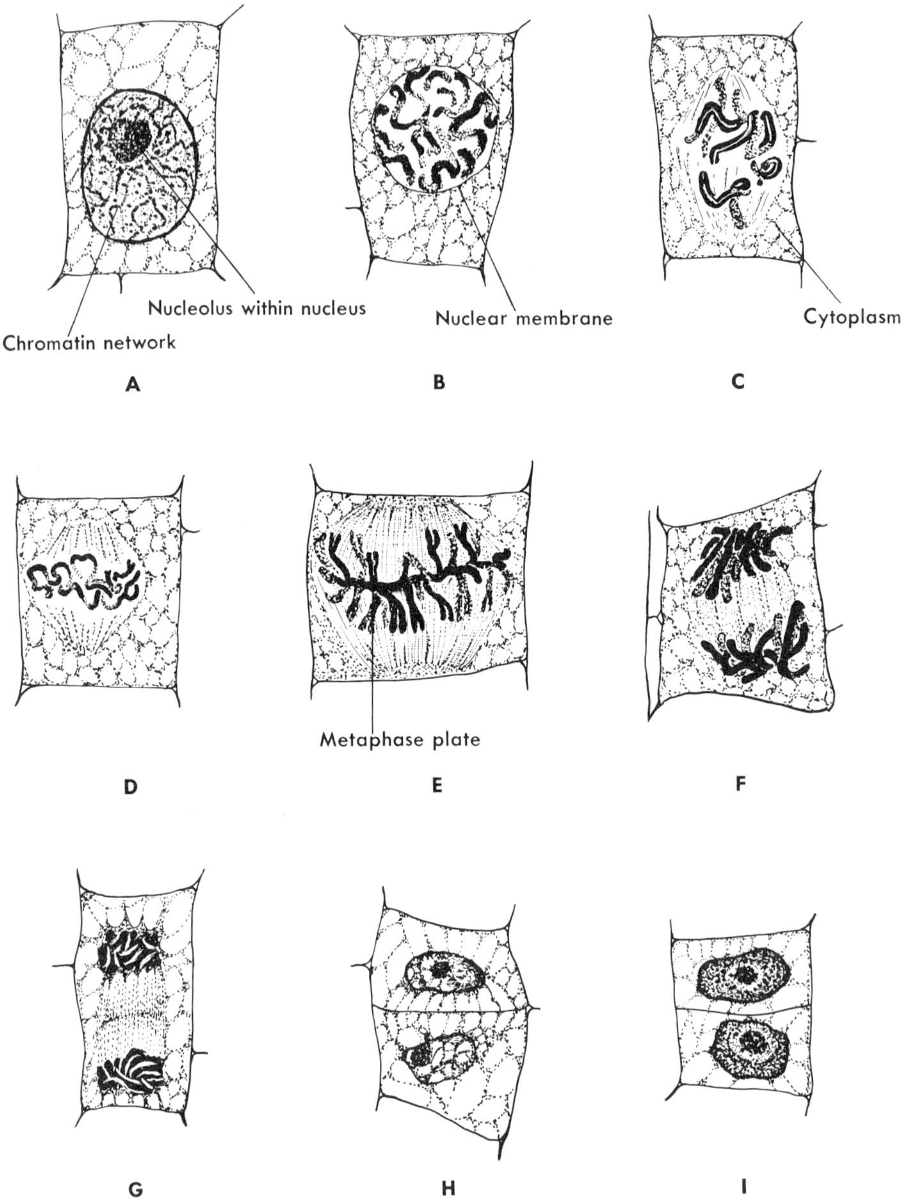

Fig. 6-6
Mitosis or cell division diagrammatically represented to show all the different stages. **A,** Early prophase (chromatin begins to coil; nucleus present). **B,** Medium prophase (chromosomes forming, becoming visible, nucleus disappears). **C,** Late prophase (nuclear membrane disappearing, spindle forming). **D,** Early metaphase (chromosomes splitting on equator of spindle). **E,** Later metaphase (chromosome showing definite longitudinal split). **F,** Anaphase (chromosomes moving toward spindle poles). **G,** Telophase (chromosomes arranged at poles; cell plate forming on spindle equator). **H,** Late telophase (nuclear membrane forming around nuclei; nucleolus appearing; chromosomes no longer visible). **I,** Two new daughter cells.

stage begins and another ends. These four stages are called, in proper sequence, prophase, metaphase, anaphase, and telophase (Fig. 6-6). Prior to any visible evidence of cell division, complete biochemical replication has taken place.

Of all the activities that can be observed in a cell, mitosis and meiosis are the most complicated, and their machinery appears to be the most elaborate on occurring in the cell. During cell division a beautiful geometric design is created out of the seeming formlessness of the cell (Fig. 6-7). This design operates with extreme coordination. It begins with the resting cell (**interphase**), resting only as far as its division is concerned, but certainly not in any other way. From chemical and genetic data it is known that chromosomes are present, although they are not visible as such.

Prophase. As the prophase begins, the previously invisible chromosomes become visible as slender threads, which thicken and become demonstrable as distinct chromosomes. Upon close examination each chromosome is actually a double filament, a single filament that has replicated during interphase. Gene replicas were formed on

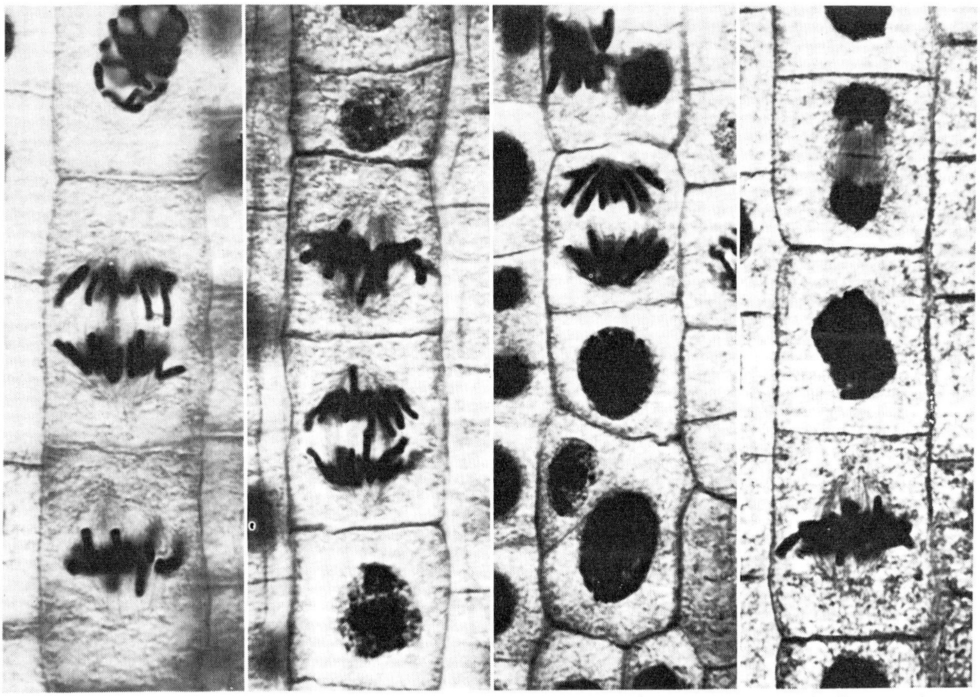

Typical prophase and anaphase — Typical metaphase and anaphase — Late anaphase — Telophase and metaphase

Fig. 6-7

Mitosis in onion root, tip cells in various stages. Compare with Fig. 6-6, to which the following letters refer. Upper left, medium prophase (**B**); middle left, anaphase (**F**); lower left, early metaphase (**D**); top of left center, two new daughters cells (**I**); next down (**D**); next (**F**); lower left center, early anaphase (**B**); right center, interphase (**A**); and another anaphase; upper right, telophase (**G**); lower right, later metaphase (**E**). (Courtesy Carolina Biological Supply Co., Burlington, N. C.)

the template of existing genes, and the new genes have been incorporated into the new chromosomes. The duplicate chromosomes are so closely apposed that they appear as one. The disappearance of the nuclear membrane has resulted in the coalescence of nuclear and cytoplasmic contents. The presence of long, thin threads of visibly distinct chromosomes in the cell and the absence of a nuclear membrane constitute the early prophase stage of mitosis. The chromosomal threads separate longitudinally into identical halves called **chromatids.** The increasing visibility is probably caused by loss of water, thereby making the stainable parts appear more compact and dense. Equally important is that the chromosomes shorten and thicken by a process that transforms the slender chromatids into a structure coiled like a spring. The chromatids continue to shorten during prophase, and the coils decrease in number as they increase in diameter. The nucleoli, which are formed by particular chromosomes and which are initially prominent, diminish in size and have disappeared by the end of prophase. The entire prophase takes, on the average, about 88 minutes, by far the longest stage of mitosis.

The complete disappearance of the nuclear membrane, which occurs in late prophase, coincides with the appearance of a new structure in the cytoplasm called a **spindle.** Chemically, the spindle consists of long-chain protein molecules oriented longitudinally between two **poles.** This introduces the next stage.

Metaphase. Although these phenomena mark the beginning of the second stage, metaphase, one should keep in mind that each stage is merely the continuation of a process begun in the preceding stage. Once the spindle has formed, the chromosomes move through the cytoplasm to a region midway between the two poles, called the equator, and become attached to the spindle by their **centromeres.** The centromere (or kinetochore) is a small, clear circular zone on the chromosome at the point where the arms of the chromosome are joined. It has a functional relation to the movement of chromosomes during mitosis and is constant in its position. The division of the centromeres occurs simultaneously in all chromosomes during metaphase. The daughter centromeres move apart and in turn separate the chromatids, which then begin their migration to the poles. Thus the chromatids cannot separate from each other later, or would the chromosomes have the characteristic V or rod shapes that identify them should the centromere be absent. The position of the centromere is visible in a metaphase chromosome as an area of constriction, and since the position is characteristic for each chromosome, the chromosome is divided into two arms of the same or varying lengths. Very few chromosomes have terminal centromeres. In the onion this stage lasts only 1.4 minutes, compared to the 88 minutes of the previously mentioned prophase.

Anaphase. In the mitotic cycle, metaphase is followed by anaphase. The centromers have divided so that each chromatid will have its own centromere. Each chromatid (chromosome half) will move away from the other so that sister chromatids will eventually be found at opposite ends of the cell. Anaphase terminates when the new chromosomes are densely packed at the two poles. The process lasts, on the average, about 3 minutes.

Telophase. Essentially, the events that occur during telophase appear to be opposite to those found during prophase. The nuclear membrane forms anew, the chromosomes uncoil and become slender threads again, the nucleoli reappear, and the nucleus again takes on the appearance found in a resting cell. A new cell wall, which forms at the equator, initially appears as a **cell plate** within the spindle and soon reaches to the outer walls of the mother cell, roughly dividing the cytoplasm into two equal parts. The spindle disintegrates, two new cells have been formed, and mitosis is complete. This final stage lasts, on the average, about 4.6 minutes.

• • •

As a result of mitosis, two daughter cells, the chromosomes of which are exactly like the parent cells in number, in characteristic shape, and presumably in genetic makeup, have been formed in approximately 97 minutes. Cells that appear to be resting between successive mitoses are said to be in the interphase stage.

A description of the mitotic process as a cell passes from one interphase to another by no means explains the complex aspects of cell division. The entire process is a delicately balanced and coordinated drama in which the nucleus, cytoplasm, and all parts of both come in on cue to play their individual roles. Above all, mitosis is a process of growth that involves the assimilation of materials from the outside and their transformation through breakdown and synthesis into new cell parts, all of which utilize energy. Not only are cells undergoing enlargement during mitosis, but also, more significantly, a continuous succession of genetically identical cells is ensured.

The control center of the cell is the nucleus, and since the nucleus is composed mainly of chromosomes, they are the regulators both of cellular metabolism and of the structural characteristics of the cell. Since the chromosomes are composed of nucleic acids and proteins that have been intricately fabricated, the two daughter cells, which are alike, must have the same amount and type of nucleic acids and proteins as did their parent cell. The duplication of chromosomes into identical chromatids and their separation toward the poles must be exact in every detail, a mechanism provided by mitosis. Even though exact lineal heredity is established from cell to cell and from organism to organism, accidents and variations, some of which are important if evolution is to take place, can and do occur, a fact established by experiment.

Mitosis in plants, as in animals, occurs at a rate characteristic for each organism, but it is affected by temperature and is governed by such factors as nutrition. A hair cell of a plant, *Tradescantia* sp., will go through one cell division at 10° C. in 135 minutes, but at 45° C. it will take only 30 minutes. Other plants, such as bacteria, will divided every 15 to 20 minutes at 37° C., whereas some root-tip cells may take 22 hours at room temperature. Among some organisms in which the mitotic cycle takes 45 minutes, more than 17 hours elapse between mitoses, indicating that the time spent by a cell in preparation for division is an important factor, a factor about which our knowledge is limited.

MEIOSIS

In mitosis the chromosome number is maintained constant during successive nuclear divisions, and the chromosomes divided lengthwise after duplicating themselves. The heredity determiners or **genes** are arranged in a linear fashion on the chromosomes. Obviously, then, there is more than one gene on each chromosome. Therefore, when a chromosome duplicates itself during mitosis, the genes are also duplicated, and each new set of chromosomes will carry identical and corresponding genes. In those organisms that originate as a result of sexual reproduction, there are two sets of chromosomes, one from the female parent and one from the male parent. Each pair of chromosomes constitutes a homologous pair, and the maternal chromosome has a corresponding paternal one that is similar in structure, size, and gene makeup. The members comprising this pair are termed **homologues.** Thus any organism arising from sexual reproduction will have the **2n** or **diploid** number of chromosomes. Two sex cells unite to form the diploid zygote, and each will contribute the **n** or **haploid** number of chromosomes to that union.

The process by which the chromosome number is reduced from the diploid to the haploid is termed **meiosis.** It is essentially the same in plants and in animals. The plant's production of sex cells is called **sporogenesis** (Chapter 9). We shall consider here only the process that results in the haploid chromosome number in the sex cell. The body cells of a plant are produced by mitosis as we have seen, but during the production of either spores or sex cells of plants a series of divisions occur in which the diploid chromosome number is reduced to one half, or the haploid number. Meiosis (Fig. 6-8) occurs in two steps, one immediately following the other; that is, there are two successive nuclear divisions in which the chromosomes divide

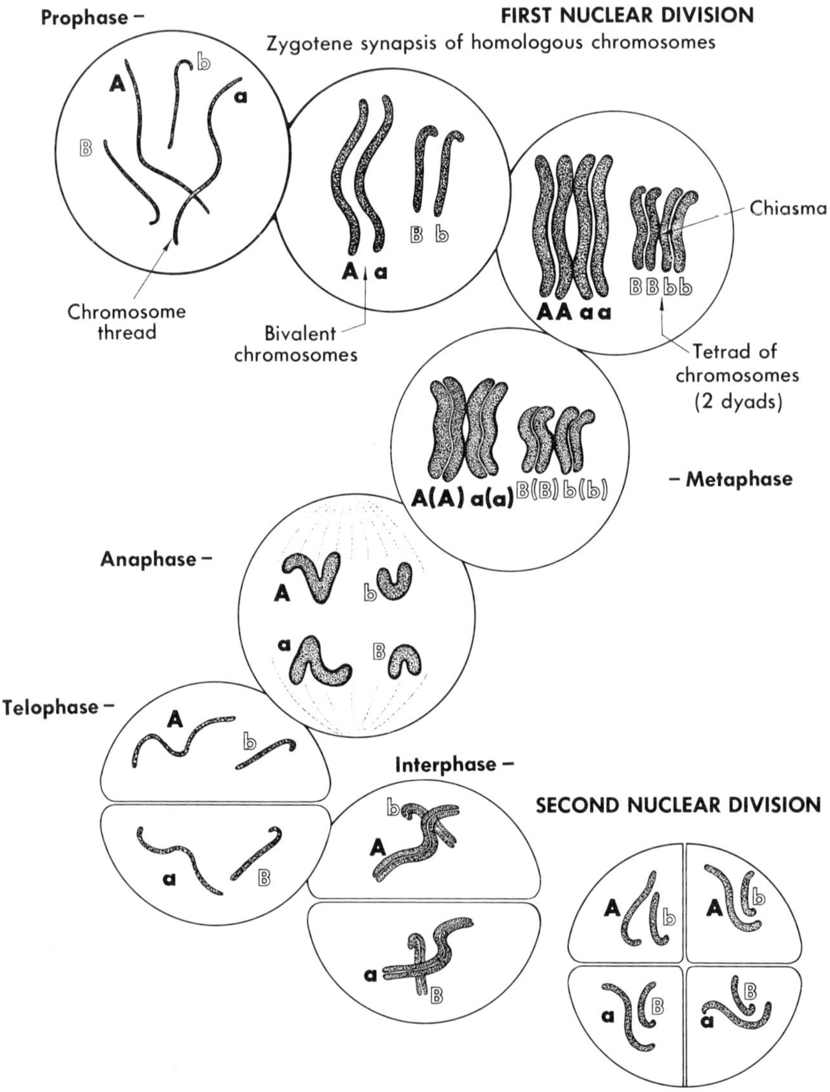

Fig. 6-8
Details of the process of meiosis showing the fate of the chromosomes.

only once. During the early prophase of the first nuclear division the chromosomes appear as long, single-strand, slender thread—the chromatids—which are made of a series of granules, the chromosomes. This is the **leptotene** stage of prophase. Only the homologous threads come together in pairs (a phenomenon known as **synapsis**), and this is termed the **zygotene** stage of the prophase. Each chromosome pair is a **bivalent chromosome** that will then duplicate itself to form a **tetrad.** The four chromatids remain united, being held together more firmly at some regions than at others. An area at which they are held closest together is called a **chiasma** and is cytological evidence that **crossing-over** (the exchange of homologous regions of a chromatid) has taken place; this is known as the **pachytene** stage. Soon after this, the attraction of the chromosomes gives way to repulsion, and they tend to separate, the stage termed **diplotene.** When the tetrads become shorter, thicker, and more compact, the division between homologous chromatids is barely or no longer visible, and the **diakinesis** stage is reached, the stage during which the loops are formed between the successive chiasmas. The half-loops lie at right angles to one another. At the end of the prophase, successive stages take place that are similar to the metaphase, anaphase, and telophase stages of mitosis. As a result, the daughter nuclei receive a dyad of each kind of chromosome. After a short interphase, a second division occurs; the dyads divide so that each of the four daughter nuclei will have one of each chromatid type present. Thus meiosis is complete.

In summary, meiosis consists of two nuclear divisions, one of which is a reduction division. The chromosome number is halved, and four nuclei (later form cells) are produced, each having the haploid number of chromosomes. The union of the haploid sex cells accomplishes two things. (1) it restores the diploid chromosome number and (2) it provides the genetic traits of the new individual. In addition, it results in the random distribution of chromosomes in the gametes or sex cells. Thus the physical basis for the new combination of hereditary material that occurs in every generation is found in this process of meiosis.

SPECIAL TERMS USED IN CHAPTER 6

anaphase
bivalent chromosome
cell
cell division
cell membrane
cell plate
cell wall
cellulose
centromere
chiasma
chlorophyll
chloroplast
chromatid
chromoplast
cristae
crossing-over
cuticle
cytoplasm
deoxyribonucleic acid (DNA)
diakinesis
dictyosome
diploid
diplotene
gene
Golgi complex
grana
haploid
homologues
interphase
lysosome
leukoplast
leptotene
matrix
meiosis
metaphase
microsome
mitochondria
mitosis
n number
nuclear sap
nucleolus
nucleus
pachytene
plasma membrane
plastid
poles
prophase
ribonucleic acid (RNA)
ribosome
spindle
sporogenesis
subfractionations
synapsis
telophase
tetrad
vacuolar system
zygotene

QUESTIONS TO CONSIDER

1. Why is it bettter to refer to a cell not actively dividing as in the interphase stage rather than as a resting cell?
2. Do all cells continually divide?
3. Active cell division is a sign of what process?
4. Locate following cellular activities: control, reproduction, respiration, food manufacture.
5. When do the two parts of the double helix DNA molecule separate?
6. What is the essential difference between the processes of mitosis and meiosis?
7. In what part of the cell does most of the growth in size take place?
8. Considering the previous discussion, in what way do you think cells "communicate"?

DISCUSSION FOR LEARNING

1. Suggest possible reasons why the nucleus of the cells of higher organisms is separate from the cytoplasm.
2. Review the structure of the DNA molecule and again discuss how the genetic code is like an alphabet that spells out instructions to the cytoplasm.
3. There is no doubt that the cell is the unit of life. Discuss, in view of the description in this chapter, the possible effect of cell size on the efficiency of operation. Include in the discussion, comments on why cells are all more or less in the same size range regardless of the organism.

Part III

Function and environment

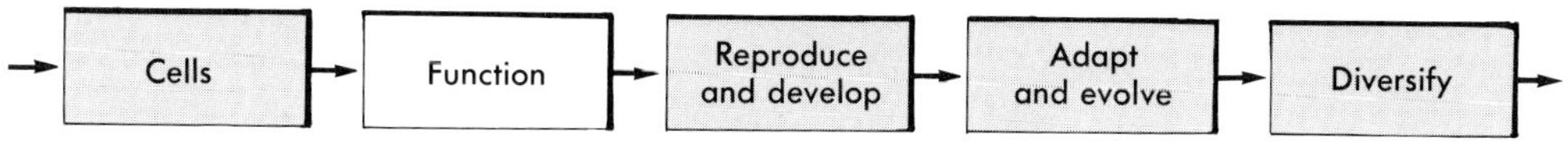

- Function is dependent upon the innate quality of the cell and its dynamic stability with the effective environment.
- Cell quality is controlled by its basic structure and its genetic program.
- Environmental effect is the remainder of cellular action less cellular reaction to the environment.

Chapter 7

Function of molecular systems in plant cells

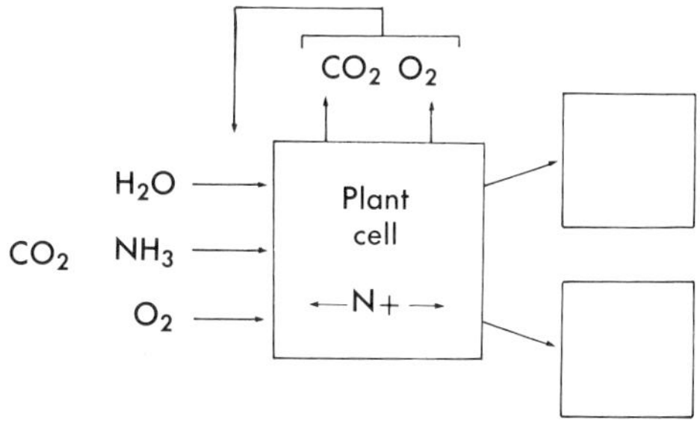

The first level of organization, the cell, is the center of all plant activity, as the system diagram above shows. The input or raw materials are processed, with feedback and growth as the output of this busy molecular system. Specialized cells of the multicellular higher plants or of certain colonial Monera and Protista need to share some function with other cells and thereby become dependent upon them. This chapter describes the basic plant functions. Detailed and special functions of the higher plants are described in Chapter 13.

It is apparent that a living organism is distinct and separate from its environment, interchanges with its environment, while remaining in a steady state, and, finally, is endowed with internal divisions even down to the molecular level.

Plant cell functions involve intake, synthesis, transfer of energy, reproduction, and excretion (or **secretion**). Nearly all cells will absorb water, simple inorganic ions, respiratory gases, simple carbohydrates, and other organic nutrients from their surroundings. Substances that are not dissolved in water or in some other media are synthesized. A plant cell will absorb through its surface raw materials dissolved in water and will eliminate various products in the same way. Substances to be eliminated are collected in vacuoles before being expelled to the outside. In some cases these substances may be formed into crystals or other insoluble precipitates and stored by the cytoplasm before elimination.

The synthesis of protoplasm and the energy-liberating processes require food, inorganic raw materials, and gases from outside the cell. The waste products of metabolism, as well as any other unusable materials, must be eliminated, and all secretions must be expelled.

A living organism uses food (basically carbohydrates, fats, and proteins) as a source of energy, which it replenishes when the food is exhausted either by absorbing more food or by manufacturing it. It replaces its worn-out parts and even generates a new organism from a part of itself when it reaches maturity. It is not possible to overemphasize the complexity of living matter. The average person not trained in biology thinks of life in simple terms, but the biologist with added knowledge appreciates its complexity, the most striking difference between it and nonliving matter.

Each living plant, as well as other organisms, is the expression if its own protoplasm. Although Huxley aptly terms protoplasm the "physical basis of life," it has been learned that protoplasm is not a single uniform substance, as we have seen from the previous discussions. The functional differences and similarities need to be considered.

PLANT CELL FUNCTIONS

- Intake (diffusion, osmosis)
  - Water
  - Oxygen
  - Carbon dioxide
  - Minerals
- Synthesis
  - Fermentation
  - Photosynthesis
- Transfer of energy
  - Respiration
  - Metabolism
  - Enzymes
- Reproduction
  - Gene action
- Excretion (secretion)

INTAKE

Water

Protoplasm is largely a colloidal system, the liquid phase of which is water containing dissociated ions and small molecules. The solid or dispersed phase consists of large protein molecules and complex fat and carbohydrate particles. Any system that consists of a liquid phase and a solid phase must be either a crystalloid or a colloid, depending upon the size of the particles (Fig. 7-1). If all the particles are very small, such as inorganic ions or small molecules, the system is a true solution or **crystalloid.** It is so named because crystals form readily from it. Ex-

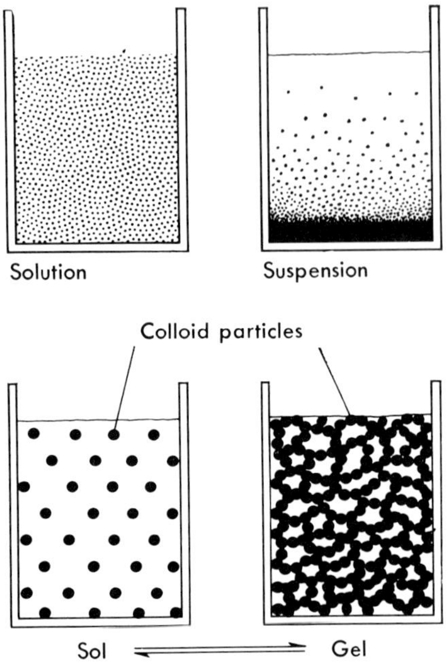

Fig. 7-1
Diagram of crystalloids, suspensions, and colloids.

tremely large particles would merely settle to the bottom by gravity, as in a suspension. On the other hand, if the particles are intermediate in size, they will neither settle out nor form a true solution. Such a system is called a **colloid.** A gas dispersed within either a solid or a liquid, a solid within a gas, liquid, or solid, and a liquid within a gas, solid, or liquid are examples of general types of colloids. For example, water droplets in air (or fog) and protein and fat particles in a liquid (milk) are examples of nonliving colloids. The dispersion of a liquid within another liquid is commonly called an **emulsion.**

Since protoplasm is a colloid, it is necessary to discuss the characteristics of colloids. The molecules that make up a liquid are constantly in motion, thereby bombarding any particles in the liquid. Large particles would fall to the bottom of a container, but smaller particles are kept suspended. The random movement or agitation of these small particles is called **Brownian movement** and can be easily detected under the microscope in the living protoplasm of an ameba or in the vacuoles of some desmids (unicellular green plants) or even in dust in a sunbeam. Although Brownian movement helps to keep these particles suspended, they are not prevented from settling out by this force alone. Electrical charges are largely responsible for their dispersion, since colloidal particles are either positively or negatively charged.

As a colloid, protoplasm may undergo a phase reversal transforming from a **sol** to a **gel** and likewise from a gel to a sol. If more colloidal particles are added or if water is withdrawn from the sol phase, the particles will become more closely associated and form a gel. This gel stage of the colloid is an intricate network of colloidal particles within which are held droplets of water. Thus there is formed a resilient semisolid characteristic of protein colloids such as gelatin. Should the water phase be increased or colloidal particles be removed, the sol or more fluid state will return. Although some protoplasm may be in a more or less permanent gel state, it is not abnormal for protoplasm to alternate repeatedly between the sol and gel phases in accord with the fluctuations in the water-organic compound ratio. Many physical and chemical conditions such as hydrogen ion concentration (pH), pressure, temperature, and absorption of energy may influence the sol-gel phase reversal. For example, the temperature tolerance of colloidal protoplasm is confined to a relatively few degrees coincident with the temperature ranges found in the areas of the earth inhabited by living organisms, that is, from a few degrees below 0° C. to approximately 75° C. above

When it is freshly formed, any colloid is enveloped by what is known as bound water. This is the molecules of water held to the surface of the particles by electrochemical attraction. With time, particles lose this capacity to bind water. For example, it is common knowledge that a gelatin or a custard will lose water when it stands for any length of time. The water lost is bound water, the loss of which causes a significant drying of the colloid.

Colloidal particles, such as ions and molecules concentrated in a certain area, are constantly agitated either as a result of Brownian movement or as a result of heat. In a concentrated region they will collide with each other and be stopped. Those moving toward a less concentrated area will be stopped less frequently through collision since the particles are farther apart. This migratory movement of particles from a region of greater concentration toward one of lesser concentration is called **diffusion.**

When any colloidal system comes into contact with a different medium such as water, air, or even a colloid of a different kind, a membrane will form where the two faces meet. This is known as the interfacial membrane, which in a living system has become known as the plasma membrane. In a protoplasmic membrane it appears that the boundary molecules of water, fat, and protein are oriented in layers or are parallel or are both, since they are subjected to a variety of electromolecular forces. It is obvious that such colloidal membranes in living systems are extremely important, since they are the medium through which all materials enter and leave the cell.

Diffusion. The principal process involved in the passage of water and solutes into, within, and out of the cell is diffusion. It is dependent upon the movement of ions, molecules, and molecular aggregates as a result of their own kinetic energies. Diffusion occurs only when there is a difference in the concentration of particles in two different regions. The particles will move from a region of greater concentration to one of lesser concentration as a result of their intrinsic energy. The force or pressure developed when the particles move is called the **diffusion pressure.** This pressure is proportional to both the kinetic movement and the concentration of particles and is dependent upon their concentration per unit volume and temperature, among other things. Therefore diffusion pressure is not only the result of diffusion but also is the cause. There are instances in which particles are moving in all directions simultaneously at the same rate, as when water molecules pass in equal numbers in both directions through a membrane. Such a system is said to be in **dynamic equilibrium,** since there are no differences in concentration nor is there any diffusion pressure and, therefore, no diffusion. Similarly, when particles are moved by the outside forces (forces other than those of the particles themselves), the process is not diffusion but mass movement. An example of mass movement is smoke carried by the wind from a stack. When a diffusion system is in balance, that is, when there is an equal concentration inside and outside the cell, it is known as net diffusion. It is often apparent that water diffusion is taking place against an activity gradient, a process known as dynamic diffusion, which requires an energy source and a barrier, in this case, a cell membrane. Several possible agencies may be the active energy for this water movement, but the only one that seems to be of importance to the cell is ionizable sidechains of protein molecules capable of unfolding and absorbing water. Of course, the nature of the cell membrane has a direct bearing on this process. This leads us to the discussion of osmosis, permeability, and absorption.

A special kind of diffusion known as **imbibition** occurs when water molecules move into the living and nonliving parts of a plant or a cell through adhesive attraction between the imbibing and the diffusing substances. The swelling of gelatin is due to the imbibition of water; that is, the colloidal particles absorb the water in tightly compressed shells around the particles. Plant materials such as cellulose, starch, and protein are capable of imbibing water, which is a factor responsible for the swelling of seeds in water. Imbibition occurs only when the diffusion pressure of the imbibed water is less than that of the water surrounding the imbibing substance and when there is an adhesive attraction between the water and the imbibing substance.

Osmosis. Diffusion obviously takes place through a membrane, as well as within the cell. This determines not only the rate of diffusion, but also the substances that enter and leave the cell. In this special type of diffusion, which is termed **osmosis** (Fig. 7-2), it is the solvent rather than the solute that is of interest. The difference between the hydrostatic pressure on the inside of the cell and the atmospheric pressure on the outside is the **osmotic pressure.**

In the living cell both the cell wall and the layer of cytoplasm lining the wall determine the permeability of the cell, but the latter is the least permeable of the two. The ability of a substance to pass through these barriers is known as the **permeation** of the substance. Since not all substances may pass into or out of the cell, the cell surface is **selectively permeable** or **semipermeable.** The semipermeability varies from cell to cell and under certain conditions. Water molecules alone permeate it freely, but most ions more slowly, and many large molecules of nonelectrolytes, for example, sugar, not at all. Yet it is known that sucrose will permeate some cells. So it appears that the cell wall will pass a wide range of molecules, which includes all those necessary for the cell's metabolism.

Heretofore, living membranes have been compared to nonliving ones such as cellophane, since that material is also semipermeable. But a living membrane is not like a window screen that allows air to pass through freely, keeping out larger particles because of its pore size. That a membrane has pores is one hypothesis of

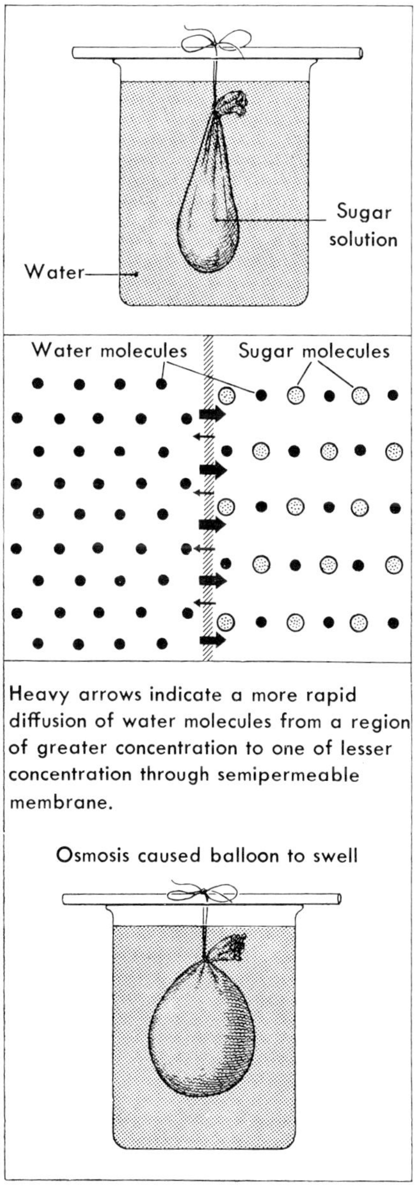

Fig. 7-2
Osmosis. The top panel shows a balloon made of a semipermeable membrane filled with a sugar solution, the molecules of which are too large to pass through the membrane. Water does pass through, however, as shown in the diagram in the middle panel. The result is the swollen balloon shown in the lower panel.

membrane permeability. It has been learned that the pore hypothesis is inadequate to explain what occurs in a living membrane, since two particles of the same size do not necessarily enter with equal facility, and one may not be able to diffuse through the membrane at all. The living membrane is therefore selective, and an expenditure of energy is involved in the passage from one side to the other. Therefore, instead of visualizing a living membrane as a passive membrane wherein there is merely a diffusion through pores from a region of greater concentration to one of lesser concentration, it appears that the membrane is a dynamic structure in which the passage of any substance is accompanied by complex chemical processes. It is thought that the fatty components of the membrane are particularly active in this transfer.

Osmotic processes characterize many well-known phenomena, such as those that result when a concentrated salt solution is placed on the leaves of an elodea plant. The sodium chloride does not pass through the cell membrane, but the water inside the cell membrane will diffuse out at a greater rate (since the salt concentration is less inside the cell than outside). As a result, the cell becomes dehydrated as the salt solutions on each side of the membrane tend to reach equilibrium. A solution in which the concentration of salt is greater than that inside the cell is called a **hypertonic solution** and will cause the plant cell to become plasmolyzed. In other words, it loses water and becomes dehydrated through **plasmolysis.** If the concentration of the salt solution outside the cell is less than that inside the cell, a **hypotonic solution,** the water will diffuse into the cell, making it **turgid.** should the **turgor pressure** become great enough, the cell membrane will burst. When the concentration of salt is the same on both sides of the membrane, the solution is said to be **isotonic.** This explains why aquatic plants must remain in water to which they are adapted

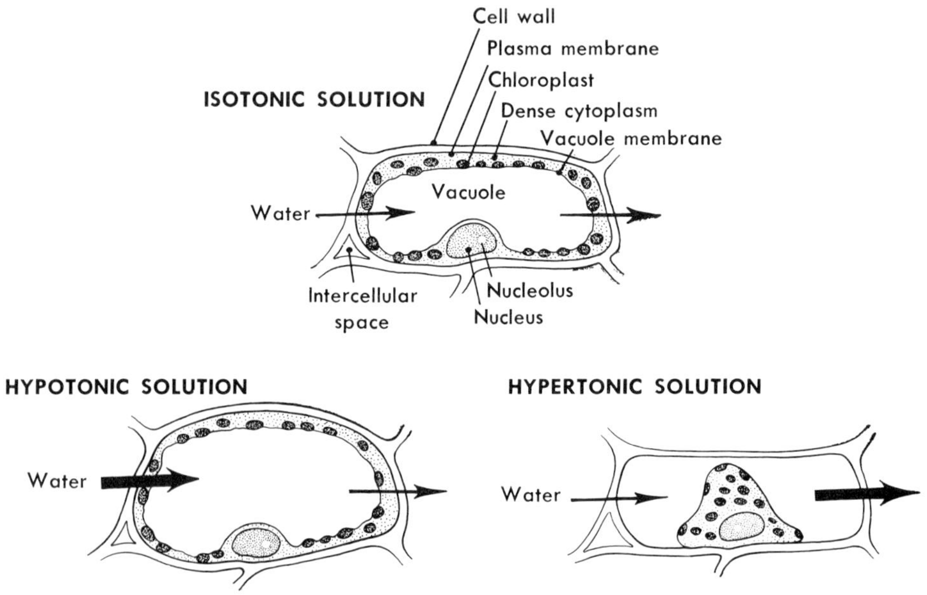

Fig. 7-3
Osmosis in plant cells showing the effect on the cell of different solutions. Normal, or isotonic, solution at the top does not distort the cell. A hypotonic solution, that is, one with a lower amount of salt in the water, causes the cell to swell as in the diagram on the left. The reverse occurs when the solution has a greater amount of salt than does the cell, as can be seen in the diagram on the right of a shrunken cell. Arrows indicate the relative amount of water entering and leaving the cell.

and cannot be changed from fresh water to salt water or vice versa (Fig. 7-3).

Active secretion takes place when the principal ions in the cell are at higher concentrations than the solutions outside. This is usually accomplished by substitution of one ion for another. This transfer or exchange of ions, as, for example, potassium ions for the sodium ion in salt water, is termed secretion.

Excretion in animals as a means of removing metabolic waste products is of little or no importance in plants due to the fact that the cells do not contract and therefore do not do work in this sense. Sometimes plant cells, especially those living in areas of high-salt concentrations, will give off, or excrete, excess salt. Others may excrete sucrose, but this is a special case involving special cells, or nectaries; still others may secrete or excrete digestive enzymes, as found in the insectivorous plants. Nitrogen from protein metabolism is either used in new protein manufacture or stored as inactive protein. **Guttation** is more general and involves the excretion of water droplets or very dilute solutions from a highly turgid plant.

Minerals

Plant cells absorb all kinds of mineral elements combined to form salts. This fact is apparent, since a great number of chemical elements are found in plant cells. However, physiologists have found that plant cells are composed primarily of carbon, hydrogen, oxygen, and nitrogen. Carbon and oxygen are obtained from the air as carbon dioxide, hydrogen from water, and nitrogen from the soil. A number of additional elements are necessary for normal plant growth. Ten elements, designated as essential **macronutrients,** are found to be necessary. The chemical elements essential for plant growth in addition to those mentioned previously are phosphorus, sulfur, calcium, potassium, magnesium, and iron. Further investigation of the mineral needs of plants reveal that chlorine, copper, boron, manganese, molybdenum, and zinc are also necessary for normal growth but in such small quanitities that they are called **micronutrients,** or **trace elements.** Each of these minerals is obtained in soil water.

Of all the macronutrients, carbon, hydrogen, oxygen, and nitrogen comprise approximately 95% to 98% of the entire plant, whereas the remaining six make up less than 5%. Micronutrients are present only in minute parts per million.

Oxygen for respiration

Whether the organism is composed of one cell or many, the production of energy is necessary to continue life. With or without the aid of free oxygen and with the help of organic nutrients, each cell produces energy through respiration. This energy may be used within the cell producing it or in the maintenance and reproduction of the cell or the entire organism.

Oxygen enters by direct diffusion from the atmosphere into the water of the cell. In the higher plants, specialized structures control the gas exchange.

Carbon dioxide for photosynthesis

Carbon dioxide is abundantly present in the atmosphere and in most water habitats due to its release as a by-product of respiration. Green plants require this compound for photosynthesis. Like oxygen, this readily diffuses into the plant cell and is easily dissolved in water.

SYNTHESIS

Fermentation

Fermentation is a complex reaction usually involving microorganisms. It takes place without the use of free oxygen and is often termed **anaerobic respiration.** Because of these two factors, microorganisms and the lack of oxygen, many believe this to be one of the first stages in the original development of life processes.

The following simplified formula diagrams the process:

$$C_6H_{12}O_6 \xrightarrow{\text{Enzymes}} 2\ \underset{\text{Ethyl alcohol}}{C_2H_5OH} + 2\ CO_2 + \text{Heat}$$

Fermentation may take place in an atmosphere of free oxygen, but without using it. Also, this process, sometimes referred to as glycolysis, is known to occur in higher plants, at least for a short period of time when oxygen is deficient. The process involves twelve chemical reactions and several different enzymes in about seven steps. Much less energy is released in this process than in aerobic respiration, and, of course, the end product, ethyl alcohol, may be further reduced. Substances other than this may be produced as well, for example, lactic acid, acetaldehyde, and glycerol. In advanced processes involving the transfer of energy this fermentation (as glycolysis) is a part of the metabolic process which is known as the glycolytic pathway.

Photosynthesis

Photosynthesis is probably the most important process found in nature. It, like respiration, is an oxidation-reduction process. With energy from the sun a green plant-cell can utilize carbon dioxide and water, build up carbohydrates, and store energy. The carbon dioxide and water necessary as raw materials for photosynthesis are among the end products of metabolism in all cells.

Investigators have proved that plants get some of their nourishment from the air and that the gaseous exchange in plants is affected by light and darkness; that is, in light, carbon dioxide is absorbed and oxygen is given off, and in darkness, carbon dioxide is given off and oxygen is absorbed. Although all the earlier experiments on photosynthesis were conducted using the leaves of land plants, recently much of the work has been done using a one-celled green plant, *Chlorella vulgaris*. One reason is that this plant is most easily kept under controlled conditions.

The reaction that takes place in photosynthesis can be stated simply in the following formula:

$$CO_2 + H_2O \rightarrow (CH_2O)_n + O_2$$

The $(CH_2O)_n$ represents some carbohydrate, but the available information does not permit the designation of any particular one. The energy source necessary to combine carbon dioxide and water for the formation of organic material is light. Thus light energy becomes chemical energy in green plants.

It is clear, then, that photosynthesis by green plant cells, in addition to being the only ultimate source of food on earth, is also the only source of animal energy. Since wood, peat, coal, and oil have been our principal sources of heat, light, and industrial power, photosynthesis may be considered responsible for fulfilling many of the energy requirements of modern civilization except, of course, those now being supplied by water power and, more recently, by nuclear reactions.

The ratio of oxygen evolved to carbon dioxide consumed is 1:1, as shown in the following formula:

$$6\ CO_2 + 6\ H_2O \rightarrow \text{Carbohydrate} + 6\ O_2$$

Water in the cell appears to be the hydrogen donor (Fig. 7-4), as demonstrated by the isotope tracer methods. The oxygen in water is labeled with the isotope O^{18}. When a culture of *Chlorella vulgaris* is exposed to light in the presence of radioactive water, only the oxygen liberated contains the isotope of oxygen. This is proof that all the liberated oxygen comes from the water and that the water functions as a hydrogen donor.

From evidence of the rate of photosynthesis under different conditions, it is concluded that photosynthesis is not a single photochemical reaction. At least one "dark" reaction is necessary. In other words, some part of the process of photosynthesis is not affected by light. It is believed, then, that a dark or thermochemical reaction follows the light or photochemical reaction; that these are consecutive reactions is known from temperature coefficients and from the use of flashing light.

When a suspension of *Chlorella* sp. is exposed to flashing light, more photosynthesis occurs per unit dose of light than

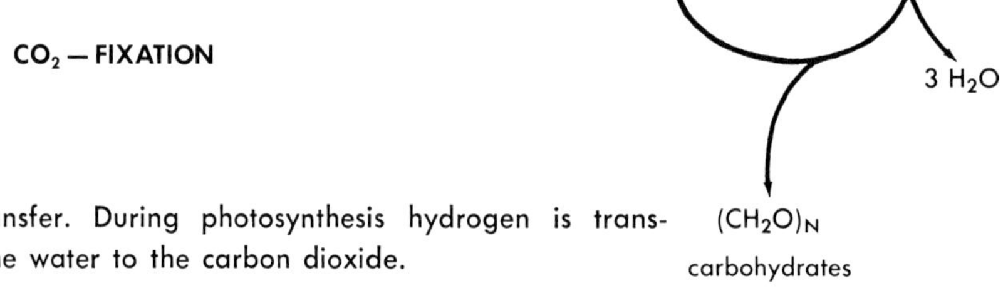

Fig. 7-4
Hydrogen transfer. During photosynthesis hydrogen is transferred from the water to the carbon dioxide.

when a suspension is exposed to continuous light. This indicates that light is used more efficiently when it is followed by a dark period. The lower the temperature, the longer will be the dark period required for the completion of the thermochemical reaction. Therefore, if the light intensity is sufficient to bring about the maximum photochemical reaction, increasing the temperature will increase the rate of photosynthesis. However, if the light intensity is low, thereby limiting the photochemical reaction, the temperature may be varied from 15° to 32° C. without greatly altering the rate of photosynthesis.

Two types of reactions are involved in photosynthesis: first, a reaction in which light is absorbed and used by chlorophyll and, second, a dark reaction during which the carbon dioxide is fixed. The photochemical reaction is known as the **Hill reaction** named after Robert Hill, who discovered that freshly ground leaves in water containing suitable hydrogen acceptors give off oxygen in the presence of light even though the cells are crushed and no carbohydrate is produced. This and other experiments indicate that a reduction pool of hydrogen is produced when light is absorbed and that this process is a basic chemical reaction. It is proposed that water, being the hydrogen donor during photosynthesis in green plant cells, is broken down in the presence of chlorophyll by light energy. A definite quantum of light will split water into the free radicals, H and OH. The OH radicals then recombine to form water, and the hydrogen is used to do work in the cell.

Plant respiration as will be seen later involves the two types of reactions; one breaks down carbon chains in the large organic molecules, and the other removes hydrogen atoms from their association with carbon, with the help of catalytic enzymes, and transfers them to oxygen, forming water. It appears, therefore, that photosynthesis is the first step in the plant's metabolism, which begins with the split off from the water molecule and ends with its reconstruction during respiration (Fig. 7-7). The reverse of decarboxylation is carbon dioxide fixation, and the reverse of dehydrogenation is reduction. Therefore the same processes operate in photosynthesis, but in reverse. The transfer of hydrogen in respiration liberates energy, whereas in photosynthesis it stores energy. This reversible process is not unique and can be illustrated by mixing a solution of the dye thionine with a solution of ferrous sulfate. In intense light the color of the dye disappears in seconds, whereas in the dark the color returns. This illustrates how an oxidation reduction can occur in one direction in the dark and in the opposite direction in the light. Ferrous iron reduces the dye to a colorless form in the light and is itself oxidized to ferric iron; in the dark, ferric iron oxidizes the dye to the colored form and is itself reduced to ferrous iron.

The reactions that occur during photosynthesis may be summarized as shown at the top of the following page.

The sum of all these reactions can be stated simply in this formula:

$$3\ CO_2 + 2\ H_2O + \text{Phosphate} \rightarrow 1\ \text{Triose phosphate} + 3\ O_2$$

Phosphorylation occurs in photosynthesis and the pathway by which carbon dioxide is reduced by the hydrogen radicals is complex. The hydrogen radicals yield the energy for the formation of high-energy phosphate bonds. This is known as the carbon assimilation cycle.

Although chlorophyll may be masked by other pigments, it is present in all plants in which photosynthesis takes place. Even though several kinds of chlorophyll have been identified, all have essentially the same structure: four pyrroles attached to a nucleus of magnesium. Chemically, the chlorophylls are water-insoluble methyl phytol esters of the chlorophyllins and differ only in their side-chain attachments. The **phytol** is a long-chain alcohol derived

Fixation of carbon dioxide

Phosphopentokinase
1. Pentose phosphate + ATP ⟶ Pentose diphosphate + ADP
Carboxylase
2. Pentose diphosphate + CO_2 + H_2O ⟶ 2 Phosphoglyceric acid
Acid kinase
3. Phosphoglyceric acid + 2 ATP ⟶ 2 Disphosphoglyceric acid + 2 ADP
4. 2 Diphosphoglyceric acid + 2 H^+ + 2 Diphosphopyridine nucleotide (reduced) ⟶
Dehydrogenase (Coenzyme I)
2 Triose phosphate + Diphosphopyridine nucleotide (oxidized) phosphate
Isomerase
5. 2 Triose phosphate ⟶ Hexose diphosphate

Regeneration of pentose

Diphosphatase
6. Hexose diphosphate + H_2O ⟶ Hexose monophosphate + Phosphate
Transfer enzyme
7. Hexose monophosphate + Triosephosphate ⟶ Pentose phosphate + Tetrose phosphate
Transfer enzyme
8. Tetrose phosphate + Hexose monophosphate ⟶ Heptulose phosphate + Triose phosphate
Transfer enzyme
9. Heptulose phosphate + Triose phosphate ⟶ 2 Pentose phosphate

Regeneration of H^+, ATP + Coenzyme I

Splitting enzyme
10. Coenzyme I (oxidized) + H_2O ⟶ Coenzyme I (reduced) + H^+ + O^- + 2 electrons
11. Coenzyme I (reduced) + 3 ADP + 3 Phosphate + H^+ O + 2 electrons ⟶ 3 ATP + 4 H_2O + Coenzyme I (oxidized)

from vitamin A by hydrogenation. The chemical formula of a typical chlorophyll is given in Fig. 7-5. Some ten different chlorophylls are known, but in the higher plants only chlorophyll **a** and chlorophyll **b** are present.

Photosynthesis occurs only in the chloroplasts of plant cells, since they are the only structure that contains the chlorophyll pigments. Photosynthesis may take place experimentally in chloroplasts isolated from living cells, showing the independence of these organoids. Some protozoans have been known to carry on the process after ingestion of plant chloroplasts.

Chloroplasts are usually disk-shaped protoplasmic bodies consisting of units called grana, which are dispersed in a matrix of protein globules (Fig. 7-6). The grana appear as piles of disklike membranes connected by fibrils in the matrix. A single granum contains several million chlorophyll molecules. Chlorophyll is synthesized within the cell from products containing carbon, hydrogen, oxygen, nitrogen, and magnesium.

One of the most characteristic properties of chlorophyll is that it shows a brilliant fluorescent band in the red part of the spectrum when illuminated. After illumination, chlorophyll will emit light in the dark for as long as 2 minutes, a characteristic called **bioluminescence.** All green plant cells grown in the dark appear white or etiolated but contain protochlorophyll, which is rapidly reduced to chlorophyll when the plant is placed in the light.

Yellowish pigments called **carotenoids**

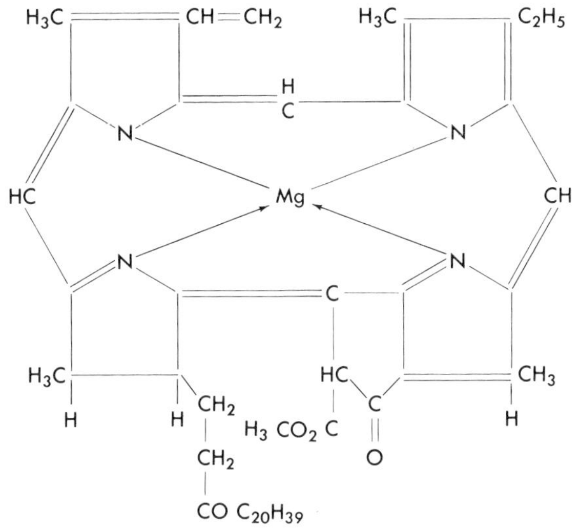

Fig. 7-5
Chlorophyll has a chemical formula very much like that of hemoglobin in animals; the significant difference is the replacement of magnesium by iron in hemoglobin.

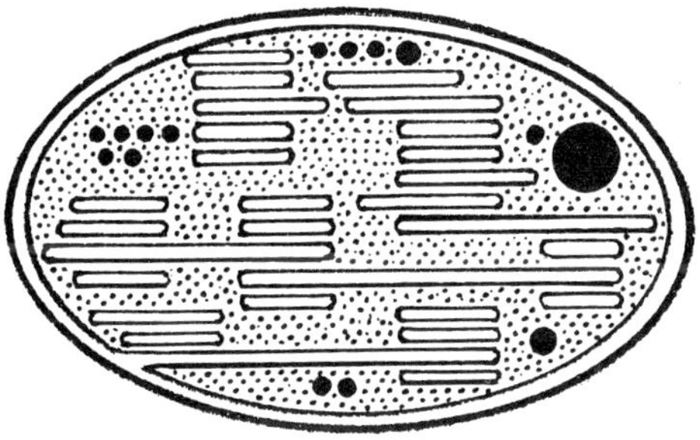

Fig. 7-6
Details of structure of a chloroplast showing diagramatically the outer membrane, internal lamellae or grana, and the matrix and globules.

are present in many cells. They are complex hydrocarbons that can be differentiated into two main groups, **carotenes** and **xanthophylls**, and are attached to the proteins in the chloroplast, like the chlorophylls. Although the chlorophylls alone absorb sufficient light for photosynthesis, it has been shown that the carotenoids can absorb light, which in turn may also be used in photosynthesis. Although the absorption of light by carotenoids is not too important in the higher plants, it is of considerable importance in certain algae and bacteria. It is not known whether they transfer the radiant energy directly to the carbon dioxide reduction process or whether it is transferred to the chlorophyll and then used in photosynthesis. Nevertheless, it is evident that light absorption by pigments other than chlorophyll is not necessary for photosynthesis in the higher plants.

The rate of photosynthesis is dependent upon light, concentration of carbon dioxide, temperature, water, and other factors in the environment, the organic and inorganic materials available. It is evident that one or a combination of these factors may decrease or increase the rate of photosynthesis and that any one factor is probably influenced to some extent by the others. In addition, the general physiological activities of the cell will likewise have some bearing on the photosynthetic rate.

Three aspects of light are important for photosynthesis: intensity, quality, and duration. Intensity depends upon the number of quanta falling on a given chloroplast per second, regardless of quality. This intensity affects the life cycle of some plants. Quality refers to the color bands making up light. A tungsten bulb produces light rich in red but low in blue, whereas a fluorescent bulb is richer in blue than in red.

The amount of light (intensity) upon which a cell can maintain itself with the sugar produced in a day is called the critical light intensity. The rate of photosynthesis will be increased if the light intensity is increased up to its critical or maximum intensity. It has been shown that maximum carbon dioxide fixation (photochemical reaction) occurs at light intensities varying from 200 to 10,000 footcandles.

Light energy is transferred by enzymes from one product to another through several intermediate chemical reactions to finally become part of the sugar molecules.

The rate of photosynthesis in cells exposed to full sunlight for 10-minute periods varies with temperature and carbon dioxide concentration. When carbon dioxide concentration is increased, the rate of photosynthesis will increase in proportion to the increase in temperature.

The water available to a cell also influences the rate of photosynthesis. Since water is one of the raw materials essential for photosynthesis, the rate of carbohydrate synthesis will be reduced according to the decrease in available water. Water is also necessary for the numerous other chemical processes in a plant cell, and any decrease will affect its metabolism, in addition to affecting those processes connected with photosynthesis.

Extremes of temperature will diminish, if not entirely inhibit, photosynthesis. However, temperature is interrelated with the concentration of carbon dioxide.

Although water, temperature, carbon dioxide concentration, and light are the most important factors influencing photosynthesis, there are also various environmental and structural factors, which are discussed in Chapters 8 and 13.

TRANSFER OF ENERGY

Respiration

Respiration can be defined as the production of chemical energy from the organic components of living protoplasm. Ultimately, this energy represents stored energy from the sun that, through photosynthesis, has become part of the organic molecules, particularly that energy required to bind the atoms of a molecule together. When these bonds are broken, as much energy will be released as was necessary to bring about their formation.

Through respiration, following a great number of steps, the carbohydrates, sometimes organic acids, and fats, are finally converted into carbon dioxide and through the reoxidation of hydrogen, to water. This process is maintained by enzyme activity, which promotes the removal of hydrogen when a carbon chain is broken down. The energy released during this process is used partly to sustain other life processes, and partly to continue growth. The stored supply of organic foods, particularly fats and carbohydrates, is the source of this energy.

More simply stated, during respiration, oxygen is being used in **oxidation** of the organic compounds, with the subsequent release of energy and the production of carbon dioxide, water, and other by-products:

$$\text{Oxygen} + \text{Organic compounds} \longrightarrow H_2O + CO_2 + \text{Nitrogen compounds} + \text{energy.}$$

Even though this is correct as far as it goes, it is not so simple as it appears. The energy does not pass freely into the surrounding area and become useless metabolically.

Respiration takes place within the mitochondria of the cell, along the surface of the cristae. The water released is used during the various steps of the process and is then diffused out of the organoid. In respiration very little of the liberated energy escapes as heat, but most of it is combined into a new kind of chemical energy having new chemical bondings. These bondings are the sources of energy needed by the cell for its metabolism; they are transferred to other areas of activity through a process described later. The energy stored by the chemical bonding of compounds may be released in several ways. Fermentation, previously described, is a simple but low-energy-producing process. This is also a part of the respiratory process. Along with a group of phosphate compounds and enzymes, first this glycolysis takes place followed by the citric acid cycle. In all cells a pathway is present whereby the energy is captured by a group of chemical components known as adenosine phosphates. At least two of these phosphates are present in each cell. They are **adenosine diphosphate (ADP)** and adenosine triphosphate (ATP). ADP can be transformed into ATP by the addition of another phosphate. Conversely, ATP can be transformed into ADP. When the phosphate is added, the process is known as high-energy phosphorylation and is one of the first steps in respiration:

$$\text{ATP} \leftrightarrows \text{ADP} + \text{Phosphate} + \text{Energy}$$

When energy is necessary, the ATP breaks down through enzyme action, phosphate and ADP reappear, and energy is released for use (Fig. 7-7). Thus the first reaction requires energy, whereas the second, or reverse, reaction produces the loss of energy.

The process of respiration, therefore, is not so simple as the original formula indicates. At first there is the oxidation of an organic compound, which means that there has been the removal of hydrogen and its attachment to the oxygen to form water. Other fragments form as a result of oxidation into 1-carbon portions to finally appear as carbon dioxide.

Energy is released also through oxidation of the organic compounds (carbohydrate),

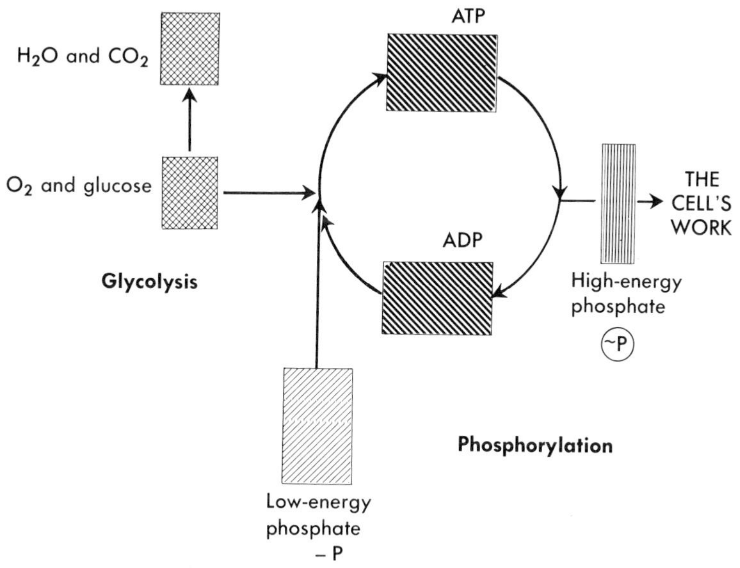

Fig. 7-7
Diagram showing how energy is captured in a cell by a group of chemicals known as the adenosine phosphates.

most of which is captured by the ADP-ATP system (Fig. 7-7). The process of respiration may now be diagrammed as follows:

$C_6H_{12}O_6 + O_2 \longrightarrow$ Energy + CO_2 + H_2O
Energy + ADP + Phosphate → ATP
ATP → ADP + Phosphate + Energy for metabolism

The net result of respiration is the transfer of energy from the organic compound to the ATP substance. This transfer is necessary, since the energy needed for synthesis of protoplasm and its components must be in a concentrated form. The energy within an organic molecule is not the same for all carbon-carbon, carbon-hydrogen bonds, and so on, but they are all low-energy ones. Therefore breaking these bonds releases small amounnts of enery. By incorporating these small amounts of energy into ATP, a supply of energy is built up of uniform high intensity available for all metabolic reactions. The creation of high-energy bonds means that an oxidation process has taken place, and, as previously noted, the most important oxidative change is the removal of hydrogen. Also involved is the removal of carbon joined to oxygen. Thus the final reduction, once again stated, is the formation of water and carbon dioxide.

The large amounts of energy in an organic molecule are never released at one time but rather in small fractions through a series of steps. In each step a catalyst is required in the form of a specific enzyme. The general scheme, as outlined briefly before, for the oxidation reduction of glucose in the cell is shown in more detail in Fig. 7-8.

With these preliminaries as a background, the entire detailed process of respiration may be reviewed. The breakdown of glucose, glycolysis, consists of a series of steps. Glucose is broken down into smaller fragments. At the same time, phosphorylation occurs and energy from ATP is expended. Therefore the glucose is converted to glucose-phosphate. A series of internal rearrangements takes place that transforms the glucose-phosphate to fructose-phosphate, a process that does not change the low-energy values of the phosphate bonds. Using ATP as the source of energy and phosphate, another low-energy phosphate group is added to fructose-phosphate, changing it to fructose-diphosphate and releasing dihydroxyacetone phosphate.

Again, without changing the energy values, fructose-diphosphate splits to form two 3-carbon fragments having slight internal rearrangements. This new compound is phosphoglyceraldehyde. A final phosphorylation occurs when a phosphate group from inorganic phosphoric acid is added to the phosphoglyceraldehyde, converting it to diphosphoglyceric acid.

When phosphoglyceraldehyde undergoes oxidation, the end product is diphosphoglyceric acid, which raises the energy level. During this process, the H^+ atoms are transferred to a hydrogen acceptor, nicotinamide adenine dinucleotide (NAD). Diphosphoglyceric acid loses phosphorus to ADP, thereby creating ATP, and the acid changes to phosphoglyceric. This acid now undergoes dehydration, losing not only hydrogen but also oxygen in the form of water. At the same time, a new high-energy bond is created in the form of phosphoenolpyruvic acid. It transfers its phosphorus to ADP, creating ATP, and the acid becomes pyruvic acid. With the production of

Fig. 7-8
Glycolysis and the citric acid cycle in schematic diagram. To determine the total number of ATP molecules resulting, it would be necessary to double the reactions and yield from phosphoglyceraldehyde on, which would give a total yield of 40 ATP.

ENERGY TRANSFER AND LIBERATION IN LIVING CELLS

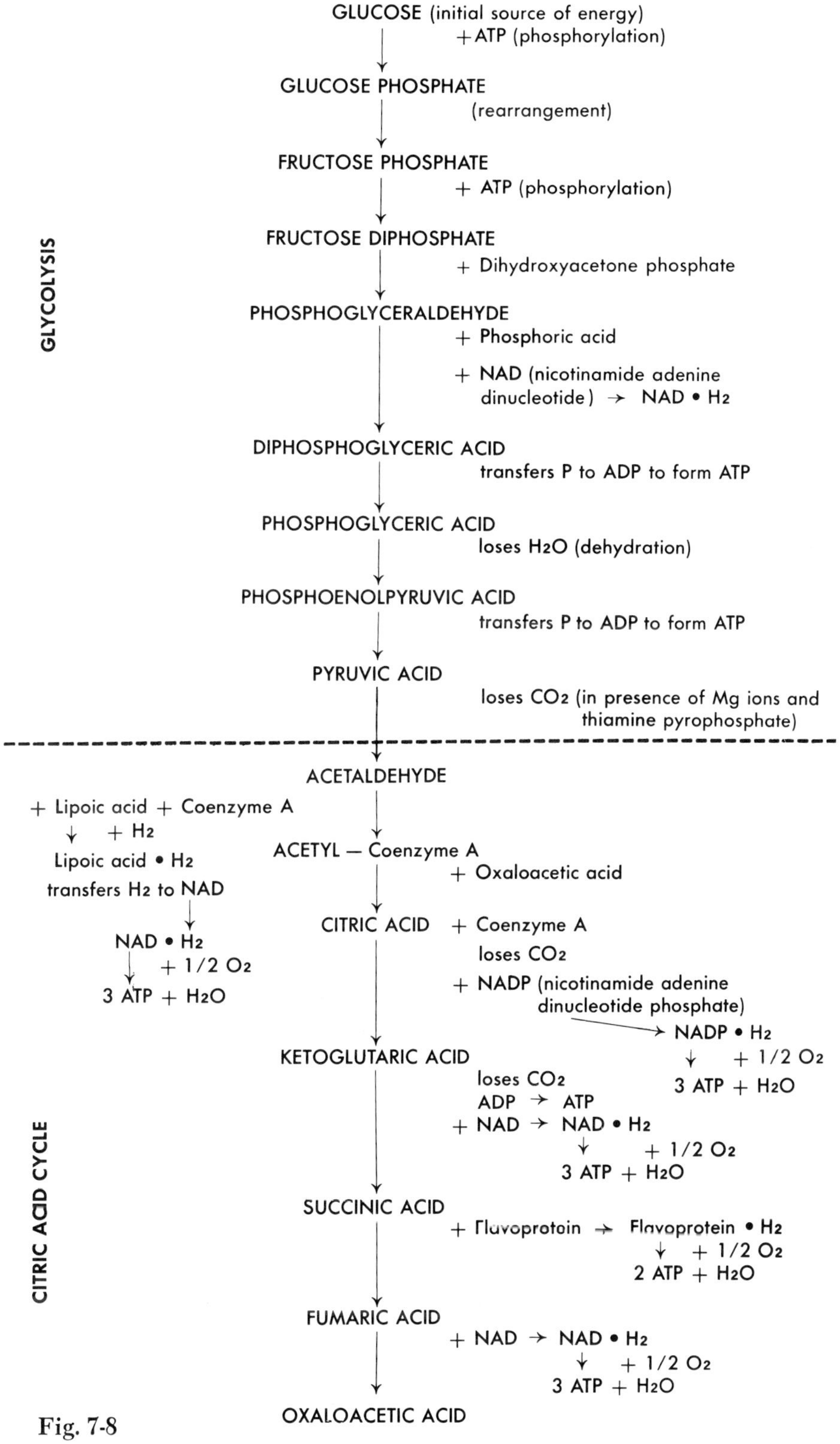

Fig. 7-8
For legend see opposite page.

pyruvic acid, glycolysis may be regarded as complete.

Under anaerobic conditions the pyruvic acid serves as the final hydrogen acceptor, forming lactic acid or alcohol and carbon dioxide. Under aerobic conditions the pyruvic acid eventually supplies many more ATP molecules.

In the presence of Mg ions and thiamine pyrophosphate (a derivative of vitamin B_1), carbon dioxide is removed from pyruvic acid, resulting in acetaldehyde, which is converted to coenzyme A. The hydrogen lost from acetaldehyde is accepted by lipoic acid, but no high-energy bonds are created. However, the H_2 transferred to the lipoic acid is further transferred to NAD, thereby regenerating the lipoic acid. From NAD the hydrogen eventually combines with oxygen to form water, and ATP is produced.

Oxaloacetic acid reacts with acetyl-coenzyme A to yield citric acid and coenzyme A. Citric acid subsequently loses CO_2 and H_2 to NADP, forming NADP·H_2. This leads to the formation of reduced compounds (NAD·H_2 and flavoprotein·H_2), which can yield additional ATP. From Fig. 7-8 it can be seen that the final product of the reaction is oxaloacetic acid, which can then be used to combine with another acetyl-coenzyme A to repeat the cycle. This is the **Krebs** or citric acid **cycle.** Whereas the final yield of glycolysis and the citric acid cycle is 40 ATP molecules, the net result is only 38 because two ATP molecules are used during phosphorylation of glucose and fructose-phosphate.

Negative feedback

Life, we may conclude from this description of cellular activity, is an **open system** of biochemical activity, but is is dependent upon feedback (Chapter 1) to keep up the inward flow of energy. Without this, life processes would run down, and life would cease or at least remain in a suspended state. There is still no evidence that life may remain in a completely suspended state; therefore it must be concluded that even in spores and similar inactive states, slow biochemical processes continue. Feedback may be either negative or positive. Most biochemical systems are negative feedback systems, as positive feedback involves replication. Negative feedback input may be through environmental influence, enzymes, or simply through circulation within the cell.

Metabolism

Growth through metabolic processes is one living characteristic. It is generally accompanied by an increase in physical size. However, that is not all, since a snowball rolling down a hill or a crystal enlarging in a solution also can be said to grow. Such growth by nonliving matter is by accretion; that is, a similar material is merely added to the outside, much as one would add a shovelful of soil to a pile of soil to make it grow larger. Therefore, increase in size but not growth is exhibited by nonliving things, such as a piece of gelatin swelling in water. A living seed, on the other hand, will grow only if it is supplied with water, oxygen, a suitable temperature, and the proper nourishment. True growth, then, centers around a process called synthesis or the putting together of nonliving material into living material. In synthesis, plants utilize foods, chiefly sugars and proteins, in building up the living matter of the plant body. This process is basic to growth in living things and is distinguished from mere increase in size, which can also occur in the world of inanimate matter.

All living systems accumulate, store, and expend energy during the processes of respiration and photosynthesis. The energy accumulation is used to build complex compounds from simpler ones in the growth of the organism, as well as in many other processes. Much of what is known about these phenomena has been previously discussed, since this energy balance is necessary to the maintenance of life itself. The term applied to the overall chemical processes of the living organism that are necessary to life and growth is **metabolism.** Some phases of metabolism, **catabolism,** break down the materials of the living system into simpler constituents. Others, **anabolism,** build up the materials and, when they exceed or predominate over the breaking down process, are responsible for growth.

In summary, the only materials needed for the growth of green plants are simple inorganic compounds. As raw materials, oxygen, carbon dioxide, mineral elements, and water are the necessary substances for plant growth. The most frequently used minerals in plant metabolism are potassium chloride, sodium chloride, sodium monohydrogen phosphate, magnesium sulfate, and calcium nitrate. Sunlight furnishes the energy to produce ATP through phosphorylation. Subsequent reactions lead to the production of sugar from carbon dioxide. The sugar is then used to build protoplasm (anabolism) or for other energy-requiring activites in the cell (catabolism). Food, therefore, is any substance that can be used as a source of energy and as a source of material for the production of protoplasm, and it is this that is taken in by the plant.

All the processes involved in bringing food to the individual cells of the plant body are collectively referred to as **nutrition.** As we have seen, the three basic foods within the cell are carbohydrates, fats, and proteins. Of the many kinds of carbohydrates, five are common: glucose, fructose, sucrose, starch, and cellulose. Glucose and fructose are simple sugars or monosaccharides ($C_6H_{12}O_6$). Sucrose ($C_{12}H_{22}O_{11}$) is a disaccharide or a complex sugar. Starch $(C_6H_{10}O_5)_n$ is a polysaccharide, a series of simple sugars hooked together into a complex molecule. However, starch is stored in plants because it is only slightly soluble in water and therefore does not become involved in other cellular function until it is converted to sugar by enzyme action. Then it can be removed from the storage site and utilized as food. Starch is not the only common type of plant storage food, since sugars, either monosaccharides or disaccharides, occur frequently. Cellulose, used primarily in the formation of cell walls, is a polysaccharide and appears to be the most abundant carbohydrate on earth. Only a relatively few organisms, such as bacteria, fungi, and protozoa, can use cellulose as food. If an efficient means for converting cellulose to simpler carbohydrates could be found, the world's food supply would become almost unlimited.

Fat is often stored by plants. **Lipid** is the more general term for substances of a fatty nature, and fat is one of the esters known as a lipid. Like carbohydrates, fats are composed of carbon, hydrogen, and oxygen, sometimes also phosphorus or nitrogen. The fats are complex molecules of two main parts, glycerin and fatty acids. Fats have a very high energy yield when metabolized because of the low quantity of oxygen in each molecule. This means a greater affinity for oxygen per unit amount, so that when oxidized it releases a greater amount of energy than does the same amount of a carbohydrate. Fats are stored in abundance in cells, sometimes composing the greatest portion of the cell. Fat is also abundant in cutin, plasma membranes, vacuolar membranes, mitochondria, plastids, and nuclei.

Proteins, unlike the other food molecules, always contain nitrogen and sometimes sulfur or phosphorus. Proteins are known only in the protoplasm of the cell; that is, no protein has been discovered as naturally occuring outside of living organisms, except as cell secretions. The details of the chemistry of proteins have been discussed previously.

Metabolic reactions taking place within the plant cell are very different from those known in animal cells. Because green plants synthesize all their own food, their process of **digestion** is quite different from digestion in higher animals. In no case is digestion **extracellular** in photosynthetic plants. Hydrolysis is a better term to apply to this process in plants because it is one that renders large insoluble molecules water soluble by breakdown reactions resulting in two or more smaller molecules. Also, this is a reversible reaction; therefore by the same process, but in reverse, termed condensation, two or more molecules may be joined. Fats, carbohydrates, and proteins held in reserve in the cell are thereby converted into soluble substances used in growth processes. As one would expect, enzymes are involved in each of these reactions. Chemically, therefore, the process of plant cell digestion is a matter of hydrolysis and is **intracellular.** When maltose, for example, is converted to glucose by enzymes, the following formula shows the reaction:

$$\underset{\text{Maltose}}{C_{12}H_{22}O_{11}} + H_2O \xrightarrow{\text{Enzyme}} 2\ \underset{\text{Glucose}}{C_6H_{12}O_6}$$

The enzyme acts as a catalyst to bring about the conversion, since the addition of water to a complex chemical compound such as maltose could not otherwise occur because maltose is a stable compound.

After hydrolysis, a transfer reaction makes these energy-yielding molecules available for growth. This is the ATP-ADP reaction described previously as a part of respiration.

Nitrogen, because of its presence in protein, is metabolized by plant cells, but in a much different manner than in animals. All twenty required amino acids are synthesized by plants, so that they, unlike animals, have no vitamin requirements. Nitrogen is absorbed as a part of the minerals used by plants. They use either nitrates or ammonia, but both must be reduced to usable form through enzyme action. The enzyme, nitrate-reductase, which requires the trace element molybdenum as its activator, reduces the nitrate to ammonium ions, which are used in the citric acid cycle.

Transport of energy

Energy is expressed as work, which falls, in cellular activity, into three major categories: biosynthesis, mechanical, and transport (Fig. 7-9). Biosynthesis or chemical work accounts for the greatest part of cell activity, since this is the way growth takes place. Tracer studies using radioisotopes of nitrogen and carbon have shown, for example, that all proteins are periodically replaced. Therefore it seems evident that growth involves renewal as well as increase in size.

Mechanical work of plant cells is usually limited to pressure exerted through growth, or in single-celled plants, flagellar locomotion, if at all.

Transport work involves osmosis or the absorption of substances against a diffusion gradient, as previously discussed. However,

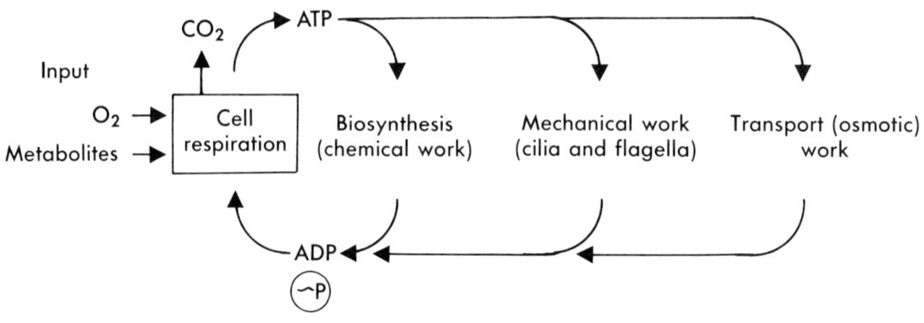

Fig. 7-9
Transfer of energy in cells by the ADP-ATP system.

electrical work also may occur when the substances are in ionic form and are absorbed against an electrical potential gradient. This is active transport and results in the accumulation of a solute in one zone from another zone of lower concentration, or chemical potential through energy input.

Environment. Influences of the environment are discussed in Chapter 8; however, it may be pointed out here that biochemical processes of the cell, such as enzyme activity, are limited to strict temperature ranges and that within these ranges, the speed of the reaction is increased or decreased with the rise and fall of temperature in logarithmic proportions.

Enzymes. Catalysts for the hydrolysis of organic compounds are the enzymes isomerases, transferring enzymes, splitting enzymes, oxidases, and dehydrogenases. All these, as was indicated in the discussion on the decomposition of glucose, take part in the synthesis and breakdown of nutrients in the cell. The isomerases are responsible for catalyzing changes, such as the shift from glucose-1-phosphate to fructose-6-phosphate in both aerobic and anaerobic metabolism. Oxidases and dehydrogenases are concerned with electron and hydrogen ion transfers, for example, cytochrome oxidase, which transfers electrons, and coenzyme A, which transfers hydrogen ions and electrons.

The transferring enzymes catalyze the transfer of the high-energy phosphates, methyl, acetyl, amino groups, and other radicals. Glucose is phosphorylated when the high-energy phosphate bond is transferred from ATP to a glucose molecule. The adding and splitting enzymes catalyze the addition to or removal of a fragment from a molecule. This occurs when carbon dioxide is either added or removed from a carboxyl group.

Although different and specific, all the cell enzymes have certain properties in common. Each oxidation enzyme is composed of a protein carrier and a prosthetic group. The latter is an organic radical, often very complex and attached to the protein. Neither alone can act as an enzyme; they are effective only in combination. Although in most cells they are associated as the enzyme, in some cases they remain dissociated until they act on the substrate. The prosthetic portion is called a coenzyme in its dissociated form. This dissociation is an advantage, since the coenzyme can associate with any protein, thereby varying the specificity of the enzyme; this specificity is actually due to the protein rather than to the prosthetic group, making catalyzing enzymes quite specific in their action.

Hydrogen is removed from a substrate and transferred by a series of enzymes until it is oxidized to water, with the release of energy as in all oxidation-reduction reactions. Through the complicated chain of events already described, the hydrogen is released into the cell fluids as the hydrogen ion, and the electrons are passed on to specific cell pigments, called cytochromes, which are universally present in cells in which aerobic metabolism occurs. These pigments receive the electrons and transfer them to cytochrome oxidase to activate the oxygen. Thus the activated oxygen unites with the hydrogen ions to form water.

Cytochrome is very much like the heme part of hemoglobin in structure, inasmuch as the former contains an iron atom surrounded by four pyrroles. When no metal ion is present, the pyrroles are called porphyrins. Chlorophyll is also a porphyrin containing the metallic ion magnesium. Cytochrome oxidase is a compound similar to cytochrome. These enzymes are located in the mitochondria of the cell, where, in fact, most of the enzymes necessary for the various steps in the Krebs cycle have been identified. The main role of the mitochondria is, therefore, the formation and liberation of compounds containing high-energy phosphate bonds that can be used within the cell.

REPRODUCTION

Gene action and biosynthesis

The most characteristic feature of living matter is the ability to form new individuals of the same kind. This ability to increase the number of new individuals so that the species can be maintained and dispersed over a greater area is called reproduction. It may be asexual involving cell division after chromosome replication (as described in Chapter 6) or it may be sexual, a process that involves the fusion of male and female reproductive cells or sex cells to form a new cell, the beginning of a new individual. All these forms of reproduction will be explained more fully in Chapters 9 and 14.

Closely associated with replication is biosynthesis through gene action, since it is in this way that specific macromolecules of polysaccharides, lipids, and proteins are produced. By means of coupled enzymatic reactions the energy from the terminal phosphate groups of ATP are used to activate a building-block molecule. Information from the genes is stored and transcribed to the protein molecules for expression through nucleic acids, and this controls all biosynthetic processes. The DNA molecules of each cell are the master informational units, and nearly all of it is present in the cell nucleus. Thus it is here that we find the genetic code, or the molecular language of life. This language specifies the amino acid sequence of the protein of the cell. The double helix of the DNA molecule provides, through complimentary base pairs, two identical strands in that each contains exactly the same information. Through replication, the stored information is preserved for thousands of generations. To convey the information locked in the nucleus to the ribosomes, the sites of protein synthesis in the cytoplasm, the information is transcribed to messenger ribonucleic acid (mRNA), a single-strand nucleic acid containing four different types of mononucleotides (Chapter 5). Messenger RNA is a copy of the DNA molecule synthesized enzymatically in such a way that the DNA is the template for the sequence of bases in the mRNA. The completed molecule then diffuses into the cytoplasm and joins the ribosomes, where it in turn acts as the template for protein synthesis.

About 90% of the cells' biosynthetic energy is used in protein synthesis. This control of proteins is fundamental to all biological expression, morphological and physiological and, therefore, is the ultimate definer of the individual and the species. All diversity is expressed by this mechanism.

The complicated process starts with the activation, on the surface of ribosomes, of the twenty individual amino acid molecules through the transfer of energy from ATP. A specific enzyme is needed to prepare each of the amino acids to enter into peptide linkage. Another RNA molecule, either soluble (sRNA) or transfer RNA, is involved in such a way that it becomes a carrier of amino acids. The mRNA attaches to the ribosome to form a stringlike molecule, the polyribosome, and forms the base sequence used to code the sequence of amino acids ready to be inserted. This involves a sequence of three consecutive mononucleotide residues called a **codon.** Thus, because of the structure of each amino acid, the activated molecules previously prepared by sRNA, the **anticodons,** seek out the proper codon and are then assembled into the proper sequence to form the protein requested.

Protein synthesized in this manner must now assemble to form the cell structure. Obviously then, the number of molecules produced and the time they are produced is regulated also. This is possible through another aspect of coding, the position of the amino acids on the protein chain. Their reaction in life processes depend upon their proximity to others, so that a single amino

Table 7-1
Life systems found in plants

Intake	Flow	Feedback	Replication
Water relation Minerals O_2 for respiration CO_2 for photo-synthesis	Fermentation Respiration Photosynthesis Biosynthesis Transport of energy	Environmental Enzymes Circulation	Gene action Reproduction Development

acid molecule may have more than one "meaning" depending upon which kinds it has as neighbors. As discussed previously, protein chains fold back and have distinctive structures discernible through x-ray crystallography. It is this structure that determines the gross features of cells and organisms.

Information content of proteins, nucleic acids, and cells

It has been estimated by A. L. Lehninger of Johns Hopkins University that a single very small bacterial cell, with a diameter of about 2 μ, contains information bits equivalent to about the information contained in 1,000 volumes, each with 1,000 pages, of the *Encyclopedia Britannica.*

LIFE SYSTEMS IN PLANTS

All the functions of plant cells and their interrelationships with the environment are summarized in Table 7-1. The remaining chapters of this text elaborate on these systems.

SPECIAL TERMS USED IN CHAPTER 7

active secretion
adenosine diphosphate (ADP)
anabolism
anaerobic respiration
anticodon
bioluminescence
Brownian movement
carotene
carotenoid
catabolism
codon
colloid
crystalloid
diffusion
diffusion pressure
digestion
dynamic equilibrium
emulsion
environment
excretion
extracellular
fermentation
gel
guttation
Hill reaction
hypertonic solution
hypotonic solution
imbibition
intracellular
isotonic solution
Krebs cycle
lipid
macronutrients
metabolism
micronutrients
nutrition
open system
osmosis
osmotic pressure
oxidation
permeation
phytol
plasmolysis
secretion
selectively permeable
semipermeable
sol
trace elements
turgid
turgor pressure
xanthophyll

QUESTIONS TO CONSIDER

1. What is the particular characteristic of water that makes it so important in biological function?
2. What are the major differences between carbohydrates, fats, and proteins?
3. What is meant by the light reaction in photosynthesis?
4. What is meant by the dark reaction in photosynthesis?
5. Write a simple word formula for the process known as photosynthesis, listing on one side of the reaction the raw materials and the conditions necessary for the operation of the system and on the other side the product. A limiting factor is one that is in such short supply that a system will cease to function. Explain the conditions under which any one of the above can be "limiting factors." Do this for each factor.
6. What is the difference between active transport and passive absorption or transport?
7. Name one way in which growth differs from reproduction.
8. Is there any time in the plant cell when metabolism is not equivalent to growth?
9. What is a more general name for processes like photosynthesis? Can you name some of the other processes that fall under this general term?

10. Briefly describe the nitrogen pathway in the plant cell. Why is this a major difference between plant and animal cells?

DISCUSSION FOR LEARNING

1. Discuss possible applications of our knowledge of photosynthesis. Will it ever be possible to utilize sunlight directly as a food source? Discuss.
2. What makes the Krebs cycle of respiration so different from burning? How might this knowledge be useful?
3. Discuss the genetic code again in terms of its action in biosynthesis. What connection does this have with morphology or plant form?

Chapter 8

Effective environment

The cell, whether singly or as a component of an **organism,** can exist only in an environment that can best be described as hostile. The flow of energy in the biological world takes place not only within the cell, and from cell to cell within the organism, but from organism to organism within the physical, nonbiological portions of our world. Only certain parts of this environment are effective, that is, cause dynamic, equilibrium-seeking reactions. As the picture above demonstrates, this environment is dramatic, hostile, and sometimes man made. The latter is evident in this picture showing one of the greatest problems in our world—air pollution—which can occur not only in cities but also, as can be seen, in remote areas, in this case, in Arizona.

The distribution of a species, the geographical area and ecological habitat inhabited by any particular plant species, is determined by the anatomical and physiological characteristics of each member of the population. The physical and biological environments of a given geographical area are not suited equally well to all species. Therefore any particular area is inhabited by characteristic species.

These areas or habitats also show progression in complexity similar to the progression in complexity of the species.

Two factors are concerned in the distribution of plants. Generically speaking, these are heredity and environment. The hereditary composition of a plant determines the degree of adaptability to a given environment; that is, the organism can live in a particular environment only to the extent of its genetic latitude. Within this framework the individual is affected by environmental factors, which may be grouped into two categories—soil and climate. The soil or substrate varies according to chemical composition, slope and drainage, and available water. The climatic factors vary with temperature, moisture, light, and wind. All these factors combined compose a nonliving environment.

In addition to the nonliving environment, there is the equally important living environment. Two situations, one, the environment of organisms of the same species, and the other, the environment of different species, have different effects. The former usually contributes in varying degrees to the welfare of the individual and the latter usually to its detriment.

EFFECTIVE ENVIRONMENT
- Physical environment
  - Light
  - Water
  - Atmosphere
  - Radiation
  - Soil
- Biotic environment
  - Plants of the same kind
  - Plants of a different kind
  - Animals

PHYSICAL ENVIRONMENT

Light. The role of light as a source of energy for photosynthesis is obvious. The amount of light available to plants is regulated by latitude, weather, and shade. Shade may be produced by land forms, for example, mountains, by other plants, artificially by buildings or similar structures, and by weather conditions.

The northernmost and southernmost latitudes have the longest growing seasons in terms of length of day but the shortest period in number of days. Therefore plants of these areas are quick to flower and set seed. Because of this and temperature factors (to be discussed later), plants are short and stunted. The same species growing in temperate latitudes are considerably taller. As the latitude approaches the equator, the length of day becomes more nearly equal throughout the year, with a gradual disappearance of the seasons until at the equator, light conditions are nearly uniform through the year, and night and day are equal in length. Consequently, plants of those areas have more uniform growth, and any variation is the result of other factors such as the amount of rainfall. The effects of latitude are apparent but are never independent of other environmental conditions.

Plants growing in mountainous areas may be shaded by land forms. The length of day is shortened by the early morning or afternoon shade of the mountain, depending on the slope exposure. Thus eastern slopes receive early morning sun and western slopes, afternoon sun. Northern slopes receive less light during the growing season than do southern slopes. Plants in canyons may receive relatively little light, whereas plants on mountain tops receive the most light but at the same time are in a cooler climate. The growing conditions in these two types of locations produce different kinds of plants, or at least plants having different characteristics.

Plants growing in the shade of larger plants or buildings are affected in the same way by light conditions. Therefore certain kinds of plants will grow best in shade and others in sun. Neither are able to survive in exchanged situations, showing that they are adapted to their particular habitat. A knowledge of light effects is fundamental in understanding the distribution of plant species.

The color of light may affect plant growth also. For flower formation, red light is more effective in some species than is blue, and green is less effective than either.

Duration of light during the day, that is, the number of hours of light to which a plant is exposed, evokes certain responses. This is called **photoperiodism.** Some plants such as corn are long-day plants and will only produce flowers when exposed to daylight for 12½ hours. Other plants such as tobacco are short-day plants and will produce flowers only when exposed to less than 12½ hours of daylight. Still other plants are neither short- nor long-day plants and will produce flowers during either photoperiod.

The number of days of sunshine varies from region to region. The average number for a given area is tabulated and is available in reference books along with other weather data. Areas having many sunny days produce more abundant plant growth than do areas having many cloudy days, unless some other limiting factor is present, for example, water. Some areas are characteristically covered by cloud or fog banks throughout much of the year. In spite of an abundance of water, plant growth is limited in such areas by the lack of bright sunshine and by the lower temperature resulting from the cloud cover.

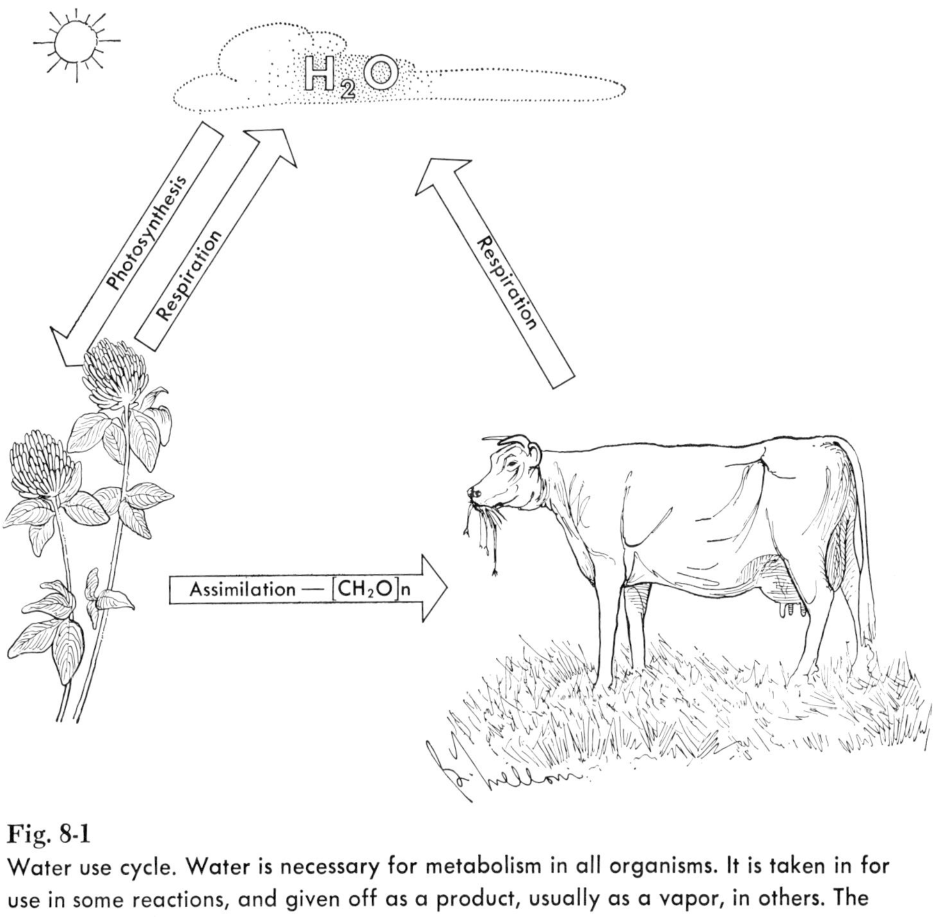

Fig. 8-1
Water use cycle. Water is necessary for metabolism in all organisms. It is taken in for use in some reactions, and given off as a product, usually as a vapor, in others. The vapor appears as clouds. (See photograph at beginning of Chapter 1.)

Water. Water is usually the limiting factor in plant growth (Fig. 8-1). Without an abundance of water, photosynthesis cannot take place even if the days are sunny and the soil is rich in minerals. The amount of rainfall (Fig. 8-3), like the amount of sunshine, is carefully recorded for all areas and is a key factor in agricultural regions. To illustrate the exchange of water, under ordinary growing conditions a square foot of leaf surface transpires 4 ounces of water each 24 hours. At that rate, a corn plant would take up 54 gallons of water during its growing season. An apple tree, however, may lose 95 gallons of water per day by transpiration during the month of July; therefore an orchard of 40 trees would have to take up 480 tons of water during a period of one month at the height of the growing season. A rainfall of 4.25 inches would be needed to replenish this water. In addition to this, large amounts of water are used by the plant for processes other than transpiration. Under natural conditions the vegetation found is indicative of the amount of available water. Plants found in a tropical rain forest (Fig. 15-11), where heavy rains occur, are characteristic of that region and are not found in drier areas. Conversely, desert plants, xerophytes, are characteristic of dry areas. The necessary water need not be in the form of rain. Sur-

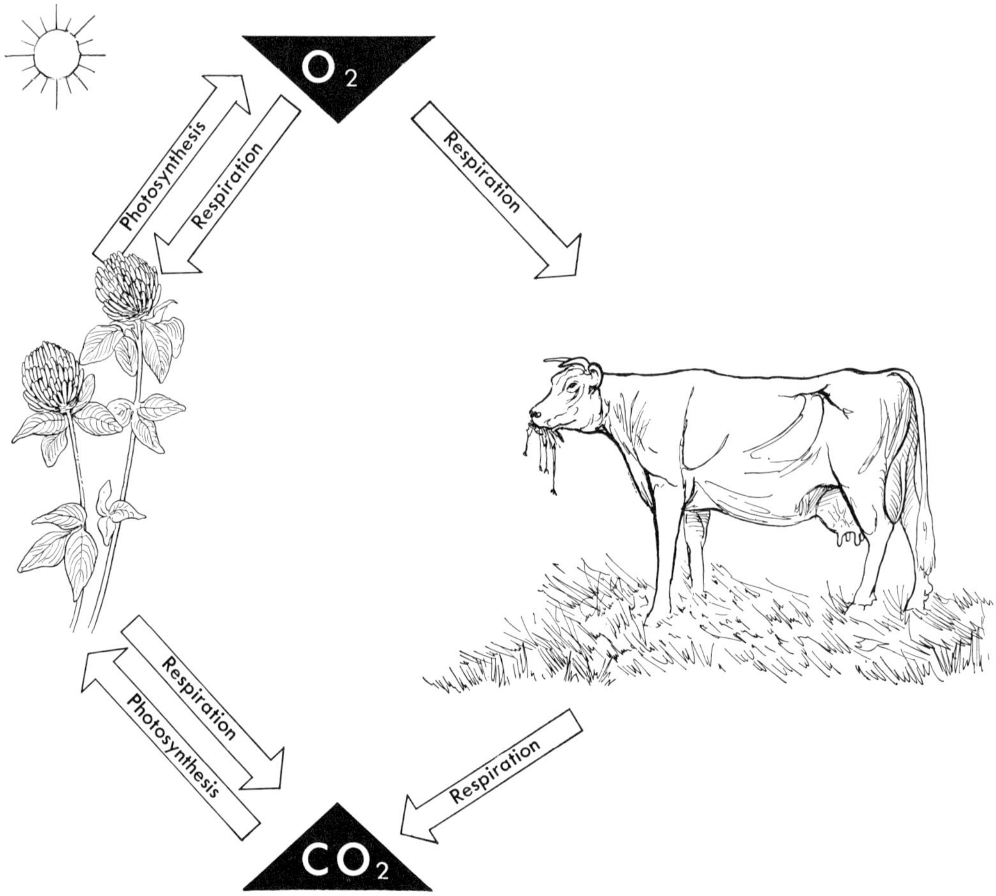

Fig. 8-2
Oxygen and carbon dioxide cycle. Both are dependent upon respiration and photosynthesis. (See Fig. 8-7.)

face or subsurface runoff and marshy soil are other sources of water for plants. A few areas are known as fog deserts in which rain is rare; yet moderate plant growth is maintained because the moisture in the air is absorbed by the plant. Finally, truly aquatic situations have an obviously unlimited supply of water, and plants are variously modified to live there. For example, air bladders may keep the plant floating on the surface of the water (Fig. 8-4). Plants with floating leaves (Fig. 8-5) have stomates located on the upper surface instead of ventrally. The extent of the root system of plants usually varies with the availability of water.

Atmosphere and oxygen. Oxygen is of primary importance in the growth of plants and in their ecology. The role of oxygen in plant life is termed the **oxygen cycle** (Fig. 8-2). Usually oxygen is not a limiting factor because plants give off oxygen during photosynthesis. Some plants growing in water have special structures to help aerate their roots. Among the best known are those of the mangrove (Fig. 8-6), with aerial roots. Cypress "knees" (Fig. 8-8), long thought to be aerating structures, are now

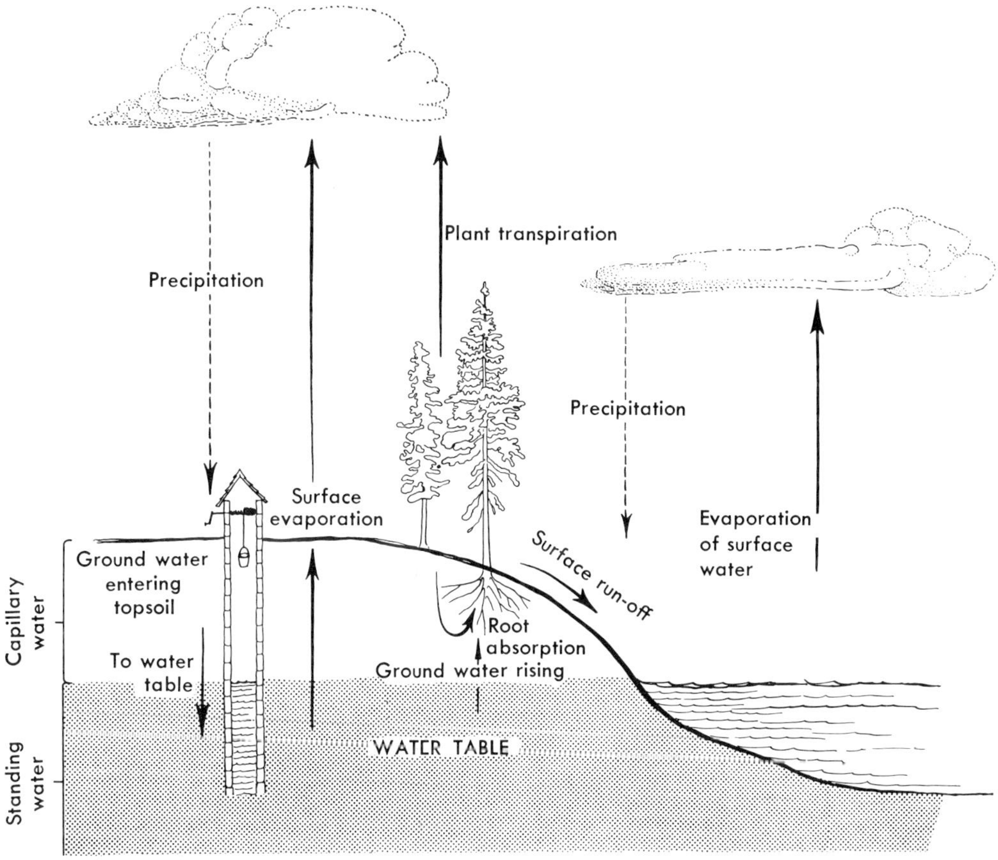

Fig. 8-3
The water cycle is a physical phenomenon and nearly independent of life. However, plants contribute water vapor from transpiration and effectively increase atmospheric humidity in local areas. (See Fig. 8-2.)

Fig. 8-4
Water hyacinths in Arbuckle creek in Florida. This plant chokes out all other life in streams and drainage ditches.

Fig. 8-5
Water lettuce. The roots of these plants are favorite sites for the breeding of certain kinds of pest mosquitoes.

known to be only a growth reaction to a particular environment.

Atmosphere and wind. Air currents in the lower atmosphere do not often grossly affect plants except by controlling somewhat the rate of transpiration. Wind will blow away water vapor that builds up near the stomates, and unless they close, the rate of transpiration will increase. In desert areas particularly, this can be an important factor. Winds that are strong, constant, and always from the same direction affect the growth pattern of plants. Trees growing in prevailing winds are bent over, and the branches are pushed to one side. Frequently this effect is evident in coastal or mountainside plants.

Carbon dioxide. The atmosphere contains approximately 0.03% carbon dioxide. However, it varies greatly near the cell's surface, especially if decomposition is occurring, as in a pond or similar area. It has been found that carbon dioxide may measure as much as 0.35% within a few inches of the cell. Other factors being equal, this increased concentration of carbon dioxide

Fig. 8-6
Mangroves (*Avicennia* sp.) in a Colombian estuary. The mud in which they grow contains little or no oxygen so that many "breathing" roots grow up into the air.

is sufficient to increase the rate of photosynthesis. Increasing the carbon dioxide concentration when other factors are favorable will increase the rate of photosynthesis, but it is doubtful that carbon dioxide is ever a limiting factor in nature. If one refers to the **carbon dioxide cycle** (Fig. 8-7), it is evident that carbon dioxide fixed in the cell must be returned to the atmosphere, or eventually the atmosphere would become exhausted of carbon dioxide.

The amount of carbon dioxide in the atmosphere may have profound effects on the climate of the earth, since it is this material in the atmosphere that effectively shields the surface from cosmic cold. In fact, through geological history evolutionary changes in the flora have been in response to changes in the CO_2 blanket. As the cover increases, the temperature increases (sometimes called the greenhouse effect because it is comparable to the same process that occurs in a greenhouse). Over the past hundred years, CO_2 has increased in the atmosphere by 13%, causing an average increase in temperature of 1° F. (= 0.56° C.). This may account for such events as the ice age and the carboniferous age (Chapter 14). The effect of the increase of carbon dioxide from industry and carbon monoxide from transportation, and other air pollutants, on plant growth has not been determined as yet.

Temperature. Heat, primarily from infrared radiation reaching the earth's surface, affects the rate of photosynthesis and other metabolic processes by increasing or decreasing the speed at which the chemical processes take place. There is little effect other than this unless the temperature is lowered to the point of freezing or raised to the point at which the protoplasm is damaged. The rate of transpiration increases with higher temperatures. The evaporation of the transpired water vapor will cause a cooling effect that is readily noticeable on hot days in forests and other shaded areas. If water is a limiting factor,

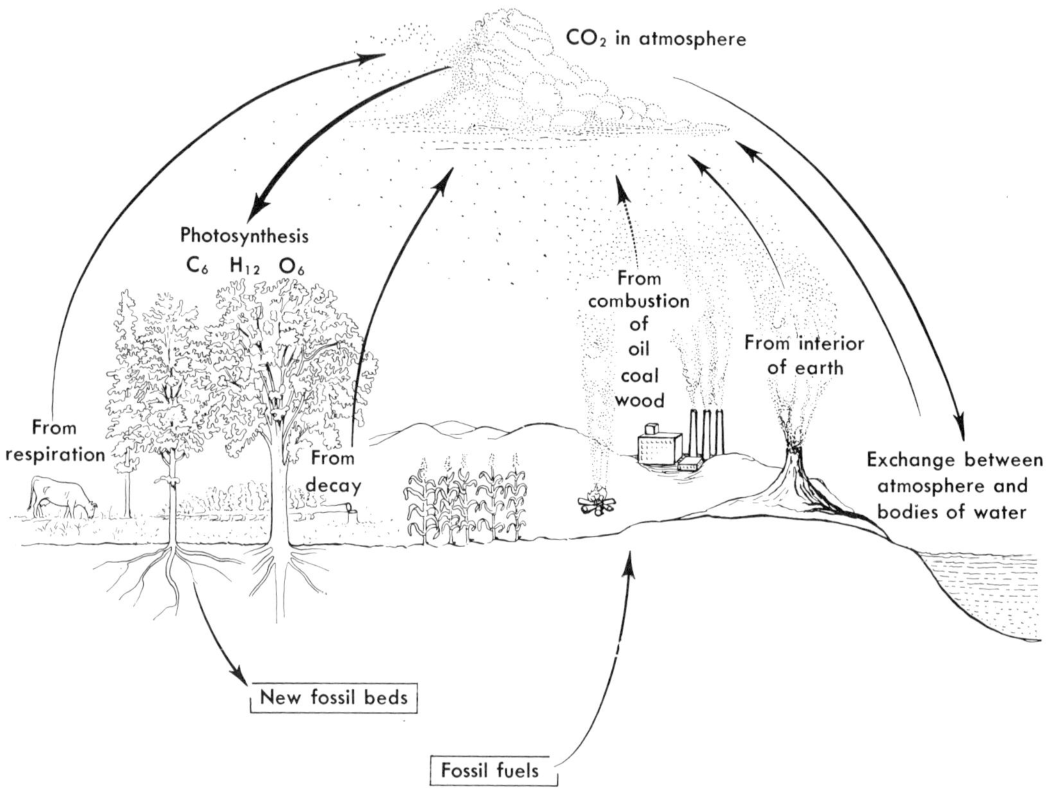

Fig. 8-7
Carbon cycle illustrates the chain of events in the utilization of carbon by plants. The broken arrow indicates one artificial process by which man adds carbon dioxide to the atmosphere.

the increased transpiration resulting from high temperatures will cause the plant to wilt. Wilting slows down transpiration by allowing the stomates of the leaves to close. If water continues to be a limiting factor for too long a period, wilting will cause permanent damage to the plant, and it may not recover.

Radiation. In addition to radiation of light as we see it, and to infrared light, there are other kinds of radiant energy affecting plant life, either already present on the earth, or reaching it from outer space. Ionizing ultraviolet, nuclear, and cosmic rays have sufficient power to damage cells under certain conditions. Fallout from nuclear reactions is now a well-known example of causes of cellular changes of many types. Ultraviolet light, usually screened by the atmosphere, may be important at extremely high mountainous habitats. Cosmic rays will be important as space travel develops, and it is now known that the amount of this kind of radiation may be responsible for many mutations. Radioactive minerals in the earth's shell cause changes in plant growth in some areas. After volcanic activity, radioactive lava has the same effect on growth as does fallout. The amount of energy exchange varies with the latitude and altitude, which in turn affects the amount of radiant energy reaching a cell surface.

Soil. Most of the plants, except some algae and fungi, are terrestrial and are rooted in soil, the substrate. Many plants growing in water are anchored to the bot-

Fig. 8-8
Cypress knees, a growth adaptation in swampy areas. (U. S. Forest Service.)

Fig. 8-9
A landslide in Ecuador. Thin tropical soils do not hold during heavy tropical rains.

tom soil. All plants are dependent upon the moisture and mineral content of the substrate, but completely submerged plants are also dependent upon the gases dissolved in the water. In terrestrial plants the soil affects growth in other ways. Most plants depend upon the soil to anchor and support the stem and leaves. For these reasons the quality and quantity of the soil are important in plant growth. Contrary to what might be expected, tropical soils are thin and subject to landslides (Fig. 8-9), so that there are many areas in the tropics where plant growth has been disturbed.

Soils of different types have varying moisture-holding qualities. A heavy clay soil will hold moisture because it is composed of very fine particles of rocks. These small particles provide a tremendous surface for holding water. A thin layer of water will cling to each particle by surface tension. Loose soils such as sand hold less water because the particles are larger. The organic matter in the soil also determines its ability to retain water. Heavy clay soils will hold water but make root penetration more difficult because they dry to a hard surface. Such soils also keep out air so that roots may be drowned, eliminating the oxygen needed for their respiration as in other plant parts. Desert soils are generally loose so that the little water available soon drains off. The best soils are loams because they are a mixture of sand and clay, with a considerable amount of humus. This combination holds the proper amount of moisture, drains well, and does not harden at the surface.

Soil is formed principally by the slow cracking of rocks from frost or root penetration or by erosion. Humus, organic matter from decaying plant and animal bodies, covers the best soil and provides roots with mineral salts containing nitrogen and phosphorus formed when the final decay process is complete. Humus is a substance added to soil and is not a soil type. Principal soil types are sand, clay, and loam, as mentioned previously. Soil is transported by a variety of means, each of which results in a characteristic deposit. Thus alluvium is formed by receding flood waters that deposit the soil they have held. This is generally a rich soil. **Lacustrine** soils form in lake bottoms that have eventually dried. Such soils are rich if they are high in organic matter, but they may also be poor because of heavy mineral and salt deposits, including sodium salts, borax, and lime. Glacial tills are stony soils deposited at the foot of receding glaciers and are generally poor for agricultural purposes. In desert regions much of the soil is in the form of loess, a deposit of fine yellowish gray loam transported by wind. Actually, loess is rich but porous, permitting rapid runoff of the sparse water supply. Usually this soil crusts over to form a hard cover, making plant growth very difficult.

The mineral content of soils is critical for plant growth. Any soil deficient in essential minerals will not support plant growth unless this lack is counteracted either by the use of fertilizers or by water running into the soil from other areas. Even the trace elements must be present. Soils deficient in any of these show characteristic growth deficiencies in the plants they support. Although organic substances are not used directly by the plant, soils high in organic materials support better plant growth by gradually breaking them down into their component parts. The organic matter is also a source of nitrogen through bacterial action.

The type of minerals and the amount of humus regulate the pH (acidity and alkalinity) of soil. Some soils, particularly in humid regions, are acid in reaction because the basic salts are leached out. Others, particularly in desert regions or in certain types of bogs, are alkaline in reaction. Most plants grow best in nearly neutral or slightly acid soils. The salt ($NaCl$) content of soil, although it does not affect the pH, is sometimes so high that plants cannot grow; they are unable to absorb water because of the increase in osmotic pressure outside the root cells (Fig. 8-10). Areas with poor drainage and high evaporation suffer from high salt content. The most notable of such areas is the Great Salt Desert of Utah.

Unstable soils result in a particular kind of dynamic ecology. Among the best examples of this are sandy areas with prevailing winds that blow and shift the soil surface into dunes. Dunes are mounds of sand that shift continually so that momentarily they cover established vegetation. The battle is endless in such regions. Only through man's efforts can the soil be held in place long enough for a firm plant cover to become established. The slightest disturbance from disease, insect pests, or fire will again release the nomadic dunes. They may be man made through faulty agricultural practices, a good example of which is part of the dust bowl.

Other than nitrogen, the minerals essential for plant growth come from the weathering of rocks. Nitrogen, however, does not, since its origin is ultimately the atmosphere. Before atmospheric nitrogen can become available to a plant in soil, it must be combined with other elements. Symbiotic and free-living microorganisms in soil are primarily responsible for combining free nitrogen from the air with other elements. A group of bacteria (*Rhizobium* sp.) that live symbiotically on certain legumes are able to utilize the free nitrogen gas from the air in their metabolism and thus fix the nitrogen in a combined form. These bacteria are called **nitrogen-fixing bacteria.** They are responsible for the formation of nodules on legumes (Fig. 18-9) such as clover, vetch, and the pea. The combined nitrogen, probably in an amino form, is absorbed by the legume and synthesized into its proteins. Other saprophytic bacteria, such as *Azotobacter* sp. and *Clostridium* sp., are also capable of nitrogen

Fig. 8-10

Effect of salt on plant growth. The grass in the drum on the right is growing in soil with a high salt (NaCl) content; the one on the left, grown under the same conditions but without salt, shows greater growth. (USDA photograph.)

fixation. The organic nitrogen of the legume after death is broken down into nitrates by still other microorganisms in the soil and can then be used by nonleguminous plants. Animals secure their nitrogen supply from plants or other animals as food, which is used in synthesizing their proteins. The nitrogen of the animal and plant proteins is returned after death to the soil, where it is broken down into nitrates. However, this breakdown occurs in several stages and is aided by the microorganisms in the soil. The organic residue from dead plants and animals is reduced first to ammonia, then by nitrification to nitrites, and finally to nitrates. Nitrates are the only form in which plants can utilize nitrogen. The diagram in Fig. 8-11 illustrates graphically the use and reuse of nitrogen. The transfer of nitrogen from the air to the soil, then to plant and animal proteins, and back again to the soil and air is called the **nitrogen cycle.**

BIOTIC ENVIRONMENT

In addition to the physical factors in the environment, the relationships of individual plants to others of the same kind and of different kinds are of equal importance. Through a study of these relationships the transition from the autecological approach to the synecological approach

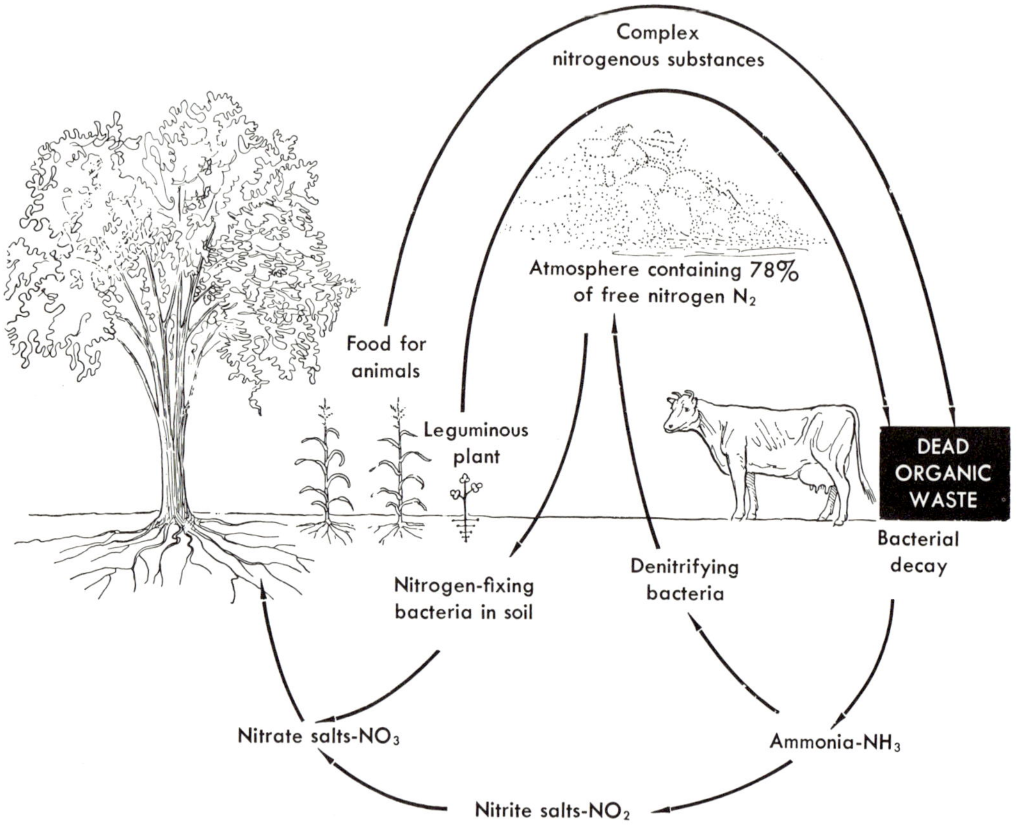

Fig. 8-11

Nitrogen cycle. There are various sources for the nitrogen needed for protein production in both plants and animals. Animals obtain all their nitrogen from either plants or other animals. Plants, however, obtain all theirs from the soil.

is apparent. Both these approaches are necessary if the complexities of biological phenomena are to be understood.

Competition with other plants of the same kind. The simplest form of competition in plant life is the need for a certain amount of space for roots in the soil and for exposure of leaves to light. As any plant grower knows, well-spaced plants grow best. The more crowded the conditions, the greater will be the reduction in the size of the plants and the number of seeds. On the other hand, beneficial effects from crowding are also possible. The colony of a single species of plant, once it occupies all available space in a given area, presents a united front against the invasion of other species. Dioecious plants must be close enough together so that pollen can easily fertilize the female plant. In many monoecious plants this is necessary also because the anthers mature at a different time from the pistils or some structural device prevents cross-fertilization; therefore fertilization may not occur in widely separated plants.

Competition among plants of different kinds. This is well known to the practical biologist. Weeds penetrate any cultivated area, and the struggle between these two types of plants is as active and dynamic in its slow, persistent way, as is competition between animals. Some plants have developed special structures to aid them in their invasion techniques. Reproduction by runners is one, the more vigorous part of the plant colony contributing to the growth of the invading shoot until it becomes established. Vines, such as poison ivy or the strangling fig, can completely choke out large trees and occupy the space once held by them. A vine growing over smaller plants can block out the sunlight that would otherwise reach them, and these plants must give way to the new growth. Any pest or disease that weakens one species and leaves another unharmed will give the latter the advantage, and it will soon occupy the space held by the former. Certain plants have the ability to colonize quickly an area destroyed by fire. However, once they have become established, other plant species will invade the area and by some superior growth habit push out the original colony to take over. Thus there is a continuous succession of plants until a climax condition is reached. Once this is achieved, conditions remain more stable until the area is again disturbed. Weeds are particularly tenacious in this respect. Some, for example, plantain (Fig. 8-12), will flower and fruit, using nutrients stored in their leaves, after they have been removed from the soil by the gardener.

Parasites of plants fall into two main classes, bacteria (Fig. 8-13) and fungi that cause disease and worms that cause nematode diseases by invading the plant tissues. External feeders such as insects and other herbivores cannot be considered parasites in the usual sense of the word because their effect on the plant is immediate. Plant diseases caused by fungus

Fig. 8-12
Plantain fruiting even after being uprooted. (USDA photograph.)

Fig. 8-13
Crown gall bacterial disease affecting plants. (Courtesy A. S. Riker, Department of Plant Pathology, University of Wisconsin, Madison, Wis.)

are many, and because of their importance to agriculture, their study and control form a separate field of biology, plant pathology. Some of these are discussed in Chapter 28. Today, nematode parasites of plants have become increasingly important. They have been brought to the attention of the lay public by importation of the golden nematode disease of potatoes. There are many other diseases of this kind, all causing serious loss and influencing competition for space by plants.

Certain symbiotic relationships among plants occur. One of the best known is the relationship between certain algae and fungi to form lichens. This is the relationship between two plants of different kinds for their mutual benefit and is called **symbiosis.** Other relationships of this nature are to be found, as, for instance, the denitrifying bacteria that grow in root nodules.

The term symbiosis is now used in a much broader sense than is implied in the preceding paragraph. It includes almost any mutual and constant relationship between two species, including parasitism. This is the more generally used meaning of the word, but a few moments' reflection will show how difficult it is to limit the term and to assign a given case to this phenomenon.

Animal relationships. Plants of many species, but not all, are eaten by many kinds of animals. Without this relationship, there would be no animals. It seems obvious, therefore, that the growth of plants is sufficient not only to maintain the plants themselves but also to provide for the entire animal kingdom, including man. This is done primarily by overproduction of the vegetative and reproductive parts of the plant in the form of accessory tissue such as fruits and nuts. Secretions such as nectar and sap also provide food for animals. These relationships are quite obvious. Less apparent is the fact that feeding on plants by animals has become an established pattern so that, in the normal balance of nature, the plant species are not harmed by this predatory action. It is only when the balance is momentarily upset that the plant population is harmed. If the plant is under cultivation by man, animal predators are called pests, and steps are taken to eliminate them. This view is anthropocentric and not meaningful in a true biological understanding of the phenomenon. Plants not cultivated or otherwise useful also are held in check by predators. So far as the plant species is concerned, being preyed upon by animals is only one factor to prevent the species from occupying the maximum possible space for which it is fitted. This means that there may be a constant struggle for the greater area occupied. Likewise, from the point of view of the animals, if the plants grew better, the animals would have more food and could occupy more space. Therefore, in any given case, the relationship is one of dynamic equilibrium, sometimes with great fluctuations of individuals. Once host specificity such as this is firmly established it will result in the extinction of both. Only man and a few other animals, for example, some ants, have learned to profit by the cultivation of plants. When this happens, the relationship is no longer one of host and predator but becomes symbiotic in the broadest sense.

Actually, the kinds of animals feeding directly upon plants are relatively few. Most animals feed on other animals as a part of a food chain that only eventually gets back to the plants. These food chains may involve many kinds of plants and animals, one eating the other until either man or some animal dominant in a given community is reached.

Most notable of the animals that are **phytophagous** (that is, feed on plant parts) are many insects and the herbivorous mammals. Both of these groups have many species that feed directly upon the plants. Aquatic animals, particularly some arthropods and some fish, are the herbivores of the sea. Most of these are filter feeders; that is, they feed by straining sea water through their mouths and eat the filtrate of microscopic plants, as well as of animals. Together, the floating plants and animals form the **plankton.** The extent of this feeding is apparent when one sees the thousands of tons of fish removed from the sea each year or when one realizes that some of the whales are plankton feeders and grow to tremendous sizes on a diet of plankton.

The following outline briefly sketches "plant" and "animal" relationships.

I. Unicellular plants
   Consumed entirely by macrozoa such as sponges, jellyfish, anemones, corals, flatworms, rotifers, entoprocts, bryozoans, sea worms, lamp shells, mollusks, worms, crustaceans, starfish, and vertebrates.

II. Multicellular algae
   A. Marine algae—eaten mainly by crustaceans and vertebrates.
   B. Fresh-water algae—eaten mainly by crustaceans, fish, reptiles, and birds, but rarely by mammals.
   C. Terrestrial algae—eaten by small insects and rarely by some other terrestrial arthropods.

III. Vasculuar plants
   A. Roots—insects, particularly the larvae of beetles and moths, are obligate root feeders, and a number of larger mammals, for example, pigs, are sometimes root feeders. The insects are primarily borers in roots, which causes considerable loss of vitality and frequently death to the plants.
   B. Stems—this part of the plant is usually the least palatable, partly because of the hard, nonnutritious supporting tissue. The majority of stem-feeders are boring insects or ruminate mammals, for example, cow and horse.
   C. Leaves—the greatest variety of plant-feeding animals are leaf feeders. Many mammals depend entirely upon the consumption of leaves of vascular plants. Unless the plant is well protected by heavy spines or poisonous secretions, it may be stripped of its leaves by these animals. In addition, the adults and larvae of beetles, moths, and some bees feed on or in the leaves of plants. These number into the tens of thousands of species, many of which are host-specific. Because of the variety of leaf-feeding mechanisms, these must be further elaborated.
      1. Feeders on the entire leaf—insect larvae and mammals.
      2. Leaf-skeletonizers—insects larvae which feed only on the softer parts of the leaf, leaving untouched the skeletons of vascular tissue.
      3. Leaf-miners—larvae of moths, flies, midges, sawflies, and beetles. The larvae and sometimes the adults of these insects bore into the leaf and feed on the parenchyma tissue, leaving a tunnel or mine in the leaf. The

mine grows larger as the insect grows, producing a characteristic winding tunnel pattern on the leaf that is different for each species of miner.

4. Leaf-cutters—some species of insects cut away parts of the leaf, using it to build nests, or as their food supply in subterranean fungus gardens.
5. Leaf-rollers—a few insects, while not feeding on the leaf, cut and roll the leaf to form a nest held together by silk.

D. Fruit- and seed-feeders—mammals, birds, and insects consume a large part of the fruit and seeds produced by many species of plants.

E. Flower-feeders—animals consuming floral parts before they mature into fruit and seeds are relatively rare. However, the sugary secretion of the flower, the nectar, is consumed by birds and insects. Usually, while these animals are feeding, they come in contact with pollen which is also gathered, either by accident or on purpose, and transferred to other plants so that pollination results. A large number of flowering plants are entirely dependent upon this method of pollination. A few insects are pollen feeders, but do not pollinate.

F. Gall-makers—a number of insects, particularly flies and wasps, and also some fungi produce galls, that is, growths on leaves, stems, and roots of plants. The shape and characteristics of the gall depend upon the insect or fungus guest living within and are specific for each guest and each host. The galls are different from those resulting from disease in that the normal function of the plant is little affected.

G. Sap-feeders—insects with sucking mouthparts, particularly the true bugs such as plant lice and some others, are able to pierce the tissues of plants to feed directly on the fluid within the vascular tissue of the leaves and stems of the plant. This not only weakens the plant but also may stunt or even kill back the plant if feeding takes place at the growing tip. The scars left by such feeding also destroy the aesthetic value of flowers and fruit.

IV. Fungi

The larger fungi such as the sac fungi and club fungi are eaten by some mammals, rarely by birds, and by a great number of insects, many of which are host-specific. Several families of flies and beetles have many species that are obligatory feeders on fungi of this type.

Other more specialized animal-plant relationships are properly reserved for detailed study.

SPECIAL TERMS USED IN CHAPTER 8

carbon dioxide cycle	oxygen cycle
lacustrine	photoperiodism
nitrogen cycle	phytophagous
nitrogen-fixing bacteria	plankton
organism	symbiosis

QUESTIONS TO CONSIDER

1. From the discussion so far and from what can be observed of life about you, list the conditions under which life can exist as we know it.
2. What are the chances of finding life like ours on other planets? In other regions of space?
3. Explain how a greenhouse is a controlled growth chamber.
4. Under what conditions of plant growth can there be, and often is, an oxygen deficiency?
5. How would you set up a balanced aquarium? A terrarium?
6. Considering the nature of protoplasm, why is extreme cold less injurious to plants than is extreme heat?
7. When is wind a factor in the effective environment? Name some other factors that are a part of the effective environment only under certain circumstances.

DISCUSSION FOR LEARNING

1. Discuss the relationship of the cell wall to the effective environment. What does it protect; what processes are slowed down as a result?
2. Considering all the factors involved in the effective environment, discuss the following agricultural practices: (a) spacing of plants; (b) weed control; (c) insect control; (d) irrigation. What factors are not ordinarily controlled?
3. Which one of the environmental factors is most likely to lead to overpopulation if it is in over abundance?

Part IV

Reproduction and development

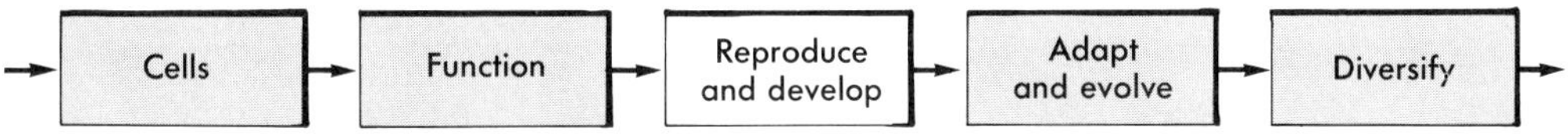

- Living processes end either in death or through replication, in birth, signifying in each the end of one series of processes and the beginning of a new series.
- Replication takes place at the molecular level and birth at the cellular level.
- Continued replication with modification occurs within the organism.
- Birth of individuals occurs within the population; birth of new populations occurs within the species; and finally even the birth of new species occurs through evolution.
- Development by growth and by changes in physiological processes facilitates replication through the formation of diverse reproductive structures suited to the environment of the species.

Chapter 9

Reproduction

Courtesy Ferry-Morse Seed Co., Detroit, Mich.

Life itself is defined by its unique ability to reproduce, and this is often symbolized by a bee or a flower such as the one pictured above. A philosopher might refer to a reproductive series as the chain of life, a maze of organisms linked together since the first replication of nucleic acid. Profound studies have been made to determine how life originated. These interesting and scholarly deductions probe deeply into biochemical reactions. It is apparent that the process was slow to develop, but it also seems obvious that the real beginning of life was at the moment of replication, through the synthesis and arrangement of amino acids to preserve an orderly structure capable of growth and further replication. It is in this sense that we define **reproduction.** The very nature of the biota, as we shall see, is based on reproductive structures and processes, and the ordinary connotation of reproduction is signified by the union of gametes, as will be seen throughout this chapter.

All plants from the simplest to the most complex have structures for reproduction and for getting food. Plants that have a complex organization are able to live in situations uninhabitable by plants lacking these complexities. Every organism, as an individual, is equipped with some means of reproduction, and that organism is occupied with growth, preservation of self, and reproduction of its kind. The most distinctive characteristics of any plant are its structures for, and means of, reproduction. This chapter presents the basic principles of reproduction. The adaptive structures and processes of reproduction are very diverse and are further discussed in Chapter 13, where twenty-two life cycles are described, in Chapter 12, and in Part VI.

BASIC PRINCIPLES

The reproduction of all plants is very similar in some respects. Means are usually provided for a **gamete** union to form a **zygote.** This is necessary to mix the genetic material and thus provide for individuality (variation). The gametes usually have the haploid chromosome number. The union of two gametes forms a zygote which is the diploid chromosome number. All the complexities of plants simply involve refinements of structure for facilitation of the union of gametes or the dispersal of spores and seeds.

Reproduction of most plants is basically an **alternation of generations,** that is, an asexual plant body alternating with a sexual plant in the succession of generations. Relatively few of the simple plants have sexual generation, and most of the higher plants (seed plants) have so reduced the sexual reproductive plant body that it is negligible. It is the asexual plant that becomes dominant in the higher plants; the gametophyte is so greatly reduced that it is, for all practical purposes, a part of the sporophyte.

Most organisms are either single-celled, **unicellular,** or divided into several to billions of cells, **multicelluar.** Considerable difference in terminology is necessary in considering these two types of organization. A third condition exists in certain fungi and slime molds that have two to many millions of nuclei without cytoplasmic walls separating the body into cellular units. Finally, the cells of bacteria and related organisms (i.e., the procaryotic organisms) lack an organized nucleus. Each of these types of organisms maintains their reproductive processes in a different manner, utilizing quite different structures.

In spite of this great diversity, a general plan is seen that is no doubt inherent in the universally present control system provided by the DNA molecule. This uniformity is seen as follows:

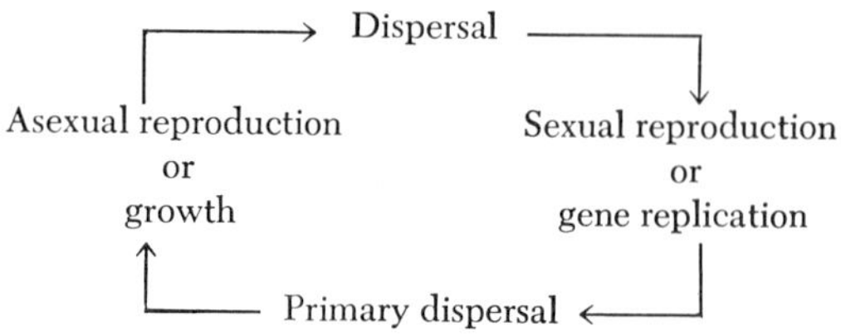

Each of these stages is considered in detail in the following discussion.

ASEXUAL REPRODUCTION

Asexual reproduction is not a single process or even a homologous process in plants. In unicellular plants, cell division or mitosis is a form of asexual reproduction. It is asexual because it does not involve a mixing of genes; that is, the organism is duplicated by the mitotic process, but each of the offspring is genetically identical with the original cell. This same process in multicellular plants is called growth. When the cells adhere to each other after mitosis, a multicellular organism is formed. This growth process is not considered to be a form of reproduction because no new organism is produced.

True asexual generation is evident by the production of **spores, sporulation.** In the lower plants this is the primary dispersal stage. The **sporophyte** produces large quantities of cells that may be specially fit to resist adverse conditions and that are at the same time capable of wide dispersal. In the lower aquatic forms these cells are provided with cilia or flagella that permit them to swim for some distance and thus populate new areas if conditions are suitable for growth. Swimming spores do not have a protective cover. They may have either the haploid or diploid chromosome number. These forms continue to produce new plants without sexual reproduction. The same plant also produces the gametes, and sexual reproduction takes place. In the higher but nonflowering plants these spores are produced by a sporophyte that has the diploid chromosome number. The spore itself is formed after meiosis has taken place. Such spores are light and airborne. They are provided with a protective covering resistant to adverse growing conditions. These spores, instead of producing new sporophytes, produce the gametophyte. True alternation of generations is then present.

There are other types of asexual reproduction. Some of the bryophytes are provided with special organs, the cupules, which develop **gemmae.** A small clump of cells breaks away from the parent plant and grows into a new gametophyte. This type of vegetative reproduction is well organized. In lower plants simple fragmentation of the **vegetative body** will result in the formation of new individuals. In flowering plants the vegetative parts of many plants will reproduce new individuals by means of **runners,** which may be either stems or roots, or even by means of rooting of leaves or parts of leaves. A clump of plants developed in this manner, each with identical genetic composition, is termed a **clone.** The airplane plant, a common succulent house plant, normally reproduces with the development of small plants along the margin of the leaves. The plantlets are budded off, fall to the soil, root, and grow into a new plant.

SEXUAL REPRODUCTION

The key to **sexual** reproduction is meiosis. The formation of gametes must be preceded by the reduction of the chromosome number. The sporophyte is usually diploid in chromosome number. If the formation of gametes were not preceded by the separation of the homologues (chromosome pairs), the fusion of the gametes would cause a doubling of the chromosome number at each fertilization. Obviously this process could not continue indefinitely. In many lower plants meiosis takes place immediately after fertilization. In these plants the cells of the **gametophyte** are now prepared for fusion without the need for reduction in the chromosome number. The true sporophyte is absent in such plants. Because of this, it is assumed that the haploid condition of the chromosomes is primitive, and that diploid plants are due to a delay in the time of meiosis, which is the result of the development of a spore-producing plant, the sporophyte.

In many of the algae the production of gametes occurs only toward the end of the growing season and cannot be regarded as the normal means of increasing numbers. With more complex development of body structure in the higher algae and the rest of the plant groups, gamete production plays a more prominent role in reproduction. As may be expected, special structures are developed in these plants for gamete production. The gamete chamber is characteristic of particular groups of plants. The gamete chamber of the algae and related plants is always unicellular, whereas the gametangia of higher plants are multicellular. The simple algae, and also many fungi, produce isogametes with no morphological distinction between the two gametes that unite. Certainly, however, in all cases there are as yet undetectable physiological differences. The gametes of the higher algae and most of the rest of the plant kingdom show distinct morphological differences. The female gamete or egg cell is larger than the male gamete or sperm cell. The **sperm** is always motile, but the **egg** remains in situ until fertilized.

In the algae the cell that produces the egg is referred to as an oogonium; in the higher plants the gamete chamber producing the egg is called an archegonium and is many celled. The male gamete-producing chamber, whether unicellular or multicellular, is referred to as an antheridium. Many sperm cells are produced by the antheridium, but only a single egg cell develops in the oogonium or in the archegonium. Collectively these structures are referred to as **gametangia.**

SPOROPHYTES

The production of spores as a form of asexual reproduction is the function of the sporophyte. In the higher forms this plant is also the vegetative or food-producing plant. In some it remains as a parasite on the gametophyte. Gradually the sporophyte evolved to become independent until in the seed plant its role as the vegetative plant is dominant, with the gametophyte remaining as a parasite on the sporophyte.

GAMETOPHYTES

The role of the **gametophyte** is much different from that of the sporophyte. The essential purpose of the gametophyte is the production of gametes, haploid cells of opposite types, that is, genetically different. Upon fusion they produce a new individual with a gene complement unlike that of the parents. The gametophytes of algae are vegetative, in which any cell is capable of producing the gametes. In others, the gamete-producing plants are distinct from the spore-producing plants.

Various types of gametophytes are known. The bryophytes have vegetative gametophytes to which the sporophyte remains attached. However, in some of the bryophytes the sporophyte becomes partially independent. The lower vascular plants have separate, independent gametophytes with relatively limited vegetative functions. This is typical of the ferns and horsetails, but not of all the others. The club mosses, for example, have partially parasitic gametophytes on the sporophytes. This is a forerunner of the seed habit of the conifers and the flowering plants, the remaining groups in which the gametophyte is completely dependent upon the sporophyte. Spore production in such plants is extremely limited, since the spores are retained in the plant and mature into a tiny gametophyte. Thus the two generations of the plant life cycle merge into one as the ultimate development of plant reproduction, the seed habit. This development best fits the conditions presently on earth and accounts for the success of the conifers and the flowering plants that have become the dominant element of our modern floras.

EVOLUTION OF THE PLANT BODY

Considerable attention has been given to the reproductive habits of plants because an understanding of these processes is necessary in order to understand the role of plant life in the balance of nature. Efficient reproductive processes are necessary for the survival of the plant species. In addition to the evolution of the reproductive organs, there is a parallel evolution of the vegetative body.

The organization of the single-celled vegetative plant is the least complex because each of the cells carries on all life processes, independent of each other. Plants organized in this way vary greatly in size but within definite limits. Although some are just visible to the unaided eye, most are microscopic in size. These plants are able to populate certain habitats in great abundance, but they are strictly limited regarding the types of habitat in which they can thrive. They are all restricted to an aquatic niche, either a body of water, such as the ocean, the most uniform of ecological habitats, fresh water, or a boggy area that rarely dries out. Therefore, as a rule, a generalized body organization demands a generalized ecological condition. Lack of specialization prevents most single-celled vegetative plants from inhabiting locations that are subject to great variation. However, the land forms have developed one feature that permits them to emerge from the aquatic habitats. This adaptation is the ability to form spores that can resist extremes of temperature and fluctuations in moisture. The introduction of sporulation in the life cycle is probably the first specialization of the plant body. The existence of this in bacteria is probably indicative of the specialized rather than generalized nature of these plants.

The next development in the specialization of the plant body is not clear, but the formation of filaments is the most logical one. A **filament** is easily formed by successive cell divisions without the breaking away of the newly formed cells. If a certain cell grows and divides rapidly and the cells remain attached to each other, the plant body becomes multicellular and filamentous. Frequently a sheath is secreted around this string of cells, affording common protection. If some of the cells in this multicellular plant become specialized for spore production, or for gamete production, a division of labor within the plant body results. The remaining cells in the body contribute food to the reproductive cells, and the filament becomes an individual. Additional specialization among the cells may be the formation of a basal holdfast cell that holds the plant to the substratum.

The difference between filamentous organization of the plant body and colony formation is slight. A true **colony,** more simply organized than a filament, is a collection of unicellular organisms clumped together and perhaps providing some mutual protection. There may be an exchange of food material, the individuals of the colony sharing in the food-getting process. However, this is not evident in photosynthetic organisms. Colony formation occurs but rarely in the plant kingdom, although it is complex in the animal kingdom, in which it is very difficult to distinguish between a colony and an organism because of the complex interaction among the individuals. The only way to test this is to determine whether the individual cells are able to survive and reproduce when broken apart from the main mass. Even this is a poor criterion at times because the process cannot always be distinguished from regeneration.

The change from a filament to a **thallus** organism, that is, a plant body made of several parallel filaments, is accomplished by the lateral division of cells as well as by the apical division, to form a sheet one or more cells thick. Once this plant body is formed, specialization within the new individual becomes complex. This type of plant body is the most common.

The plant body does not always develop from cells in the usual sense. The third type, mentioned previously as occurring in some algae, slime molds, and fungi, often show a multinucleated organism composed of only one or only a few very large cells. In these forms only the nucleus divides; the cytoplasm remains as a common material throughout the large cell. The individual nucleus controls the cytoplasm in its immediate vicinity, but the cytoplasm carries on the common functions of the plant body. Multinucleated cells are called **syncytia** and may be filamentous, as are the hyphae of bread mold, or they may be in the form of a sheet, the **plasmodia,** as in the slime molds, in which individual cells unite to form the plant body.

Cell specialization soon leads to tissue formation in the higher algae, the bryophytes, and the fungi. Several tissues then becomes associated to form the major parts of the higher plants. These plants have functional tissues that serve much the same purpose as the integumentary, reproductive, excretory, regulatory, skeletal, and transporting systems of the higher animals. Higher plants never have even a simple receptor system, but some of the lower forms have the same structures for reception as those found in the Protozoa, and some of the higher plants react to touch and changes in light intensity. Plants always lack digestive, muscular, and nervous systems. Therefore, compared to animals, plants are more simply organized. The movement of fluids in the vascular plants, although somewhat analogous to the circulatory system of animals, is poorly developed. Even in trees it is still restricted to a thin cylinder of living tissue. The ability of most plants to manufacture their own food seems to have eliminated the necessity for developing the complex body systems found in animals, and yet at the same time plants are well adapted for life on this earth. Plants can easily survive without animals, but animals cannot live without plants.

SPECIAL TERMS USED IN CHAPTER 9

alternation of generations
asexual
clone
colony
egg
filament
gametangium
gamete
gametophyte
gemma
multicellular
plasmodium
reproduction
runner
sexual
sperm
spore
sporophyte
sporulation
syncytium
thallus
unicellular
vegetative body
zygote

QUESTIONS FOR CONSIDERATION

1. Distinguish between reproduction and growth.
2. Three terms are rather close in meaning: filament, colony, and clone. How are they distinguished? What is the difference between multicellular and thallus?
3. Distinguish between plasmodium and syncytium.

DISCUSSION FOR LEARNING

1. Alternation of generations seems to take place in many plants. Is there any reason, considering the basic principles of reproduction, why this is an advantage?
2. Sex to the human mind usually implies maleness and femaleness. Is this a basic feature of reproduction?
3. As we have seen, dispersal is usually closely associated with reproduction, yet it has no direct part of reproduction as such. Why, however, is dispersal considered to be a link between the two generations?

Chapter 10

Plant genetics and evolution

Charles Darwin gazed upon the same scene in the Galápagos islands that is pictured above. As he did so, he was impressed by two things: the weird landscape, where common plants and animals were so changed as to be barely recognizable—sunflowers and cactus became trees—and the continuity of forming through change. His thoughts crystallized many years later. His ideas along with those of Gregor Mendel have led the way in developing biology from a descriptive to a quanitative science.

The details of both the genetic and the evolutionary process are the most intricate of all biological phenomena, and some aspects are still little known. The combined knowledge of all genetic, biochemical, and physiological functions is needed to interpret the phenomenon termed evolution. There is no doubt that it is the greatest conceptual scheme of science ever devised. In it we see a progression of biological systems from the replication of DNA to cell division, to reproduction of the organism, and finally to the multiplication of species. This seems to be the theme common to all aspects of biology.

In previous chapters the chemistry and physiology of cells, cell division, and reproduction have been discussed. The mechanism for the control of the development of the individual plant body as well as its variation will be considered first in this chapter, followed by a discussion of how these same mechanisms are involved in the multiplication of the species, and its adaptation to its environment.

The most striking characteristic of living things goes farther than mere reproduction —it is the ability to produce offspring essentially like themselves in their fundamental structures, yet at the same time,

Fig. 10-1
Varieties of strawberries produced by selection and crossbreeding. (USDA photograph.)

individuals with slight differences. Like begets like is a common phrase. An organism will inherit the characteristics peculiar to its species. However, despite the fact that parents and offspring are unmistakably similar to each other, they will usually differ in certain respects. Fig. 10-1 shows a number of strawberry fruits. Although each is recognizable as a strawberry, it is apparent that they are not all alike. In fact, they are an example of several varieties of fruit that differ as much in taste and hardiness as they do in size and shape. Not only does one plant differ from another, but also variations can be found on the same plant. For example, on the same hibiscus plant can be found several differently shaped leaves (Fig. 10-2). This tendency of organisms to differ slightly, or to a greater degree, from their parents is called variation. Variation is the one characteristic that makes possible the adaptation of an organism, and eventually a species, to various habitats. If all offspring were exactly alike, environmental changes would cause

Fig. 10-2
Variations in hibiscus leaves from the same plant.

rapid extinction instead of evolutionary change.

VARIATION

Variation within a particular species of plant is due either to environment, hybridization, or mutation. Plants, particularly, often vary as a result of their environment. A plant growing in poor soil that lacks a certain essential mineral will not grow as large or produce as much food or as many flowers and seeds as the same plant growing in fertile soil in which the necessary minerals are present. Changes in plant growth due to environment alone are not inheritable and will not be transmitted by the variant plants to their offspring. Organisms can be modified through the action of external influences, but such acquired characters are a modification that arises as a direct response to an external stimulus and are not inherited because there is no mechanism that will change the coding of the DNA, and hence permit transmission of the new trait.

Although the structure of organisms and the way in which their parts function are determined primarily by heredity, the environment is also important in providing an opportunity through selection for modification of inherited characteristics and the way in which they will be expressed. If a plant is grown in the cold, it fails to grow large, the leaves remain small, and the stem short. When the plant is exposed to higher temperature, it will develop and increase in size. When grown in the shade, many plants develop larger leaves than when grown in sunlight. The cuticle of the leaf becomes thicker if a plant is growing in dry soil, thereby reducing transpiration. On the whole, plants exhibit great versatility in response to their environment.

Another type of variation is due to **hybridization,** which is the result of crossing two varieties of the same species. The offspring of such a cross is likely to have some of the characteristics of each parent and hence is called a **hybrid.** Since they are heritable, the variations resulting from hybridization differ from those induced by the environment.

Because of the ability to produce offspring that resemble the parents, living organisms can be classified into groups of similar and presumably related individuals. At the same time, because offspring are not always identical to the parents, in other words, do exhibit variation, it is possible to study the way differences are inherited. This science is termed heredity. The physical basis for the laws of heredity is to be found only in the behavior of the chromosomes.

The only physical bridge between two generations of sexual organisms is the sex cell or gamete, and more specific than that, it is the individual chromosomes. The science of genetics deals with this aspect of heredity and is, therefore, concerned with the study of the behavior of genes and chromosomes and their effect on the form and function of plants and animals. Thus we see that heredity deals with the **phenotypes,** or external appearance of an organism, and genetics treats the **genotype,** or internal inherited components.

MENDELIAN HEREDITY

As described in Chapter 2, Mendel was the first to experiment with heredity. His may experiments extended over a period of eight years. The methods he used and the conclusions he reached will be explained and illustrated in the following pages.

MENDEL'S LAWS

1. Law of segregation of characters
2. Law of independent assortment of characters

Although Mendel experimented with crossing seven pairs of characteristics in peas, the one that is most familiar deals with their tall and dwarf habit. He crossed a variety having tall stems with one having dwarf stems (Fig. 10-3) and found that all the offspring of these two parents were tall. After allowing the tall offspring to undergo self-pollination and hence self-fertilization, he realized that there were two kinds of offspring from this cross. Most of the offspring resembled their original tall parents, whereas a few were dwarf like their dwarf parent. Upon counting the numbers of each and carefully recording them, he discovered that 787 of the plants were tall and 277 were dwarf. Thus, by crossing a tall plant and a dwarf plant, he learned that the hybrid progeny of the first filial generation

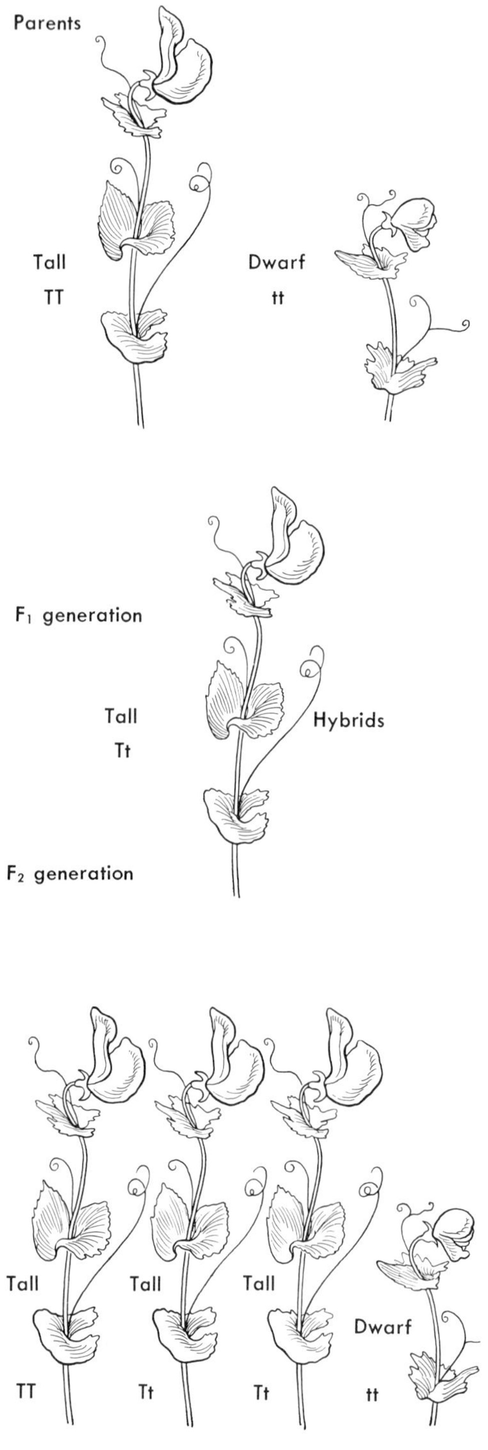

Fig. 10-3
Illustration of cross between tall and short pea plants (monohybrid cross).

(F_1) resembled one of the parents and not the other. The characteristic that remained apparent in F_1 was **dominant** over the other, which was **recessive.** Whether a characteristic was dominant or recessive could be determined only by breeding experiments. The F_1 hybrids breeding among themselves produced progeny that were a mixed group, occurring in a ratio of 3 tall plants to 1 dwarf plant.

Mendel then allowed these offspring, the second filial generation (F_2), to pollinate themselves, and from accurate records of their numbers learned that the F_2 dwarf plants produced nothing but dwarf progeny; one third of the tall F_2 plants produced only tall offspring, whereas two thirds of the tall plants produced both tall and dwarf plants, again in a ratio of 3 tall plants to 1 dwarf plant, like the F_1 hybrid tall parents. From the data of this experiment and from similar experiments, Mendel discovered and formulated what has become known as Mendel's first law of heredity, the law of **segregation.** According to this law, when unit characters such as tallness and dwarfness are combined in the F_1 hybrid generation, the bearers of these hereditary characters (now called genes) remain distinct. However, at the time hybrid plants produce spores and subsequently sex cells, the bearers of the hereditary characters segregate or separate from one another unmodified by their association in the hybrid and are distributed unchanged to different sex cells. Consequently, in the next generation, the F_2 after self-pollination, the chance combination of the sex cells containing either a characteristic for tall or dwarf will give rise to offspring that are either pure tall, hybrid tall, or pure dwarf. Although the phenotypic ratio of these different kinds will be 3 tall plants to 1 dwarf plant, the genotype will exhibit a ratio of 1 pure tall plant, 2 mixed tall and dwarf plants, and 1 pure dwarf plant in the F_2 generation (Fig. 10-3).

In modern terms, the law of segregation

states that two genes of a pair do not contaminate each other but are segregated during reduction division and pass into different gametes or sex cells. Since knowledge of the gene is available, it is now a common practice to let a letter symbolize a gene, such as **T** for tallness and **t** for dwarfness. Since the genes are on the chromosomes, and chromosomes occur as homologous pairs, each cell before meiosis will then have two genes for each characteristic and each sex cell only one. The genes that occur as pairs in a zygote may either be alike, for example, **TT,** which is homozygous tall, or be one of each allele, **Tt,** which is heterozygous for tall and dwarf.

When only one pair of genes is involved in crossing two individuals that differ, the resulting offspring is called a **monohybrid.** Although the phenotypic ratio in a monohybrid cross is 3:1 when one allele is completely dominant over the other, the genotypic ratio is 1:2:1. However, if the monohybrid individual is crossed with the double recessive individual, half the offspring will be heterozygous like one parent, and half will be homozygous like the other parent. For example, **Tt** × **tt** will result in half the offspring being **Tt** and the other half **tt.** This is called a **backcross.** It is used to determine whether a particular organism is phenotypically and genotypically homozygous or heterozygous. Thus a 1:1 ratio is obtained if one parent is homozygous recessive and the other heterozygous.

By experiments using other pairs of contrasting characters in peas, Mendel discovered that round seeds were dominant over wrinkled seeds (Fig. 10-4). In his experiment he found 5,474 round-seeded plants and 1,850 wrinkle-seeded plants in the F_2 generation. Again, it was an approximate ratio of 3:1, such as he had observed for the height of stems. Flower color in peas was also inherited in the same way. Mendel's records show that he got 705 purple-flowered plants and 224 white ones in the F_2 generation when he crossed homozygous purple and white parents (Fig. 10-5). By crossing yellow cotyledonous plants with green ones, Mendel found that yellow was dominant over green and that this characteristic behaved as did the others. All the aforementioned crosses involved alleles in which one characteristic was completely dominant over the other, emphasizing a principle established by Mendel's work. This is the principle of dominance. In other words, in a pair of alleles one gene may be completely dominant over the other, the recessive gene.

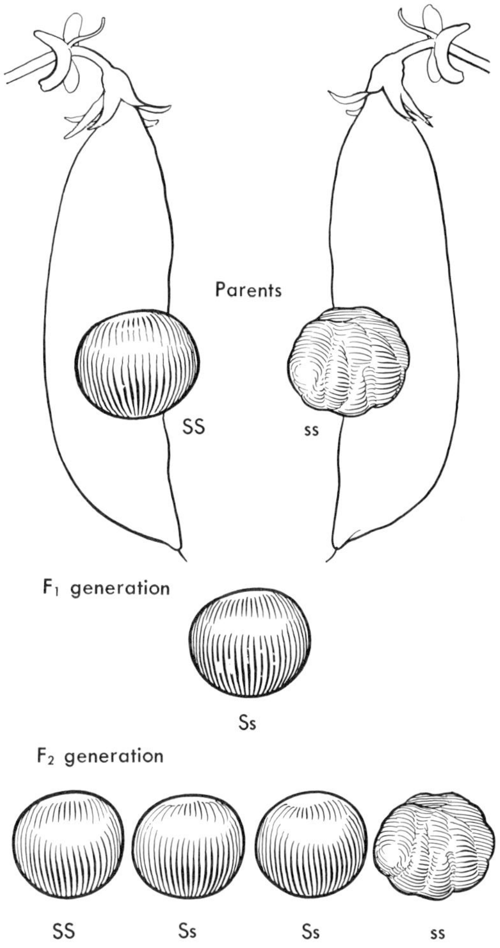

Fig. 10-4
Chart of round and wrinkled peas (complete dominance).

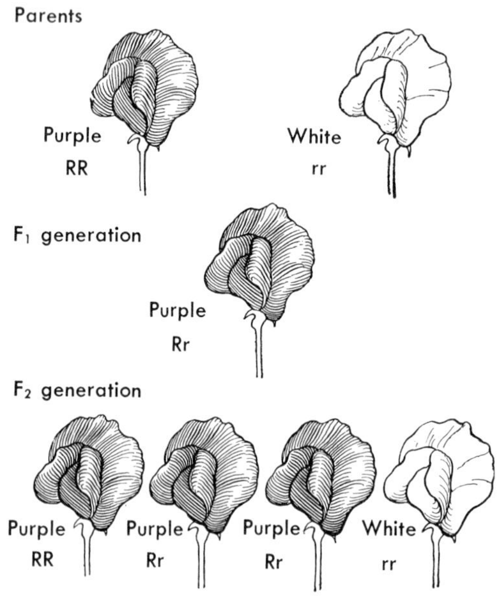

Fig. 10-5
Cross between purple- and white-flowered pea plants (complete dominance).

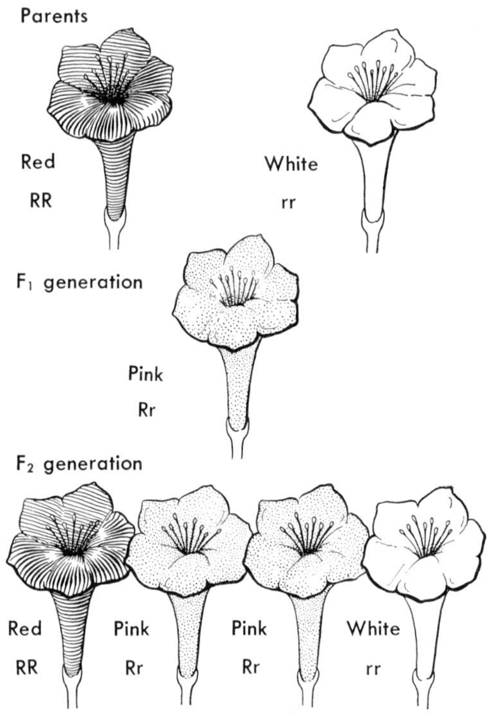

Fig. 10-6
Incomplete dominance in four-o'clocks.

However, dominance is not always the rule in plants. For example, when a red-flowered four-o'clock is crossed with a white one, it may be expected that the F_1 generation would be either all red or all white. Such is not the case, since the plants are all intermediate between red and white—pink (Fig. 10-6). When the hybrid pink plants are self-pollinated, they do not produce all pink offspring, but one fourth are red (like the original parent), one fourth are white (like the other parent), and one half are pink (like the hybrid F_1 generation). If this cross is diagrammed using letters for the genes, it is found that the phenotypic ratio of the F_2 generation is 1:2:1 and that it corresponds to the genotypic ratio. Again, as in Mendel's crosses involving complete dominance, the genes do not lose their identity in the hybrid pink plant but segregate normally in a subsequent generation. This type of inheritance illustrates what can be referred to as **incomplete dominance** or blending inheritance. Initially, it may appear to refute Mendel's first law of inheritance, but actually it strengthens it.

It should be emphasized, since assuming that the genes are located on the chromosomes, that it is the chromosomes that are assorting at random in each generation. Each of the seven pairs of factors used by Mendel to demonstrate the segregation of unit characters was actually on seven separate chromosome pairs. Thus he found that when he crossed pea plants differing in two characteristics (such as cotyledon color and appearance of the seed coat) the characters assorted independently of each other in the hybrid progeny (Fig. 10-7). From the **dihybrid** cross of a yellow-smooth and a green-wrinkled seed plant, he found that all the F_1 plants had yellow-smooth seeds (dihybrids) and that in the F_2 generation each character segregated independently of the other; as a result, there were 9 plants with yellow-smooth seeds, 3 with yellow-wrinkled seeds, 3 with green-smooth seeds, and

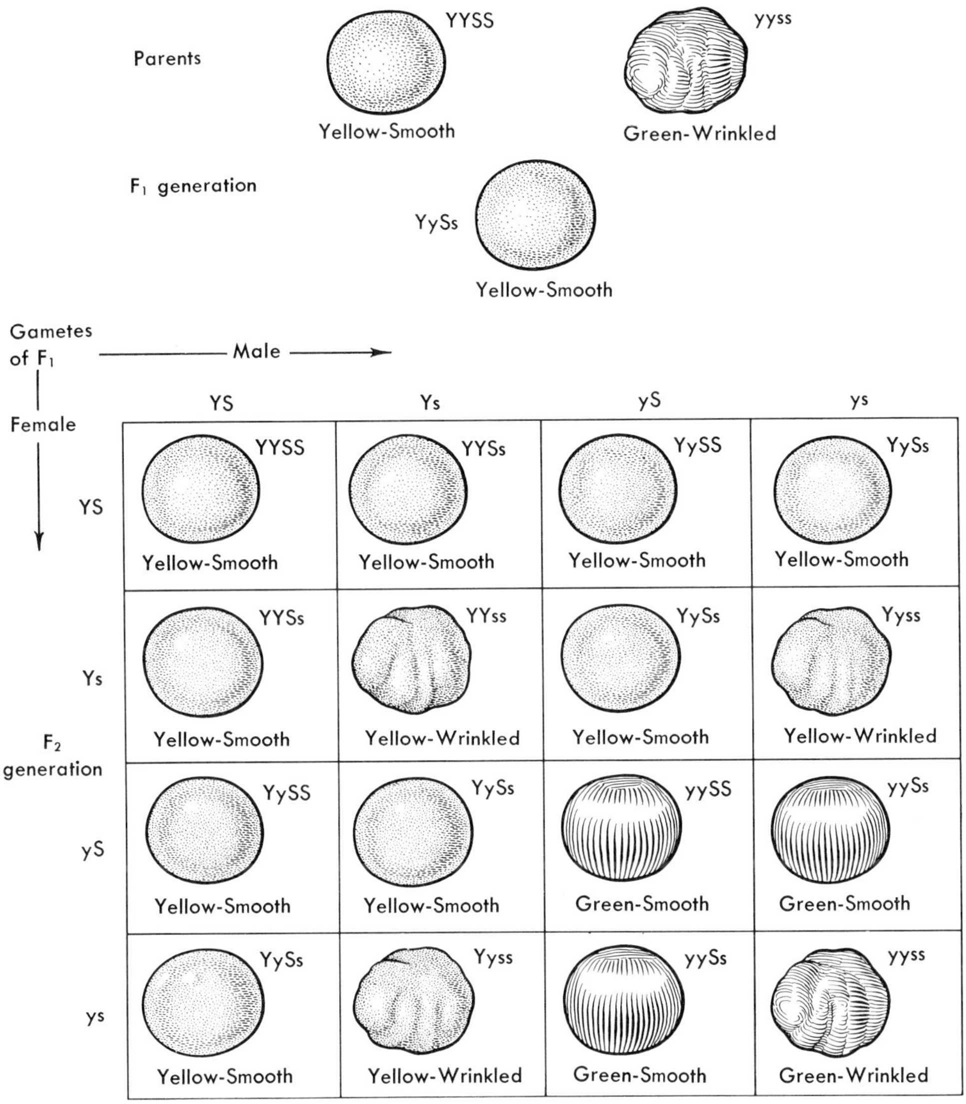

Fig. 10-7
Dihybrid cross. Yellow and smooth seeds × green and wrinkled seeds.

1 with green-wrinkled seeds. If the inheritance of each characteristic is determined separately, it is found to follow a simple monohybrid ratio of 3:1. No zygote will have more than four genes (two pairs), and no gamete will have more than one of each of the two pairs. If the four genes are allowed to assort independently, then 9 plants will exhibit both dominants; phenotypically, 3 will exhibit the one dominant but the other recessive trait, 3 will exhibit the other dominant and the other recessive trait, and 1 will exhibit both recessives. The dihybrid cross thus illustrates Mendel's second law, the law of independent assortment, which holds true if the genes are located on separate chromosomes, but will not be true if they are located on the same chromosome, a phenomenon known as linkage. Therefore, when plants are crossed in which

two characteristics differ and two of the alleles are completely dominant over the other two, the characteristic ratio of 9:3:3:1 will occur. Should one pair of alleles be incompletely dominant (such as the red and white four-o'clocks), that ratio will be modified into 3:6:3:1:2:1, giving six phenotypes instead of four. Should both be incompletely dominant over the other two, the number of phenotypes will then be increased even more.

All these ratios can be obtained by simple arithmetic. For example, in a 3:1 ratio, three fourths of the plants have yellow seeds in the F_2 generation, and so also will three fourths of the plants have smooth seeds. Three fourths of the plants with smooth seeds also have yellow seeds, and three fourths of that three fourths, or nine sixteenths, will have smooth seeds as well. Of the three fourths having yellow seeds, only one fourth of those, or three sixteenths, will have wrinkled seeds; one fourth will have green seeds, and three fourths of those, or three sixteenths, will have smooth seeds; and of the one fourth of the plants having green seeds, only one fourth will have wrinkled seeds, in other words, one sixteenth. Thus the phenotypes can be calculated, or they can be worked out as in Fig. 10-7.

The ratio of the genotypes can be calculated in the same way. In those individuals having YY, the ratio is 1 out of 4; in those having Yy, 2 out of 4; and in individuals having yy, 1 out of 4. The same is true for SS, Ss, and ss. Therefore a genotype YYSS will be $\frac{1}{4} \times \frac{1}{4}$ or $\frac{1}{16}$; YySS will be $\frac{2}{4} \times \frac{1}{4}$ or $\frac{2}{16}$; YYSs will also be $\frac{2}{16}$; and YySs will be $\frac{4}{16}$. Ratios can also be calculated in the same way for crosses involving one or more incompletely dominant characteristics by assigning the correct fraction to the genotypes.

In all the crosses cited, it does not matter which parent produces the pollen or the ovules. A cross of YYSS × yyss will give the same results as a cross between yySS × YYss. The F_1 generation will be the same (YySs), but the gametes from the first cross will be YS and ys, whereas in the second generation they will be yS and Ys.

A cross involving three pairs of contrasting factors is called a **trihybrid** cross. If the parents are homozygous for three characters and differ in all three, and if three characteristics are completely dominant over the remaining three recessive characteristics, the normal phenotypic ratio is 27:9:9:9:3:3:3:1. This can be demonstrated by a checkerboard diagram, or it can be calculated mathematically as before. Suppose we cross AABBCC and aabbcc, in which the capital letters indicate dominance. Then the F_1 hybrid will be genotypically AaBbCc. Again, A occurs in three fourths of the offspring, B in three fourths, and C in three fourths; the small letters occur in one fourth of the offspring. The number of possible gametes produced by AaBbCc will be eight: ABC, ABc, AbC, aBC, aBc, Abc, abC, and abc. Therefore the possible combinations of the eight gametes from each parent are 64. In the F_2 generation $\frac{3}{4} \times \frac{3}{4} \times \frac{3}{4}$ or $\frac{27}{64}$ of the offspring will have all three dominant characteristics; $\frac{3}{4} \times \frac{3}{4} \times \frac{1}{4}$ or $\frac{9}{64}$ of the offspring will have two dominant characteristics and one recessive; $\frac{3}{4} \times \frac{1}{4} \times \frac{3}{4}$ or $\frac{9}{64}$ will have two other dominant characteristics and one recessive; $\frac{1}{4} \times \frac{3}{4} \times \frac{3}{4}$ or $\frac{9}{64}$ of the offspring will have two other dominant characteristics and one recessive; $\frac{1}{4} \times \frac{1}{4} \times \frac{3}{4}$ or $\frac{3}{64}$ of the offspring will have one dominant and two recessives; $\frac{1}{4} \times \frac{3}{4} \times \frac{1}{4}$ or $\frac{3}{64}$ of the offspring will have two recessives and one dominant; $\frac{3}{4} \times \frac{1}{4} \times \frac{1}{4}$ or $\frac{3}{64}$ of the offspring will have the other two recessive characteristics and one dominant; and $\frac{1}{4} \times \frac{1}{4} \times \frac{1}{4}$ or $\frac{1}{64}$ of the offspring will have all three recessives. Should one or more of these three characters be incompletely dominant, the ratio may become more complicated, but it can also be calculated mathematically.

If the original parents differ in **n** pairs of characters, their hybrid progeny consists of 2^n groups, the numbers of which occur in the ratio of $(3:1)^n$ when dominant and recessive characters are involved, as shown in Table 10-1.

Every added character pair multiplies the number of possible combinations by 4, the number of genotypically different combinations by 3, and the number of visibly different combinations (when dominance is complete) by 2, as shown in Table 10-2.

To determine if a phenotypically dominant dihybrid or trihybrid is homozygous recessive as was done for the monohybrid: if a dihybrid is heterozygous for two pairs of characters, four different phenotypes will be expected; if it is heterozygous for only one, then two phenotypes occur, as illustrated in the following diagram:

P_1	YySs	×	yyss	
gametes	YS Ys yS ys		ys	
F_1	YySs Yyss yySs yyss		yellow-smooth yellow-wrinkled green-smooth green-wrinkled	
		or		
P_1	YsSS	×	yyss	
gametes	YS yS		ys	
F_1	YySs yySs		yellow-smooth green-smooth	

Table 10-1

Phenotype groups and F_2 ratio associated with pairs of characters when dealing with complete dominance in each pair

Pairs of characters	No. of phenotype groups	Ration of F_2
1	2	3:1
2	4	9:3:3:1
3	8	27:9:9:9:3:3:3:1
4	16	81:27:27:27:27:9:9:9:9:9:9:3:3:3:3:1
•	•	•
•	•	•
•	•	•
n	2^n	$(3:1)^n$

Table 10-2

The way a number of F_2 phenotypes and F_1 gametes increase with pairs of characters involved; number of genotype combinations in F_2 and possible F_1 combinations of gametes

Pairs of characters	No. of F_2 phenotype groups	No. different kinds gametes in F_1	No. genotypically different combinations	No. possible combinations F_1 gametes
1	2	2	3	4
2	4	4	9	16
3	8	8	27	64
4	16	16	81	256
•	•	•	•	•
•	•	•	•	•
•	•	•	•	•
n	2^n	2^n	3^n	4^n

In the first cross, in which one parent was heterozygous for two characteristics, the ratio is 1:1:1:1, whereas in the second it is 1:1 just as in a monohybrid backcross. Only independent assortment could give results such as these.

LINKAGE

The pairing of homologous chromosomes and their subsequent separation at meiosis implies that the genes are arranged in linear order and that corresponding genes, whether they are identical or not, also pair. The position they occupy on the chromosome is fixed whether they are or are not alike. This position is termed its **locus.** Either member of a pair of genes that occupies the same locus on two paired chomosomes is called an **allele.**

Any organism that exhibits the characteristic (heterozygous) of one or more alleles is a hybrid, which is the result of crossing two individuals that differ in one or more genetic characteristics, as we have seen. Natural hybrids occur wherever there is population variation, which always is the case under wild conditions. But because of the position of the genes, hybridization cannot go on indefinitely due to linkage.

The Mendelian ratios discussed thus far depend upon the independent assortment (random assortment) of the genes at meiosis and upon their recombination at fertilization. Such a chance assortment will occur only if the genes are located on separate and different chromosomes. Plants, like animals, contain hundreds and probably thousands of genes, whereas there may be only a limited number of chromosomes (*Secale cereale*, rye, has only seven pairs). Since the number of chromosomes is limited, each chromosome must of necessity carry more than one gene. If two or more genes are located on the same chromosome, these will tend to be transmitted together to the progeny. As a result, independent assortment of these genes cannot take place. When genes are located on the same chromosome and hence tend to be inherited together, they are said to be linked. If two plants are crossed, one having two pairs of linked dominant genes homozygous for them (AB) (AB) and the other being homozygous recessive (ab) (ab), the F_1 generation will then be heterozygous for both (AB) (ab). Four kinds of gametes would be produced if they were not linked, for example, AB, Ab, aB, and ab. However, if they are linked, only two kinds of gametes can be produced, AB and ab. As a result, the F_2 generation will have a monohybrid ratio. Three times as many plants will exhibit the two dominant characteristics as exhibit the recessive: (AB) (AB), (AB) (ab), (AB) (ab), and (ab) (ab). Therefore, instead of the expected 9:3:3:1 ratio characteristic of independent assortment in a dihybrid, the ratio will be 3:1. A **linkage group** is the number of genes that tend to be inherited together. Although more than 360 different sets of alleles in corn are known, there are only ten linkage groups, which represent the ten pairs of chromosomes common to corn.

SEX DETERMINATION

A special kind of linkage in certain chromosomes occurs in both plants and animals, generally referred to as sex determination. Most higher animals are bisexual and have a special pair of sex chromosomes that determine the sex of the individual. Some plants have a similar condition, but instead of being called bisexual, they are referred to as dioecious. Actually, since only the gametophyte produces sex cells, or gametes, only these plants can be termed bisexual, and in many cases they will produce both kinds of gametes. The lower plants, for example, the bryophytes, have dioecious individuals. In the flowering plants where the sporophyte is dominant, the plants that show bisexuality or dioecism are better termed either staminate or ovulate plants, that is, those that produce anthers or pistils, but not both. However, regardless of the situation, sex is determined by X and Y

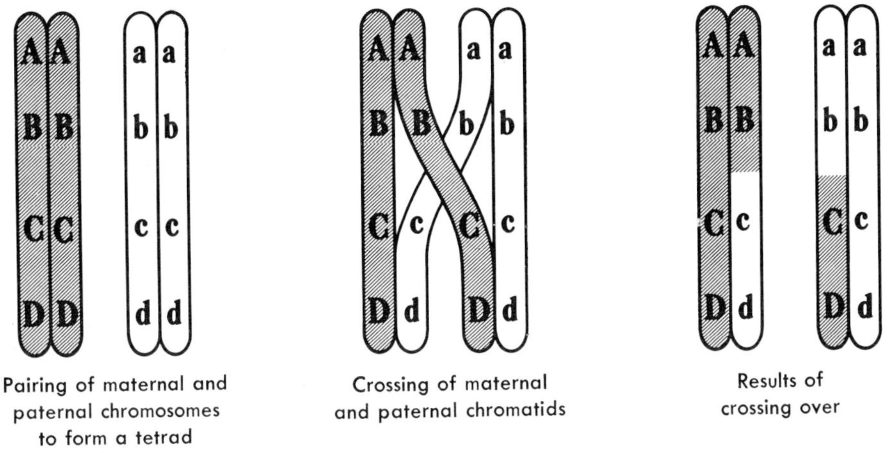

Fig. 10-8
Diagram of crossing-over.

chromosomes. If the cells contain only X chromosomes, only ovules are produced. However, if the cells contain both X and Y chomosomes, the pollen that is produced is of two types, some with a X chromosome, and some with a Y chromosome.

CROSSING-OVER

Even though complete linkage is known in some organisms (male *Drosophila* sp., for example), the genes originally found linked on the same chromosome do not always remain that way. There is a tendency for the genes on the same chromosome to become separated during the tetrad stage of meiosis as a result of thce process of crossing-over which is evident cytologically by the chiasma.

The process of crossing-over is illustrated in Fig. 10-8. In the tetrad stage of meiosis four linked genes may be on the same chromosome. If we assume that they are four dominant ones, **A**, **B**, **C**, and **D**, and that the homologous chromosome has the four recessive alleles, **a**, **b**, **c**, and **d**, then during the diplotene stage of the first prophase each has divided to form a tetrad. At diakinesis there is evidence that they may cross over so that, as a result, one chromatid will have the four dominant genes, one the four recessive, one two dominants and two recessives, and one the other two dominants and two recessives. The genetic consequence of crossing-over is the transfer of hereditary material from one chromosome to its homologue. Two of the chromosomes (in either spores or sex cells) will have the genes that were formerly linked on the other pair. Should the linked genes on one chromosome be those that are responsible for disease resistance and late seed production and the others responsible for no resistance and early seed production, it is possible, through crossing-over to get plants that are disease resistant and early seed producers. Phenomena such as these are of interest to the plant breeder who is ever alert to the improvement of crop yields.

Although crossing-over between two or more genes may occur, it does so in only a small percentage of cases. The percentage of crossing-over between two genes is proportional to the distance between them. If genes are close together, the frequency of crossing-over is low; if they are far apart, the percentage of crossing-over is high. Thus it is possible to determine how far apart genes are from their crossover values and at the same time to define their positions on the chromosome relative to other genes.

The percentage of crossing-over between two genes is determined by backcrossing the F_1 generation to the double recessive parent. If genes B and C are linked in a homozygous individual and the parent containing them is crossed with another homozygous individual having the recessives b and c, the F_1 generation will then have the genotype (BC)(bc). When an organism in which linkage is suspected is crossed with the homozygous recessive, the offspring having the linkage groups of the parent will occur in numbers greater than 50% (if all were to appear in equal numbers, it would be an independent assortment). The following diagram illustrates how this phenomenon occurs:

Parents	(BC)(BC)	×	(bc)(bc)	
F_1	(BC)(bc)			
Backcross	(BC)(bc)	×	(bc)(bc)	
F_2	(BC)(bc)		noncrossover	35%
	(Bc)(bc)		crossover	15%
	(bC)(bc)		crossover	15%
	(bc)(bc)		noncrossover	35%

If the genes were on separate chromosomes, independent assortment would prevail, and the offspring in the F_2 generation would appear as a ratio of 1:1:1:1. When this ratio does not occur, it is known that the genes are linked. If only noncrossovers resulted, then complete linkage would exist. In a hypothetical case, let us assume that most of the offspring resemble the orginal parent type (70%), whereas only 30% are the result of crossover gametes. The distance between **B** and **C** is 15 units and not 30 because if crossing-over occurs between **B** and **c**, it must obviously also occur between **b** and **C**.

HYBRID VIGOR

It has often been recognized that crosses between two supposedly different species or varieties of the same species produce offspring more vigorous than either parent (for example, a mule in the animal kingdom). Although it is not always true in hybrids, there are some outstanding examples of **hybrid vigor** or **heterosis.** Among those plants exhibiting heterosis is hybrid corn (Fig. 10-9), which is widely grown on farms throughout the country. There are several explanations (theoretical, however) for this increased vigor of hybrids. It may be that through a chance combination of genes, the hybrid has the better genes from the two parents in which minor homozygous defects appeared. In the hybrid these defects are covered up by normal, dominant genes from each parent. Another explanation is that a heterozygous makeup is in itself an advantage, conferring greater survival rate on the bearer. A variety of genes may contribute to the plant's greater versatility in thriving under different environmental conditions.

MUTATION

Mutations are also responsible for variations in an organism. They are the result of sudden heritable changes that are not explained by either hydridization or environmental factors. Mutations may be slight and insignificant or so marked and striking that they often seem to be new varieties or even new species.

Anything that will disturb the precise duplication of genes during mitosis and meiosis will also have its effect on heredity. A mutation is any heritable change in the genetic material. Most genes have a defined mutation rate; in some this may be as high as 1 in 2,000 or as low as 1 in several million gametes in plants. Anything that alters the mutation rate is referred to as a **mutagenic agent.** The mutation rate can be increased by chemicals such as mustard gas, extremes of heat and cold, or high-frequency radiations ranging from ultraviolet light to X rays and the radiations produced by radioactive substances. Many mutations are distinctly harmful, others have little or no obvious importance in survival, and a few are obviously beneficial. Finally, most mutations are recessive; therefore they may not appear for several generations.

LETHAL GENES

One of the most harmful characteristics a green plant may have is the inability to produce chlorophyll. Seedings without chlorophyll will die since they cannot manufacture their own food. Inability of green plants to produce chlorophyll is caused by the presence of a **lethal gene.** It is a recessive characteristic so that plants heterozygous for it are phenotypically normal. However, when they are self-pollinated, they will produce seeds in a ratio of 3 nomal to

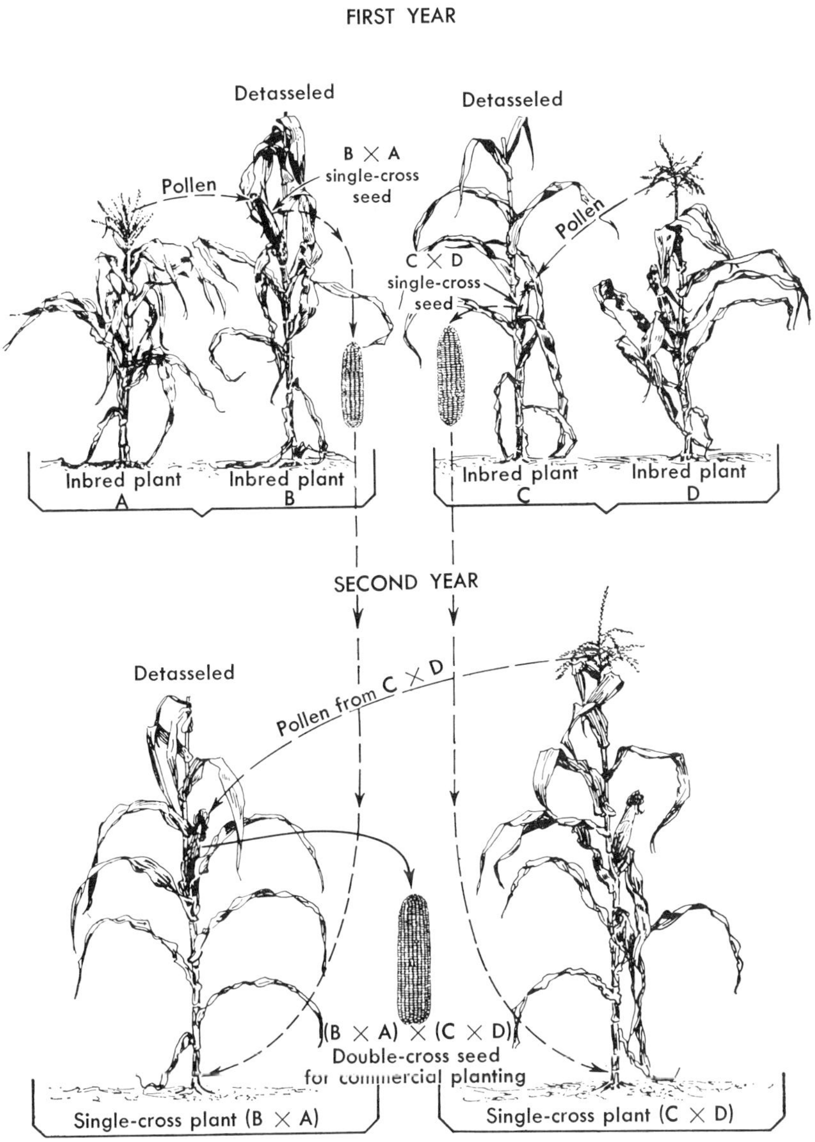

Fig. 10-9
Method of producing new hybrids of corn by controlling fertilization. The first year 2 plants are crossed, each with desirable characteristics, and the second year, another cross is made using offspring of the 2 first year's plants. Such hybrids show heterosis. (USDA photograph.)

1 albino. The albinos die early from starvation.

MULTIPLE GENES

In all the genetic crosses illustrated, the expression of the characteristic was governed by a particular pair of genes. Many phenotypic expressions are the result of pairs of genes known as **multiple genes,** in which each gene has its own effect, which is frequently cumulative. The most striking effect of multiple gene inheritance is that some of the offspring may be more extreme than either parent. If a cross is made between two varieties of wheat, one having red kernels and the other white, the F_1 offspring may be intermediate in color. The F_2 generation, however, will vary in intensity of color from red to white, but most of the offspring will be intermediate. The range may be from red, like the original red parent, to lighter red, to pink, to pale pink, and to white. The number of classes will be dependent upon the number of genes involved. Should one sixteenth of the offspring be red, four sixteenths be light red, six sixteenths be pink, four sixteenths be pale pink, and one sixteenth be white, it then means that two pairs of factors are involved, as shown in the following diagram:

Parents	$R_1R_1R_2R_2$ red	×	$r_1r_1r_2r_2$ white	
F_1	$R_1r_1R_2r_2$ pink			
F_2	$R_1R_1R_2R_2$		red	1/16
	$R_1R_1R_2r_2$		light red	
	$R_1r_1R_2R_2$		light red	4/16
	$r_1R_1R_2R_2$		light red	
	$R_1R_1r_2R_2$		light red	
	$R_1R_1r_2r_2$		pink	
	$r_1r_1R_2R_2$		pink	
	$R_1r_1R_2r_2$		pink	6/16
	$r_1R_1r_2R_2$		pink	
	$R_1r_1r_2R_2$		pink	
	$r_1R_1R_2r_2$		pink	
	$r_1r_1R_2r_2$		pale pink	
	$r_1R_1r_2r_2$		pale pink	4/16
	$R_1r_1r_2r_2$		pale pink	
	$r_1r_1r_2R_2$		pale pink	
	$r_1r_1r_2r_2$		white	1/16

Each gene contributes to the ultimate color of the seed. The more red genes present, the darker will be the final color of the seed. If one plant out of 64 were as extreme as either original parent, it would indicate the action of three pairs of genes; if one plant out of 256 were extreme, four pairs would be indicated, and so on.

CONVERTER GENES

In 1959 a gene was discovered in corn that governed the color of the plant. Ordinarily, the gene for color retains its ability to produce the color through an unlimited number of crosses. However, a gene capable of changing the color gene from one color to another color, a **converter gene,** was discovered. The results of such a cross involving a converter gene as compared with a normal color gene are shown in Fig. 10-10.

MODIFYING GENES

In plants some genes are not responsible for a specific character in themselves but can modify the expression of other genes. These are called **modifying genes.** For example, the color of a flower may be diluted when modifying genes are present in the homozygous state.

INTERACTION OF GENES

When two or more pairs of genes produce a specific effect in combination only, the process is referred to as the **interaction of genes.** Simply, this effect can be explained by the example of two-factor crosses In a two-factor cross involving two dominants and their recessive alleles, the normal ratio is 9:3:3:1, and the number of different phenotypes is four. If the genes interact with one another, the number of phenotypes is fewer than four even though the number of recombinations is still 16. To illustrate: when two flowers having yellow centers are crossed, the offspring all have purple centers. The F_2 generation will be composed of both purple- and yellow-centered flowers in a ratio of 9:7, which in-

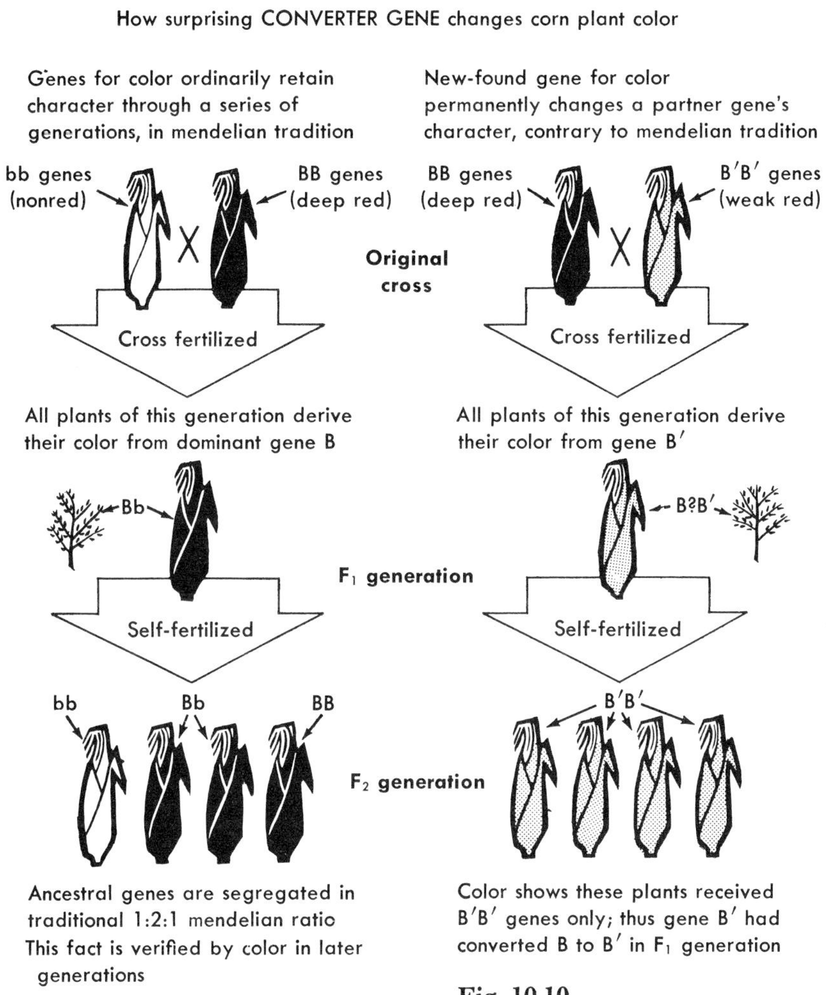

Fig. 10-10

Converter genes used to change the color of corn ears. By adding a gene of this kind to the breeding stock, color and other traits may be controlled. (USDA drawing.)

dicates that two pairs of genes are involved. The presence of the homozygous recessive of either pair produces yellow centers, as shown in the following diagram:

Parents	ppRR yellow	×	PPrr yellow	
F_1	PpRr purple			
F_2	P–R–		purple	9/16
	ppR–		yellow	7/16
	P–rr		yellow	
	pprr		yellow	

In other types of gene interaction in which only one homozygous recessive inhibits the expression of a dominant gene, the ratio is modified to 9:3:4. When one dominant gene prevents the expression of the other dominant pair, the ratio will be 12:3:1. When both dominant genes are inhibitory to the recessive, the ratio becomes 15:1, and when one dominant gene and one recessive gene are inhibitory, the ratio is 13:3. Thus the interaction of genes in combination can alter the normal phenotypic ratio of a dihybrid. Whereas it is presumed that genes are responsible for phenotypic characteristics, a single gene can control both form and function. This is known as the **pleiotrophic effect** of genes.

CHROMOSOMAL VARIATION

All the cells of a plant body have, as a rule, an identical number of chromosomes, and the genes are arranged in linear order as shown by studies of linkage groups. However, there are some exceptions. Some genes are known to mutate, and the chromosomes undergo certain rearrangements. To do this, the chromosomes must break. Sometimes this occurs without outside influence; other times chemicals or ionizing radiation causes the changes. The various kinds of variation are summarized in the following outline:

I. Changes in numbers of chromosomes
   A. Changes involving whole sets of chromosomes
      1. Haploid (n) in which n is monoploid or basic number; each chromosome represented singly
      2. Polyploid in which each chromosome represented by more than two homologues (Fig. 10-11)
         Triploid (3n)
         Tetraploid (4n)
         Pentaploid (5n)
         Hexaploid (6n)
         a. **Autopolyploid** (multiples of diploid but of single homozygous or pure strain)
         b. **Allopolyploid** (multiples of hybrid derived from two different diploids)

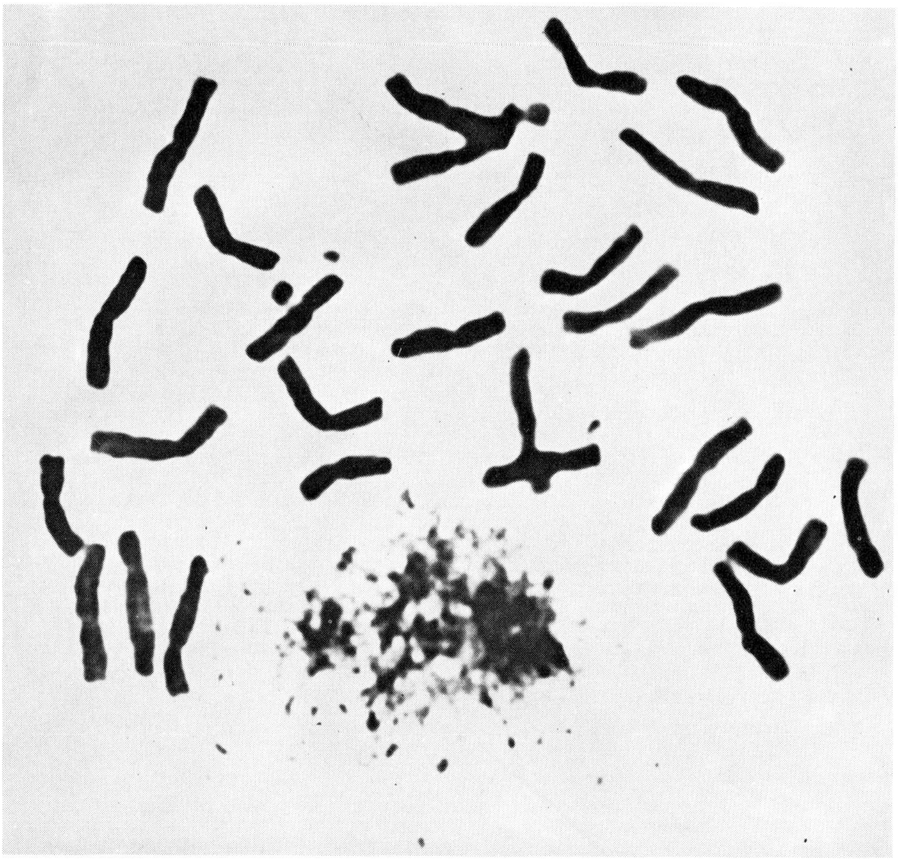

Fig. 10-11

More than two homologues for each chromosome are apparent in this polyploid oat. Polypoid crosses increased resistance to crown rust in oats. (USDA photograph.)

- B. Changes involving numbers of chromosomes within set, **heteroploidy**
    1. **Monosomic** (loss of one chromosome from set; in diploid, 2n − 1)
    2. **Polysomic** (addition of one or more chromosomes to set)
       Trisomic = 2n + 1
       Tetrasomic = 2n + 2
       and so on
    3. **Nullisomic** (loss of both chromosomes of pair)

II. Changes involving number of genes or their arrangement and position on chromosome
- A. Changes involving number of genes
    1. Deficiency or deletion (loss of one or more genes)
    2. Duplication (addition of one or more genes so that same gene is repeated in haploid complement)
- B. Changes involving arrangement of genes
    1. Translocation (exchange of parts between nonhomologous chromosomes to form two new ones)
    2. **Inversion** (180° rotation of block of genes within chromosome)

Of all the chromosomal aberrations, polyploidy is the most adaptable for the improvement of economic plants. The most common polyploids are tetraploids (Fig. 10-12). Polyploidy may influence the phenotype of the individual as well as its genetic and evolutionary potentialities. One of the most common effects of polyploidy is increase in cell size and frequently in size and vigor of the plant. However, the direct effects of increasing the chromosome numbers are not always predictable. Some annual plants may become perennials, the polyploids may be less fertile than diploids, and other physiological factors may be changed. Occasionally it is possible to make fertile plants from sterile hybrids through polyploidy. Blueberries are a well-known polyploid complex (Fig. 10-13).

Fig. 10-12
Chromosome polyploidy may sometimes be recognized by flower size, as in this case. The tetraploid (4n) lily flower on the right is larger than the diploid (2n) lily flower on the left. (USDA photograph.)

Fig. 10-13
Fruit may be larger as a result of polyploidy. For example, several species of the blueberries *(Vaccinium* spp.) of eastern United States are well-known polyploid complexes in which enhanced fruit has resulted. (USDA photograph.)

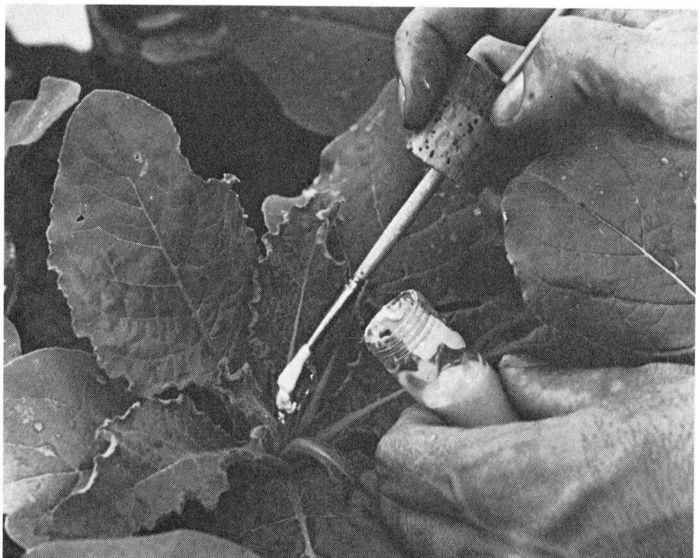

Fig. 10-14
Colchicine, used by plant geneticist to induce hybridization after crossing a tobacco plant having 48 chromosomes with one having 24 chromosomes. (USDA photograph.)

Polyploidy has been studied extensively by plant geneticists, who have learned that it originates through the suppression of the meiotic process. In other words, the gametes have the unreduced number of chromosomes. Polyploidy may also arise through failure of cell division but not of chromosome division. In nature these phenomena are accidental, being favored by extremes of temperature, particularly low temperatures. Polyploids are more frequently found at high altitudes than at low ones. Sometimes scientists use a drug called colchicine to produce polyploidy in plants. Colchicine is an alkaloid derived from the autumn crocus *(Colchicum autumnale)* and is commonly used by physicians to alleviate gout. It is a deadly poison except when used in minute amounts. It can be applied to plants as a paste (Fig. 10-14) or in some other form to inhibit the formation of spindles in mitotic cells, thus inducing polyploidy.

CYTOPLASMIC INHERITANCE

In plants some characteristics are inherited through the cytoplasm instead of through the nucleus. Since an egg contains much more cytoplasm than a sperm, **cytoplasmic inheritance** is usually maternally dominated. Inheritance through the cytoplasm is recognized by its maternal effect, particularly when plastids are involved. There is a yellow-leaved variety of primula that contains less chlorophyll than the normal variety. The seeds of this yellow-leaved form transmit the characteristic but the pollen does not. There are other instances supporting the interpretation that cytoplasmic inheritance is involved in maternal cytoplasm.

EVOLUTIONARY THEORY

Evolution deals with the changes in the kinds of organisms from the first life on earth to the present time. The history of plant life is described in Chapter 16. Here we are concerned with the actual mechanisms or processes that take place to bring about these changes. This is an interaction of chromosomes and the environment.

We must assume that life originated only once. By this we mean that macromolecules capable of replication are of one type only, and that they are the surviving life forms that eventually provided the great diversity of life as we know it today. This is so important to the evolution theory that any other idea would make all other aspects of the theory logically untenable because the uniformity of cellular biochemistry cannot be coincidental. There is absolutely no scientific evidence against this; also there is little direct evidence to support it. The premise can be assumed only because there is no way to prove it. Not to assume this premise would prevent the organization in any other manner of the available data. Great effort has been expended to explain in a most logical way the origin of life. Attempts have been made to create life in the laboratory, which may be done sooner or later. As soon as this is done, one way that life could have originated will be known. There will be other ways demonstrated as time goes on, but none of these will show how life was created, and none will prove that it was created only once. At the same time, it should be made clear that this does not overrule the possibility of numerous abortive origins. As far as is known at the present time, living substance may be produced every day at that ragged edge between the living and the nonliving. When it was proposed that there was a single origin of life, the supposition applied to life as it is now on the earth and as it was according to the paleontological evidence at hand. Any other new life appearing now or at any time in the past was either immediately consumed or was unable to maintain itself independently and merged back into the environment.

Assuming, as we must, a common ancestor for all life, it then becomes obvious that all life comes from life. This took a long time to prove, but it is now a scientific fact. It naturally follows then that if life originated only once and if all life comes from life, that all life must be related in an unbroken chain from the present back to the time of its origin. All organisms that ever lived had parents, which in turn had parents, and so on back in an ever-narrowing genealogical line to those first macromolecules that hovered on the brink between living organisms and nonliving chemicals. This is sometimes referred to as a vast sea of protoplasm that forms a network through time to the present. It is described as a network because it is not a series of absolutely integrating units but is organized into supraindividual units called kinds. In the case of bisexual kinds the unit of operation is the breeding population, and in the case of asexual organisms the unit is generally a colony. One or more such morphologically and physiologically similar units capable of producing additional similar units constitutes the species. Based on this is the theory of homology by inheritance. Similarity of form and structure is obvious in orderly arranged units of all kinds. Paper clips can be classified (or arranged in orderly units) by grouping like with like. It is validly assumed that parts are homologous, and, therefore, possessors of homologous parts have a common ancestor. This is the theory of homology and the essence of the theory of evolution. Similarity of parts, but not necessarily similarity of function of those parts, indicates relationships. All the information in Chapter 14 on life cycles in this book is based on this assumption and relates the various stages in the life cycle of plants. Sporophytes, for instance, are homologous throughout, even though in the primitive condition they are limited to single cells producing spores and in the higher plants they form the entire vegetative body. The pitfalls in dealing with the theory of evolution come only in the interpretation of the concept of similarity-by-homology as opposed to the concept of similarity-by-accident. Unrelated similarity must be recognized and accounted for in evolution. So far there is no adequate empirical way to demonstrate whether similarity is related or unrelated.

The true nature of species remains an unsolved biological problem. It seems certain that the term species does not necessarily refer always to comparable entities, although our present means of dealing with these phenomena by naming them seems to imply this. We do see species generally as units set apart by a similarity **gap.** That is, there are species' boundaries—morphological, physiological, and geographical. It is this gap concept that delimits species. When any group of organisms are examined, both those now living and those represented by fossils, it is possible to arrange them in an orderly fashion from the least complex to the most complex and from the simple to the specialized. These are first arranged linearly. One of the principal features of this arrangement is the presence of gaps between one group of individuals and another group. Gaps allow the recognition of species. Suppose one were to go into the field and gather all the plants he encountered, taking several specimens of each kind. Upon returning to the laboratory, without consulting a book or making use of any previous knowledge, he sorts and arranges the specimens, in the way most logical from the data at hand, to form an orderly sequence of specimens. It would be obvious at once that some individuals clump in one pile because they have similar features; other individuals clump in other piles for the same reason, and so on. This process is continued until all the specimens are piled, like with like. In other words, the ones that are alike are separated from those that are different. This classification can be made by anyone with any group of objects, regardless of what knowledge the collector may have about them. Any number of characteristics may be used to make this classification, but no matter how it is done, it will usually be possible to make assignments, and there will usually be discrete piles, with degrees of variation within any given pile. It may be necessary sometimes to unite piles and other times to split piles, but regardless, gaps between piles will usually be evident, some narrow and some wide. These groups may be called species, both philosophically and biologically, since the logical process is the same in both fields. Thus far, only the logical process has been discussed.

As one learns more and more about the specimens of plants collected, one will find that some of the things used to separate the piles are significant and others are not. The latter are called "unnatural" because a deeper knowledge of the characteristics shows that the features used conflict with greater evidence in favor of another arrangement. For instance, if the color red is used as a criterion and all the red-flowered plants are put into a single pile, it will soon be noted that there are great differences in the flower structure, the leaves, the size of the plants, and other features. If the plants are rearranged on the basis of flower structure, these other features will agree, and there will be several piles with red flowers, but all of them will be very different otherwise. Therefore the color red is insignificant. The new arrangement will be much more logical. Even though this is a hypothetical case, it embodies all the principles of classification that eventually caused the proposal of the modern theory of evolution. This process and this reasoning led to the discovery of a natural process in biology, speciation or the reproduction of species. Species are first recognized by the presence of gaps and are then defined so that they can be recognized again. The similarities are studied, and each species is assigned to higher categories. Finally, relationship to ancestors is proposed by studying all the data, but particularly the fossil evidence and the genetics. If this is possible, a sound phylogeny can be developed. Most of these studies result in a logical arrangement, however, and such a classification remains as a theoretical proposal changeable as new evidence becomes available.

SPECIATION

In Chapter 16 is a brief summary of the actual evolution of the plants, showing the change of species from period to period. All the plant groups mentioned are represented by fossil species, but these are not necessarily ancestors of modern plants. The actual mechanism of speciation, the formation of new species, is a complex process of genetic mutation and hybridization. The new forms may be either slightly different or substantially different from their parents. Changes such as described previously are taking place all the time apparently and result in variation so that no two are exactly alike. This situation obtains in any ecological habitat. Despite the fact that habitats may be characterized in general terms, no two locations are exactly alike, and a given location does not remain constant from year to year. As environmental conditions change, given individuals vary in their adaptability to this environment. At the same time, every individual has a maximum, a minimum, and an optimum tolerance for these factors. Thus, some individuals can survive at lower temperatures than others. During a particularly cold period many members, but not all, of a population may be killed. Those that survive have a genetic aptitude for low temperatures, and some of their offspring will inherit this ability. The gene pool of the population is then shifted toward a tolerance for cold in this case, and this same result follows for every condition that prevails in the habitat. As time passes, the gene pool changes in this manner. If conditions become too severe, however, the population may be entirely destroyed. If the situation is general throughout the range of the species, in other words, in each population, the species then becomes extinct—the usual fate. On the other hand, if some of the many mutations that are always taking place happen to produce a combination of effects which enable the offspring to survive, a new species may then be born.

As a result of this, a great amount of change is going on at the population level. Several factors, such as isolation of a population from the rest of the populations throughout the range of the species, may bring about change. Isolation may be caused by changes in land form, by changes in climate in part of the range of the species, or by invasion by predators in part of the range. As soon as gene exchange is cut off, either temporarily or permanently, the entire **gene pool** is changed because not all the genes available to the species are then available to each population. These breeding inhibitors are termed **isolation mechanisms.** The new gene pool plus the mutations that may have occurred change at varying rates. Eventually there may be change sufficient to be detected in the phenotype, and when it is recognized, a new species has been formed.

The rate of this evolutionary change at the species level varies considerably. It may take a million years or much more, but apparently considerably less when conditions are right. Never, however, has this change to a new species been rapid enough to be observed, except under laboratory conditions. The plants pictured in the most ancient of the herbals look the same as their descendants today. However, under controlled laboratory conditions, these processes can be speeded up to a point at which it is very likely that new species can be created, and in a few cases natural species can be duplicated.

DATA USED TO SHOW THE EVOLUTION OF ORGANISMS

Biologists continue to look for evidence of the evolution of organisms, but the already accumulated mass of information is so overwhelming that most groups can be related and at least a working phylogeny given. Taxonomists use the following kinds of evidence to make a classification and to show the evolution of plants and animals.

Fossils. Before the theory of evolution was proposed, **fossils** were believed to be abortive attempts of the Creator to fashion kinds of plants and animals that were discarded because of imperfections. By the eighteenth century the real nature of these specimens was understood, and fossils were collected, classified, and studied in detail. When these specimens were arranged in a classification sequence in accordance with the classification of existing species and the theory of homology of parts was applied, the ancestry of several groups of plants and animals could be seen. When the rocks in which the fossils were embedded were dated, a time sequence was worked out, and the story of evolution unfolded, making fossil evidence the strongest proof of relationship.

Anatomy. A comparative study of the parts of living organisms is the most common means of showing relationship. Living groups that have a well-preserved fossil series show their pattern of development. If the same patterns of evolution are applied to other groups lacking a fossil history, logical and acceptable hypothetical phylogenies may be constructed. Such reconstructions are called phylogenetic trees and serve a very useful purpose in devising a sound and useful classification based on the principles of evolution. Although a strictly artificial classification can be useful without such procedures, it lacks the meaning and scientific interest of a natural classification.

Genetics and development. The development of the embryo and the plant body may offer important comparative features useful in relating groups that are otherwise very different. Comparative embryology of plants, however, is seldom possible but is used in the flowering plants to a certain extent. Most studies of development in plants are confined to life cycles. Such studies have been mentioned already and may be seen in Chapter 14.

Physiology. Comparative physiology offers evidence in support of evolution in the same manner as does comparative anatomy. Specific physiological processes are under the control of the gene, just as the formation of anatomical structures is controlled by genes. Furthermore, recent work has shown that many genes, perhaps all, have pleiotropic effects; that is, a gene will affect a number of characteristics, both structural and physiological. Therefore it is reasonable to believe that if comparative anatomy is a legitimate means of studying evolution, then the comparative study of processes is valid and useful and is especially so as a means of determining the fitness of a plant to its environment and of registering environmental tolerance. Another way is to study the effect of plant juices on the formation of antibodies. Comparative serology has recently become a very active field for both plant and animal studies.

Distribution. If any single form of evidence for evolution of a group is the most convincing, is is probably the random distribution pattern. Each kind of plant has a specific and particular geographical and ecological (microgeographical) distribution. No organism is universally distributed, despite the frequent use of the word **cosmopolitan** in describing the range of a given species. The so-called cosmopolitan species are restricted to a certain type of habitat and are therefore limited in range. The majority of species have a much more restricted distribution. They are confined to certain very definite areas and to definite conditions in those areas. In evolutionary studies when one species is compared with other species, the range of the species and its adaptation to particular conditions will yield many valuable clues about the origin of a given taxon and about the possible mechanism of evolution in the group because a study of the various members of a taxon will show a variety of distribution patterns. The range of a species, coupled with anatomical changes, furnishes information for speculations on the origin of the taxon and, finally, will show the **radiation** of the group, that is, the development of many species as they evolve and spread. When fossil evidence and knowledge that certain basic distribution patterns exist are available, some idea of the age of the taxon can be determined. These same principles can be applied to extant groups. From these kinds of data a great many very logical phylogenies can be constructed.

Genetics and populations. A change in species can be fixed in a population only by genetic change. Therefore evolutionary process is principally a process of genetics, and it is difficult to separate the two as separate subjects. The concept of the gene pool, that is, the total available genes of a breeding population, is the explanation of the mechanics of evolution. These mechanisms were described earlier in this chapter. In all natural populations, that is, populations occupying a place in nature unaffected by man, there is great genetic variation in the individuals that comprise it. However, much of this variation is concealed in the phenotype because it occurs in the form of heterozygotes.

When it is proposed that a species became adapted to a certain situation, the implication is that mutations occurred which permitted certain individuals to live in a particular place better that can another individual. Because of this, they survived and propagated, whereas the less well-adapted individuals died out. This is what the statement "the survival of the fittest" implies, and this is the process of natural selection.

Whenever a phylogenetic tree is proposed for the evolution of any group, it is based on the supposition that the various species and other taxa evolved in varying degrees into the spectrum of kinds that occur today and that occurred in the past by means of mutations in the gene pool (Chapter 16).

Polymorphism. Some populations may contain two or more distinct types of individuals and are, therefore, **polymorphic.** Several factors may contribute to this condition. It may be simply a matter of the presence in the population of certain genes as homozygous recessives that are neutral so far as selection is concerned; that is, they are neither beneficial nor unfit. This is known as the Hardy-Weinberg equilibrium. Other cases of polymorphism occur because the rate of mutation is in balance with the rate of selection. Other conditions involving various selective factors may account for polymorphism.

Genetic drift. The Hardy-Weinberg equilibrium applies to random breeding of large populations of plants, especially those that are wind pollinated. In smaller populations, especially those that are isolated, a different phenomenon may take place. Here gene frequency fluctuations may be large, and from year to year the character of the population may show great change. This is termed **genetic drift** and takes place under conditions where a small sample of the genes of the population are drawn upon for the next generation. This might occur among herbaceous plants where a large amount of seed may be produced, but only a very small number of seeds germinate for the following year's plants. Such a condition may obtain until one or the other allele, in the case of a dominant and recessive gene complex, becomes homozygous and fixed. After that, the population will remain stable for the particular character, but other changes may occur through mutation.

Introgressive hybridization. Hybridization undoubtedly plays an important role in speciation, but it probably is not as important as mutation. One may expect to find in any area a very small percentage of naturally occurring hybrids because isolating mechanisms, especially between species of newly evolved groups, may not be irrevocably fixed. Populations of species that are geographically isolated are said to be **allopatric.** Sometimes such populations may be only ecologically isolated and still be allopatric. Populations of two separate species that live in the same geographical area or ecological habitat are said to be **sympatric.** As one might expect, sympatric species will have stronger isolating mechanisms than do allopatric species. Several things may happen to cause two allopatric, and weakly isolated species to come in contact. They may, for example, enlarge their range because of optimum growing condition or absence of pests. Or they may come in contact through disturbance of the land by man. When this happens and hybrids form in the area of contact, **introgressive hybridization** may occur. Hybridization in such cases does not result in intermediates between the two species because crosses between the hybrides (F_2) are less likely to occur than a backcross with one or the other of the parent types. In cases of survival of this stock through introgression, the newly formed population may be better fit for the new ecological conditions, and not only is survival possible, but also actual range extension into the new habitat takes place.

SPECIES

Attempts to write a scientific, all inclusive definition of the category species have been unsuccessful over the past two hundred or more years. Since evolutionary theory has emphasized the change and multiplication of species, it seemed imperative that the category be defined because, if the theory proposed that species evolve, then the biologist must know what a species is.

The reason the category has never been defined to the satisfaction of all systematists working in many different groups is simple: the category is an unnatural one and, therefore, cannot have a scientific definition. For example, if it is defined to fit one group of sexually reproducing forms, say the members of the maple tree family Aceraceae, then it will not apply to asexually reproducing bacteria, and so on. Therefore it seems reasonable to think of the category species as a unit that serves only to handle certain types of information about organisms.

The importance of distinguishing between a category and a taxon now becomes apparent. The information reference, species, is a category; therefore, when a name is applied at this level, one understands that the information is limited to facts about certain individual organisms. The organisms themselves form a taxon. The assignment of this naturally existing taxon to the category species is merely a process of information storage and retrieval. This permits writing of descriptions and the identification of specimens in an efficient manner. In fitting these units into classification schemes designed to show relationships and evolution, one must understand that it is the taxon that evolves and not the category.

There are several features of species that are universal. First, all species are composed of individuals collectively referred to as a species. These are the taxa and they are named. Sometimes it may be difficult to determine exactly the limits of an individual, particularly if it exists as a member of a colony, or an individual may be a variant that can not be immediately assigned to a taxon. These are specific problems, real enough, which can be solved eventually after sufficient study. The individuals form populations, that is, the total number of living individuals in a geographically limited area. All the individuals of a breeding population are related to each other. The area may be microscopic or macroscopic, changing or static, but it is definite at any one moment. The members of this population may or may not be interbreeding, depending on the kind of species and their stage of development. The total area occupied by these living populations constitutes the range of the species. As a final criterion, the former populations, most of which are unknown but which are a part of the species, should be taken into consideration whenever possible. In a few cases, former population representatives may be fossilized, but this is rare.

A species, to be fully understood, should be traced back in the fossil record, but rare is the case where this is possible. However, it is known that fossil series, like extremely variable and close living species, have intermediate forms that fit neither one species nor the other. These may represent transitional populations, or they may be hybrids. In fossil species, because of the time differential, carefully worked out series can be logically pieced together and can truly show the changes that have taken place in

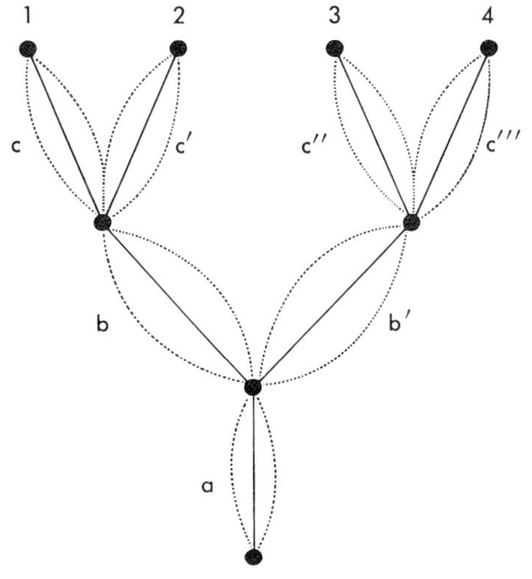

Fig. 10-15
Gentes, as illustrated in this diagram, provide the third dimension—time—of a three-dimensioned concept of species that includes time, space, and form. Each dot represents a species. **1**, **2**, **3**, and **4** are extant species, whereas the units surrounded by dashes represent gentes, **a**, **b**, **b′**, **c**, **c′**, **c″**, and **c‴**, giving the species concept its needed third dimension.

the species. All such series have limits. The gaps are the result of the incomplete fossil record. As paleontological research continues, more evidence is gathered to explain the gaps. A three-dimensional concept of species, we can easily see, is necessary, one that includes time, space, and form. Such a concept of the species is called a **gens** (plural, **gentes**). A gens is defined in such a way that time and change are included so that the species is represented not only by its extant members but by its ancestors as well, going back to the point of fracture from the previous species (Fig. 10-15).

These criteria characterize all species. The determination of which populations are species and which are not is the real problem, and a definition of the category species is now of more historical interest than of scientific interest.

EXPERIMENTAL SYSTEMATICS

Modern systematic biology concerns itself more with the study of present and past populations than it does with the description of new species. It does not, however, neglect the many thousands of new species of plants and animals remaining to be discovered. But with little doubt, the major features of the classification of plants have been recognized. Changes in classification details can be expected, but it is unlikely that further major changes are forthcoming unless there are radical discoveries in plant physiology and genetics, something that is not impossible.

Increasing attention is being given to the study of population variations, and to the mechanisms that cause these changes in an attempt to find the underlying principles of evolution. This is done by extensive field observations and laboratory and field experimentation. Investigations and studies constitute a division of biology, experimental systematics, which is a very new field.

As mentioned previously, one of the unsolved problems of biology is the true nature of the units called species and how they are formed. The full story of evolution will be told only in the future when this question is answered.

SPECIAL TERMS USED IN CHAPTER 10

allele
allopatric
allopolyploid
autopolyploid
backcross
converter gene
cosmopolitan
cytoplasmic inheritance
dihybrid
dominant
fossil
gaps
gene pool
genetic drift
genotype
gens
heteroploidy
heterosis
hybrid
hybrid vigor
hybridization
incomplete dominance
interaction of genes
introgressive hybridization
inversion
isolating mechanism
locus
linkage group
lethal gene
modifying gene
monohybrid
monosomic
multiple genes
mutagenic agent
mutation
nullisomic
phenotype
pleiotropic effect
polymorphic
polysomic
radiation of species
recessive
segregation
sympatric
trihybrid

QUESTIONS TO CONSIDER

1. What is meant by a "blood relative"? "Blood lines"? And so on?
2. Many "kinds" of cultivated plants are selected variations of a single species. Under what circumstances might these "breed true"?
3. Describe what is meant by "backcrossing" and give an example of where this might be done.
4. Why do polyploids tend to have large flowers? Do you think that complicated flowers like those of the cultivated rose are polyploids?
5. Describe how you might find out if a particular feature in a plant, for example, bright red flowers, was a dominant or a recessive character.
6. There are two meanings to the word "related"; one is used in genetics, and the other is used in discussions of evolution and phylogeny. What are these different meanings?
7. Compare a population of plants with a human population. What are the similarities? The differences?
8. Why is not a horticultural variety considered to be a separate species by taxonomists?
9. What is meant by a new species? Is it "new"?

10. Describe how you think a new species might arise, keeping in mind the terms allopatric, sympatric, isolating mechanisms, and gene pool.

DISCUSSION FOR LEARNING

1. Discuss some possible effects of atomic warfare on the future of plant and animal life on the earth. What kind of changes might take place?
2. Discuss some of the practical applications of genetics; some of the theoretical applications.
3. The idea of evolution has disturbed many people in the past. What are some of the misconceptions that have brought about this fear of scientific knowledge? What are some actual misconceptions or limitations of the current knowledge of the processes of evolution?

Chapter 11

Growth and differentiation of roots, stems, and leaves

Whether a plant is a single cell or is made of billions of cells, as is the complex vegetation pictured above, the functions of obtaining raw materials, internal transportation, elimination and secretion, synthesis, liberation of energy, support and protection, locomotion, and, finally, self-perpetuation are necessary to sustain life. A one-celled organism must carry out all these complex processes within the confines of its cell membrane, whereas in a many-celled organism certain cells become specifically adapted to carry on one or more processes exclusively; in other words, in a many-celled organism the cells become specialized. Thereby, through a division of labor among the cells, it appears that the vital functions are performed more efficiently in a multicellular plant than in a single-celled plant that perhaps may be functionally more versatile.

Whereas there may be varying degrees of specialization among cells, there is very little specialization in the embryonic tissues of plants, the cells of which differentiate as the requirement arises. All in all, there appears to be great versatility as well as great economy in the resources of a multicellular organism. (Similar cells specialized to perform a particular function are grouped together to form **tissues.** The cooperative specialization of a number of tissues leads to the formation of an **organ.** Finally, the close cooperation of several organs develops into the **organ system.** Several of these organ systems compose the organism in higher plants.) The structure and function of the cells constituting the tissues, the organs, and the organ systems in large measure determine the organism. However, the difference between organisms is more than that, since it lies in structural organization rather than in the sum of each organism's parts. For example, a mass of carbon atoms may form a diamond, graphite, or soot. The secret of each lies in its structural organization. In much the same way this holds true for the variety of plant life found on the earth today.

Living organisms, whether one-celled individuals or complex multicellular ones, increase in number and size by a series of interrelated processes that are common to all living matter. This is recognized as some form of **growth.** Growth can be expressed quantitatively; if plotted, it shows a curve, the shape of which is typical of all forms of growth. It can be measured by such things as numbers of cells, height or weight in relation to time, or volume of a cell. Regardless of the criterion, the growth curve is usually a typical S shape. Such a curve shows that all life is characterized by three stages of growth: an early period when growth is slow, a middle period when it is rapid, and a final slow period. Therefore growth rates do not continue at a constant velocity, and this inconstancy is probably due to hormone inactivation or inhibition. It has also been suggested that the rate of growth may be influenced by the inactivation of the phosphate energy-transferring system. Whatever the causes, S-shaped curves are indicative of the growth of entire plants, parts of plants, or individual plant cells, from daughter cell to maturity.

CHEMICAL REQUIREMENTS OF GROWTH

The various elements necessary for the normal growth of plants were determined by growing plants in nutrient solutions either containing or lacking certain mineral salts dissolved in water. The growing of plants in nutrient solutions without soil is known as **hydroponics.** The role of each of the known essential elements is given in Table 11-1.

Although some of these essential ele-

Table 11-1

Role of the essential elements that form compounds used in plant growth

Macronutrients	Role
Carbon	Entire cell and parts
Oxygen	Entire cell and parts
Hydrogen	Entire cell and parts
Nitrogen	Protoplasm (especially proteins)
Phosphorus	Protoplasm, enzymes, fats, and nucleoproteins
Sulfur	Protoplasm, synthesis of chlorophyll, and proteins
Calcium	Constituent of cell wall, permeability
Potassium	Enzyme, citric acid cycle, respiration, and photosynthesis
Magnesium	Formation of chlorophyll, enzymes, and photosynthesis
Iron	Enzyme (cytochrome, formation of Chlorophyll)
Micronutrients	**Role**
Boron	Enzymes, synthesis of glutamine
Copper	Activation of enzymes, synthesis of ascorbic acid
Manganese	Formation of chlorophyll, photosynthesis, activation of enzymes
Molybdenum	Change nitrates to nitrites, synthesis of tannin
Zinc	Formation of auxin, enzymes
Chlorine	Unknown

Table 11-2
Effects of the lack or deficiency of mineral compounds containing the essential elements needed for proper plant growth

Deficiency	Characteristics
Nitrogen	Pale green; older leaves become yellow and dry; stems short and slender
Magnesium	Chlorotic leaves; edges of leaves curled
Phosphorus	Dark green; older leaves become yellow and dry; stems short and slender
Potassium	Chlorotic leaves; tip of leaf dies; dead, dry spots in leaf between veins as well as on edges
Zinc	Large, dead, dry spots over entire leaf; leaves become thick
Copper	Wilting of young leaves; tip of stem droops
Calcium	Terminal buds die
Boron	Terminal bud light green; stalk withered near apex
Manganese	Leaf becomes spotted by dead, dry areas
Iron	Chlorosis of young leaves; stems short and slender
Sulfur	Light green young leaves

ments are necessary only in minute amounts, any deficiency or absence of them will have a marked effect on the appearance of the plant. Lack or deficiency in the available amounts of nitrogen, phosphorus, and potassium causes reduction in the growth of the plant. Because the symptoms for a deficiency of more than one element may be the same, it becomes difficult to diagnose exactly which element is lacking. In a general way, how the lack or deficiency of each of the better understood elements taken from the soil by plants affects their growth is listed in Table 11-2.

From Table 11-2 it can be seen that several mineral deficiencies cause similar abnormal growths and dwarfing of the plant. The yellowing of leaves, which may occur in a variety of patterns, is called **chlorosis.** It appears to be a lack of chlorophyll in various parts of the leaf and gives them a variegated appearance. A deficiency of minerals such as boron and calcium affect the meristem, resulting in distorted growth of the tip of stems and buds.

All the essential elements have a critical level, amounts above which there is no increase in growth or yield. However, when these essential elements are below this critical level in the soil, increase in growth and yield can be produced by increasing the concentration of the elements. The critical level of elements associated with maximum growth must be determined for each species of plant. Furthermore, the concentration requirements can vary with the season, and anything above that critical level is termed luxury consumption.

SIMPLE VEGETATIVE GROWTH

All the processes discussed in preceding chapters work together in a plant to produce a complex phenomenon called growth. For growth to occur, all the parts of a vascular plant must be correlated. The plants are composed of a system of roots, stems, and leaves. The roots must be able to absorb water and minerals, and the stems must be able to conduct these up to the leaves, which in turn furnish carbohydrates and other organic materials necessary for the development of the stems, flowers, other plant parts, and more leaves. In other words, a plant grows or increases in size, and growth, in turn, is regulated by minute quantities of organic substances called plant hormones. Research in plant physiology is able to explain growth only in part, even though it is known that growth involves coordination of both hereditary and environmental factors. In its broadest sense, growth is concerned with three phases of cellular development: cell division, cell elongation, and cell differentiation.

Growth can be defined as an increase in mass through either cell enlargement or increase in number of cells by mitosis. In either case the cell takes the raw materials it needs from its environment and converts them to more substances or more cells. In any sexual organism in which two cells unite to form a new individual, the fertilized egg may produce millions of cells through a series of successive divisions. Higher organisms are not a mass of similar cells comparable in size and character to the original cell. Instead, they are composed of cells having different shapes, sizes, and functions. Growth, then, is not merely multiplication of cells but a more complicated pattern of multiplication that must be coordinated to produce the form that distinguishes one organism from another.

The unique structure and function of each cell is a result of **differentiation.** Through a process of progressive changes a generalized cell becomes transformed into a specialized one. Thus variation is introduced into a functioning organism. In a tree, for example, growing cells are transformed into the different cells that make up the leaves, the stems, the roots, the flowers, and the seeds. What little is known about differentiation has been learned from living systems. In a sense it is creative as life itself is creative. Differentiation is limited only as far as the genetic information in the nucleus determines these limits. Whereas growth is a process of multiplication of similar units, differentiation is the extracting of these units from the general mass, thereby making them unlike the rest. The more differentiated a cell becomes, the less likely it is to divide. Thus differentiation tends to restrict further multiplication. A cell is destined to have a certain structure and function, even though no obvious change has taken place. At a specific time, certain cells, even when grafted to another organism, will develop into those specialized cells whose destiny has already been determined. Consequently, differentiation begins long before any visible change takes place in the cell. Obviously, any visible changes must be preceded by chemical alterations. As differentiation takes place, some potentialities are accentuated, whereas others are suppressed or even lost; therefore it is essentially a change in cellular protein, a change brought about by the nucleus, which has in its genes the DNA containing coded information that exercises control over the cell and directs its destiny.

Although growth and differentiation are two of the processes by which development is achieved, development, when considered in terms of cells, is found to be a progression from simple and uniform types to more complex and diverse ones. When the growth and differentiation of an egg, a seed, or an embryo is studied as a functional, living entity, it is found that the organism develops as a whole and not as a collection of individual cells. In other words, development is integrated by chemical stimuli, such as hormones, cell movements, and interactions. To understand how this takes place would necessarily demand a complete understanding of organization at the molecular, cellular, tissue, and organ levels—an understanding that is possessed in only a fragmentary way.

The capacity of cells to grow, in the sense that they add to the cell contents chemical substances that do not exist in the inorganic world, appears to be limited. Although there are exceptions, a cell can only double its mass. Because of this fact, any significant growth requires the production of new cells, and this involves the process of cell division or mitosis. The amount of living matter can be added to continuously, but only by multiplication, and the multiplication occurs only by division. Of all the activities that can be seen in a cell, mitosis appears to be the most complicated and its machinery the most elaborate. It is now understood why cells must always arise from preexisting cells and by a process that seems to be so complicated. The reason is evident when one realizes that the history and capabilities of a cell are determined by the threadlike chromosomes in its nucleus. The various forms of life can maintain their historical continuity only if the chromosome patterns are transmitted from one cell to another.

The act of division is impressive enough to demonstrate the powers of dynamic organization possessed by living matter, but even more impressive is the appearance of two identical sets of chromosomes where only one existed before.

THE SEED AND ITS GERMINATION

The embryo of dicot plants is differentiated into three main parts: the **hypocotyl,** which is the primitive stem and root; two seed leaves, the **cotyledons,** which are attached to the upper portion of the hypocotyl; and the **plumule,** which is a bud located between the cotyledons (Fig. 11-1). In the dicot, such as the garden bean, all the stored food is in the tissues of the embryo, particularly the cotyledons. Therefore many dicot cotyledons appear thickened and have lost their leaflike characteristics. The plumule of dicot embryos is large and readily develops a leaf-bearing shoot upon germination.

The processes occurring from the time the embryo resumes its growth until the seedling is established are called, collectively, **germination.** The first noticeable evidence of the germination of a seed is the bursting of the seed coat and the protrusion of the root tip (Fig. 11-2). As a result of the absorption of water, the embryo and seed coat swell to rupture the seed coat. No matter how the seed is oriented in the soil, the root always grows downward into the soil (positive geotropism) and rapidly develops root hairs. The initial or primarily root readily produces branch or **secondary roots,** thus anchoring

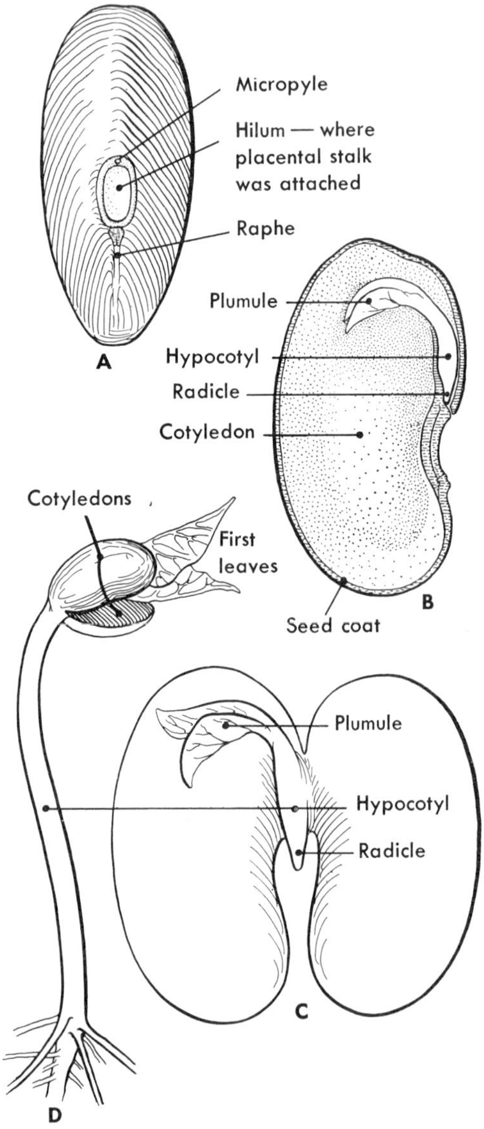

Fig. 11-1
Bean seed (dicot), its parts and development.

the young plant firmly in the ground. In the bean the cotyledons emerge from the seed, turn green, and serve as the first leaves of the seedling plant (Fig. 11-2). Eventually the cotyledons wither and drop off when their reserve supply of food is depleted. The more or less permanent leaves of the plant develop on the shoot located above the cotyledons. The young

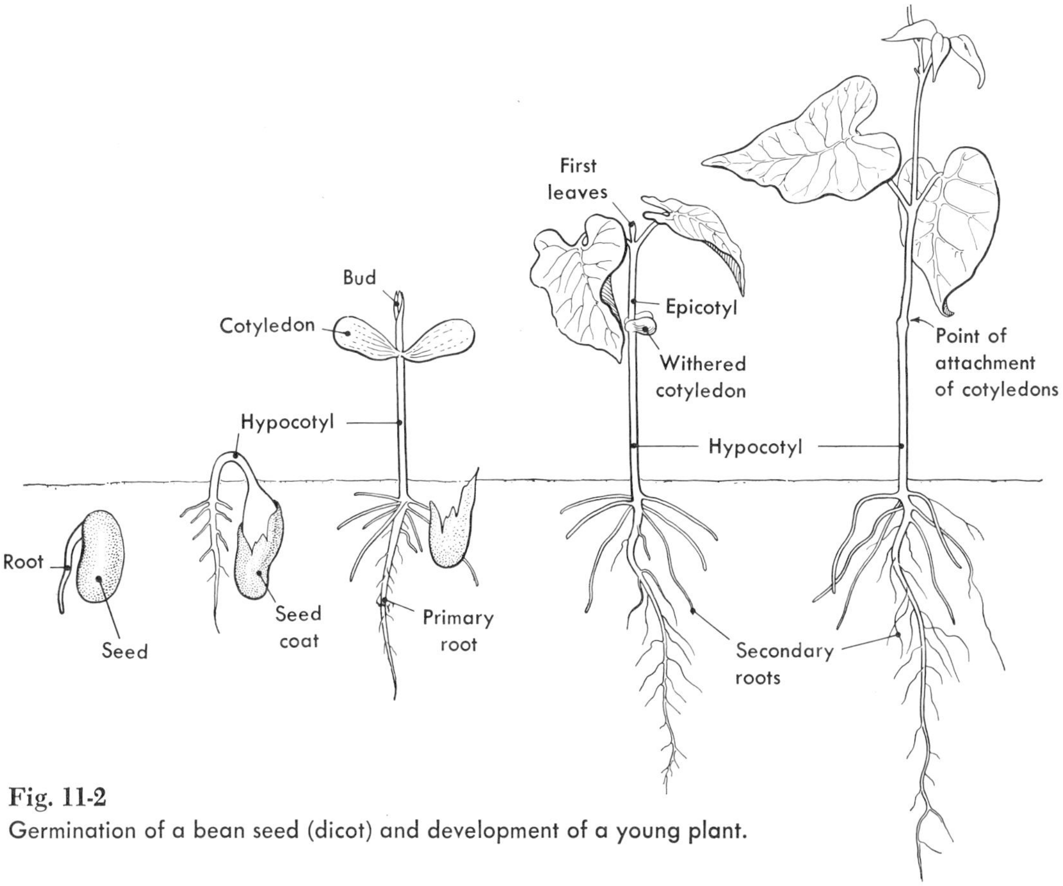

Fig. 11-2
Germination of a bean seed (dicot) and development of a young plant.

plant is now capable of sustaining itself through food manufactured by photosynthesis.

The seeds of all plants have embryos, but not all are without an endosperm. In a grain of corn the endosperm persists and is not absorbed until germination begins. The endosperm of the corn seed is a prominent tissue, rich in accumulated food in the form of oil, starch, or protein. The embryo of the corn is embedded in the endosperm (Fig. 11-3), which developed from the polar nuclei in the embryo sac (life cycle 16). The milk of the coconut consists of a pint or more of fluid containing numerous nuclei and is the free nuclear stage in the development of its endosperm. As the seed matures, these nuclei migrate toward the outer side, cell walls form around them, and the mature endosperm or coconut meat develops. The embryo of monocots develops in much the same way as that of dicots, except that only one cotyledon (Fig. 11-3) is formed, rather than the two described for the bean.

The primary function of this one cotyledon is the digestion and translocation of the reserve food in the endosperm to the growing parts of the seedling. During germination of the corn seed (Fig. 11-4) the **radicle** grows downward and the plumule grows upward to produce the stem and leaves of the new plant. The corn grain, with its endosperm and cotyledon, remains in the soil. Protecting the plumule from injury as it grows upward through the soil is a pointed sheath called the **coleǫptile.** This is regarded as the first leaf of the seed-

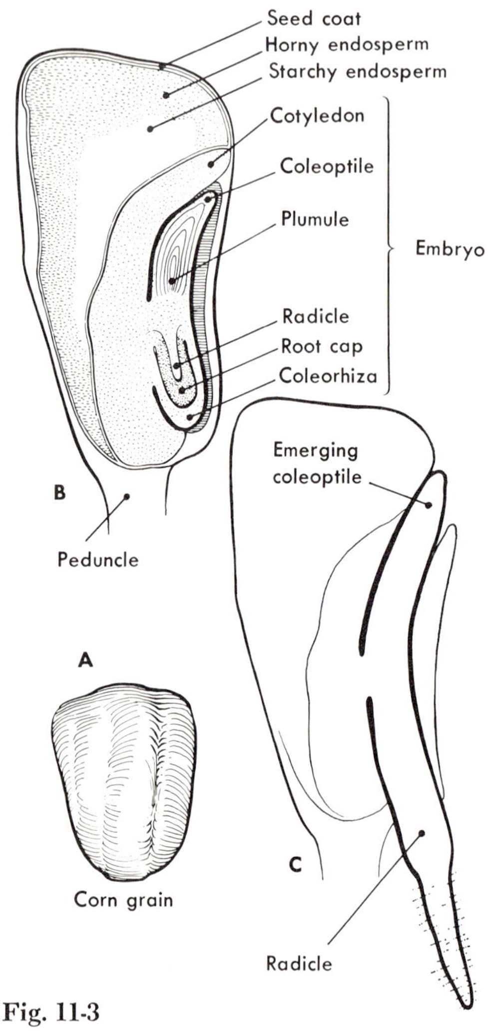

Fig. 11-3
Grain of corn (monocot seed) showing parts and early development of the embryo.

ling, but one that does not expand to become a leaf characterististic of corn. The radicle does not persist in monocots, although it develops into the primary root and even produces branches or secondary roots. It is supplemented by a stronger root system that is adventitious in origin. The adventitious roots develop from the lower nodes of the stem just under the soil, and the plumule, the first node, and the coleoptile are pushed from the seed principally by elongation of the first internode of the seedling's stem. Elongation in an embryo is due to the translocation of auxin, a growth hormone, from the tip of the coleoptile downward to the stem (Fig. 11-5). Auxin inhibits the growth of roots and leaves from the first node while stimulating the elongation of the internode. Once the coleptile leaves the soil and enters the light, the production of auxin is reduced, and leaves and roots will develop. This makes certain that no matter how deeply a seed is planted, neither adventitious roots nor leaves develop until the coleoptile reaches the surface (Chapter 12).

The radicle regularly becomes the primary root, and the **epicotyl** becomes the primary shoot in both dicots and monocots. However, the functions of hypocotyl and cotyledons are not the same in all seeds. In some, such as the peach (Fig. 11-6), the hypocotyl remains short, and the cotyledons never emerge from the seed. The epicotyl will appear above the ground only by its own growth. This type of seed exhibits hypogeal germination, in contrast to epigeal germination, in which the hypocotyl elongates, bringing the epicotyl and the cotyledons with it.

PARTS OF A SEED—THE POTENTIAL PLANT
Plumule—leaf bud
Epicotyl—upper stem
Cotyledons—Food storage for embryo tissue
Hypocotyl—lower stem
Radicle—root

The requirements for germination of all seeds are adequate amounts of moisture and oxygen, as well as a suitable temperature. There are some seeds, however, that require light for germination, but light inhibits the development of most seeds. The water content of most seeds is very low, and the water requirements for the physiological processes necessary for germination are high. Germinating seeds respire rapidly; therefore an adequate supply of oxygen is

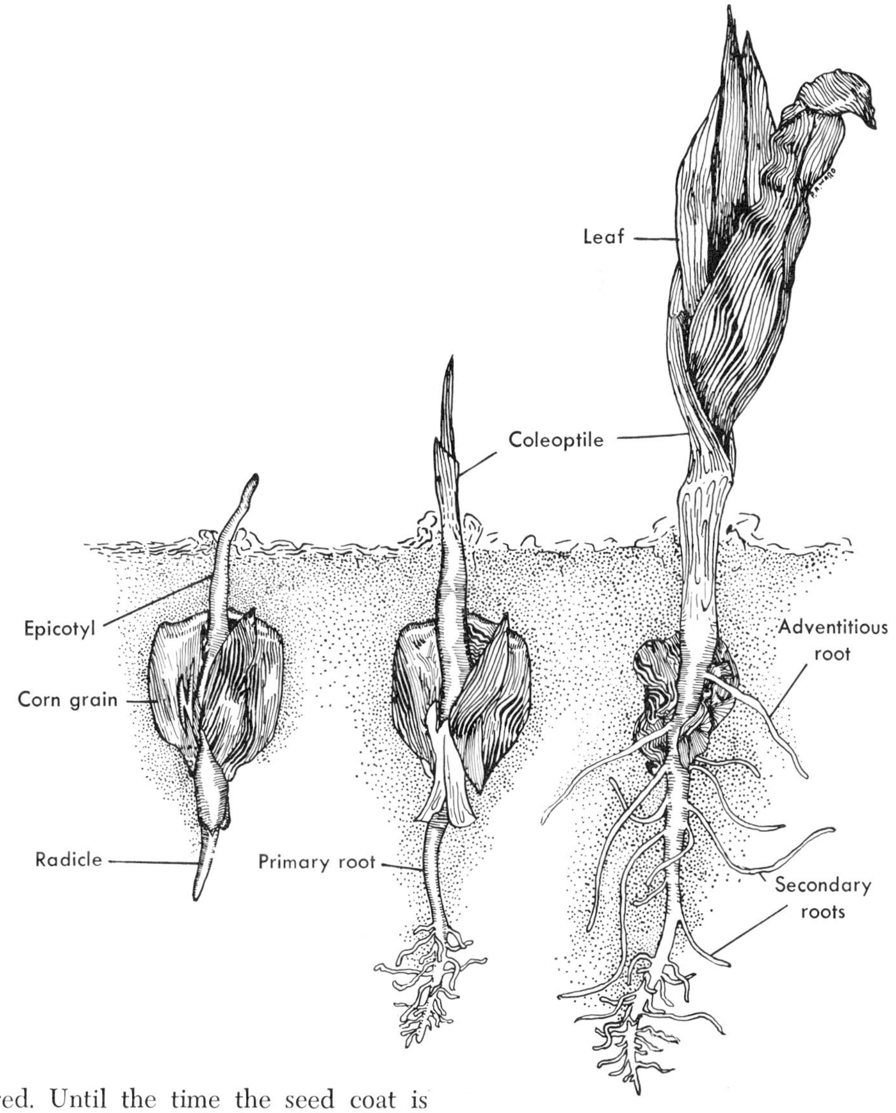

Fig. 11-4
Germination of a corn seed (monocot) and development of a young plant.

required. Until the time the seed coat is ruptured, respiration is almost entirely anaerobic. Aerobic respiration is more prominent as soon as the embryo ruptures the seed coat. The optimum temperature for germination of seeds varics. For all, there is a maximum and a minimum, above or below which germination will not take place. The effect of light upon germination also varies since some seeds require light, whereas others are retarded by it.

Some seeds will not germinate when initially produced, even under favorable environmental conditions. Seeds that require a certain period to elapse before germination begins are said to be dormant. This may be due to several factors. The following are the most common: (1) seed coat is permeable to neither water nor oxygen nor both, (2) seed coat is so tough that it resists rupture, (3) embryo is not mature, or (4) chemicals inhibit germination. The seed coat may have to decompose partially before oxygen or water can enter the seed and the embryo and rupture the seed coat wall. This is true of very hard, stony seeds, such as the coconut. Seeds with hard, stony coats generally remain viable for longer times than do those with softer coverings. Most of our common crop plants produce seeds that remain viable only one year or at most a few years. However, others, such as the Oriental lotus, have been reported viable for what appears to have been 800 to 900 years.

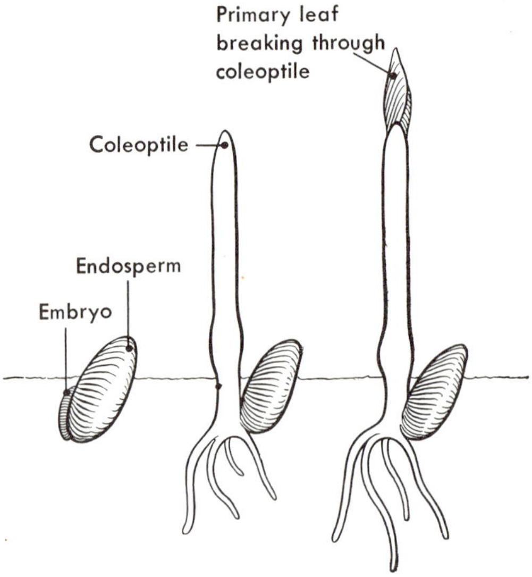

Fig. 11-5
Elongation of the coleoptile of a seed and the formation of the first leaves.

DIFFERENTIATION IN HIGHER PLANTS

The term **differentiation** is used to describe any kind of change in a cell, either morphological or physiological or both, that occurs after the cell has ceased elongating. Originally, it was used to describe the morphological changes occurring in cells as they developed to maturity. Therefore, today, differentiation is used to designate the formation of cell substances as well as the structures produced by the cells.

It is appropriate to begin the discussion of differentiation with those cells capable of indefinite division or mitosis. Those tissues made up of cells undergoing division are called **meristem.** Two types are recognized: meristematic tissues at the tip of roots and stems and tissues in which cell division occurs laterally. However, any living cell may become meristematic if injury or some other abnormal condition arises.

The terminal or apical meristem of roots

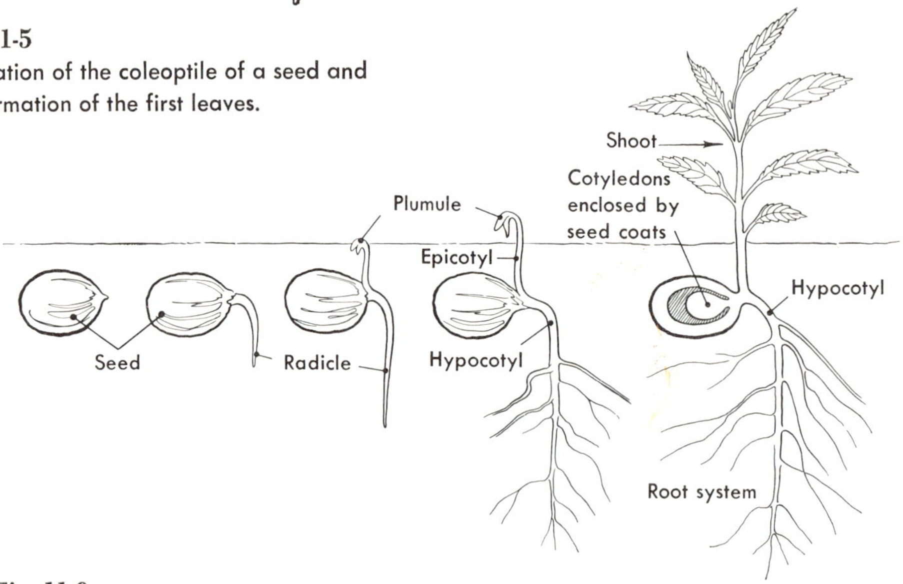

Fig. 11-6
Development of a peach seed (endocarp removed) into a young tree.

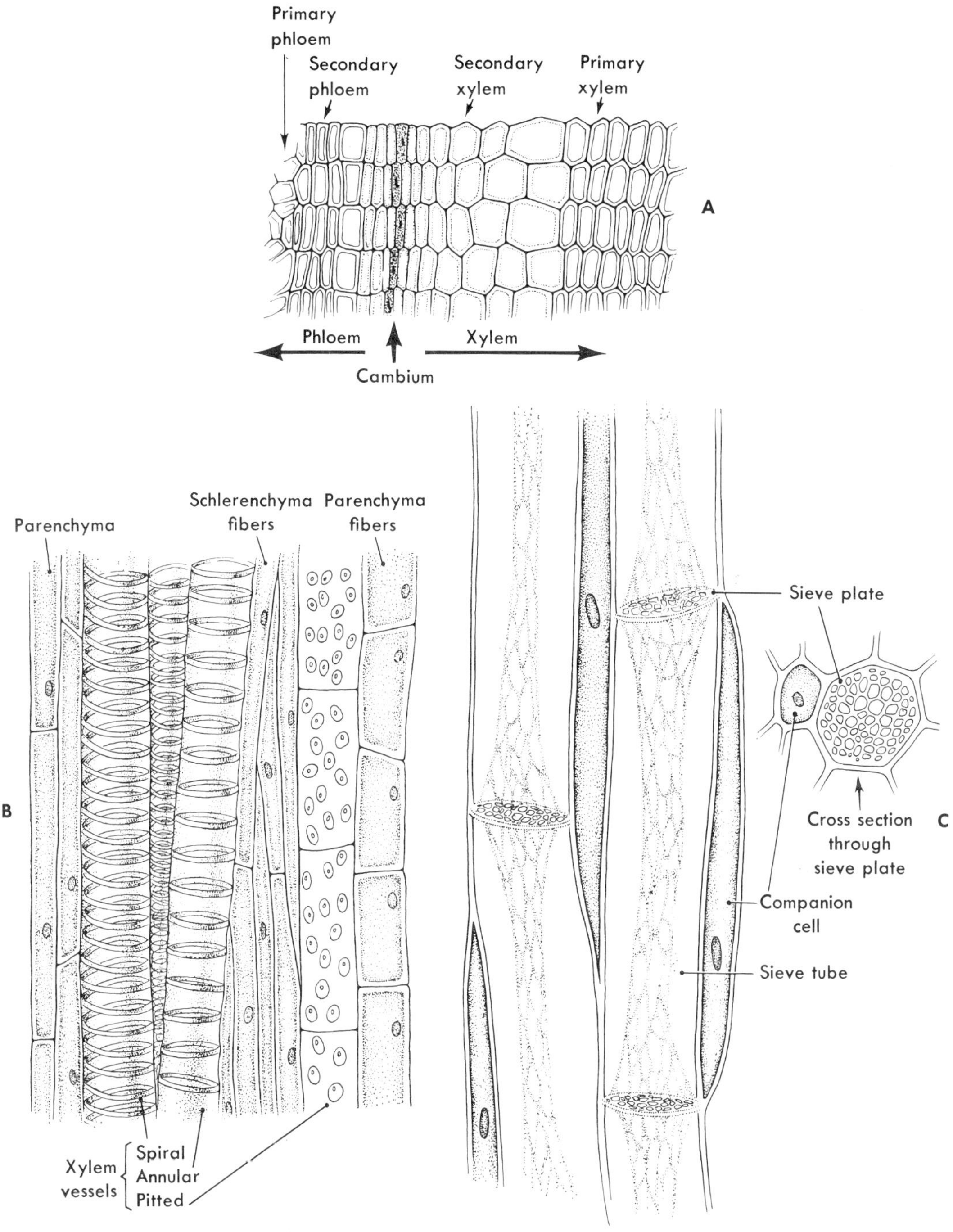

Fig. 11-7

Details of the cells of the vascular system. **A,** Vascular cambium with phloem on the left and xylem on the right, both specialized cells produced by the growing cambium. **B,** Xylem vessels in detail. **C,** Phloem cells in detail. Note the remains of the cytoplasm in the sieve tubes of the phloem.

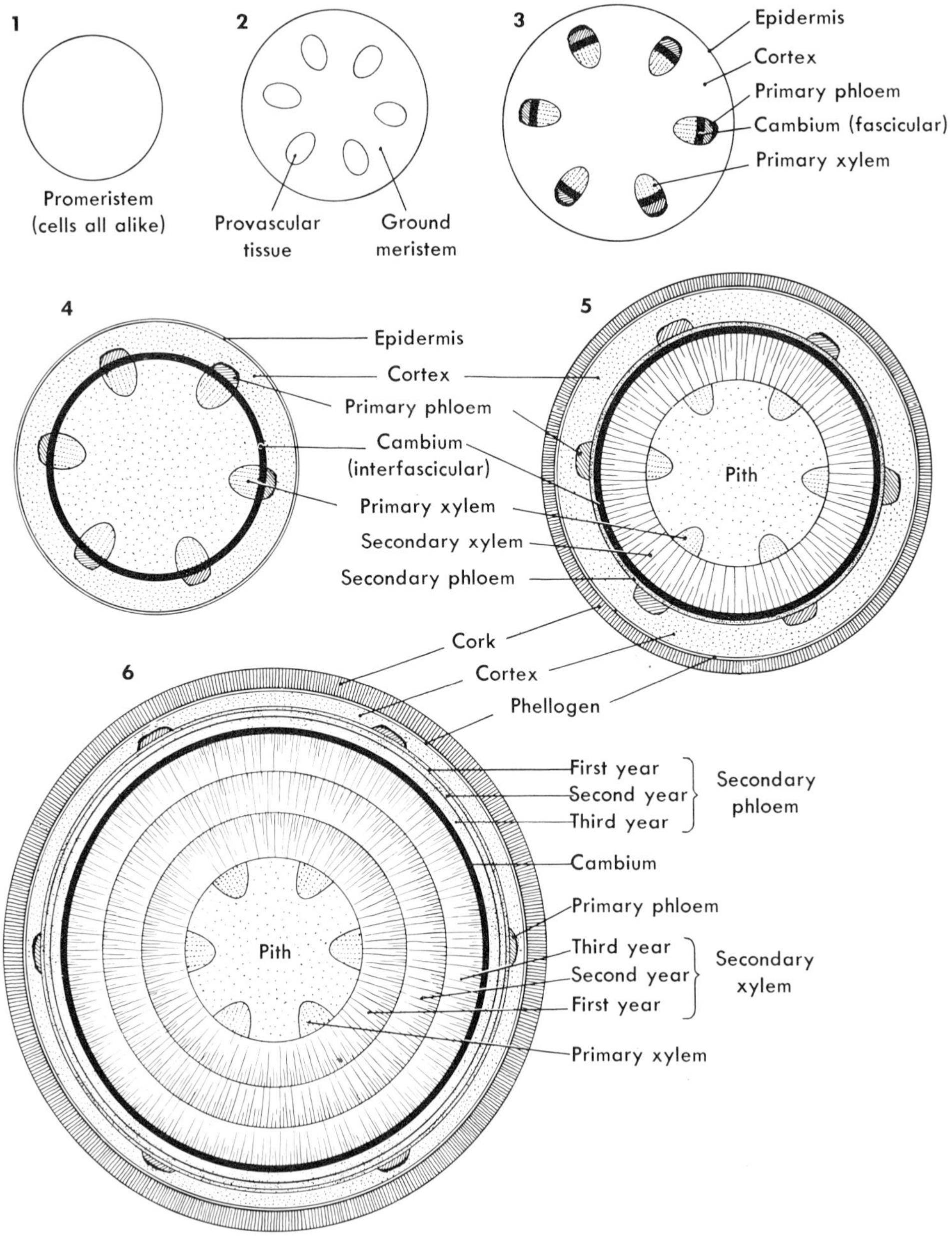

Fig. 11-8
Development of new cells in young woody stem. Various cross sections of the stem are numbered to correspond to the regions in the longitudinal section. (See Fig. 11-22.)

and stems is commonly called a growing point. The cells that make up the meristem are immature and have walls composed of a few thin layers of cellulose. There are no thickened secondary walls. The nucleus is large and surrounded by dense cytoplasm, in which may be found small and scattered vacuoles. The cells are isodiametric in shape. In other words, the cells have approximately the same length, width, and depth. Few, if any, intercellular spaces can be found in meristems. The most striking characteristic of the meristem is its capacity for repeated cell divisions.

The terminal meristem of stems possesses the potentialities for producing both vegetative and reproductive organs. The formation of leaves or flowers depends upon either heredity or environmental factors, such as photoperiodism and temperature or both. It is also known that important growth regulators are also produced in the terminal meristematic tissue. The terminal meristem in both stems and roots is alike and is responsible for the increase in length of these structures. However, in roots the meristem produces cells that form the protective root cap, whereas in stems the growing point is protected primarily by the bud scales and leaf primordia.

The meristematic tissues responsible for lateral growth in a plant are called **vascular cambiums** (Fig. 11-7, *A*). This type of tissue consists of a single layer of cells, which produce, by division, new cells toward the center of a root or stem called the **xylem** (Fig. 11-7, *B*) and other cells toward the outer surface, **phloem** (Fig. 11-7, *C*). The xylem is a water-conducting tissue, and the phloem a food-conducting tissue. Since the cambium consists of one layer of cells, it is obvious that one daughter cell of the division remains meristematic. The other daughter cell will develop into either a xylem or a phloem cell. Why a cell will develop one time into xylem and the next into phloem is a matter of speculation. It has been suggested that when vascular cambium is acid, xylem is produced; when it is not, phloem is produced. Regardless of the reason, during the active growing season more xylem cells are produced than phloem cells, thereby accounting for the greater amount of xylem than of phloem in gymnosperm and dicot roots and stems (Fig. 11-8). The xylem cells have walls heavily impregnated with lignin. Thus they are capable of withstanding crushing. The softer-walled cells of the phloem are easily crushed and destroyed from year to year.

Any lateral growth of monocot roots and stems is largely due to the enlargement of cells, since the vascular cambium is either poorly developed or not present at all. In perennial plants, in which a vascular cambium does exist, the cambium increases in diameter as the root or stem enlarges. This is accomplished by the division of the cambium cells in a raidal plane and the tangential enlargement of the daughter cells. Wherever vascular cambium exists, it is derived from the terminal meristem and remains meristematic to form xylem and phloem.

Another meristematic layer of cells is the **cork cambium,** which develops between the phloem and the outer surface of the plant organ. It usually originates in the cortex layer just below the surface of a stem or root. Cork cambium cuts off cells on the outer surface, which then develop into cork cells. The walls of cork cells become impregnated with **suberin,** a fatty substance, which makes the cell impermeable to food and water. Therefore mature cork cells are dead and form a major part of the bark of older stems and roots of trees and shrubs. The death of the cork layer also results in the death of all tissues outside the cork layer. Subsequent growth in diameter of the stem or root will cause the dead tissues to crack and peel off, resulting in the deep furrows so characteristic of the bark of some trees.

The meristem is embryonic tissue that by cell division and enlargement will differentiate into various groups of tissues. A **simple tissue** is one in which the cells are essentially alike in structure and function. On the other hand, there are **complex tissues,** which consist of two or more different kinds of cells that perform a common function.

There are five kinds of simple permanent tissues in plants: parenchyma, collenchyma, sclerenchyma, cork, and epidermis.

The **epidermis** is a tissue covering the external surface of leaves, young roots, and stems. Primarily a protective tissue, it consists of a single layer of cells in stems compactly arranged and uniform in thickness. Epidermal cells are without chloroplasts, except in the cells that surround the openings (stomata) through which an exchange of gases takes place in leaves. Ordinarily the outer surface of epidermal cells is covered with a layer of a waxlike substance known as cutin. This waxlike secretion of the epidermis prevents excessive loss of water through evaporation.

A common tissue found throughout a plant is the **parenchyma,** cells that are large, thin walled, and relatively undifferentiated in shape and contain conspicuous intercellular spaces. If they do have chloroplasts, they are classified as **chlorenchyma.** Parenchyma cells are living and often serve to store foods.

Collenchyma (Fig. 11-9) is a supporting tissue frequently found in the outer layers of stems and in the petioles of leaves. Like parenchyma, the collenchyma cells are living cells, but elongated, and the cell walls are thickened with cellulose, particularly at

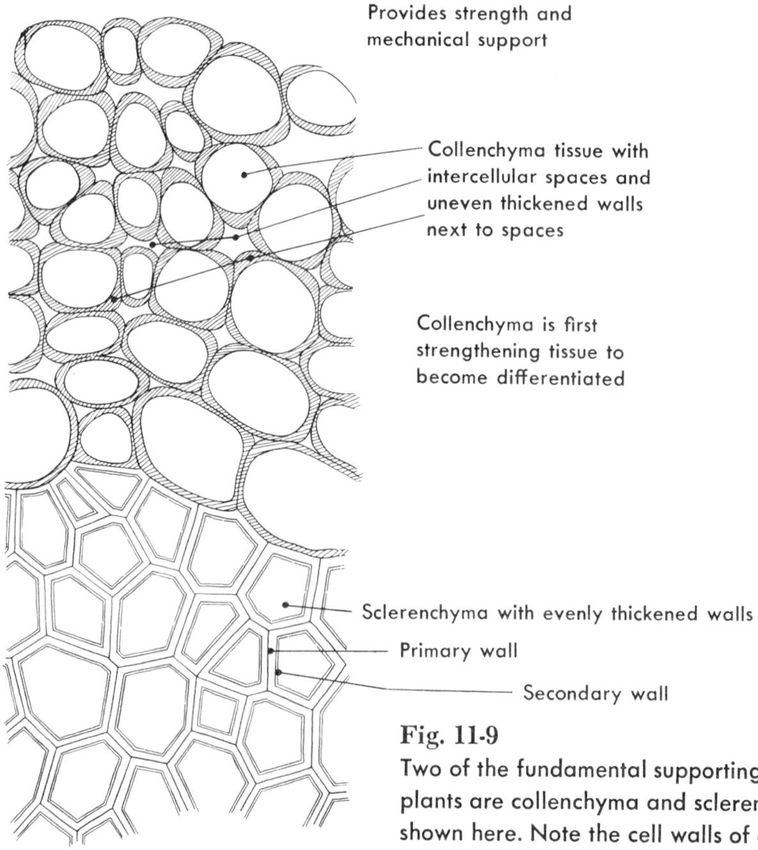

Fig. 11-9
Two of the fundamental supporting tissues in plants are collenchyma and sclerenchyma shown here. Note the cell walls of each.

the corners. Collenchyma cells are found in the stems of herbaceous plants and in the younger stems of woody plants.

Like collenchyma, **sclerenchyma** cells (Fig. 11-9) are supporting tissues that give rigidity, toughness, and hardenss to the plant, but unlike collenchyma, sclerenchyma cells are dead when they are mature because the entire wall is heavily and uniformly thickened. The wall of the mature cell is composed largely of cellulose and often, in addition, a hard substance called **lignin.** Thus sclerenchyma cells are said to be dead, lignified cells. Some sclerenchyma cells are long, slender, and pointed at both ends. These are called fibers and have great tensile strength and elasticity. Plants with very long fibers are useful in manufacturing cordage, linen, and cloth. Another type of sclerenchyma cell is the **stone cell.** They are similar to fibers, but much shorter, and sometimes form isolated groups of cells in the outer parts of stems, the hard shells of nuts, and the stones in cherries and peaches. They also account for the gritty texture of a pear.

Finally, there is the **cork tissue** previously mentioned. This tissue consists of cells that, when mature, are dead because of the fatty substance in their walls (suberin). As a protective layer, cork eventually replaces the epidermis in older stems and roots.

Xylem and phloem are **complex permanent tissues.** Both are conductive in function, forming the greater part of the vascular system of plants.

Xylem consists of several kinds of cells, most of which are elongated dead cells, with walls composed largely of lignin. Lignin is deposited on the inner surface of the cell wall, replacing most of the cellulose. The cells become hard and strong, but this does not interfere with the passage of water upward from the roots to the stems and leaves. The simplest of these xylem cells is the **tracheid** (Fig. 11-10). On the basis of the manner in which lignin becomes deposited, the secondary walls may appear as

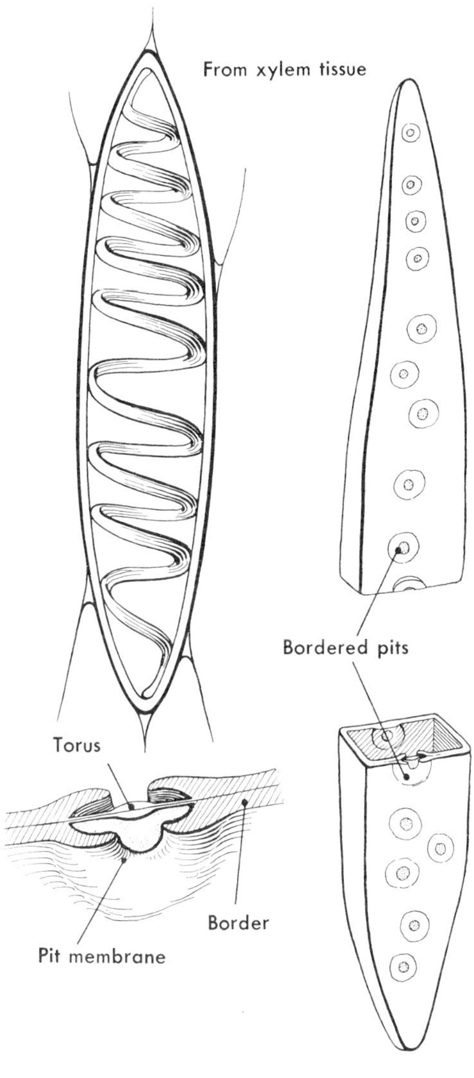

Fig. 11-10
Tracheids showing details. Note the cell walls, pits, accessory structures, and absence of cytoplasm.

spirals, rings, parallel bars, irregular networks, or pits. Pitted tracheids are the most common. For example, in gymnosperms the wood is made up almost entirely of pitted tracheids. In some plants, such as angiosperms, long **vessels** (Figs. 11-11 and 11-12) develop from a longitudinal row of cells, the end walls of which break down to form conducting tubes. Similarly, they may be

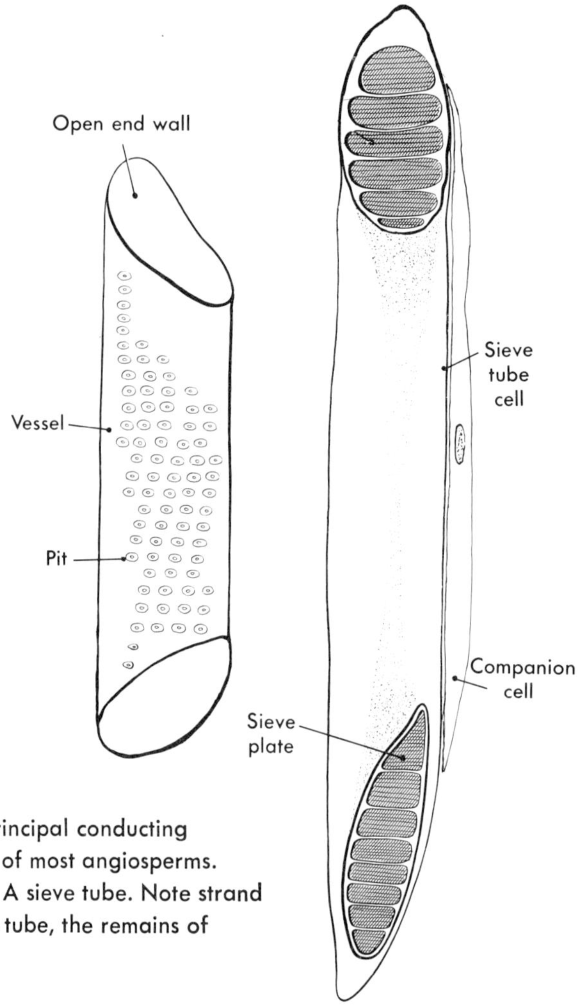

Fig. 11-11
Conducting vessels, principal conducting structure in the xylem of most angiosperms. **A,** An open vessel. **B,** A sieve tube. Note strand of "slime" in the sieve tube, the remains of cell contents.

classified as spiral, annular, or pitted. Fibers are like tracheids but have greater length, more pointed ends, and few pits. Both fibers and vessels are absent in gymnosperms, but the fibers do give the hard quality to the wood of angiosperms. Parenchyma cells may also be found in xylem.

As a part of the conducting tissue in plants and associated with the xylem are those tissues called phloem. Phloem conducts food from the leaves to the stems and roots. The principal cells in phloem are **sieve tubes**—elongated cells, the end and side walls of which may be perforated with sieve plates. They have thin walls of cellulose, cytoplasm, but no nucleus. Because they have no nucleus, other living cells, **companion cells** (Figs. 11-11 and 11-12), are usually associated with them. These are narrow cells that have a nucleus and cytoplasm. Companion cells are not present in gymnosperms. In addition to sieve tubes, some fibers and parenchyma cells may be present.

Roots

The root of a flowering plant always begins in the seed. It starts its development

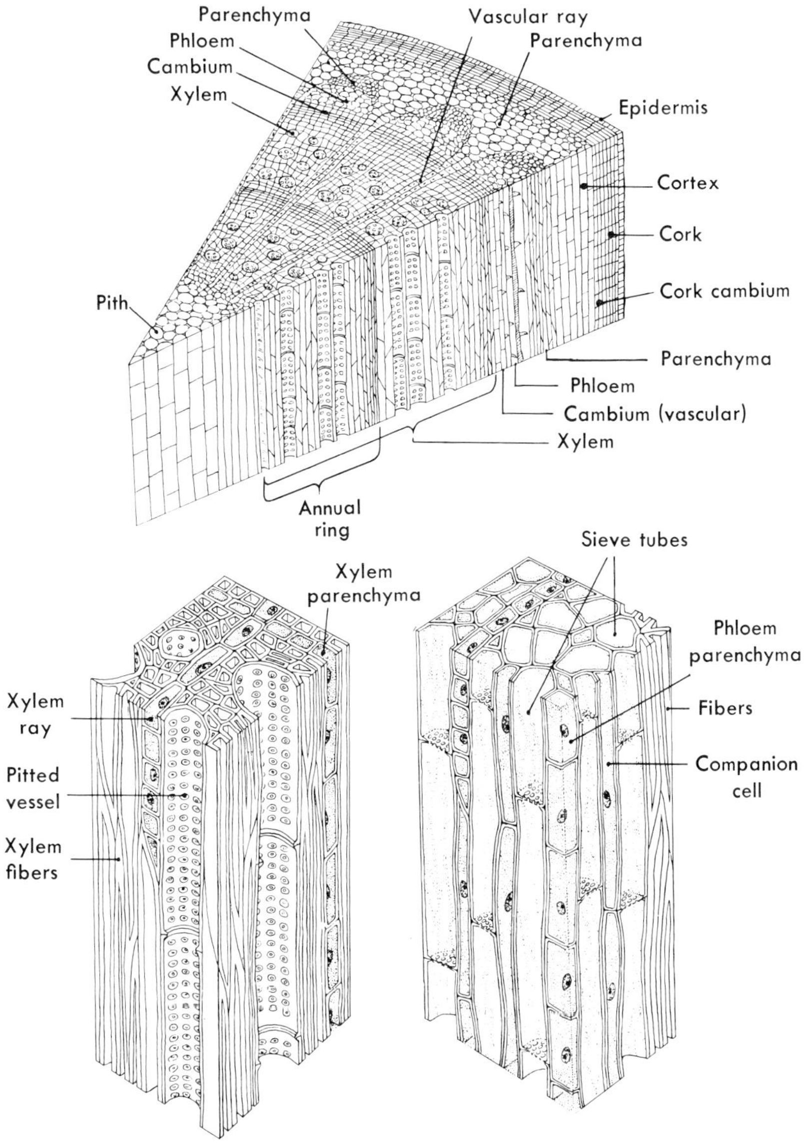

Fig. 11-12
Details of a woody stem (*Tilia* sp.) showing arrangement of cell types (see text).

from the hypocotyl of the embryo within the seed (Fig. 11-6). Once a seed germinates and absorbs water, the radicle grows out of the seed, producing the first or **primary root** of the newly developing plant. Soon the primary root produces branch or secondary roots, which in turn give rise to branches of their own. Roots ordinarily grow downward since they are positively geotropic (**geotropism**), even though the secondary roots may at first grow out horizontally. By turning a potted plant over on its side, one can demonstrate that the roots will turn downward (Fig. 12-7). Occasionally roots will develop from leaves or stems (Fig. 13-9). Roots that develop from parts other than the primary root or one of its branches are called **adventitious roots,** such as the prop roots of corn. Other examples are roots that develop from bulbs or rhizomes (underground stems). Nevertheless, all root systems, with the exception of aerial roots, find their way into the ground to form a complex root system.

If we examine any root, it will invariably be cylindrical, although tapered toward its free end. This narrowed end of the root is called the **root tip** and is that area of a root (primary, secondary, or adventitious) in which unlimited growth and elongation take place (Fig. 11-13). Growth in diameter occurs only in the older parts of a root. Each root tip is covered by a cap of epidermal cells, the function of which is to protect the more delicate dividing cells that are found immediately behind it. Thus the **root cap** fits over the end of the root tip like the short finger of a glove. A region of actively dividing cells undergoing mitosis may be found just back of the root cap. This area of meristematic cells is called the **embryonic region,** an area in which the number of cells is being increased. It is possible to see cells in all stages of mitosis in this embryonic region when the root tip is properly prepared, sectioned, and stained to show chromosomes. Since these cells are continually undergoing mitosis, they remain

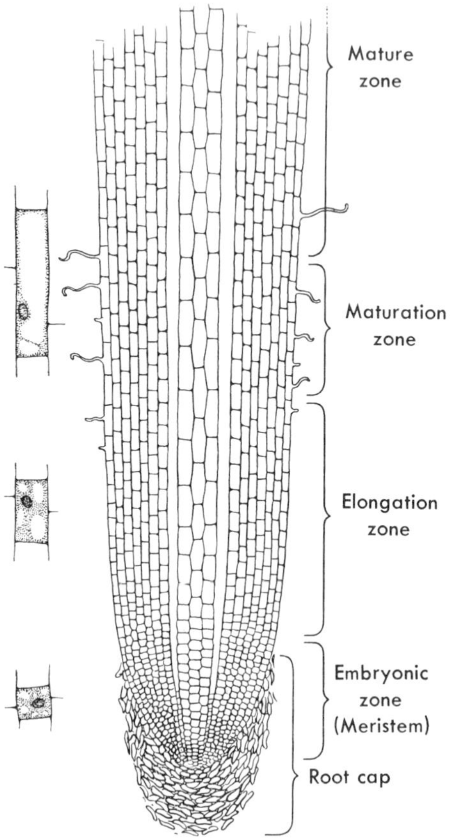

Fig. 11-13
Root tip and root cap with the major growing regions indicated.

the undifferentiated **apical meristem** of the root. The cells in the embryonic region are, therefore, the origin of those cells that contribute to the permanent tissues of the root, as well as of the root cap cells which are constantly being sloughed off as the root tip pushes its way through the soil. Because the cells are rapidly dividing, they are thin walled, are mostly cubical in shape, and have a large nucleus within a dense cytoplasm This region of the root tip appears dense in stained preparations and, in the living condition, has a yellowish color.

The cells of the embryonic region cease dividing as they become farther removed from the tip and gradually merge into a

region of elongation. In this part of the root tip the cells, typically columnar in shape, mature and increase in length. Because of the absorption of water, the cells develop large vacuoles so that the cytoplasm appears less dense than in the mitotic cells of the embryonic region. However, the cells in the region of elongation have slightly thicker walls.

A more extensive area of the root tip is found behind the region of elongation. It is an area in which the cells are becoming differentiated and can easily be detected because of the appearance of numerous **root hairs** (Fig. 11-13). This area of the root tip is called the **region of maturation** or the **root hair region.** It is in this region of the root tip that the cells become specialized in structure and function, thus becoming differentiated into the permanent tissues of the root. The root hairs first appear as small swellings on individual epidermal cells. These swellings increase in length with the age of the cell, forming slender outgrowths of the epidermal cells of the root tip. Root hairs are easily demonstrated on grass seedlings grown on moist paper. They appear as a cottony mass just behind the region of elongation, and under the microscope they can be seen as tubular extensions of a single cell. Root hairs have thin walls lined with a thin layer of cytoplasm that surrounds a large central vacuole (Fig. 11-14). Usually a nucleus can be seen lying in the cytoplasm near the tip of the root hair. Root hairs are short lived, lasting only a few days or weeks. Therefore they are not present on the older parts of a root. As the root tip grows, new root hairs arise behind the region of elongation, thus assuring a supply on every growing root.

The growth of adventitious roots on the cut end of a stem is stimulated by auxins (growth regulators). Excessive amounts of auxins check cell elongation in roots, causing a root to react positively geotropic if it is placed in a horizontal position (Fig. 12-7). Roots produce a growth regulator

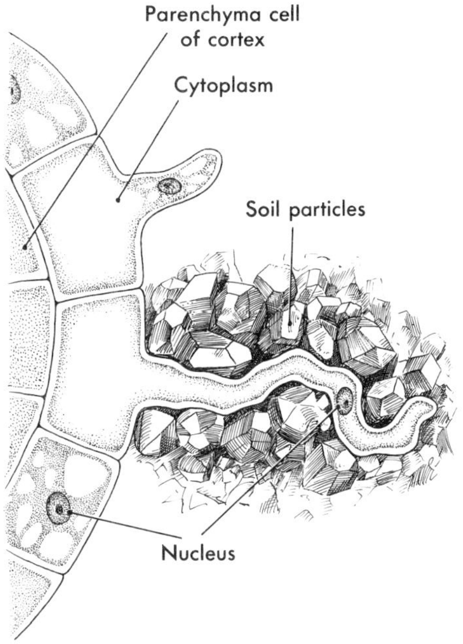

Fig. 11-14
Root hair cells showing growth of root hairs.

called caulocaline, which is needed for stem growth. Finally, it has been shown that excised roots will grow only if supplied with certain B vitamins, such as thiamine (B_1), pyridoxine (B_6), and nicotinic acid. Although the root requirements of different species vary, most have been found to need thiamine. Plants synthesize these vitamins in their leaves and translocate them to the roots.

The root hair region gradually merges into that part of the root in which cell differentiation is complete and in which the permanent tissues of the mature root have formed. This area of the root reveals three distinct regions: (1) the epidermis, which is a superficial layer of cells completely surrounding the root, (2) the **cortex,** which lies inside the epidermis and occupies an extensive regtion of many layers of cells, and (3) the **stele** or conducting portion, which occupies the center of the root (Fig. 11-15).

The epidermis is a single layer of cells,

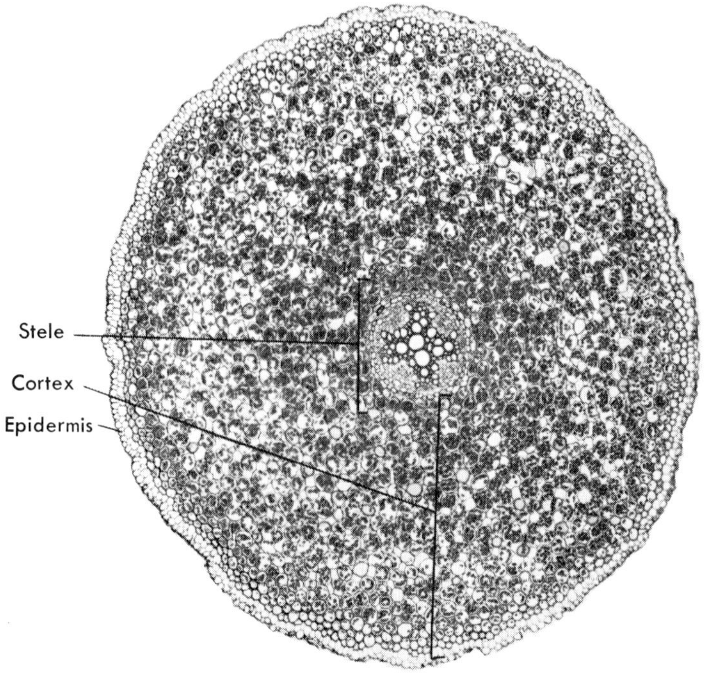

Fig. 11-15
Cross section of an older root of a buttercup showing the three distinct regions of a root. (×40.) (Courtesy George H. Conant, Triarch Products, Ripon, Wis.)

the walls of which are thin and composed almost entirely of cellulose. There is no cuticle on the outer surface of the epidermis of a root, as appears in leaves and stems.

The cortex is composed of many layers of large parenchyma cells (living and thin walled), between which are intercellular spaces (Fig. 11-16). The cortical cells transport water and mineral nutrients in a young root from the epidermis to the stele. In an older root the cortex serves primarily to store food and water. The innermost layer of the cortex consists of a layer of smaller cells, the walls of which become thickened with suberin and form the **endodermis** (Fig. 11-16). The endodermis may be a ring of thick-walled cells around the stele or may appear as a broken ring in which only some of the endodermal cells have thick walls. The thinner-walled endodermal cells provide a pathway for the radial transportation of water and other nutrients. Although its function is not precisely known, the endodermis probably prevents loss of water from the stele.

Two kinds of conducting tissue, xylem and pholem, are found in the stele. The xylem appears as a star-shaped group of thick-walled cells in the center of the dicot root, whereas in a monocot there are numerous xylem ridges (Fig. 11-16). The larger cells in the center are called **metaxylem,** whereas the smaller cells between the large ones are **protoxylem** (Fig. 11-17). Between the star-shaped arms of the xylem in dicots are small group of phloem cells. A layer of parenchyma cells, the **pericycle,** lies between the phloem and the endodermis to form the outermost border of the stele. The pericycle is sometimes several layers thick. As the cells of the pericycle divide, they push their way outward, dissolving and crushing the cells of the cortex to form **lateral roots** (Fig. 11-18). Lateral

MONOCOT DICOT

Fig. 11-16
Comparison of monocot and dicot roots showing stele inside cortex and endodermis (cross section).

roots always arise from the pericycle of the stele, unlike branch stems, which arise from the outermost layers of the stem. Lateral or secondary roots arise from the points of the star-shaped xylem and therefore produce as many vertical rows of lateral roots as there are points to the xylem. Dicot roots commonly have three, four, or five protoxylem points, but there can be more. Most roots have a solid mass of conducting tissue that consists of a center of xylem and a peripheral area of phloem but no pith, as is found in some stems (Fig. 11-19). In roots this type of stele is called a **protostele.**

Most fibrous root systems will have only the **primary permanent tissues** just described. These are the tisues that have arisen from the embryonic cells of the tip that are undergoing elongation and differentiation. However, in older roots and tap-root systems a special layer of meristematic cells develops from the pericycle, the cambium. A dicot root develops a cambium; a monocot rarely forms one. The cambium forms **secondary permanent tissues** in older roots, and, as a result of their activity, the root is able to increase in diameter. The cambium arises between the initial xylem and phloem (Fig. 11-16). It is a layer of thin-walled cells that retain the capacity for division. Initially, the cambium consists of a broken band of cells, between the xylem and phloem, that eventually becomes continuous around the entire protoxylem to form a cylinder. It will form new xylem and phloem (secondary, in contrast to the

Cross section of the [redacted]
(Courtesy George H. Conant, Triarch Products, Ripon, Wis.)

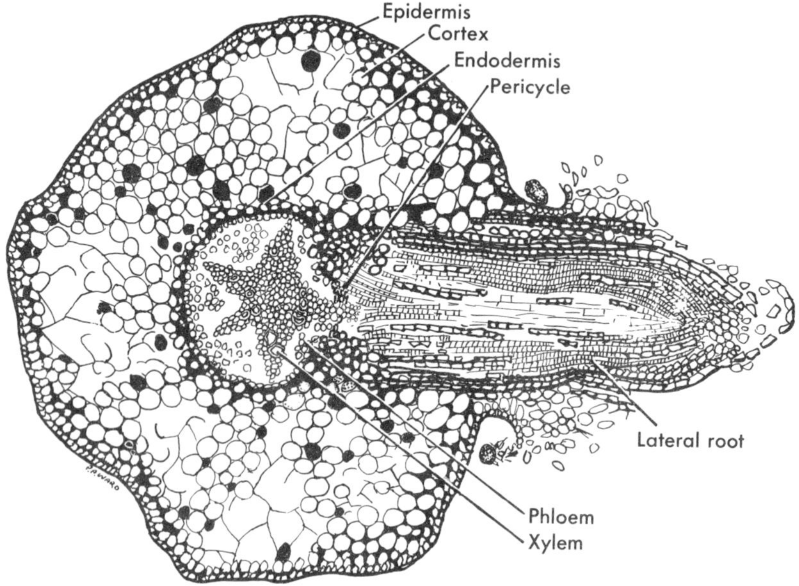

Fig. 11-18
Cross section of primary root showing origin of lateral root.

MONOCOT
(corn)

DICOT
(sunflower)

Fig. 11-19
Cross sections of monocot and dicot (typical) stems.

original primary xylem). The secondary xylem forms toward the center of the root and the secondary phloem toward the outside. As the layers of xylem increase, the cambium is pushed farther toward the periphery. The xylem layer becomes much thicker than the phloem layer, and both are traversed by thin radial layers of parenchymal cells called **vascular rays.** The thick xylem layers differentiate into seasonal growths recognizable as **annual rings.** However, they are not defined as clearly in woody roots as in woody stems. In some roots, for example, old trees, another cambium develops in the pericycle to form cork. A thick layer of cork causes the cells of the cortex and epidermis to die and slough off, leaving the corky layer as a protective covering on the outside of the root. It is this cork la[illegible] on thick and old roots of trees that makes them so difficult to cut with an axe.

Absorption through roots. Young secondary roots have a diffusion pressure deficit of water greater than that in the soil solution. Therefore there will be a new movement of water molecules from the soil solution into the vacuoles of the cells of the functional secondary roots. The movement of water molecules is from a region of lesser diffusion pressure deficit to a region of greater diffusion pressure deficit and not from a region of greater osmotic pressure to one of lesser osmotic pressure. The pressure resulting from the net movement of water molecules from the soil to the vacuzing of fluids from th plant or from specialized structures in leaves called **hydathodes.** Each hydathode will have one or more pores that resemble stomata, but unlike stomata they are permanently open.

The movement of water into a root produces what is called **root pressure** (Fig. 11-20) and may cause the formation of droplets of water on leaves, a phenomenon known as guttation. Both root pressure and guttation are a result of active absorption when transpiration (loss of water by evaporation) is at a minimum and water is being absorbed by the roots in greater amounts than is being lost at the leaf surface.

In the past the cause of active absorption of water was considered to be purely physical in nature as a result of osmosis. Today, enough evidence has been accumulated to show that it is also physiological in nature, since guttation can be affected by growth hormones, respiration, and temperature. In addition, exudation has been

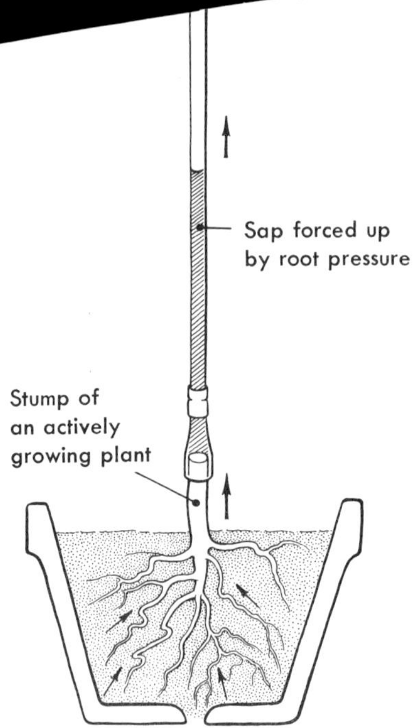

Fig. 11-20
Test to show the effect of root pressure on a column of fluid (sap).

the root hairs actually become firmly united to soil particles that are surrounded by films of water. This contact of the root hair with the soil particle enables the root to absorb the maximum amount of moisture and minerals dissolved in the water. In addition to the secondary roots, elongated conducting cells can be seen developing in the center of the root.

Stems

The stem or **shoot** of a flowering plant has its origin in that part of the embryo within the seed known as the epicotyl, which is a continuation of the hypocotyl (Fig. 11-6). The epicotyl is a cylindrical mass of meristematic tissue that is narrower at its apex or stem tip. Frequently, as in the bean embryo, a pair of tiny leaves can be seen at the tip of the young stem (Fig. 11-1). When the seed germinates, the root emerges first so that it can absorb the water and nutrients necessary for further development before the food stored in the seed is entirely consumed. The young stem emerges from the seed to grow upward

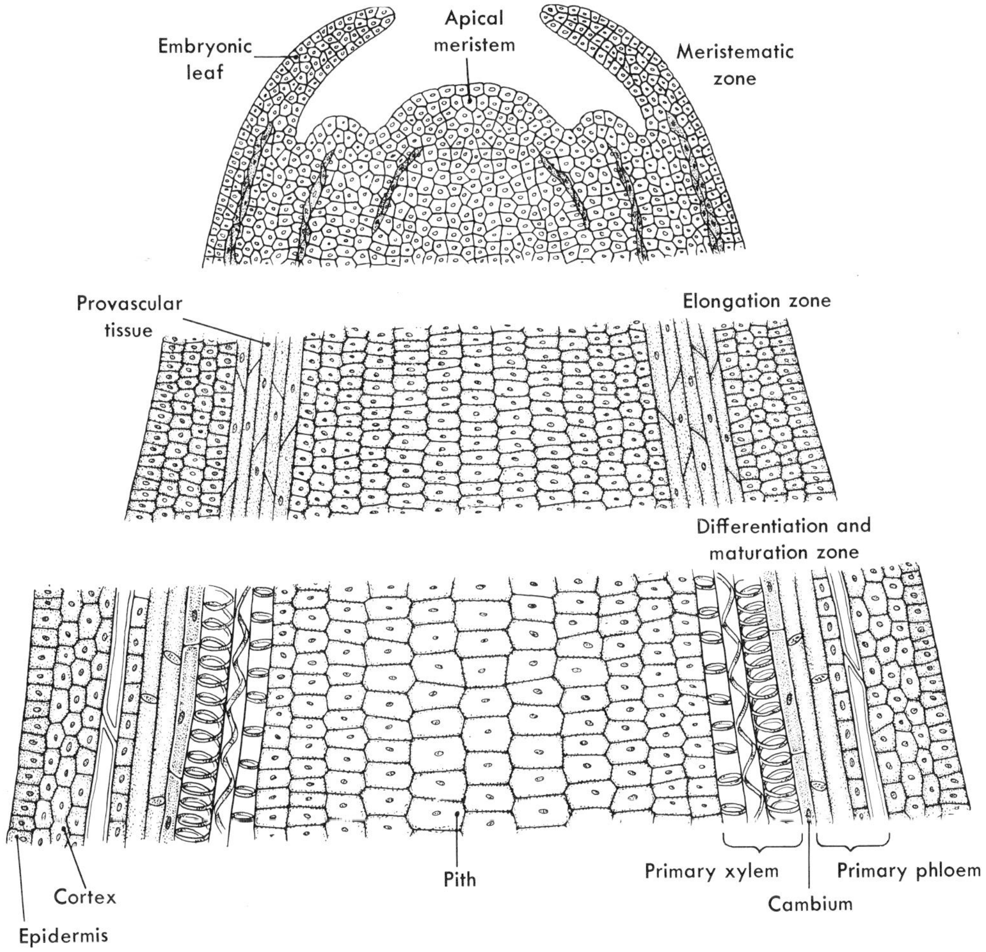

Fig. 11-21

Longitudinal section through the tip of stem to show regions of growth. (Redrawn from Wilson, C. L., and W. E. Loomis. 1957. Botany. Rev. ed. The Dryden Press, New York.)

(Fig. 11-1) because most stems react negatively to gravity.

The growth of stems, like that of roots, is unlimited. However, the length of a stem, when growth is taking place, is greater than that of a root. Furthermore, the arrangement of the permanent tissues is different. The regions in the growing point of a stem are very much like those of a root, except that a stem tip has no protective cap of cells comparable to the root cap; instead, the stem is covered by the leaf primordium. At the tip of the growing stem is a peak of actively dividing cells called the apical meristem (Fig. 11-21), the cells of which are thin walled and many in active mitosis. Immediately below this are lateral outgrowths or leaf primordia that also consist of undifferentiated embryonic cells. The youngest and smallest leaf primordia are nearest the apex of the stem; the oldest and largest are near the base of the stem tip. Behind the apical meristem is the **zone of elongation** where certain cells are undergoing some modification but where most are increasing in size. This region may extend for several inches behind a terminal bud. It merges into the **zone of differentiation,** a region in which differentiation occurs more rapidly in leaves and internodes of the stems than at the nodes. It is in this region that the primary permanent tissues are being formed (Figs. 11-8, 11-21, and 11-22). Three kinds of primary tissues can be initially recognized as developing from the undifferentiated cells or **promeristem** at the tip of the stem: (1) protoderm, a layer of surface cells, (2) procambium, strands of cells that retain the capacity for division, and (3) **ground meristem** (Fig. 11-8), cells that surround the procambial strands. The protoderm is a single layer of cells on the outer surface of the developing stem, which will form the future epidermis. Procambial strands become xylem and phloem, but a single layer of this tissue may retain the ability to produce new xylem and phloem. This layer is called the

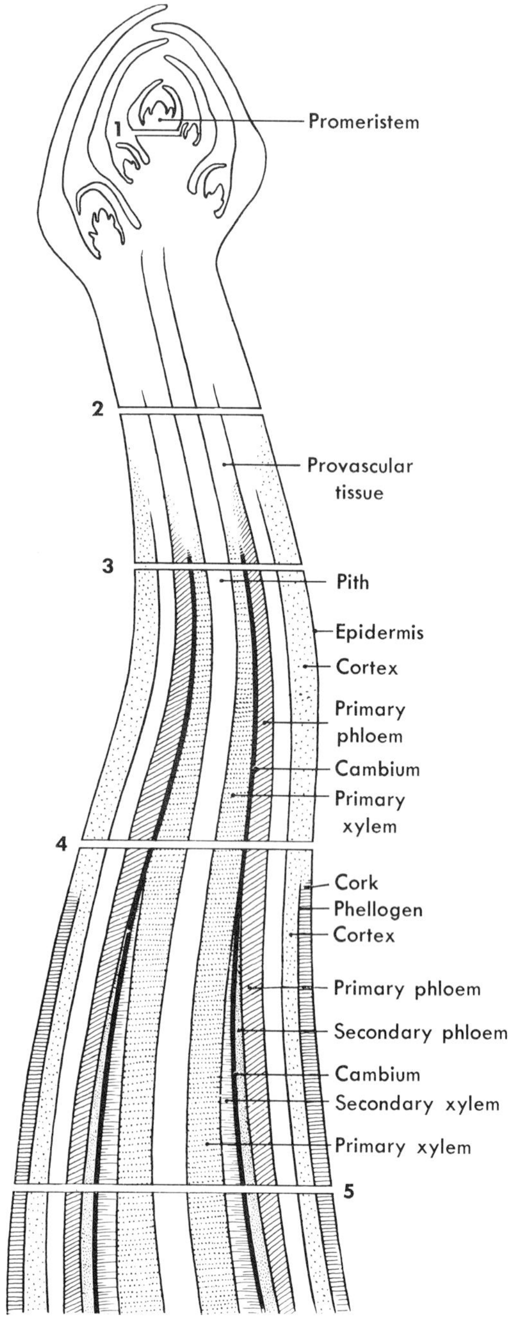

Fig. 11-22
Longitudinal section of growing stem. (Numbers refer to sections shown in Fig. 11-8.)

cambium (vascular). It functions in stems as it does in roots—to produce secondary xylem and phloem—thereby being responsible for the increase in the diameter of the stem. The ground meristem develops into a number of layers of cells located between the epidermis and the phloem. These layers of cells are called the cortex. The ground meristem is also responsible for the formation of a group of cells in the center of the stem that becomes completely surrounded by xylem. This central group of cells is known as **pith** and may be connected with the cortex by strands of cells between the vascular tissue (xylem and phloem). The strands are called **pith rays** or **medullary rays.**

The epidermis is usually one layer of cells in thickness. The stem epidermis may last throughout the life of the plant, or it may be replaced by other protective tissues, such as cork. The epidermis consists of relatively unspecialized cells with living protoplast and plastids but rarely any chlorophyll. The outer epidermal cell wall is usually thicker than the inner one. The most characteristic feature of the epidermis is the presence of **cutin,** a waxlike substance that forms as a separate layer on the outer wall of the cell, the cuticle. The cuticle appears to be highly impenetrable and thereby protects the stem against excessive loss of water.

The cortex of stems is largely composed of parenchyma cells, which occasionally may have chloroplasts. Intercellular spaces are prominent. The outer layers frequently contain collenchyma cells in groups or in a more or less continuous ring. The delimitation of the cortex from the vascular region is an important morphological criterion in identifying both stems and roots. In roots the innermost layer of cells of the cortex is called an endodermis, which only rarely is found in stems. The stems of gymnosperms and angiosperms lack a morphologically differentiated endodermis, but lower vascular plants may have one. In some young stems the innermost layer of the cortex contains large amounts of starch and becomes recognized as a **starch sheath.** The innermost group of cells in stems is the pith, which is composed of parenchyma cells. In many stems the central pith is destroyed during growth. In some stems this destruction is limited to internodes, but the nodes retain their pith as a diaphragm (for example, corn and bamboo stems). The cells of the pith have the prominent intercellular spaces (Fig. 11-19) characteristic of parenchyma.

The **vascular tissue,** composed of xylem and phloem, commonly appears as a cylinder between the cortex on the outside and the pith in the center of the stem. In young dicots the vascular tissue is separated by interfascicular regions into smaller or larger units commonly referred to as **vascular bundles.** The xylem and phloem ordinarily occur in a collateral arrangement, with the phloem located outside the xylem.

The vascular system of the stem is always continuous with that of the leaf. The vascular connection of the leaf and stem at the node may be one or more strands of vascular tissue that lead from the stem to the leaf. This close association of the stem with the leaves makes this part of the plant more complex than the root. However, like the root, the stem consists of three tissue systems: the epidermal, the ground meristem, and the vascular. The variation of these primary tissues in stems of different species or in the larger groups of plants is based chiefly on differences in the relative distribution of these fundamental tissues. The stems of gymnosperms and woody dicots have a vascular system that appears as a cylinder, with cortex on the outside and pith in the center. In monocots and herbaceous dicots the vascular tissue does not appear as a cylinder. It appears as a single ring or as more than one ring (Fig. 11-19). If it is a single ring, the bundles are widely scattered (Fig. 11-19).

Secondary growth increases the amount

of vascular tissue in a stem. However, it contributes only to the thickness of the axis, which is characteristic of gymnosperms and woody dicots. On the other hand, most monocots and herbaceous dicots ordinarily have no secondary growth, except that herbaceous dicots have limited lateral growth and a cambium. If secondary growth does occur in monocots, it is of a special type. The term secondary growth includes the formation not only of secondary vascular tissues but also of the **periderm,** which is a protective tissue that replaces the epidermis of many woody plants.

In a cross section of a young woody stem, the vascular bundles can be seen arranged in groups around the stem. The vascular cambium will eventually form a complete cylinder. It arises in part from the procambium within the vascular bundles, where it is known as the **fascicular cambium,** and in part from the parenchyma between the bundles, where it is called **interfascicular cambium.** Therefore, in the most common type of secondary growth the vascular cambium, while becoming a complete cylinder, will produce continuous cylinders of secondary xylem and secondary phloem to ultimately increase the diameter of the stem.

The development of secondary vascular tissues between the primary xylem and the primary phloem exerts a great amount of stress on the internal tissues of the stem, particularly on those located outside the cambium. As the cambium produces secondary vascular tissues, the primary phloem becomes isolated from the primary xylem. The primary xylem and the pith are enveloped by the secondary xylem so that in an older stem the small amount of remaining pith may eventually lose its identity.

Slowly the primary phloem is pushed outward and becomes nonconducting because its thin-walled elements are thoroughly crushed. The cortex may persist for many years, increasing in circumference through cell enlargement and division. The increase in circumference may also be aided by the growth of the parenchyma ray cells that, in some stems, can be seen expanding or fanning out into the cortex. The epidermis may persist; its cells enlarge and divide as the diameter of the stem increases.

In some stems a protective tissue of secondary origin called periderm may develop. This eventually replaces the epidermis in stems (and roots) where secondary growth takes place. The best examples of periderm development can be seen in gymnosperms and woody dicots. However, it does occur in the oldest parts of the stem (and roots) of herbaceous dicots, as well as in some monocots.

Periderm must be distinguished from the bark of a stem. **Bark** is used most appropriately to designate all those tissues outside the vascular cambium (for example, phloem, cortex, periderm, and epidermis when present). The development of the periderm in the cortex isolates its outer cells so that they become nonliving. Consequently, it is convenient to differentiate between the outer, nonliving back and the inner, living bark of a stem.

The first periderm commonly appears during the first year's growth of a stem (or root). In some plants only one periderm developes, whereas in others several may be found. The first periderm usually originates in the outermost layer of the cortex just under the epidermis. Subsequent layers of periderm arise in successively deeper layers of the cortex beneath the first periderm. Periderm consists of a layer of meristematic cells called the cork cambium or **phellogen,** the cells of which produce the cork or phellem, a protective tissue of dead and suberized cells. They are formed by the phellogen and the **phelloderm,** which is a living tissue of parenchyma cells formed inwardly by the meristem. The presence of cork outside the cortex accounts for the death of any cells outside it.

The vascular cambium of stems is inactive during the winter months in temperate climates. In the spring it becomes active again to produce new xylem and phloem. This activity is determined by the seasonal activity of the vascular cambium and has been found to be closely correlated with the growth of buds in the spring. This relationship was easy to comprehend when it was learned that the auxin produced by buds was transferred toward the base of the stem, thereby inducing its growth.

Seasonal growth of xylem is more readily demonstrated than that of phloem. A larger number of xylem cells (wood) is produced in late summer than in spring. Cells are smaller and more numerous in summer wood, whereas in spring wood the cells are fewer and larger. Hence it is easy to differentiate one year's growth from another, since the differences in xylem growth appear as annual rings in a cross section of a stem (Fig. 11-8). The oldest part of the stem is nearer the center and the youngest nearer the outside. The width of the annual ring will vary from year to year, giving some indication of growth or the environmental conditions existing at the time of growth. This fact has been used to determine changes in climate during the years before weather records were kept. Even the climate during prehistoric times can be determined in this way by examining the annual rings of petrified wood.

Since stems differ in their primary and seconadry structure, it becomes convenient to distinguish between types of stems. Therefore it is customary to identify them as gymnosperms, woody and herbaceous dicots, monocots, and vines.

The stem of the pine is a good example of a gymnosperm stem. Initially the vascular tissue is in discrete bundles, which later will form a continuous cylinder of secondary xylem and secondary phloem. The rings of annual growth are easily recognized in an older stem. Some pith remains in the center, the cortex and xylem contain resin ducts, and the first periderm arises beneath the epidermis and is not replaced by deeper ones for many years.

In all woody dicots the secondary vascular tissues form a continuous cylinder. Rings of annual growth are present through which many wide rays of parenchyma (vascular rays) can be found, as well as narrow rays. The secondary phloem appears as alternate bands of fibers and bands of sieve tubes, companion cells, and parenchyma. The periderm originates just beneath the epidermis and persists for years, and the cortex remains intact. The pith is composed of parenchyma cells in which cavities appear, and the pith cells next to the xylem act as storage tissue.

A herbaceous dicot (Fig. 11-19, *B*) resembles a young woody stem in which the vascular tissue occurs as a ring of separate vascular bundles. However, in some stems a continuous and narrow cylinder may form, which in turn may exhibit some secondary growth. The buttercup genus *Ranunculus* has little, if any, secondary growth, and there is no perioderm. The pith, in which a large central cavity may be present, makes up the major part of the diameter of the mature stem.

A typical monocot has widely spaced vascular bundles that are not restricted to a single ring (Fig. 11-19) but are scattered throughout. Other monocots, such as *Avena sativa* (oats), have bundles in only two rings. In both these monocots there is a peripheral ring of sclerenchyma cells near the epidermis, with which some vascular bundles are associated. Stomata appear in the epidermis; the pith may be broken down to form large or small spaces, and all the vascular bundles are enclosed in sheaths of sclerenchyma. Although monocots lack secondary growth, they may develop thick stems through primary growth, as in the palms.

Knowledges of the basic principles of stem structure and physiology is indispensable to practical botanists such as horticulturists, floriculturists, and orchardists. One of its applications is in the pruning of trees and shrubs. The chief purpose of pruning is to remove dead and diseased branches to prevent the entry of parasites and their subsequent spread to the main trunks. In addition, pruning is necessary to give the desired shape to ornamental plants, to increase side branches, or to induce flowering. Unless it is properly done, it may result in improper healing of the cut surface and encourage rot or parasitic attack.

Grafting is another practical application of stem structure and physiology. It is a process in which two freshly cut stem surfaces are bound together so that the cambium layers of both are in contact. The usual type of grafting involve the union of a base stem having a root, the **stock,** and a cutting, the **scion** (Fig. 11-23). There are many ways to do grafting. Depending upon the way that scion and stock are united, the types of grafting are called saddle grafting, cleft grafting, tongue grafting, or side grafting (Fig. 11-24).

Grafting is of use in propagating seedless varieties of plants such as navel oranges and hybrid plants, in inducing more rapid fruiting since plants may produce fruits earlier if grafted than if propagated from seeds, in accommodating a species of plant to a new environment, and in checking or eliminating parasites (particularly of roots).

Plants are sometimes propagated through stems by a process called **layering.** A stem is bent down and buried partly under the ground. That part of the stem under the ground will develop adventitious roots; after it has started to grow, it can be separated from the parent plant (Fig. 11-25). Another example of plant propagation is the use of cuttings. A short stem is removed from the parent plant and its cut end placed in moist sand or soil (or in water for some plants). It will usually develop adventitious roots at this lower end. Many horticultural plants are propagated in this way.

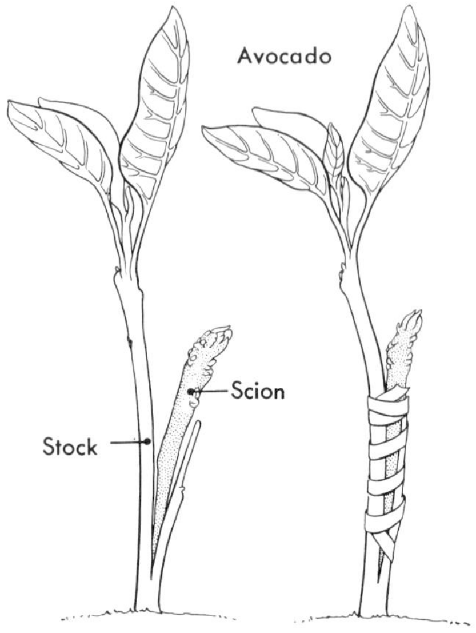

Fig. 11-23
Method of side grafting.

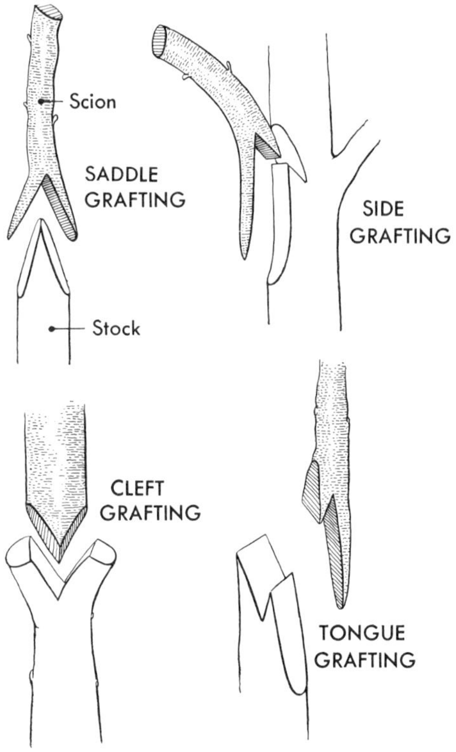

Fig. 11-24
Diagram of several methods of grafting.

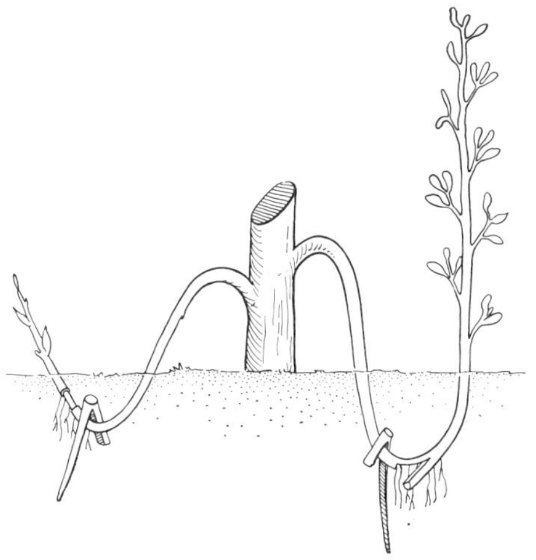

Fig. 11-25
Vegetative propagation of plants by the layering method (see text).

Physiology of stems. One of the functions of stems is the conduction of materials upward to the leaves, downward to the roots, and transversely. The **translocation** of materials in stems is not completely understood, since the mechanisms behind the rise of sap in a stem and the path it follows are very complex.

The term **sap,** as used by a botanist, is usually applied to the water and minerals that move upward in the younger annual rings of woody plants, particularly the xylem. The plant physiologist can prove by experiment that the water and ions moving upward in the plant are largely in the xylem. However, to explain the cause of this ascent is not easy. It is due partly to transpiration and partly to root pressure. It has been suggested that **capillarity** is involved in the rise of sap, as well as are the rhythmic expansions and contractions of the living parenchyma cells associated with xylem. Most recently, a more satisfactory explanation to account for the rise of liquids in a stem is based on the idea that liquids are pulled up rather than pushed up from below. The pulling action, due to evaporation of water from the leaf cells, and the cohesive action of water appear to be able to account for the rise of liquids to great heights in trees. Thus the **cohesion** theory, although accepted by most botanists, depends chiefly upon several physical forces to explain the rise of sap. However, it is likely that evaporation and cohesion are not the only forces, but that root pressure and capillarity play an important part also.

The food manufactured by the leaves is transported through the living sieve tubes and, to a lesser degree, through the companion cells of the phloem. Therefore, translocation of food is chiefly downward through the phloem in summer. The vascular rays (Fig. 11-12) transport dissolved materials and water transversely across a stem by diffusion and by the protoplasmic streaming of the ray cells. Transverse conduction is extremely important because it makes possible the movement of foods from the sieve tubes into the dividing cells of the cambium. Carbon dioxide moves outward through the lenticels and oxygen moves inward, thereby aiding respiration of cambium cells and other living cells in both xylem and phloem.

Leaves

The leaf develops during localized meristematic activity on the side of the apex of a stem. As a leaf progresses in its development, it grows outward from the leaf base, revealing the final shape and form characteristic of the species.

Leaf tissues and structures. The blade of a leaf is protected by an upper and a lower layer of cells called the epidermis. Between these two layers of protective cells is the **mesophyll.** It consists of several layers of cells containing chloroplasts. The mesophyll is transversed by numerous vascular bundles (veins) of different sizes (Fig. 11-26). The monocot leaf vein has a greater amount of mechanical tissue (sclerenchyma fibers) and xylem fibers than does the vein of the dicot leaf (Fig. 11-27).

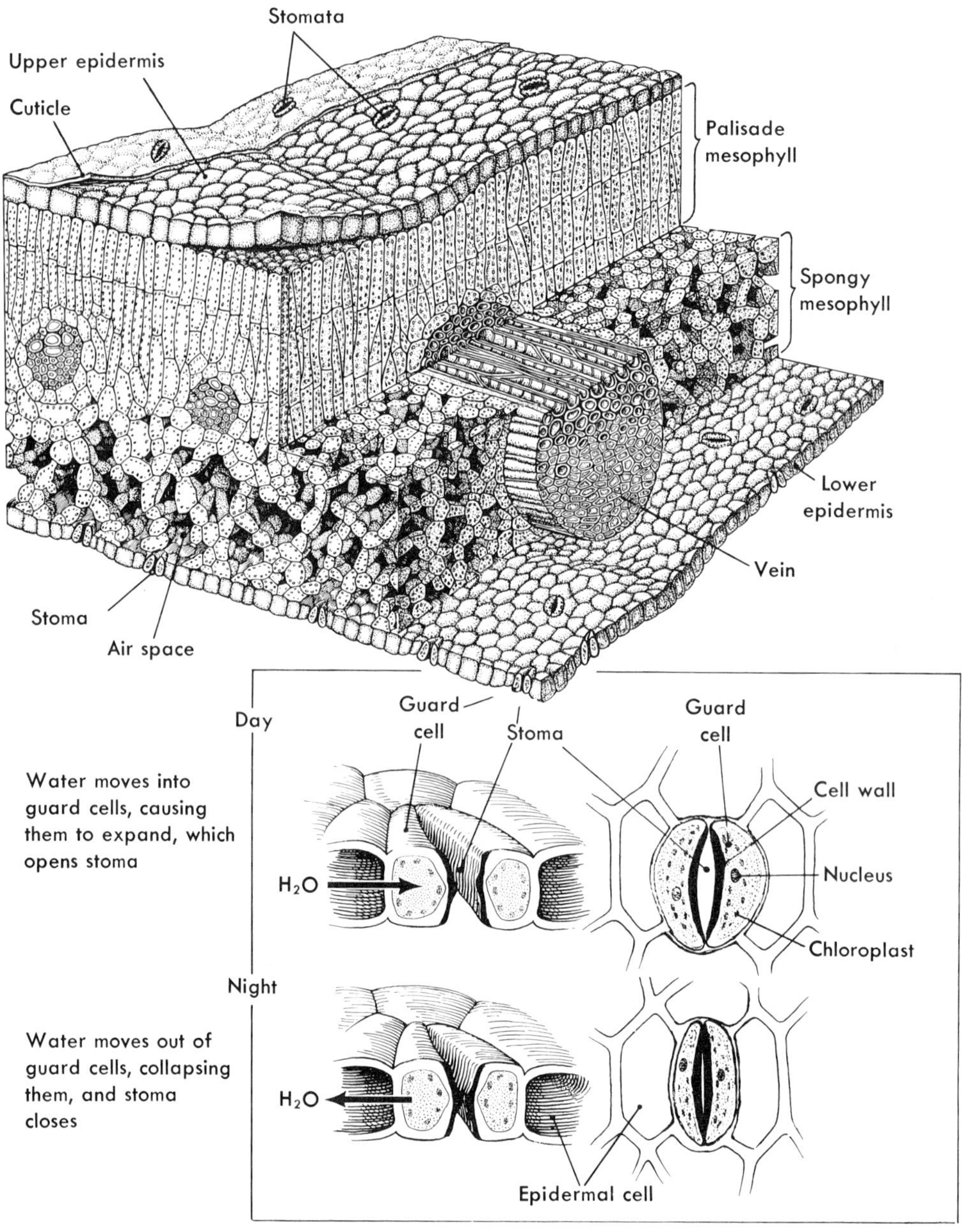

Fig. 11-26

Details of the cells and tissues of a leaf. Top, cross section of a leaf showing structural details; bottom, an enlargement of the stomata showing functional details.

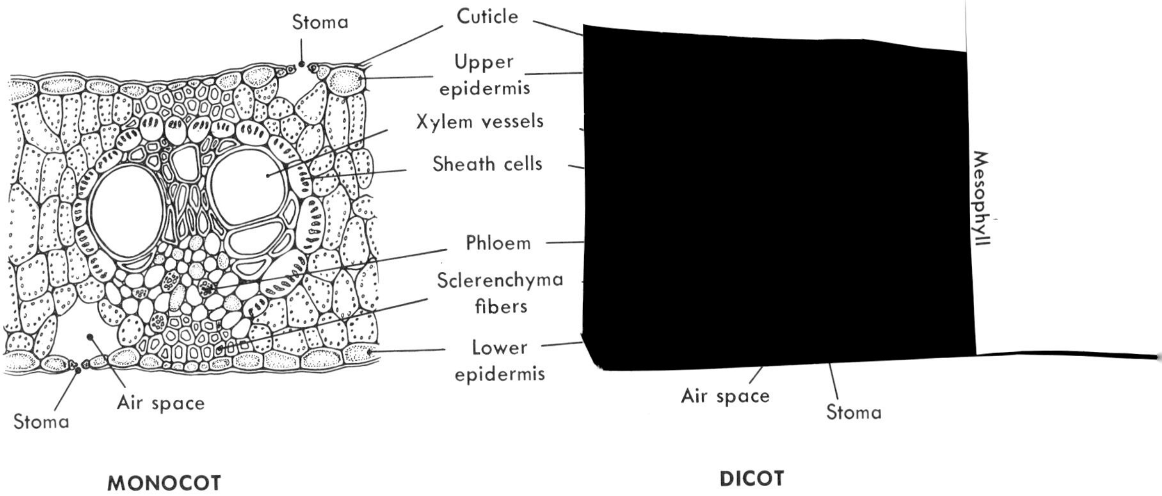

Fig. 11-27
Cross sections of typical leaves, monocot and dicot, showing differences between the conducting (vein) tissue and the mechanical tissue surrounding them.

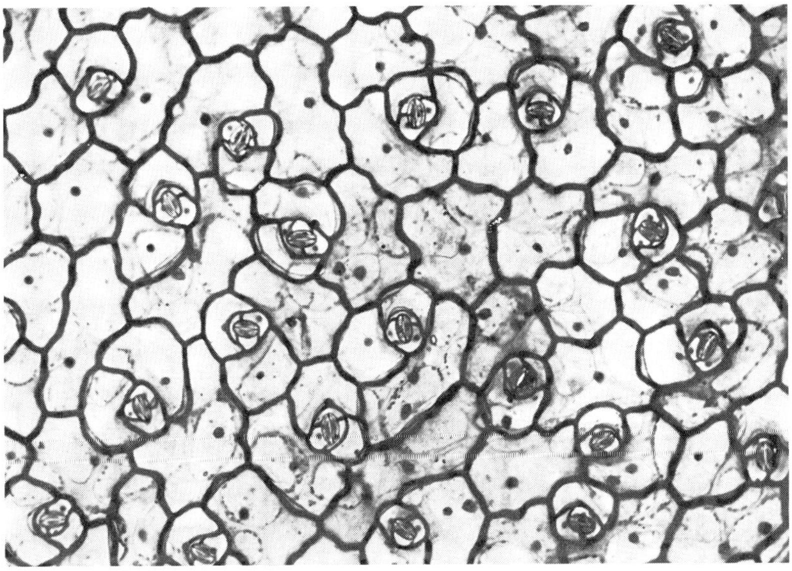

Fig. 11-28
Surface view of the lower epidermis of a leaf to show the guard cells surrounding the stomata. (Courtesy General Biological Supply House, Inc., Chicago, Ill.)

The epidermis of a leaf protects it from excessive loss of water, mechanical injury, and parasitic infection. Epidermal cells are usually covered with a definite cuticle that aids in [redacted] ces or chloroplasts, and include guard cells. Scattered among the epidermal cells of most leaves are numerous openings called **stomata** (Fig. 11-28). Each stoma is surrounded by two specialized cells known as **guard cells** (Fig. 11-26), which regulate the size of the opening. Frequently the cells adjoining the guard cells are ordinary epidermis, but in some leaves they may be differentiated into supporting cells. Although the term stoma is sometimes used to apply to the entire stomatal apparatus, it is more correctly restricted to the opening.

The guard cells are usually kidney shaped, and their concave surfaces form the intercellular space or stoma. When guard cells are turgid (Fig. 11-26), they are more bowed apart, causing the opening to enlarge; when they lose their turgidity (Fig. 11-26), they become more closely approximated throughout their length, and the opening closes. The mechanism of the opening and closing of the stomata is discussed more fully later. The structure of the guard cells and the presence or absence of supporting cells, as well as their arrangement, are two of many criteria used by the taxonomist to aid him in classifying plants.

Usually stomata are more abundant on the lower surface of a leaf than on its upper surface. However, in some leaves they are equally numerous on both surfaces. In other plants the upper surface may lack stomata entirely, whereas floating leaves such as those of water lilies have stomata only on the upper surfaces. In most dicot leaves the stomata are scattered uniformly over the leaf surface, whereas in monocots they are arranged longitudinally along the veins. The number of stomata per unit area varies in different species from as low as 10 to as high as 1,300 per square millimeter. The larger the stomata, the fewer they are in number, and vice versa. The stoma, relatively small, varies from 7 to 40 μ in length and from 3 to 12 μ in width when completely open. It is an advantage for a leaf to have numerous small stomata rather than a few large ones, since diffusion of gases through many small openings is much greater than through a few large ones.

The epidermis may develop a variety of appendages called **trichomes** (Fig. 11-29). These may include secretory (glandular) and nonsecretory hair, scales, and papillae. Trichomes are useful as morphological structures in classification.

Between the two layers of the epidermis are a number of loosely arranged cells that comprise the mesophyll (Fig. 11-26) of the leaf. This layer varies in thickness with the thickness of the leaf. The cells of the mesophyll are primarily parenchyma cells containing chloroplasts (chlorenchyma). The mesophyll can be differentiated into two easily recognizable layers; one is called the **palisade layer** and the other the **spongy layer.** The palisade layer ordinarily will consist of one or two rows of elongate cells just beneath the epidermis. If they are present just beneath the upper epidermis and not the lower epidermis, they cause the leaf to appear darker green on its upper surface and lighter green on its lower surface. If a palisade layer is present on both surfaces of the leaf, it will then be the same color on both upper and lower surfaces. A leaf, the blade of which is broad and structurally different on its upper and lower surfaces, is called a dorsiventral leaf. If it is structurally the same on both sides, it is known as an isobilateral leaf. The cells of the palisade layer are closely set together and contain many chloroplasts.

The spongy layer of the mesophyll is composed of irregularly shaped parenchyma cells containing fewer chloroplasts than those of the palisade layer. There are many intercellular air spaces in the spongy layer. Therefore, in a dorsiventral leaf the lower epidermis is easier to strip off than the upper. Near each stoma is an extra large air space. The cells of the spongy layer are not arranged in rows, and it appears porous. Mesophyll may vary in different species of plants; for example, monocots (Fig. 11-27) have a more compact mesophyll layer than do dicots.

Since the primary function of leaves is photosynthesis, it appears that the mesophyll is well adapted for this function. The densest tissue is near the upper surface in a dorsiventral leaf and contains the greatest concentration of chloroplasts. This is the part that receives most of the light. The air spaces are largest near stoma that connect to the atmosphere in order that carbon dioxide can easily enter the intercellular spaces and be rapidly distributed throughout the leaf. The compact palisade layers somewhat retard the loss of water by evaporation because only a small area of their surface is exposed to air.

Many interconnected and branching veins are present in the leaves of flowering plants. A typical vein (Fig. 11-27) is composed of xylem and phloem surrounded by a layer of sheath cells. The xylem is closer to the upper surface and the phloem is closer to the lower surface of the leaf. The smaller veins consist of only primary vascular tissues, whereas the larger veins of dicots may show secondary growth. The amount of xylem and phloem decreases with the size of the vein; for example, the smallest veins may have only a single tracheid as their xylem tissue, and a modified parenchyma cell will represent the phloem.

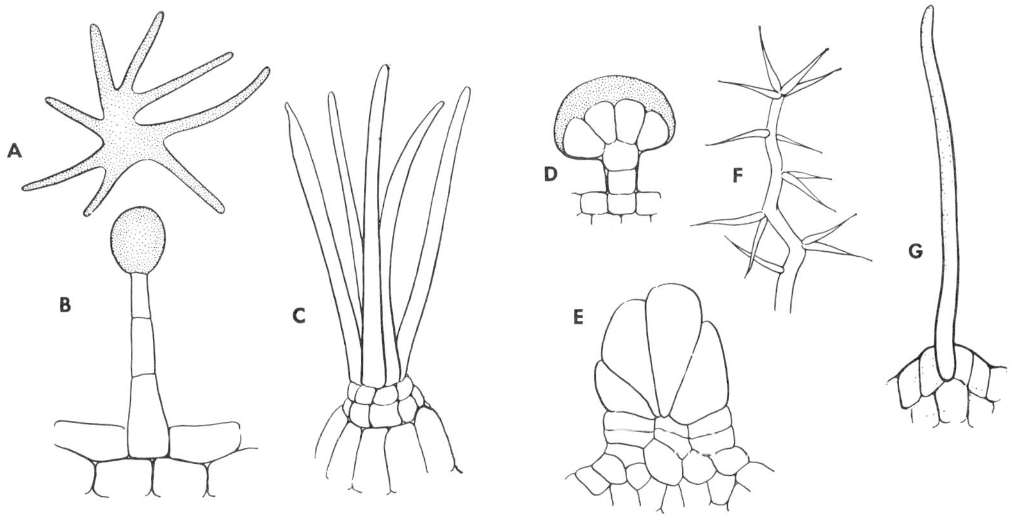

Fig. 11-29
Epidermal appendages and trichomes of leaves. **A,** Stellate hairs in *Alyssum* sp. **B,** Glandular hairs in *Pelargonium* sp. **C,** Floccose hairs in *Malva* sp. **D,** Short glandular hairs in *Lavandula* sp. **E,** Glandular hairs in *Solanum* sp. **F,** Urticating hairs in *Verbascum* sp. **G,** Stinging hair in *Cistus* sp.

The largest vein in a net-veined leaf is usually located in the center and is called the **midrib.** Large veins such as the midrib are embedded in a mass of compact parenchyma cells, rather than in a sheath. To strengthen the vein, collenchyma tissue may be present, as well as sclerenchyma tissue. It is these that make the vein stand out on the leaf.

Physiology of leaves. Primarily leaves are organs of photosynthesis. At the same time, they provide the site for **transpiration** or the loss of water through evaporation. It is this effect that makes a forested area cooler in the summer than is an open area with little vegetation. Somes leaves store water and food. All foods used by the plant originate here in most cases, and are either stored here or moved into storage organs. As the shapes and size of leaves vary from species to species, they provide one of the characteristic features of their kind.

Passive absorption (passive transportation) of water may occur through the root cells as a result of forces in the shoot system of the plant—the leaves. As the water evaporates from the cells between the upper and lower epidermis of leaves, or as water is used in their growth processes, or both, an increasing amount of water is withdrawn from the xylem tubes. This increased movement of water (Fig. 11-30) increases the diffusion pressure deficit in the cells, which in turn reduces the deficit in the leaves, stems, and root cells. The water is more or less pulled through the plant by transpiration. In other words, it is the result of evaporation of water from leaves. Almost 95% of all the water taken in by a plant is lost through transpiration. The remaining 5% is used in cell growth, photosynthesis, or other life processes.

In transpiration the loss of water vapor is through the stomata of leaves and young stems. When the vapor pressure is less in the atmosphere than it is in the spaces of the leaves, water molecules diffuse into the atmosphere. This loss of water in the leaf spaces causes water to diffuse from the mesophyll cells, which in turn causes diffusion from the xylem tubes of the veins of the leaf that extend down into the roots. The cohesive force of the narrow column of water extending from the leaf to the root is so great that there is actually a pulling force on the column. This force originating in the leaves from transpiration is responsible for the passive absorption of water from the soil. This, together with active absorption, is enough to cause water to move to the tops of our tallest trees.

The rate of transpiration is affected by many factors. The energy of the sun will affect the internal temperature of the leaves, and it is not unusual for a leaf to be 10 degrees warmer than the air around it. This increased temperature as a result of radiation from the sun will cause greater vaporization of the water in the leaf. Therefore transpiration will be greater on sunny days than on cloudy days. Circulation of air also affects transpiration. When air movement increases up to approximately 15 miles per hour, the rate of transpiration increases. When the circulation of air is above that rate, transpiration will be reduced because the temperature of the leaf is reduced. Another factor affecting the rate of transpiration is the amount of moisture in the soil. Decreasing the amount of water in the soil will decrease transpiration.

In addition to all the factors mentioned that affect transpiration, the principal one is the opening and closing of the stomata and their number and position on the leaf. Stomata are found on the upper surface of a leaf only, on the lower surface only, or on both surfaces. Most trees have leaves in which the stomata are found only in the lower epidermis and in large numbers. For example, more than 70,000 stomata per square centimeter can be found on the lower epidermis of an oak leaf.

Each stoma or opening is surrounded by two guard cells. Although the epidermal cells are normally without chloroplasts, they

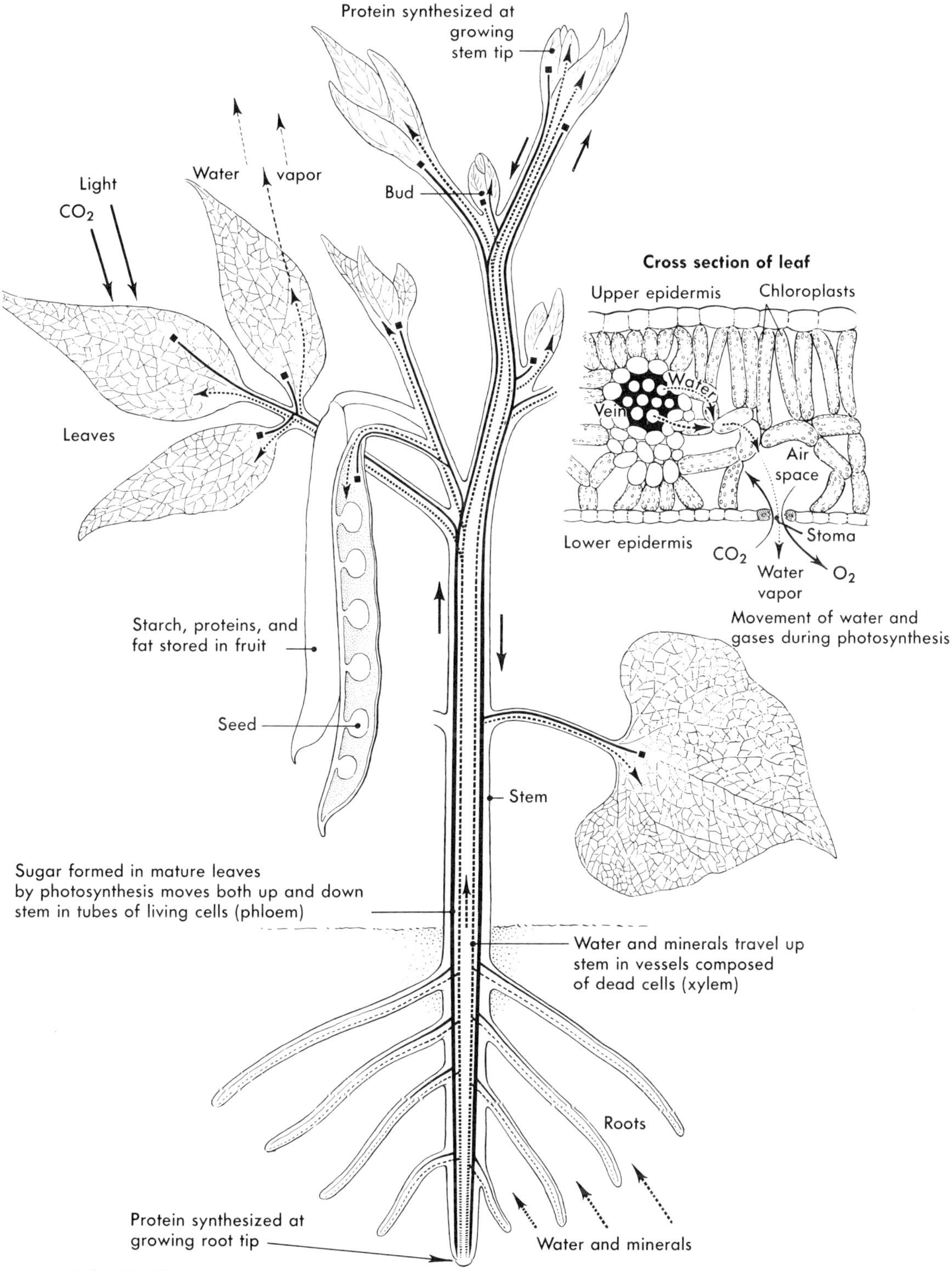

Fig. 11-30

Path of circulation of gases and liquids in plants. Upper right, details of interior of a leaf.

are present in the guard cells. Several explanations are given for the manner in which stomata are opened or closed. The walls of the guard cells are thicker toward the center of the stoma. Turgor pressure, as a result of osmosis, may increase and cause the guard cells to become more turgid, thus enlarging the stomatal opening. When fully distended, each guard cell is C-shaped. When the guard cells lose turgidity, their centers tend to meet, and the size of the opening is reduced. Pressure exerted by the surrounding epidermal cells may also aid in pushing the relaxed guard cells closer together.

One explanation for stomatal behavior may be that as night descends, the sugar manufactured by the chloroplast-bearing guard cells is converted to starch. Since starch is insoluble in water, the water content of the guard cell is almost pure, and the diffusion pressure will be greater in the guard cell than in the surrounding epidermal cells. Water passes out of the guard cells by osmosis, the guard cells lose turgidity, and the size of the stoma is reduced. In sunlight, starch is hydrolyzed to sugar, and the diffusion of water reverses, moving from the epidermal cells back to the guard cells. As a result, the guard cells become more turgid, and the stoma is opened (Fig. 11-26). It has also been suggested that an accumulation of carbon dioxide in the guard cell may cause it to lose turgidity. Whatever the cause, it is apparent that turgidity of the guard cells widens the stoma opening and loss of turgidity closes it. The longer and wider the stomata are open, the greater will be the rate of transpiration. The number of stomata makes the surface of a leaf resemble a sievelike membrane. It is known that evaporation from such a surface is many times greater than that from a perfectly flat surface. Evaporation of water from the leaf surface is less during the night than during the daylight hours for the reasons already given. At night, since absorption exceeds transpiration, the turgidity of the plant is replaced. When this does not occur, the plant becomes permanently wilted. Excessive soil, water, or salts can be responsible for preventing the uptake of water to balance the loss of water by transpiration. Why this occurs is not fully understood. One reason may be that a decrease in permeability of a root cell to water could be caused by a lack of oxygen and, hence, reduced respiration.

Factors such as the dust in the air, leaf damage, and available oxygen, influence the rate of food synthesis, the primary function of the leaf. Carbon dioxide intake may be reduced if the stomata are partially closed by dust particles; insects may feed on leaves thereby reducing leaf surface and resulting in lowered photosynthetic rate; fungus growth and disease may limit photosynthesis; poor aeration of roots, causing interference with movement of water and essential minerals to leaves, will reduce photosynthesis. The lack of oxygen to roots will indirectly affect the synthesis of food by light.

Finally, the structure of the leaf itself influences the rate of photosynthesis. Although most carbon dioxide enters a leaf through the stomata, some can enter through the epidermis if the cutin is not too thick. In some plants the stomata may be closed during hot, dry daylight hours and open only at night, whereas in others they may be open only during early morning hours or, as in water plants, open all day. However, it is doubtful whether stomata are ever closed completely to shut out all carbon dioxide. At any rate, the behavior of stomata greatly influences the rate of photosynthesis by regulating the passage of carbon dioxide. Other factors worthy of mention are the number of stomata per square centimeter of leaf and their location, the thickness of the leaf, the number of chloroplasts per cell, the size and number of intercellular spaces, and, as in all physiological processes in living systems, the enzyme activity. Enzyme activity is affected by temperature, water, acidity, and light. Finally, the accumulation of waste products plays an important role in photosynthesis.

SPECIAL TERMS USED IN CHAPTER 11

active absorption
adventitious root
apical meristem
bark
cambium
capillarity
chlorenchyma
chlorosis
cohesion
coleoptile
collenchyma
companion cells
complex permanent tissue
complex tissue
cork
cork cambium
cortex
cotyledons
cutin
differentiation
embryonic region
endodermis
epicotyl
epidermis
fascicular cambium
geotropism
germination
ground meristem
growth
guard cells
hydathodes
hydroponics
hypocotyl
interfascicular cambium
lignin
layering
lateral root
medullary ray
meristem
mesophyll
metaxylem
midrib
organ
organ system
palisade layer
parenchyma
passive absorption
pericyle
periderm
phelloderm
phellogen
phloem
pith
pith ray
plumule
primary permanent tissue
primary root
promeristem
protostele
protoxylem
radicle
region of elongation
region of maturation
root
root hair
root hair region
root pressure
root tip
root cap
sap
scion
sclerenchyma
secondary growth
secondary permanent tissue
secondary root
shoot
sieve tube
simple tissue
spongy layer
starch sheath
stele
stock
stoma
stone cell
suberin
tissue
tracheid
translocation
transpiration
trichome
vascular bundle
vascular cambium
vascular ray
vascular tissue
vessel
xylem
zone of differentiation
zone of elongation

QUESTIONS TO CONSIDER

1. What is the function of a seed? What are the factors needed for germination?
2. What happens to the cotyledons of a bean seed? How does this differ from that of a corn seed?
3. Why are the shoots of grasses (like corn) protected when the young stem emerges from the ground and those of beans are not?
4. What is the function of root hairs? Of secondary roots? Of prop roots?
5. How can a seed planted bottom side up grow properly? Does this happen in nature?
6. What are the principle functions of the stem? Of the leaves?
7. Describe the stem in cross section of: an herbaceous plant; a woody plant.
8. Describe the pathway of water through a plant; carbon dioxide; oxygen.
9. Describe growth in a woody plant.
10. What protects a plant?

DISCUSSION FOR LEARNING

1. What are some of the many factors that determine the shape and size of plants? Discuss both the individual variation and the features that differentiate the species.
2. Discuss some of the advantages to a plant of the following growth habits: tall tree; vine; broad shrub.
3. In this particular chapter we have dealt only with the morphology of the higher vascular (seed) plants. Since these plants all have similar parts we believe them to be homologous parts. Discuss the structure of some of the common plants about the home, for example, food plants, in terms of modifications of these parts from the standpoint of homology (not special adaptations, which will be discussed in a later chapter). For example, compare the parts of a carrot with those of a cabbage and point out the same parts in each.

Chapter 12

Growth and differentiation of flowers, fruits, and seeds

The flowering plants produce special stems whose leaves are modified for the production of seeds, such as the leaves of the pea plant pictured above. These modified leaves comprise the flower. Roots, stems, and leaves are strictly vegetative organs concerned with the nutrition and growth of the plant body. The **flower,** however, is a group of modified leaves specialized for reproduction and is responsible, either directly or indirectly, for bringing into existence new plants to perpetuate the species.

FLOWERS

The flower develops sex cells and accessory structures that indirectly contribute to the process of reproduction (life cycle 16).

A **complete flower** (Fig. 12-1) has four groups of parts arranged symmetrically and

FLOWER

Stigma, Style	Pistil (female) →	Gynoecium
Anther, Filament	Stamen (male) →	Androecium
Petals → Corolla		Perianth
Sepals → Calyx		Perianth
		Receptacle

in whorls at the end of an enlarged stem called the **receptacle.** The flower stalk is a stem that supports the floral parts in a position favorable for eventual pollination. When the stem bears a single flower or is

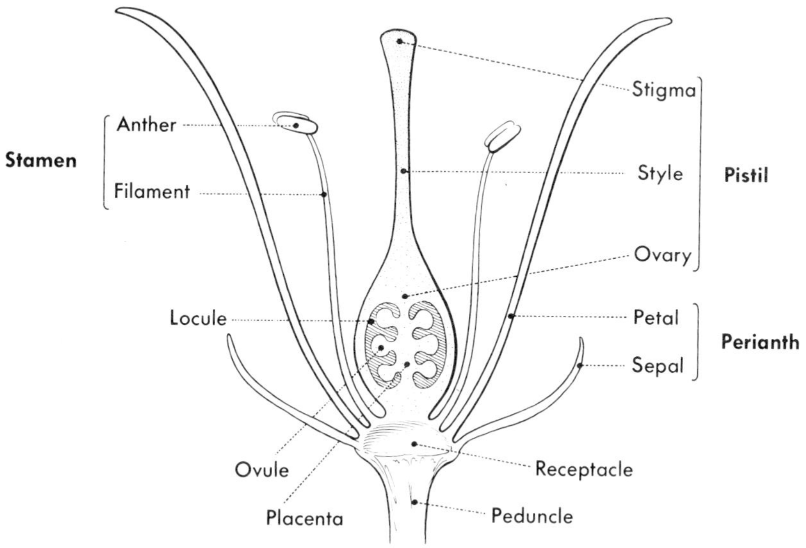

Fig. 12-1
Parts of a typical flower.

the main stalk of a cluster of flowers, it is known as a **peduncle.** In clusters the individual flower stalks are called **pedicels.** The flower develops as a branch from a bud in the axil of an ordinary leaf or in a much reduced leaf, a bract. The presence of four circles of parts makes up a complete flower, as illustrated in Fig. 12-1. The outermost circle is composed of individual leaves, which in most cases are green. These are the **sepals** and collectively are called the **calyx.** Their function is to protect the inner and more delicate parts of the flower as they develop from the bud. In some plants sepals may be colored like the petals. Inside the calyx are a group of more delicately textured and often brightly colored **petals.** In their attachment to the receptacle, petals usually alternate with sepals. Collectively, the petals are known as the **corolla.** The petals have attractive odors and colors and secrete a sugary solution, **nectar.** The glandular hairs at the base of the petal, called **nectaries,** produce this substance. Petal color ranges through the entire spectrum. Green is an uncommon floral color, but white, yellow, pink, red, and blue are common colors, and even black flowers are not unknown. The various pigments that give color to petals may be dissolved in the cell sap, such as the reds and blues of the anthocyanins, or they may be contained in special plastids in the cells, such as the yellow and orange colors (carotenes). When the plastids are colored other than with chlorophyll, they are known as chromoplasts. A petal appears white because the light is reflected from many air-filled spaces in loose cellular tissue. The **perianth** (Fig. 12-2) of the flower includes both the calyx and corolla.

The third and next innermost whorl of leaves in a flower consists of the stamens. All the stamens of the flower, collectively, are known as the **androecium. A stamen** is composed of a stalk or filament that supports at its terminal end an organ known as an **anther,** containing two **pollen sacs.** Within the pollen sacs are a large number of **pollen grains,** which are liberated when mature. Pollen grains vary in size and shape for each species of plant. They are readily identified as belonging to a particular kind of plant by a person who is trained in pollen morphology, a **palynologist.**

In the center of the flower is one to sev-

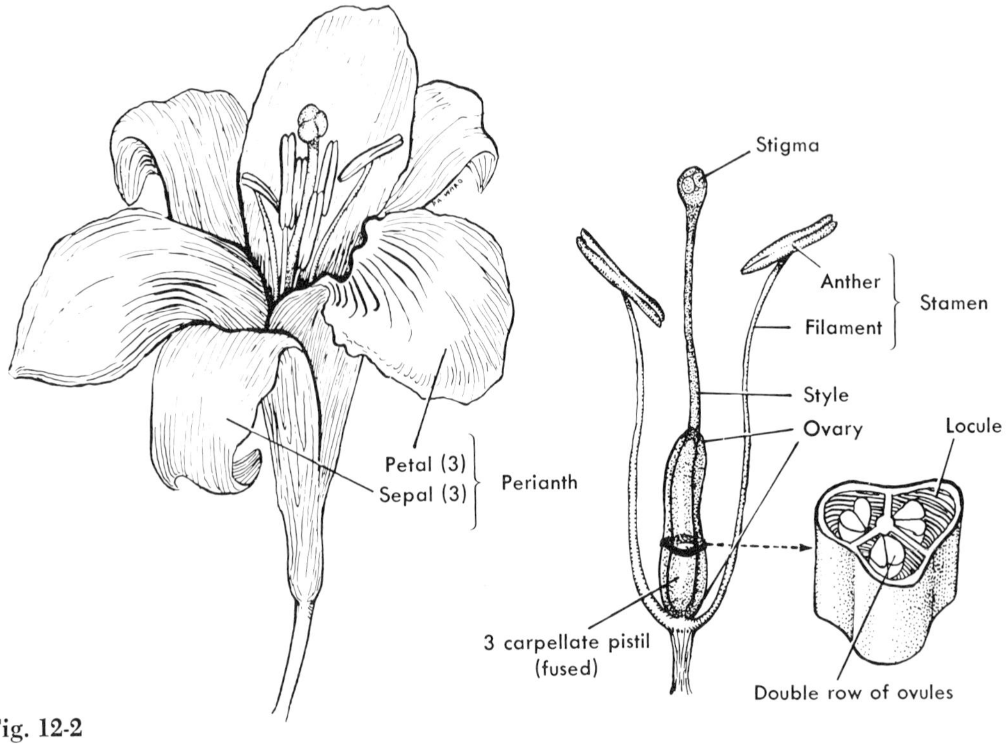

Fig. 12-2
Lily flower, a complete flower, showing details of its parts.

eral **pistils** in which **ovules** or potential seeds may be found. Collectively, the pistils are termed the **gynoecium.** Each unit of the pistil is a **carpel,** surrounding ovules located in a cavity. The base of the pistil is enlarged to form an **ovary** from which extends a slender filament, the **style.** The ovary is a hollow structure in which there are spaces called **locules.** One or more ovules, which will develop into seeds, are located in each locule. At the terminal end of the style is located an organ, the **stigma,** which is either sticky, roughened, or otherwise modified to catch and hold pollen.

For fertilization it is necessary for the pollen tube to grow down the style and reach the micropyle (Fig. 12-3). Further details of this process are discussed in life cycle 16 (Chapter 14).

FRUITS

Although we commonly think of apples, oranges, grapes, and the like as fruits, to the botanist a fruit is a fully developed and ripened ovary. Fruits are so varied that it is difficult to fit them all into even a most detailed system of classification. Fig. 12-4 illustrates one kind of fruit. Fruits are important not only to plants for seed dispersal and survival, but as a source of food for other organisms. The various types are discussed in Chapter 13.

The development of a fruit from an ovary is a complex process that involves many physiological activities. It can be shown that auxins and growth hormones play an important part in a fruit's development. Various sugars, soluble proteins, and other foods accumulate in the ovary walls during its maturation. The increase in sugars accounts for the sweetness of some fruits. In others, fats and oils accumulate in large amounts (olives, for example). Pulpy fruits accumulate large amounts of water. The

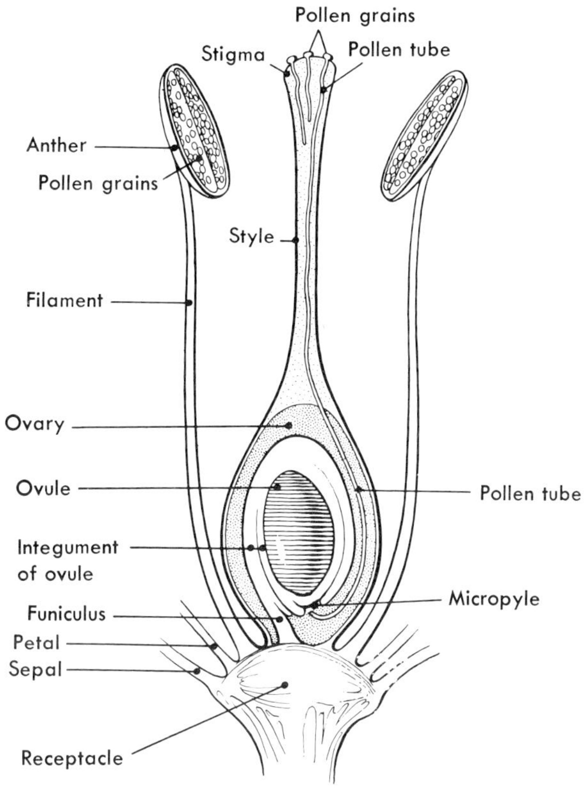

Fig. 12-3
Cross section of a fully developed flower showing the growth of the pollen tube in preparation for fertilization.

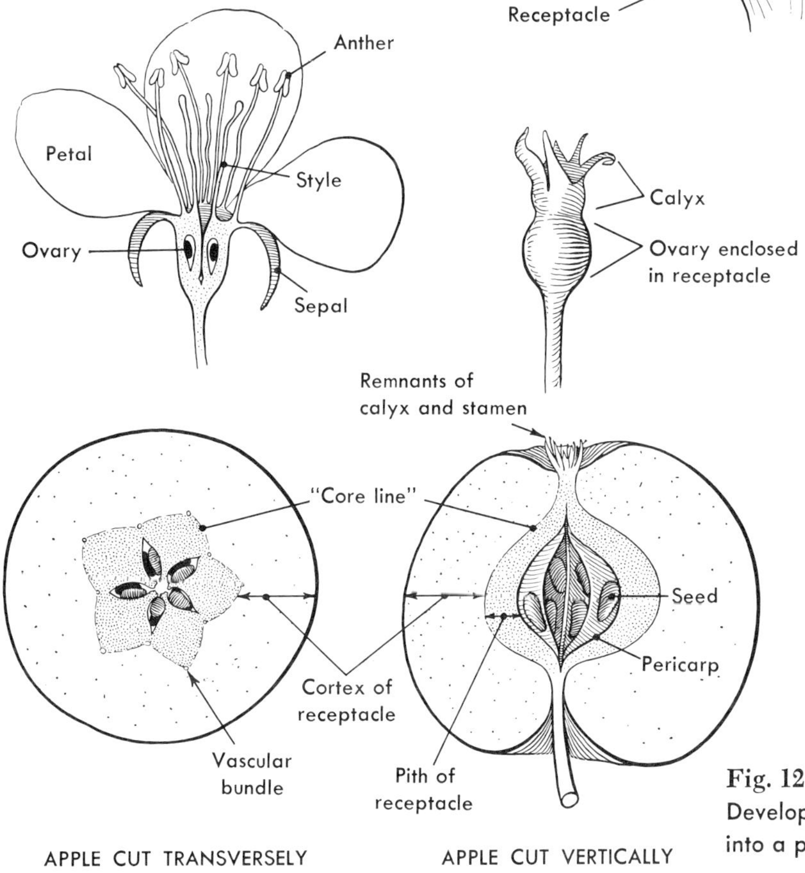

Fig. 12-4
Development of the apple flower into a pome fruit.

growth of the ovary is stimulated by auxins that are either present in the pollen or produced by it in the ovary tissue. Other changes, such as pigmentation, may accompany the physiological activities involved in developing fruits. The chlorophyll in apples (Fig. 12-4) disappears when the fruit ripens, and is replaced by other pigments. Anthocyanins accumulate in the ripening of fruits such as apples or plums to give them a charactistic red or blue hue.

The growth of the vegetative parts of a plant is frequently inhibited by the development of fruits. This could be due partly to the translocation and accumulation of foods from other parts of the plant, partly to hormones, to the fruits, or to some other physiological process. It was discovered that ethylene gas would hasten ripening, particularly of citrus fruit, as well as development of flowers and leaves. It is apparent that this artificial ripening process is due to acceleration of otherwise normal physiological processes.

The ovary wall or pericarp usually thickens when it develops into a fruit and becomes differentiated into three distinct layers known as the exocarp, mesocarp, and endocarp. The outermost covering (exocarp) consists of one or more layers of parenchyma cells. The middle layer (mesocarp) may vary in thickness from one to several layers. The innermost layer (endocarp) varies in structure, texture, and thickness in the fruits of different species. In some, parts other than the ovary contribute to the formation of a fruit.

SEEDS

A seed is a ripened ovule. Seeds vary in size from those that are barely visible to the naked eye (tobacco) to those that are several inches in diameter (coconuts). The **seed coat** and the **embryo** are the two essential parts of seeds. The seed coat develops from the integuments of the ovule, and the embryo from the fertilized egg. Sometimes the embryo is embedded in the endosperm (Fig. 11-3), which furnishes food for it. In most mature seeds the nucellus is no longer recognizable unless it has developed as food storage tissue. It is evident that a seed represents three generations—the old sporophyte (**nucellus**), the female gametophyte (**embryo sac**), and a new sporophyte (embryo).

REGULATION OF GROWTH BY HORMONES

It has been established that growth is controlled in plants by small quantities of specific organic compounds called growth regulators. These have been named **hormones**, growth substances, **phytohormones,** and **auxins.** Many of these organic materials do not occur in plants naturally, but others are actually synthesized by living plants. The American Society of Plant Physiologists has proposed a guide to these regulatory substances to avoid confusion in the use of terms describing them.

1. **Plant hormones,** also called phytohormones, are growth regulators synthesized by plants. In low concentrations they regulate certain physiological processes in the plant. They are substances that are translocated from the site of production to an area where they act in a way not unlike the hormones of animals.

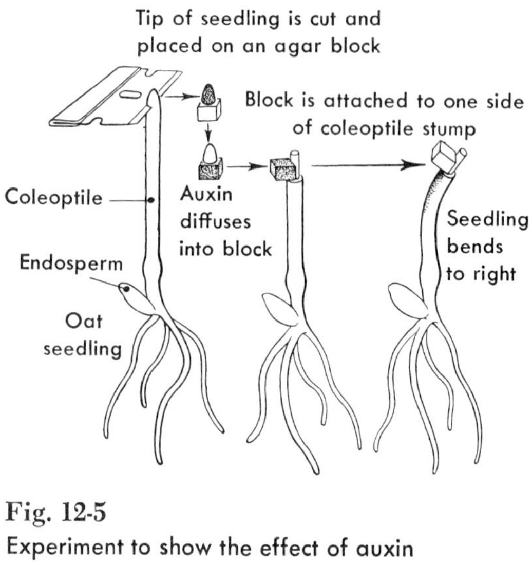

Fig. 12-5
Experiment to show the effect of auxin on plant growth.

a. Growth hormones are hormones that regulate growth.
b. **Flowering hormones** are hormones that initiate flower primordia and their formation and development.

2. Plant regulators are organic compounds, not including minerals and others, that promote, inhibit, or modify any physiological process when present in plants in small amounts.
   a. Growth regulators (or substances) are substances that affect growth.
   b. **Flowering regulators** are substances that affect flowering.
3. Auxins are compounds that induce elongation of the cells in the apex of plants. They are generally acids with an unsaturated cyclic nucleus or derivative. In 1934 a substance was isolated called indoleacetic acid, which is now accepted as naturally occurring auxin in plants. Auxins may affect other processes in the plant besides elongation, but elongation is considered the most important.
   a. **Auxin precursors** are substances that can be converted by a plant into auxin.
   b. **Antiauxins** are substances that inhibit the action of auxins.

Auxins. Auxins (from the Greek word, *auxein*, to increase) were the first growth regulators discovered. It was long known that some chemical substance diffused from the coleoptile tip of the oat to the region of cell elongation, thereby causing a phototropic response. When this substance is collected and added to small agar blocks, which in turn are applied to coleoptiles in which the tip is removed, it can affect the growth of the plant (Fig. 12-5) in various ways. It is now known to be necessary for cell elongation. Auxin is indoleacetic acid, which has the following formula:

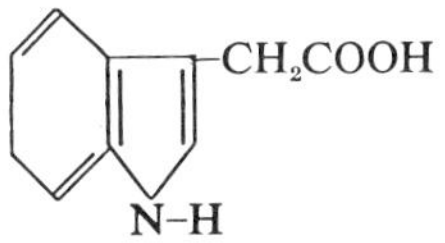

Indoleacetic acid

Although indoleacetic acid primarily appears to cause cell elongation, it does have

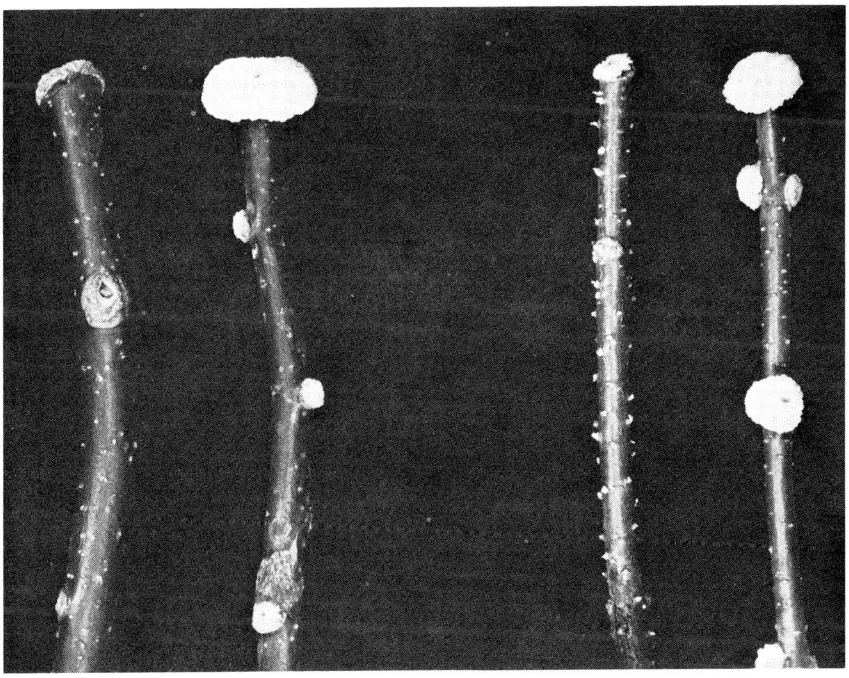

Fig. 12-6
Callus formation produced by an auxin (indoleacteic acid). (Courtesy Boyce Thompson Institute, New York, N. Y.)

other influences on plant growth, some of which stimulate or inhibit growth. It is responsible for cell elongation in oat coleoptiles, stems, veins, and petioles of leaves, for cell division in adventitious roots, for parthenocarpic fruits, for callus tissues (Fig. 12-6), and for cambium growth. It influences the inhibition of growth in root cell elongation, as well as apical dominance, and suppresses the separation of the abscission layer.

Auxin both inhibits and stimulates growth according to its concentration in the tissue. It requires much greater amounts of auxin to initiate optimum growth in a stem than in a root. The same amount necessary to promote optimum growth in a stem would inhibit growth in a root. One of the tests for auxin and its effects is the curvature test, in which the coleoptile of an *Avena sativa* (oat) seedling (Fig. 12-5) is used. Initially, the seedlings are grown in the dark and are then exposed to red light for 2 to 4 hours before the coleoptile tip is removed. A block of auxin-filled agar is placed on one side of the decapitated coleoptile, and the young stem bends toward the side opposite to that on which the agar block is placed. If the tip of a coleoptile is removed, there is no response to either gravity (geotropism) or light (**phototropism**). The auxin in small amounts will stimulate cell elongation, and it is found to be produced in very small amounts at the extreme tip of stems and roots. From these tips, the auxin diffuses backward from the region of production to the region of response, that is, only in one direction, farther away from the tip. When a seedling is exposed to light on one side only, light causes a decrease in the amount of auxin on that side. Therefore cells on the darker (opposite) side will elongate more rapidly than those on the light side, causing the stem to bend toward the light. The oat seedling is very sensitive to light, particularly the sheath or coleoptile surrounding the shoot of the young seedling.

The auxin is responsible for geotropic responses in plants. If a young plant is placed in a horizontal position, the stem tends to bend upward because gravity causes the auxin to accumulate on its lower side, thereby stimulating cell elongation on that side (Fig. 12-7). However, in roots the auxin inhibits cell elongation; thus the upper side grows faster than does the lower side (Fig. 12-7). In addition, auxins stimulate cambium growth, especially in

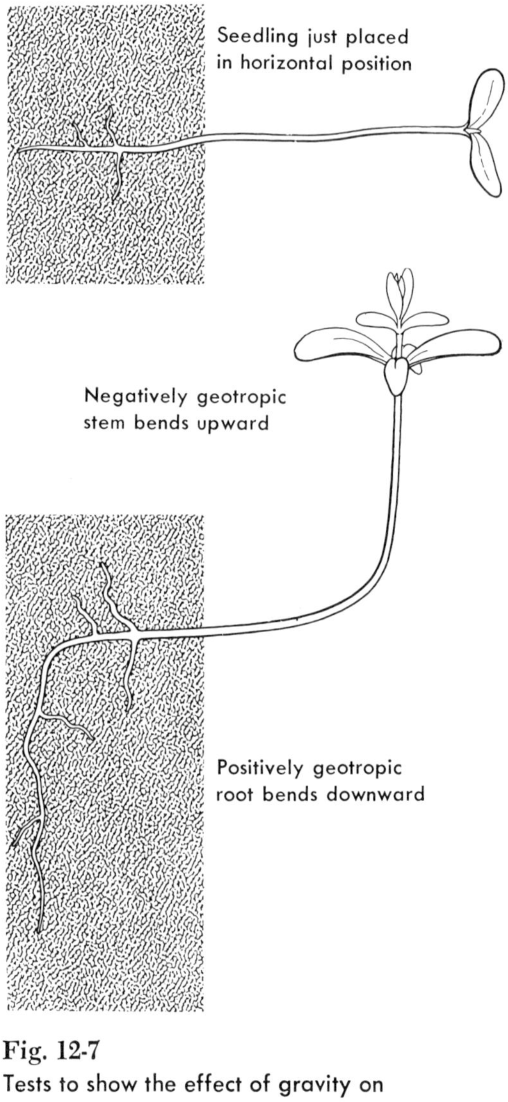

Fig. 12-7
Tests to show the effect of gravity on plant growth (geotropism).

those regions farthest away from the developing buds. It inhibits the growth of lateral buds, especially those that form branches. Furthermore, the auxin stimulates the formation of roots on cuttings (cut end of stem). If the cutting is immersed in a solution of naphthaleneacetic acid for several hours and the stem is then planted in soil, many more adventitious roots will appear on the cut end of the treated stem then on an untreated one (Fig. 12-8). Another type of plant-growth substance stimulates the development of a protective layer of tissue over a cut or injury in plants. This is generally designated a wound hormone and produces wound calluses on stems and roots (Fig. 12-6).

Other growth regulators, called **calines,** have been demonstrated in plants. One kind of caline called **rhizocaline** is believed to be synthesized in leaves but is used, along with the auxin, for the initiation of roots in cuttings. Another, **caulocaline,** is synthesized in roots but is necessary for stem growth. **Phyllocaline** is required for the growth of the mesophyll in leaves. This last substance is thought to be a purine called adenine, a substance found in many living plant cells.

A number of synthetic growth regulators

Fig. 12-8
Effect of napthaleneacetic acid. Both tomato plants have the tops removed. The plant on the right was treated with the growth substance, and the one on the left was not. Note the growth of adventitious roots on the stem of the treated plant. If this stem had been placed in soil after treatment, a new plant would have developed. (Courtesy Boyce Thompson Institute, New York, N.Y.)

Fig. 12-9
Growth-regulating substance isolated from corn pollen was applied to the bean plant on the left and not to the one on the right. (USDA photograph by Purdy.)

(Fig. 12-9) have been discovered in recent years. Indolebutyric acid is one that is applied to the stigma of flowers of tomatoes (Fig. 12-10), peppers, squash, and other species and results in seedless fruits (Fig. 12-11). Naphthaleneacetic acid has been used to prevent potatoes from sprouting during storage (Fig. 12-12). Other chemicals are used to prevent fruit from falling off trees and to delay or hasten the ripening of fruit. Weed killers are a type of growth regulator. All these are important commercially.

Gibberellins. In addition, potent growth regulators known as gibberellins have been identified. The application of one of these growth regulators will make normal plants out of dwarfed ones. Gibberellins are synthesized by a fungus that causes a disease in rice. They will produce an increase in cell elongation, promote cell division in callus tissues, cause leaf expansion, aid germination of seeds, promote flowering of plants, and, in addition, stimulate bud de-

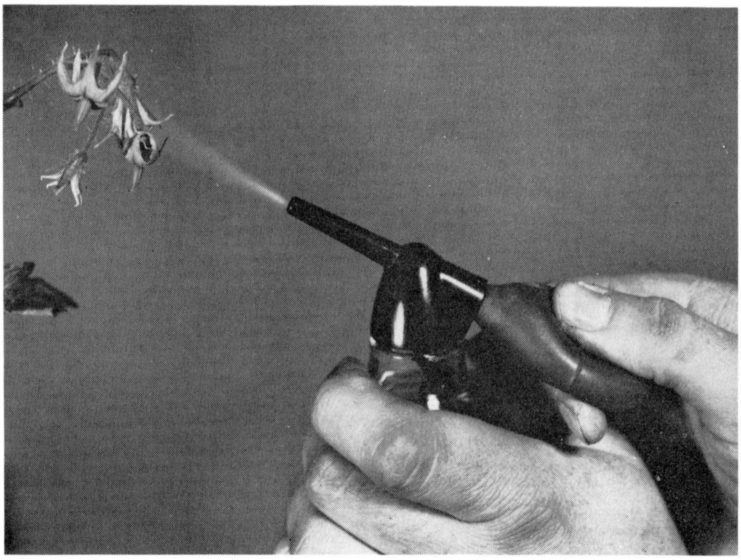

Fig. 12-10
Testing for increased fruit-set by spraying a tomato plant flower with a synthetic growth-regulating substance (indolebutyric acid). (USDA photograph.)

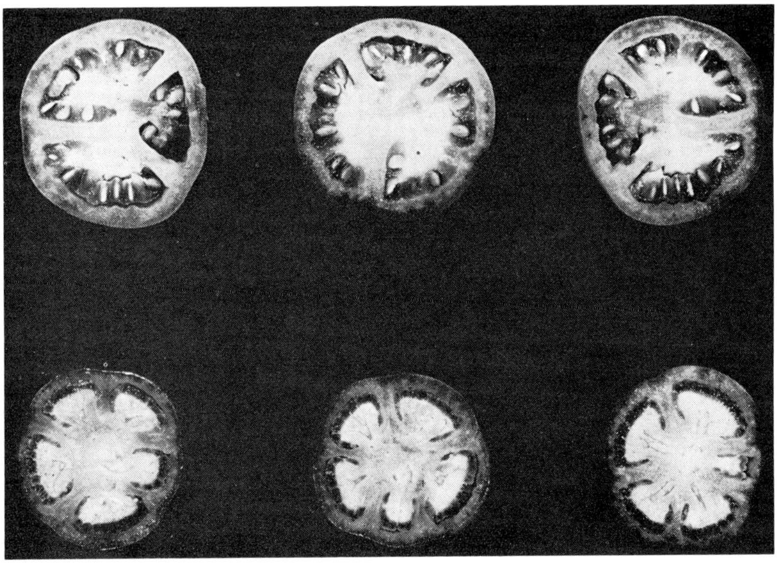

Fig. 12-11
Result of spraying tomato flowers (Fig. 12-10) with indolebutyric acid. Upper row, untreated, with seeds; lower row, treated, no seeds. (Courtesy Boyce Thompson Institute, New York, N. Y.)

Fig. 12-12
Potatoes on the right were treated with napthaleneacetic acid; those on the left with many sprouts were untreated. This indicates that this growth-inhibiting substance retards sprouting and provides for better storage. (Courtesy A. S. Riker, Department of Plant Pathology, University of Wisconsin, Madison, Wis.)

velopment. They are used commercially to make plants flower in off seasons of the year and to force the premature opening of buds.

Cytokinins (phytokinins, kinins). Various substances inhibit plant growth. These are called **cytokinins.** Although there is some confusion over their name, since they are called phytokinins also, and were first called kinins, they are becoming well-known substances affecting plant growth. One such substance is 6-furfurylaminopurine, which interacts with auxin; the former usually induces cell division, and the latter cell elongation. In various proportions they have other effects on growth.

Growth inhibitors. Various substances serve to induce dormancy or to keep plants and seeds dormant. Usually these are a mixture of auxins, gibberellins, cytokinins, and antiauxins. The latter comprise a wide variety of chemicals, many of which remain poorly known.

Summary. Differentiation and growth appear to be mutually exclusive. Whereas growth is for the most part an endless process of multiplication of individual units, differentiation is the segregation of these units from the rest, thereby making the former different and tending to prevent the cell from further multiplication. Therefore the more highly specialized the cell becomes, the less likely it is to divide. Thus growth and differentiation are the two processes by which development of a living organism is achieved.

When development is considered in terms of cells, it is found that initially simple cells of a uniform type gradually progress to more complex and varied ones. The actively dividing cells of the cambium are plastic and versatile, eventually becoming stabilized and having an unalterable structure with highly specialized functions, such as the permanent tracheids of the xylem, and the sieve tubes of the phloem. The whole course of development in a plant (or an animal) from the time of its beginning until its death is characterized both by unity and harmony in its structural unfolding and functional behavior. The completed organism is functionally no more an individual than is the sex cell, embryo, or seed that produces it. In all instances each cell behaves as a separate, living, functional entity and develops as such regardless of whether it is or is not an aggregation of cells making up a whole organism.

This phenomenon is called integration, but it is neither easy to explain nor to define. To adequately comprehend such a complex phenomenon would require more detailed knowledge of the organization of living plants at the organ, tissue, cellular, and molecular levels than is currently available. Various biochemical factors, such as hormones, are known to regulate this integration. But the exact reason why a cell stops dividing, or ceases growing, or even begins again to divide after it has differentiated cannot be simply stated. What determines how long a cell, tissue, organ, or organism will live? What is responsible for determining the size of one structure on a plant in relation to the others? All these are questions regarding the **morphogenesis** or growth form processes of plants, problems for the biologist to solve in future research, problems that perhaps hold the key to understanding the nature of cancer.

SPECIAL TERMS USED IN CHAPTER 12

androecium
anther
antiauxins
auxin
auxin precursors
caline
calyx
carpel
caulocaline
complete flower
corolla
cytokinins
embryo
embryo sac
flower
flowering hormone
flowering regulators
fruit
gibberellin
gynoecium
hormone
locules
morphogenesis
nectar
nectaries
nucellus
ovary
ovule
palynology
pedicel
peduncle
perianth
petal
phototropism
phyllocaline
phytohormone

pistil
plant hormone
pollen grain
pollen sac
receptacle
rhizocaline
seed
seed coat
sepal
stamen
stigma
style

QUESTIONS TO CONSIDER

1. What are the various growth stages? Define each.
2. What is the relationship of the flower to the roots, stems, and leaves?
3. Describe the growth process from flower to seed dispersal.
4. Define flower, fruit, and seed.
5. When are growth regulators used?

DISCUSSION FOR LEARNING

1. Since the growth regulators used commercially are chemicals that can be synthesized in many cases, discuss the possibilities of regulating abnormal growth by the use of chemicals. What are some of the problems?
2. From the standpoint of the basic parts of the vascular plants and their anatomy that has been reviewed so far, diagram and discuss the simplest or least complex plant you can and still have a complete plant.
3. From the discussion and basic plan developed in answer to the above problem, develop a mechanical plant, that is, a system that will do all the things that your simple plant above will do, except use artificial materials instead of tissues. What is the most difficult function to design?

Part V

Adaptation and evolution

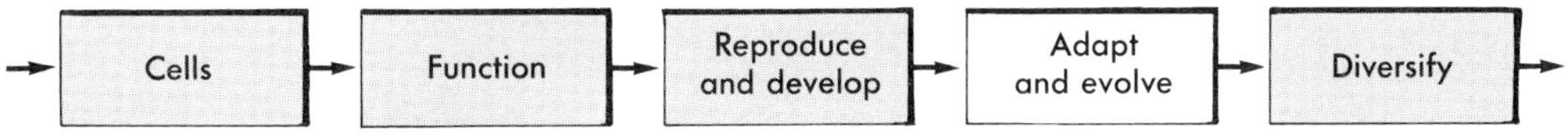

- Programming at the cellular level through genetic processes determines the survival for reproduction of each member of a population.
- Those not programmed to cope with their effective environment do not survive to reproduce.
- This determines the proportion of variants that are able to reproduce—the natural selection of those adapted to their environment.
- Once gross shifts in the programming have reached the point of reproductive isolation, species reproduction, or evolution, has taken place.

Chapter 13

Adaptive structures in plants

USDA photograph.

All living organisms in the world around us, and we ourselves, are the result of adaptation. In Chapter 10 the mechanisms of adaptation were discussed. This chapter treats the results of this process as they affect the vascular plants. The particular adaptations of the lower plants are discussed in the section on diversification.

Programmed growth results in the relatively uniform habitus of the species. At the same time, the great variation possible in the arrangement of the DNA and in its genetic processes provides for the vast array of plant life of the present and in the past. Certain features, however, are common to many kinds of plants, and these features may be grouped together to demonstrate the phenomenon of adaptation.

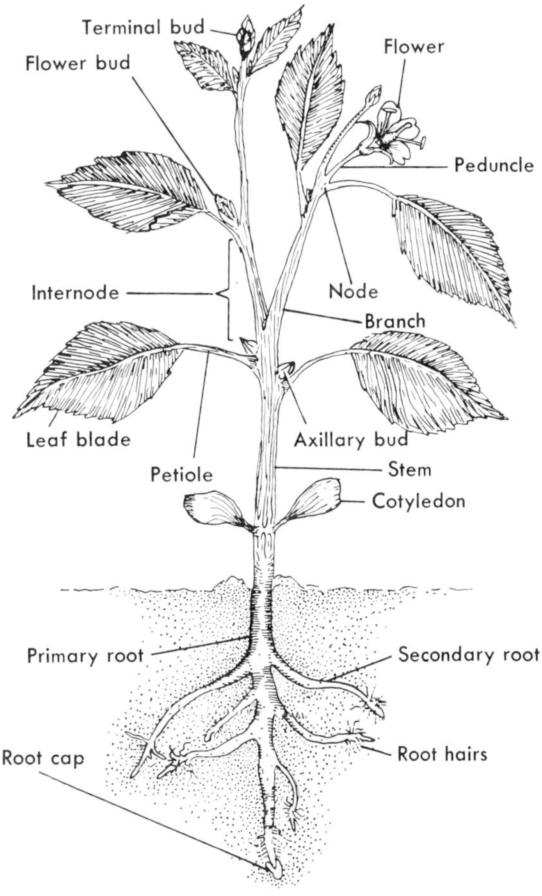

Fig. 13-1
Organs and systems of a flowering plant.

Also, many of these growth structures are useful in classification schemes. In fact, it is only after a recognition of the adaptive structures and their distinction from the nonadaptive structures that we are able to construct a classification based on evolutionary results. On the other hand, it is mostly due to these interspecific adaptive features that we are able to recognize the species of plants. All the main growth structures—roots, stems, leaves, and flowers—and the resulting fruit and seeds are involved. These are typified by the tree shown at the beginning of this chapter and by the diagram in Fig. 13-1. Form and adaptation are not restricted to the external shape of a plant. Cells show many different shapes according to their function and position (Fig. 13-2).

Seeds develop into a plant with the growth habits of its species. Some plants grow from seed, produce flowers, fruit, and new seed in one growing season. These plants are termed **annuals** and are characteristic of herbaceous plants. At the end of their growth period they die down. The same cycle of seed to seed is repeated the next year. Other plants produce seeds that germinate and produce a rosette of leaves and a taproot the first growing season. The

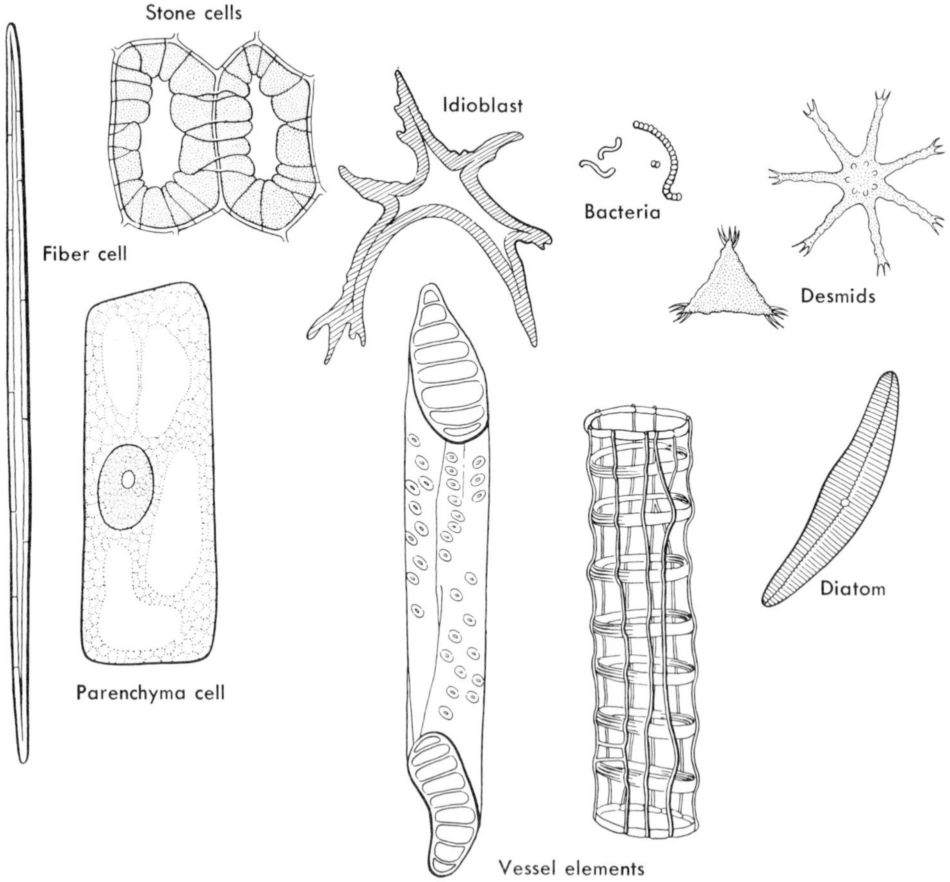

Fig. 13-2
Cells of plants show many different shapes according to their function and position. Those figured above show some of the many adaptations at the cellular level.

second year they produce a flower stalk and set seed, using the food reserves in the tap root. This type of plant growth is called **biennial.** Many of our common vegetables, carrots, beets, and parsnips for example, are biennials. They are harvested, of course, before the second year of flower and seed production. The final type of growth habit is termed **perennial.** Plants of this type are usually woody, and the plant body persists for several or many years, producing flowers and seeds only after a few years of growth. After the initial growth, the shrub, bush, vine, or tree continues to grow year after year, but more slowly. Once they have reached their maximum height these plants continue to replace limbs and twigs that may die back through disease or insect damage or that are broken during storms. Both evergreen and deciduous trees replace their leaves as they are lost.

ROOTS

Roots are adapted to function as anchors (Fig. 13-3) for plants and in some cases as food storage reservoirs (the biennials mentioned above). All the roots produced by the vascular plant spread through the soil not only to anchor the plant, but also to reach water (Fig. 13-4), much the same as the stem grows toward light. They have developed two major forms of root systems. A **fibrous root** system (Figs. 13-5 and 13-6) is composed of many slender roots, most of which are nearly equal in size. In grasses the primary root is lost in the maze of adventitious roots that have developed from the stem, whereas in other plants the

Fig. 13-3
The anchoring ability of the root system of plants serves a two-way purpose—to anchor the plant and at the same time to hold the soil, which also helps to prevent erosion. In this photograph young pine trees are planted in badly eroded mountain areas. Some will survive and eventually hold back the soil, thus preventing further erosion. (U.S. Forest Service.)

secondary roots have grown so large as to be indistinguishable from the primary root. Therefore the mass of roots, whatever its origin, constitutes the fibrous root system. In the other type of root system, called a **taproot,** the primary root is enlarged and the secondary roots are few and slender (Fig. 13-6). Common examples of taproots have been mentioned previously. The type of root system and size vary with different species of plants and also with environmental physical factors such as moisture and soil composition.

Root formation is the result of environmental stimulation. It is well known by the gardener that "rootings" can be made from slips or cuttings from the stems or even leaves of many kinds of plants (Fig. 13-7). Hormone secretion is induced by placing the vegetative part in moist soil or sand, and this same process often occurs in nature when growth conditions permit it. The gardener may speed up the process by adding commercially prepared rooting hormone to the water.

STEM MODIFICATIONS

The stem system of a flowering plant may be a distinctive mark of the plant; for instance, a pine tree, a cornstalk, and an asparagus are obviously different. The primary function of a stem is to support the leaves and distribute them in space so that they receive adequate light. In addition to the conduction of materials to and from the leaves, it often stores food. Sugar cane and corn stalks are two of the best examples of this.

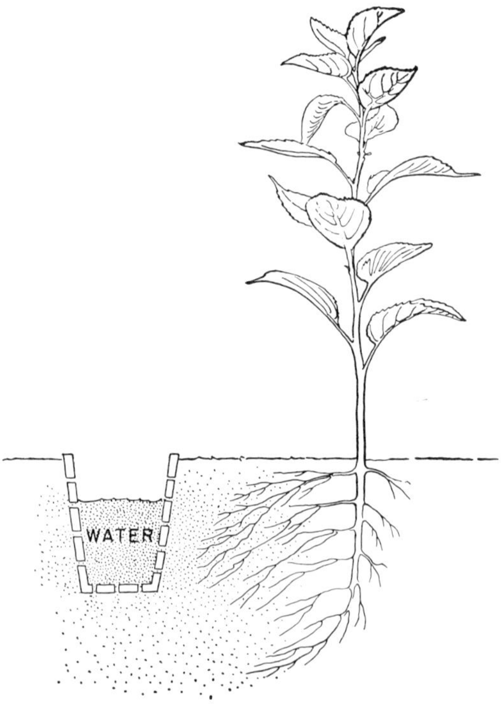

Fig. 13-4
Roots in the test shown here grow toward water, a demonstration of their hydrotropism.

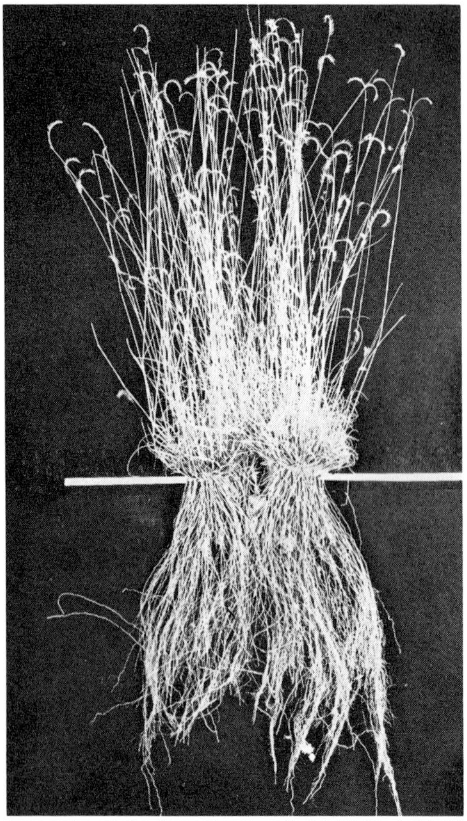

Fig. 13-5
Fibrous root system of grass. (USDA photograph.)

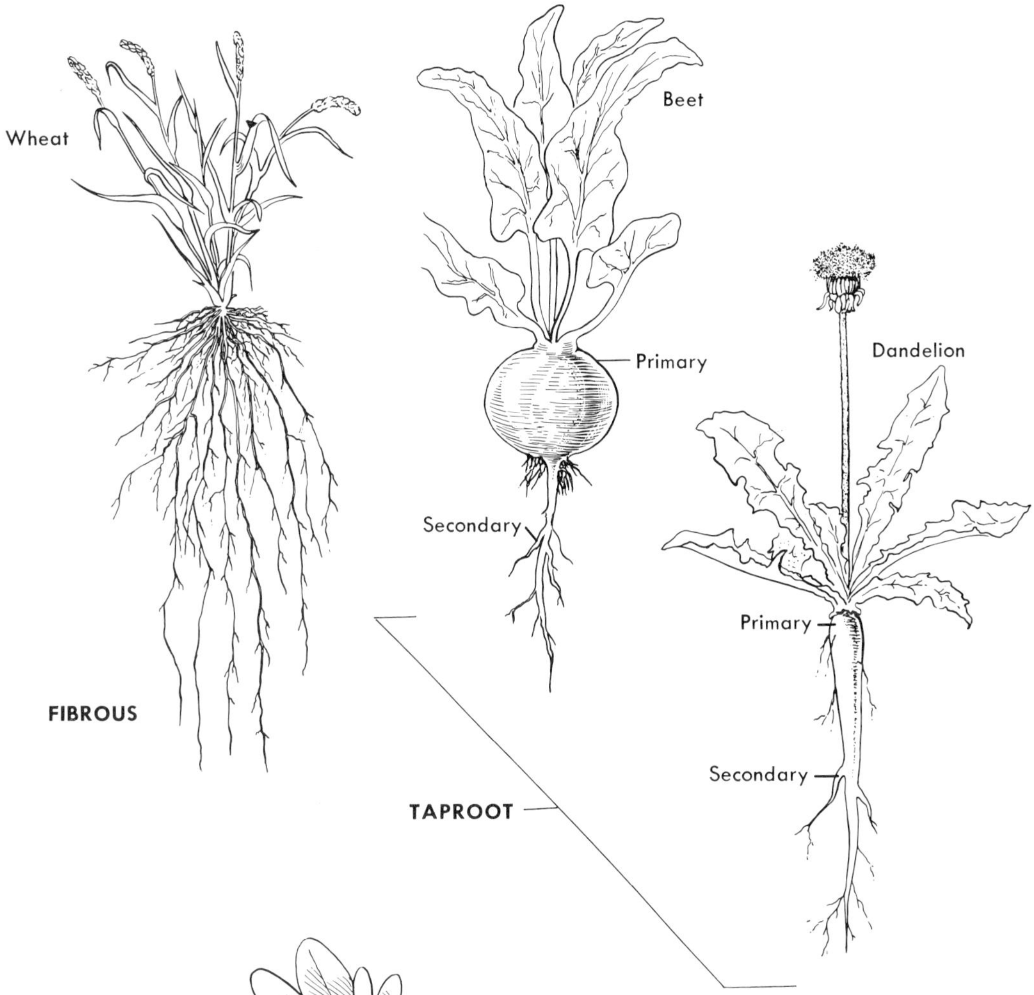

Fig. 13-6
Root systems of two types. The one on the left is fibrous (see also Fig. 13-5). Taproots, shown in the middle and on the right, are used for food storage by biennials during their first year of growth.

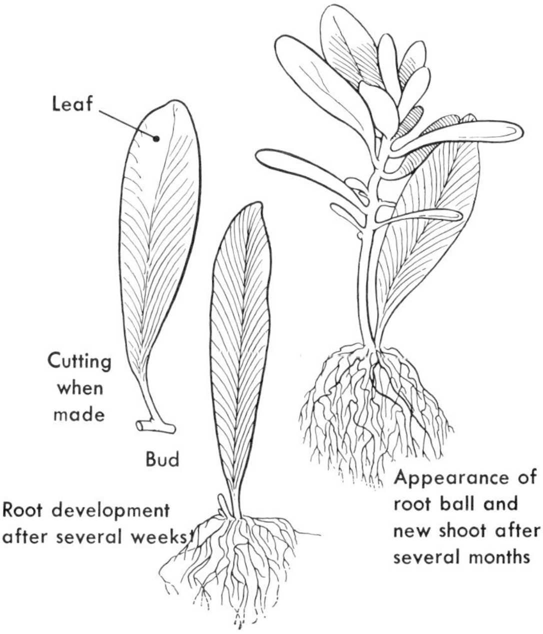

Fig. 13-7
Method of rooting rhododendron leaves.

Some stems are underground, but most typically they start at the soil level. Those stems that grow above the soil are called **aerial stems** (Fig. 13-8) to distingish them from those that grow underground as **rhizomes** (Fig. 13-9), as in the iris. Therefore an aerial stem, characteristic of most plants, is erect, but some may be more flexible in there growth habits and form climbing stems or **vines** (known as **lianas** in tropical regions), and still others are prostrate and run along the surface of the ground (Fig. 13-10). The stems of plants may vary greatly in form, size, structure, and even in the length of their life as we see in the **herbaceous stem** in contrast to the **woody stem.** The growth form distinction between **shrubs** and **trees** is difficult to make. Both are woody stem, and, in general, shrubs have many stems, all ap-

Fig. 13-8
Adventitious roots on English ivy stem.

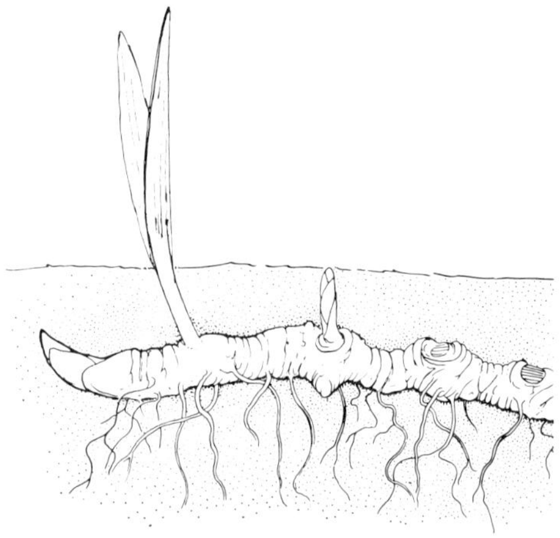

Fig. 13-9
Underground stem of iris.

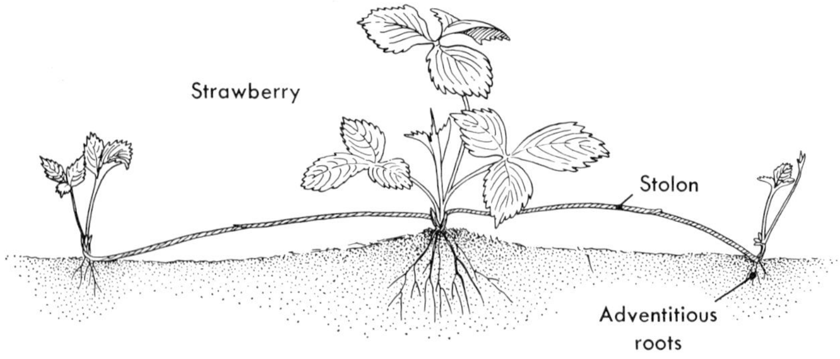

Fig. 13-10
Vegetative reproduction runners or stolons. Horizontal stem of the strawberry.

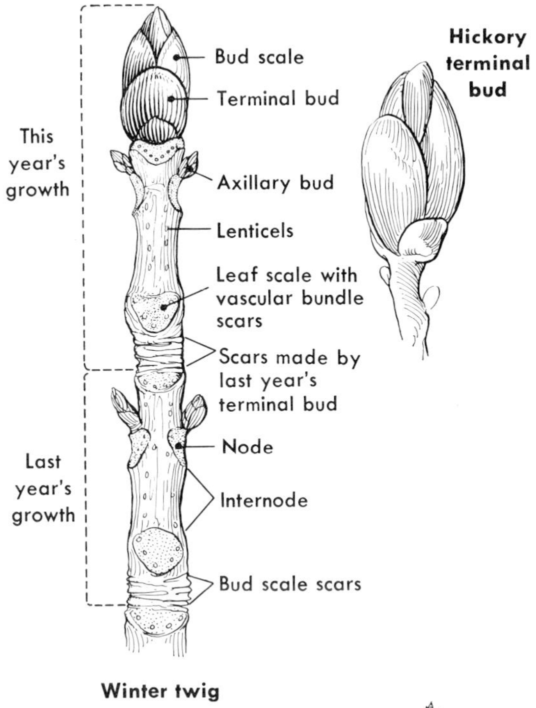

Fig. 13-11
Left, nodes and internodes of a stem; right, opening of a terminal bud.

proximately the same size. All the main stems of most shrubs branch profusely at about the ground level. However, plants may grow as shrubs in one locality and as trees in another. Unlike a root, the stem is made up of distinct areas to which leaves or buds are attached. These areas are called **nodes,** and the area between two successive nodes is called an **internode** (Fig. 13-11). The number of leaves or buds found at each node, depending upon the season of the year, will help to identify the stem. If only one leaf or bud appears at a node, the stem is described as having **alternate** leaves or buds (Fig. 13-12). When two leaves or buds are located at each node on opposite sides of the stem, it is said to have **opposite** leaves or buds (Fig. 13-12). The stem is designated as having leaves or buds in **whorls** if three or more are located at each node (Fig. 13-12). Elms and

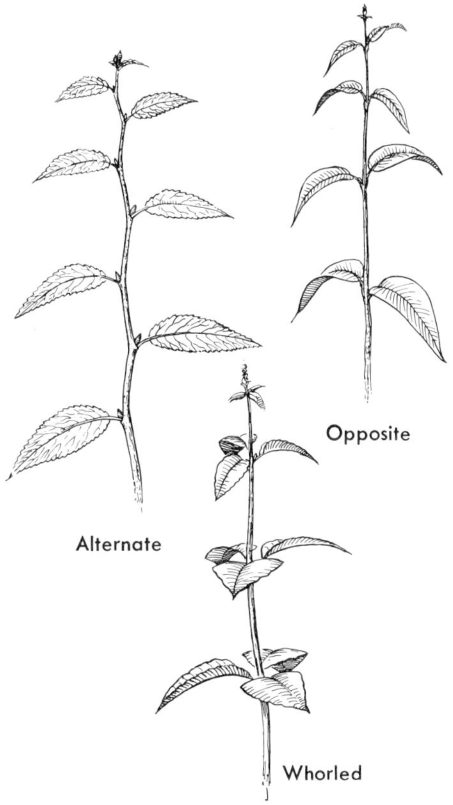

Fig. 13-12
Various leaf arrangements on stems.

apples have alternate arrangements of leaves, maples and dogwood have opposite arrangements, and catalpas are whorled. Whorled leaves are the least common arrangement.

When leaves are arranged alternately, they occur in spiral succession. In other words, each successive leaf arises slightly to one side of the one below it so that they form a spiral around the stem. When leaves are arranged on the stem in whorls or opposites, each group is not directly above the one below it. Each successive pair of opposite leaves is arranged at a right angle to the previous pair, whereas those in whorls are one third the distance around the stem from those below. This pattern of leaf attachment to the stem orients all the leaves in such a way that the maximum amount of light reaches each leaf. With such an arrangement, the greatest amount of leaf surface is exposed to light to form the **leaf mosaic.**

The surface of an active, growing stem reveals a number of structures. Most conspicuous are the leaves and buds. The leaves may be broad and flat, narrow and elongate, or needlelike and are attached at the nodes. A **bud** is usually found in the angle of the leaf attachment to the stem. This angle between the leaf and the stem is called the **leaf axil,** and the buds located there are called **axillary** or lateral **buds** (Fig. 13-11). A **terminal bud** may be located at the very apex or tip of the stem. The distinction between lateral and terminal buds is primarily one of position, since they are usually alike in structure and function. Although buds may be located in the axils of leaves, occasionally they appear on other parts of the stem and are then called **adventitious buds.** Structurally and functionally, adventitious buds are like other buds but sometimes appear as the result of an injury, for example, when a branch is cut.

A bud is an embryonic or young shoot that has the capacity to produce leaves. It is essentially a cone-shaped mass of meristematic cells that produces leaf primordia laterally as the bud grows. Therefore a bud possesses nodes, internodes, and leaf primordia. It is a miniature stem on which all the leaves of the following season are already formed. Most buds have overlapping **bud scales** that cover them and protect them from drying, injury, and cold. Buds, like leaves, can be classified as alternate, opposite, or whorled, depending upon the number located at each node. They may also be classified as to their activity and the kinds of structures they produce. Buds that grow are called active buds to distinguish them from the occasional ones that are inactive or dormant. A dormant bud may remain embedded in the bark, only to develop years after it has formed. Some buds produce only flowers, such as daisies and phlox. These are called flower buds to distinguish them from buds that produce leaves and branches, the leaf or branch buds.

When the leaves drop off in the fall of the year, **leaf scars** will remain (Fig. 13-11). In those plants that lose their leaves each year, buds are quiescent during the winter season and become active again in the spring. New buds appear in the spring in the axils of the leaves in preparation for each following year. By proper pruning of the stem, a bushier growth of the plant can be induced. The removal of buds causes the remaining ones to form sturdier branches or larger flowers. Also by pinching off lateral buds, one can induce a plant to grow tall and erect.

Besides the leaves and buds present on a stem, there may be leaf scars, **bud scars,** and often blisterlike lines or dots called **lenticels** (Fig. 13-13). These broken areas in the bark or epidermis of the stem are for the entrance of air necessary to the living tissues beneath.

It is, of course, the stems that serve as food in the case of asparagus, and lumber is made from the stem of a woody

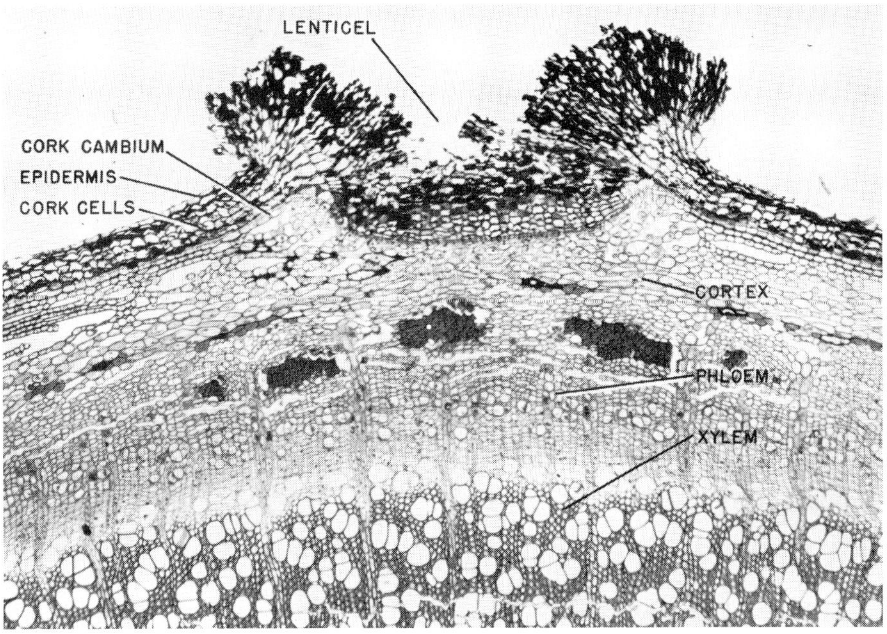

Fig. 13-13
Cross section through lenticel of stem. (×35.) (Courtesy George H. Conant, Triarch Products, Ripon, Wis.)

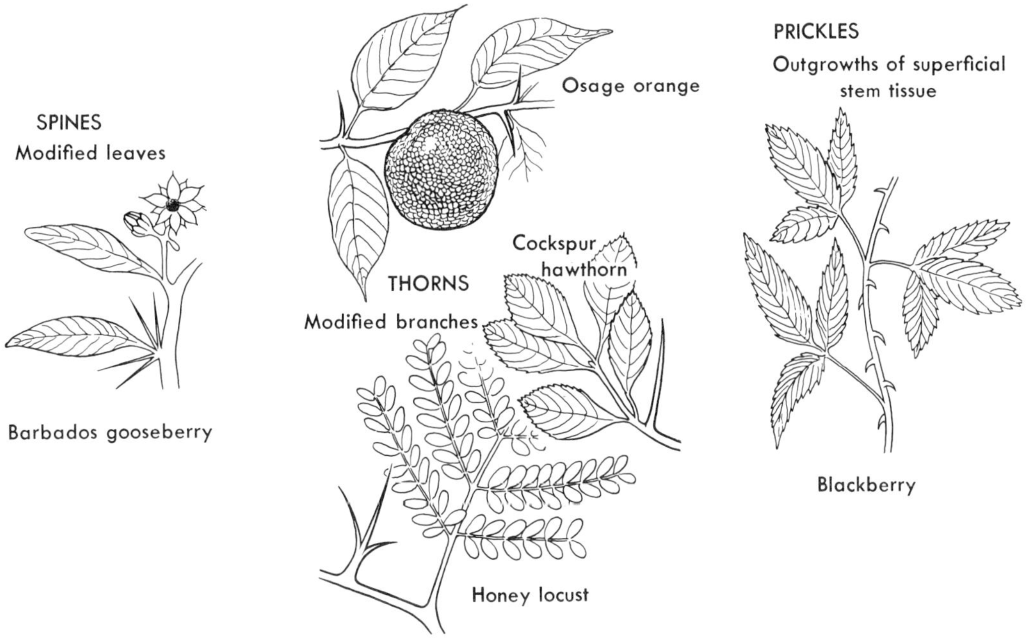

Fig. 13-14
Examples showing the different stem adaptations. Thorns and prickles are modified stems or stem parts; spines are modified leaves.

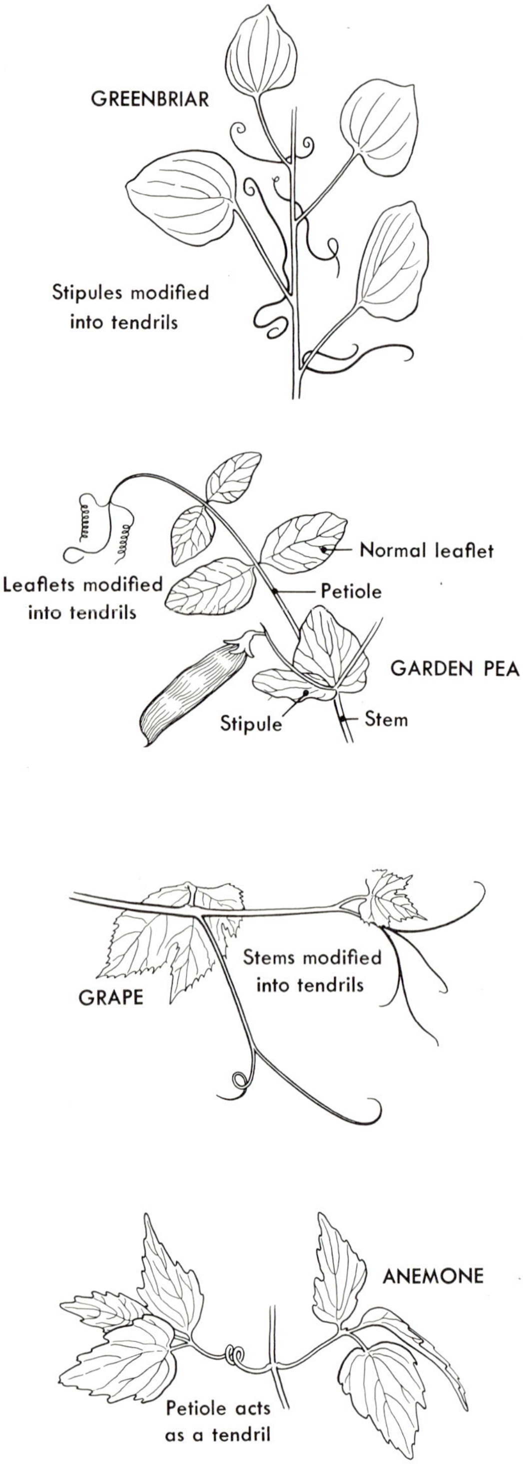

Fig. 13-15
Stem modifications that form tendrils.

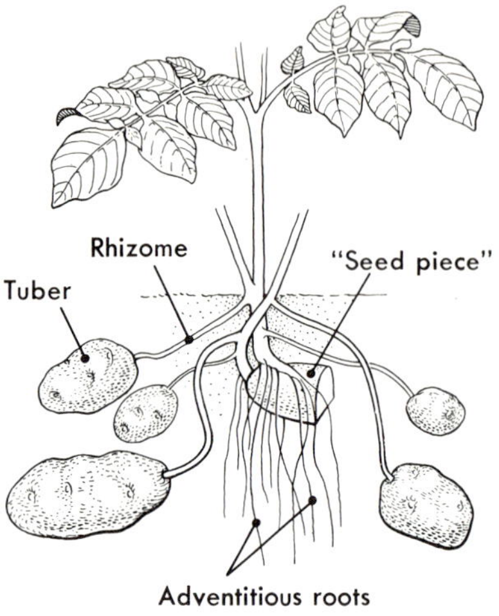

Fig. 13-16
Tuber, rhizome, and adventitious roots of potato.

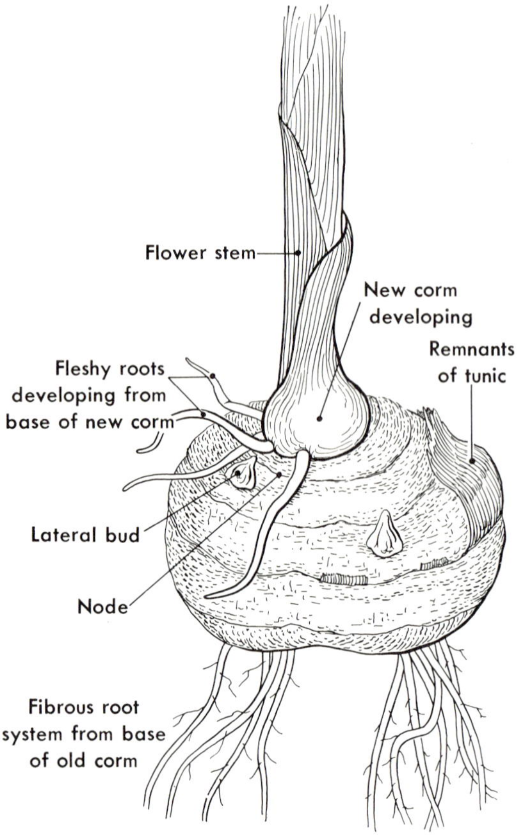

Fig. 13-17
Corm of a gladiolus.

plant. Stems sometimes undergo a particular modification and appear as **thorns, prickles,** or **tendrils** (Figs. 13-14 and 13-15). Some stems occur as **stolons,** producing at intervals erect stems and adventitious roots as a modification of the hortizontal or prostrate stems mentioned previously. If they are long, slender, nearly leafless stolons, they are known as runners (Fig. 13-10). The rhizomes of some plants, as for example the potato, may be thick and fleshy or partially thickened and are called **tubers** (Fig. 13-16). Some plants are propagated by a short, stout, erect underground stem in which food is stored. Such a stem, usually broader than it is long, is known as a **corm** (Fig. 13-17). At the end of a growing season the plant dies, but the corm sends out adventitious roots, and the terminal bud develops into the new stem, which grows and eventually produces a flower. The few dry leaves surrounding a corm are the remains of leaf bases, which do not store food. However, one of the principal storage organs is a **bulb** (Fig. 13-18), in which layers of thickened leaves (bulb scales) surround the short, erect stem. The tulip and onion are examples of bulbs.

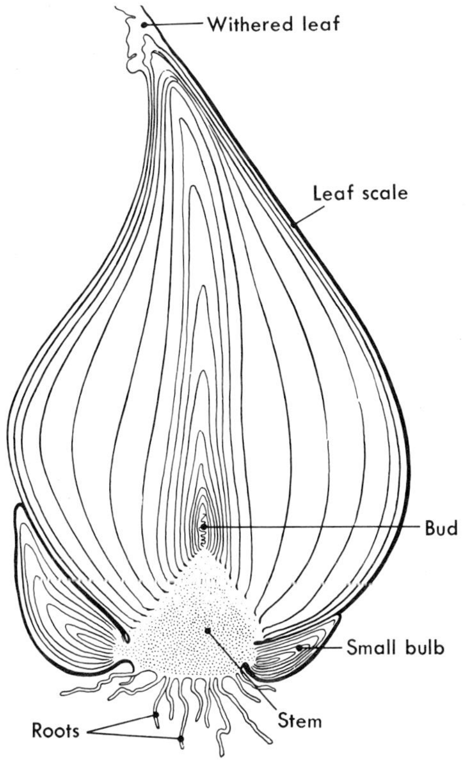

Fig. 13-18
Onion bulb and stem.

ADAPTATION OF LEAVES

Although the principal function of leaves is photosynthesis, they may have other functions, such as storage of food and water, reproduction, root formation, climbing, insect-catching, protection, or flower formation. Therefore some leaves are not photosynthetic at all but are highly specialized in other ways.

Many of our economic plants are useful because they store water, salts, food materials, and vitamins in their leaves, for example, spinach, onion, and cabbage.

The blade of the leaf of flowering plants is usually flat and expanded in form. It arises from the stem at a node and ordinarily has a bud located in the **axil.** Leaves are arranged on stems in the same way that buds are—alternate when one leaf occurs at a node, opposite when two leaves arise at a node, and whorled when there are three or more leaves at a node (Fig. 13-12).

Leaves are appendages of the stem that carry on photosynthesis; they usually have a distinct stalk or **petiole** and a thin, flat, expanded portion called a **blade.** When the blade is attached directly to the stem and the petiole is absent, the leaf is said to be sessile. In some leaves there may be a pair of small leafy structures at the base of the petiole called **stipules.**

The form and structure of the leaf blade may vary in different plants or even on the same plant (Fig. 10-2). A number of different terms have been coined to describe the kinds of leaves. These terms, while useful in recognizing and distinguishing differ-

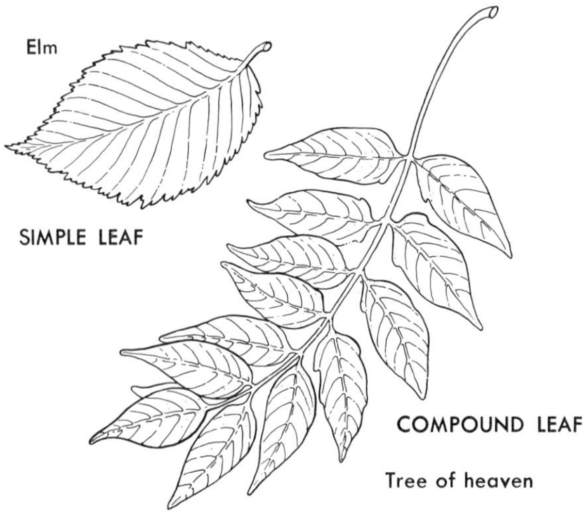

Fig. 13-19
Simple and compound leaves.

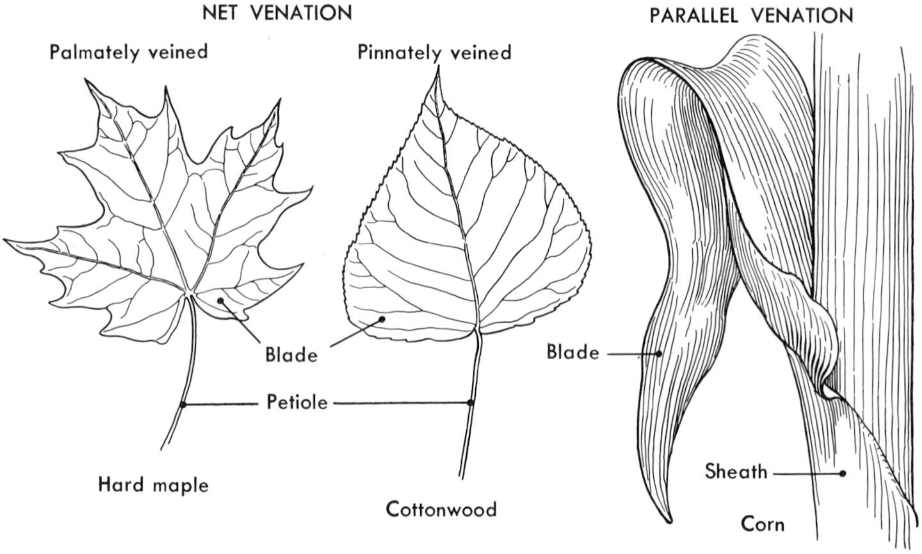

Fig. 13-20
Examples of leaf venation (net and parallel venation) in simple leaves.

ent species, are very numerous. Therefore only the most common ones will be mentioned. If the blade of the leaf is all in one piece, the leaf is called simple, and if it is composed of several leaflets, the leaf is compound (Fig. 13-19). The part of the **compound leaf** to which leaflets are attached is known as the **rachis,** and each leaflet is attached by a petiole.

The simple leaves of angiosperms have a characteristic vascular system made up of vascular bundles called **veins.** The arrangement of veins in a blade may vary from species to species. Venation alone

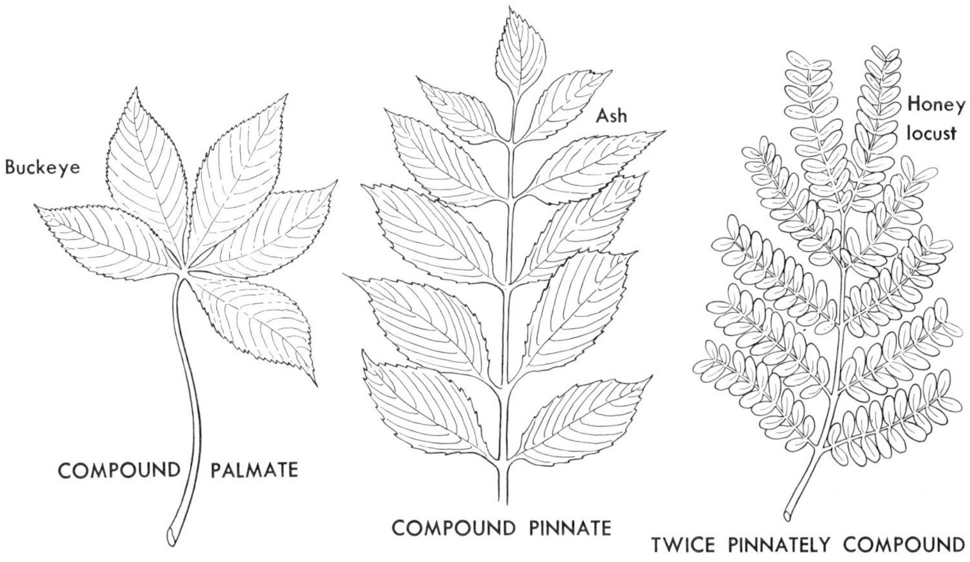

Fig. 13-21
Types of compound leaves.

generally will distinguish the two classes of flowering plants. Most dicot leaves have net venation, whereas monocots, such as corn, will exhibit parallel venation (Fig. 13-20). A net-veined leaf may have a single main vein with numerous side branches, as does the cottonwood leaf. Vein arrangement can be more specifically identified, for example, that of the cottonwood or poplar leaf as **pinnate** venation. Another type of net venation is **palmate,** typical of maple leaves, in which a number of prominent veins radiate from the petiole. Between the larger veins in the maple leaf is an obvious network of smaller ones. A compound leaf can also exhibit either pinnate or palmate venation. Compound leaves in which the leaflets are attached on opposite sides of the rachis, as in walnuts, hickories, ashes, and roses, are termed **pinnately compound** leaves. Those in which the leaflets are attached at the same point to the petiole, as in the horse chestnut and buckeye, are palmately compound (Fig. 13-21). Pinnately compound leaves of the honey locust are **pinnately bicompound,** since the primary divisions are themselves compound. Leaves in which three leaflets are attached to the petiole are **trifoliolate.** They may be pinnately compound, as in poison ivy, or palmately compound, as in clover.

Many terms derived from plane geometry and other sources are used to describe the forms of leaves. Some that are self-explanatory are linear, oblong, elliptical, and orbicular. If it is heart shaped, the leaf is called cordate; if kidney shaped, reniform. An ovate leaf is shaped somewhat like an oval, with one end narrower than the other.

The edges of leaves may vary anywhere from a smooth or entire margin to a deeply lobed one. Those having sharp, toothlike edges are called serrate, and deep indentations of the margin make the leaf lobed.

Leaves usually have a limited life span, since most grow only one season; however, others persist for several years. Woody plants that have aerial stems which survive year after year are classified as either deciduous or evergreen, according to the duration of the leaves. If the leaves formed during one season remain on the plant until the leaves of the following year are produced, the plant is called an evergreen. Nearly all cone-bearing trees are evergreen, as well as many angiosperm trees in moist tropical climates. Only a few angiosperms in temperate regions are evergreen—the holly, live oak, and sagebrush.

Trees and shrubs that lose their leaves at the end of each growing season, leaving the branches bare during the winter months, are called **deciduous.** In a temperate climate leaves form in the spring and are shed in the fall. The fall of leaves in tropical or warm areas coincides with the wet and dry seasons of the year, rather than with warm and cold temperatures. Leaves fall from most angiosperms (and gymnosperms) because of the development of a specialized transverse region at the base of the leaf called the **abscission zone** (Fig. 13-22). This region generally contains a large number of parenchyma cells and a few thick-walled cells, making

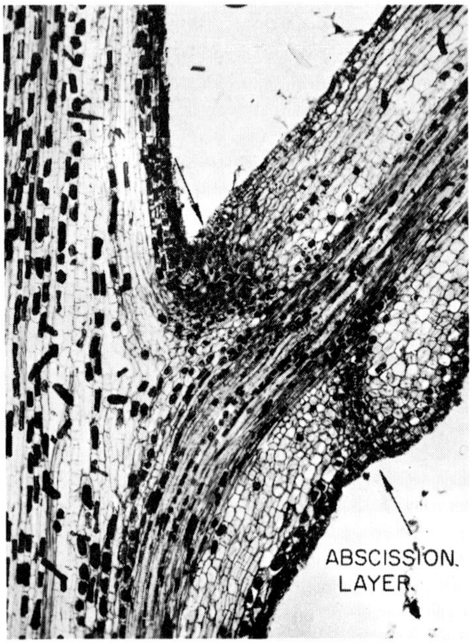

Fig. 13-22
Cross section of abscission layer in petiole.

it the weakest part of the petiole, or of the blade if the leaf is sessile. This zone sooner or later becomes differentiated into two layers. One is a protective layer, several cells thick on the side nearer the stem. The other, a separation layer, is the layer in which changes occur that eventually result in the fall of the leaf. The cells of this layer swell and become gelatinous, and the cells may or may not divide. Regardless of what occurs, the weight of the leaf and the force of the wind or rain will break the leaf away from the stem along the separation layer. Strictly speaking, the term abscission layer is used to designate the separation layer after it has begun to undergo the changes described.

Although these changes are taking place in the separation layer, other changes are occurring in the protective layer. The cells of the protective layer become suberized, making them waterproof. A layer of lignin permeates their cell walls, as well as the intercellular spaces, and the healing process is complete. Eventually, the periderm of the stem becomes continuous with the periderm formed over the abscission zone.

The falling of leaves may be brought about by any of several environmental factors. These could be a lack of moisture, low temperature, reduced light, or shorter daylight periods. It can also be induced by destroying the greater portion of the leaf blade or by treating it with a chemical such as carbon tetrachloride. Some chemicals, for example, indoleacetic acid, which is a growth regulator, tend to inhibit the fall of leaves when applied to them. There is some indication that under normal conditions a hormone produced by the blade of the leaf is translocated to the base of the petiole, thereby inhibiting abscission. Should this supply of the hormone be stopped entirely or interrupted for a prolonged period, the abscission layer will begin to form, and the leaf will fall.

Some leaves, such as those of the rhododendron (Fig. 13-7), can be induced to produce roots, as previously mentioned. Other plants may have some of their leaves modified into slender, twining tendrils, also mentioned earlier, for example, the pea or climbing plants such as the grape.

There are several families of plants with leaves that have become modified so that they can trap and digest insects. These plants, such as the Venus' flytrap or the pitcher plant, are commonly found in marshy or boggy habitats, where the supply of available nitrogen is limited.

The leaves of Venus' flytrap (*Dionaea muscipula*) are hinged along the midrib. Along the leaf are a number of trigger hairs that, when touched, cause the leaf to snap shut so that the fringed leaf margins interlock (Fig. 13-23). The closure of the leaf appears to be the result of a combined enzyme-and-turgor pressure mechanism that is initiated by chemical and physical stimulation from the presence of the insect. The pitcher plant (*Sarracenia purpurea*) has vaselike leaves lined with stiff, downward pointing hairs so that insects seeking the water at the bottom of the leaves are trapped. The water contains digestive enzymes capable of destroying the insect and of making available the nitrogen necessary to the plant.

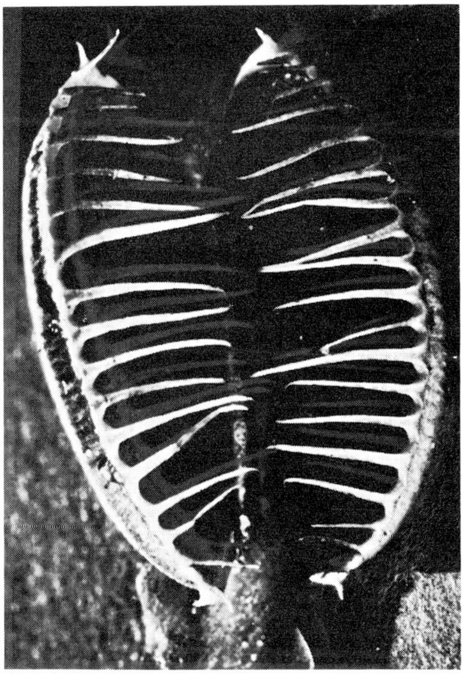

Fig. 13-23
Close-up of partly closed leaf of Venus' flytrap. (Courtesy Carolina Biological Supply Co., Burlington, N. C.)

Small modified leaves called bud scales cover the winter buds of most deciduous trees. Bud scales do not have chlorophyll and usually have more lignified cells than do ordinary leaves. Thus they are capable of protecting the developing leaves and stem they surround. When the bud opens up in the spring, the scales fall off and leave a ring of bud scale scars on the twig. Other plants, such as cacti, have leaves modified into slender, sharp spines that protect the plant from grazing animals.

We are most familiar with those leaves that add beauty to our homes or gardens. Common flowering plants, such as pansies and asters, are attractive only because the leaves have become modified into attractive shapes and colors. However, even uniformly green leaves can change their color in the fall of the year, especially those of deciduous trees in a temperate zone. The leaves of many woody plants take on yellow and red colors before they fall, particularly if the days are warm and the nights cool. In the chloroplasts of green leaves are two yellow pigments, the carotenes and the xanthophylls, in addition to chlorophyll. They cannot be seen in a green leaf because the more intense green pigments mask them. However, if the green pigments disappear, as they do in the leaves of many plants in the fall, the yellow pigments become visible. It is then that the leaves of birches, poplars, and elms appear golden over the landscape. Other pigments, such as **anthocyanins,** are dissolved in the cell sap and are responsible for the red coloring of leaves. In most leaves anthocyanins are not present during the summer but are formed in the autumn. The same kinds of trees can be different colors in different localities. In one area the leaves may appear crimson; in another area they may not. Soils low in nitrogen, bright skies, drought, and low temperatures favor the formation of anthocyanins so that under these conditions the leaves appear bright red in the autumn. The development of anthocyanins is therefore influenced by environmental conditions that favor a high concentration of sugar.

FLOWERS

The greatest variation and adaptation of the flowering plants is seen in the flowers, which are, as we have seen, really modified leaves. The arrangement of the floral parts is used extensively in classification. Some flowers, for example, the lily, are radially symmetrical in the arrangement of sepals, petals, stamens, and pistils. Others are asymmetrical, and others may lack some parts (**incomplete flowers**). If both stamens and pistils are present in the same flower, it is a **perfect** flower. If a flower lacks either stamens or pistil, it is called **imperfect.**

Many variations may appear in the number, shape, size, color, texture, and relative position of the floral parts. Monocot and dicot flowers show certain differences (Fig. 13-24). For example, monocot floral parts commonly occur in threes or multiples of three, whereas dicot floral parts are usually in fours or fives or their multiples. These variations are morphological characteristics that are important in distinguishing one species from another. More often than not, there is only one pistil consisting of several

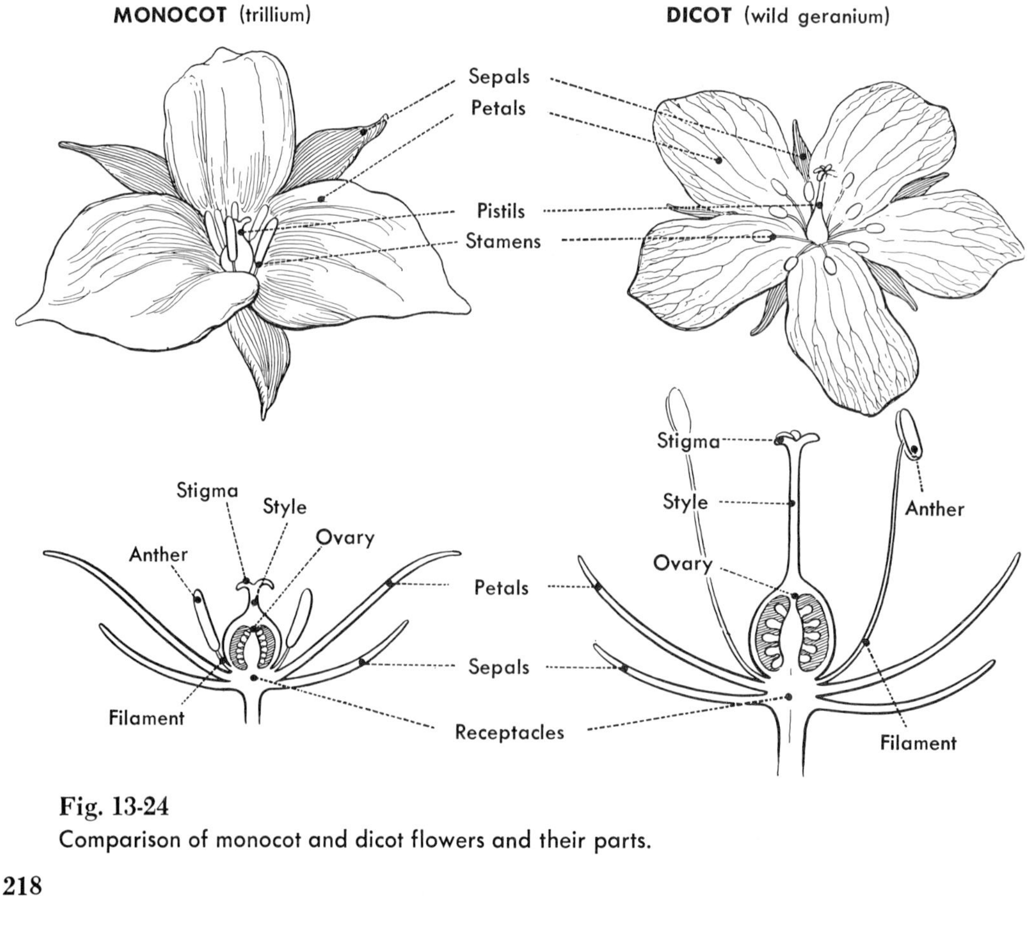

Fig. 13-24
Comparison of monocot and dicot flowers and their parts.

carpels (Fig. 13-25). The stamens are frequently the most numerous of floral parts and, in general, are separate from one another. When flower parts fuse, they are said to be **coalescent.** However, even in a bilaterally symmetrical flower such as the sweet pea (Fig. 13-26), the parts may be partially fused. The sepals are generally the same in number as the petals, but a flower is **gamosepalous** when the sepals are fused. The petals are usually fewer in number than are the stamens, and they rarely exceed ten. In some flowers the petals are fused to form a continuous corolla, a characteristic termed **gamopetalous.** In others they may be fused to form a keel (Fig. 13-26) or a similar modification. Some of the floral parts of the sweet pea are different; that is, the parts are not symmetrical. This makes the pea flower **irregular,** a characteristic of most legumes. Sometimes the sepals, petals, or even the stamens will fuse to one another. When two or more series of parts are united they are referred to as **adnate.** The lack of sepals makes the flower **asepalous;** the lack of petals makes the flower **apetalous;** and the flower is said to be **naked** if both are missing. A flower that has stamens but no pistil is said to be **staminate;** when it has a pistil but no stamens, it is **pistillate.** If both staminate and pistillate flowers are distinct from one another but on the same plant, as in corn,

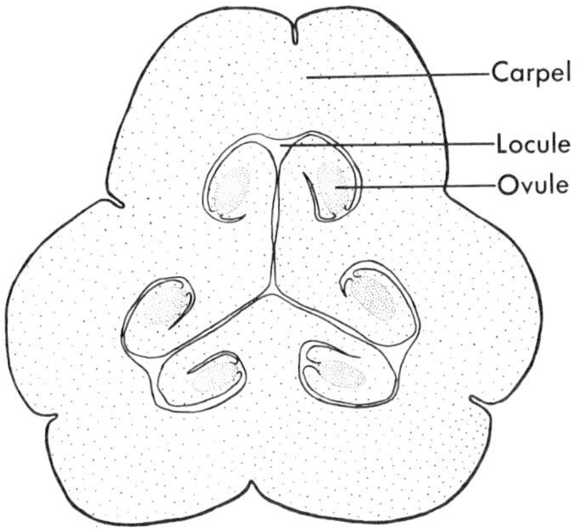

Fig. 13-25
Cross section of an ovary showing the relationship of the parts. This ovary has three locules.

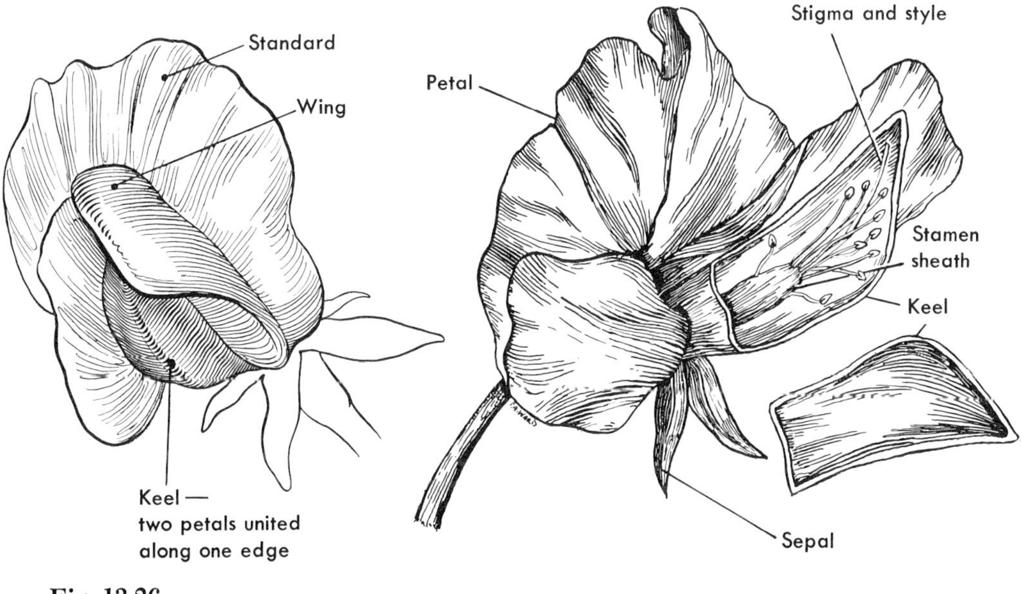

Fig. 13-26
Sweet pea flower, an example of an irregular flower.

the plant is called **monoecious.** If the flowers are on separate plants, the plant is called **dioecious.**

If the flowers arise directly from the base of the plant or are produced singly in the axil of a leaf, they are said to be solitary. When flowers are grouped together into clusters, they are called an **inflorescence.** An example of an open inflorescence is the dandelion. The shape and arrangement of inflorescences vary considerably. The inflorescence is **indeterminate** if the first flowers to open are at the base or the outside of the cluster (Fig. 13-27, *A*). If the first flowers to open in the cluster are at the tip or the center of the inflorescence, it is called **determinate.** Typical examples of both determinate and indeterminate inflorescences may be found in Fig. 13-27.

Other ways to classify flowers are by the variations in the relationship of the ovary to other flower parts and by the type of placentation. In some plants all the floral parts arise below the pistil. If the

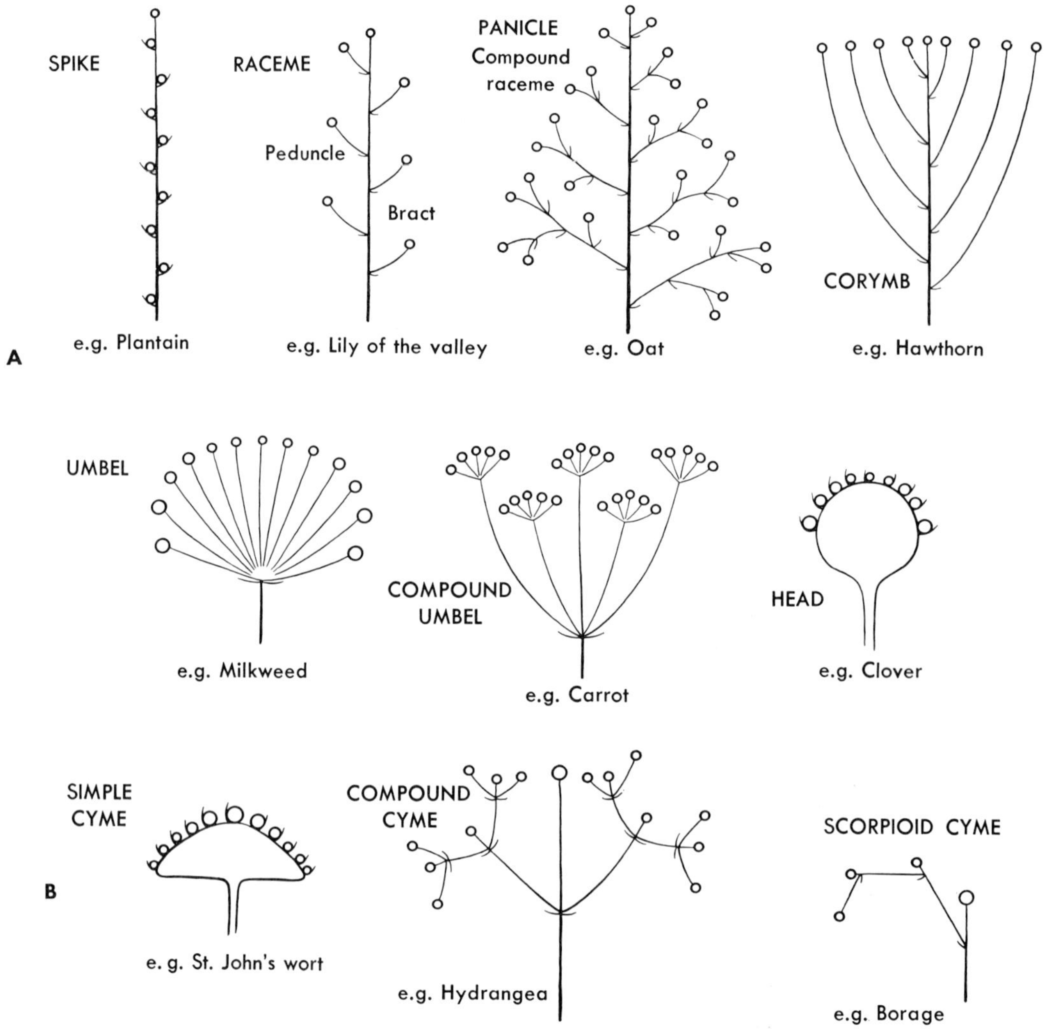

Fig. 13-27
Types of inflorescences. **A,** Indeterminate (racemose). **B,** Determinate (cymose).

calyx, corolla, and stamens originate under the ovary of the pistil, the flower is **hypogynous** and the ovary is superior; if the floral parts arise above the ovary, which is embedded in the receptacle, the flower is **epigynous** and the ovary is inferior; or if the floral parts are fused at their base and arranged around and above a "free" ovary, the flower is **perigynous** (Fig. 13-28). The ovule is attached to the **placenta** of the ovary wall by a short stalk known as a **funiculus** (Fig. 13-29). If the placenta is on the ovary wall, the placentation is **parietal;** if it is on a central axis, the placentation is **axial;** and if it is borne on a stalk that extends from the base of the ovary into the cavity, the placentation is **central** (Fig. 13-29).

Pistils vary in different plants. It is called a **simple pistil** when it consists of only one carpel such as in the bean. When a pistil represents the fusion of two or more carpels, as in the lily, it is called **compound** (Fig. 13-30). A compound pistil may have more than one style or stigma (Fig. 13-30).

The number of carpels and the arrangement of seeds vary according to the species of flower. The carpels may coalesce to form a single chamber as found in bean and pea pods, or they may remain separate as in the apple.

All these variations of flower parts are used to classify as well as to identify plant orders, families, and species.

Pollination. Except on rare occasions, ovules will not develop into seeds unless pollen is transferred from the anther of the stamen to the stigma of the pistil. This transfer of pollen is called **pollination.** It cannot occur until after pollen is produced in the pollen sacs of the anther.

The first step in the accomplishment of reproduction in seed plants is the transfer of pollen from the anther to the stigma. If the pollen is transferred from the anther to the stigma of the same flower, the process is called self-pollination. If it is transferred from the anther of one flower to the

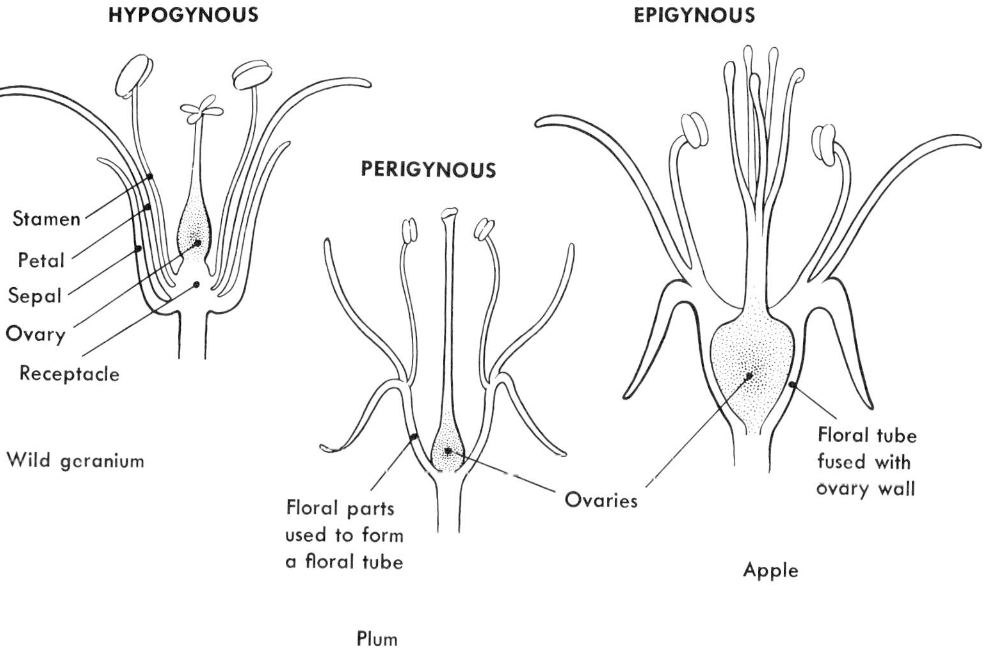

Fig. 13-28
Types of ovaries (diagrammatic).

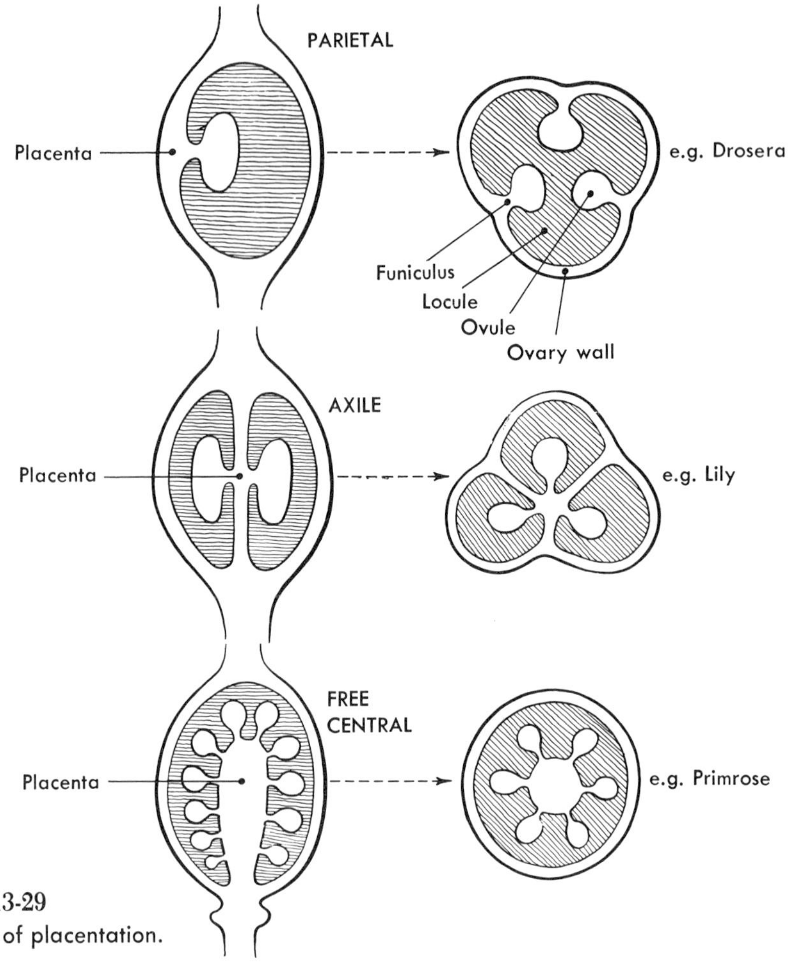

Fig. 13-29
Types of placentation.

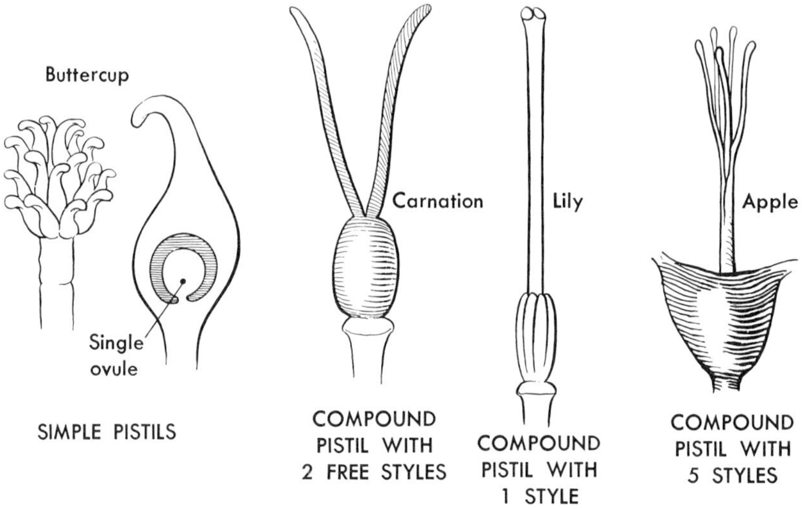

Fig. 13-30
Different kinds and arrangements of pistils.

stigma of a flower of another plant, the process is **cross-pollination.** When **self-pollination** occurs, the anther usually lies so close to the pistil that the pollen is transferred directly and even before the flower opens. In the majority of plants the transfer requires some external agent. As a result, pollen is frequently carried from one flower to another.

Wind and insects are the two most important pollinating agents. Wind-pollinated flowers are inconspicuous, are exposed prominently on the plant, and have poorly developed perianths. Their pollen is light and dry and is produced in tremendous quantities. In the spring the yellow dust (pollen) from wind-pollinated plants can be seen covering automobiles and forming a yellow blanket over ponds. It can be carried hundreds of miles by the wind and is one of the principal causes of hay fever and asthma.

The flowers of insect-pollinated plants are usually conspicuous and brilliantly col-

Fig. 13-31
Nectaries on flower petals attract insects that often act as pollinators. (USDA photograph.)

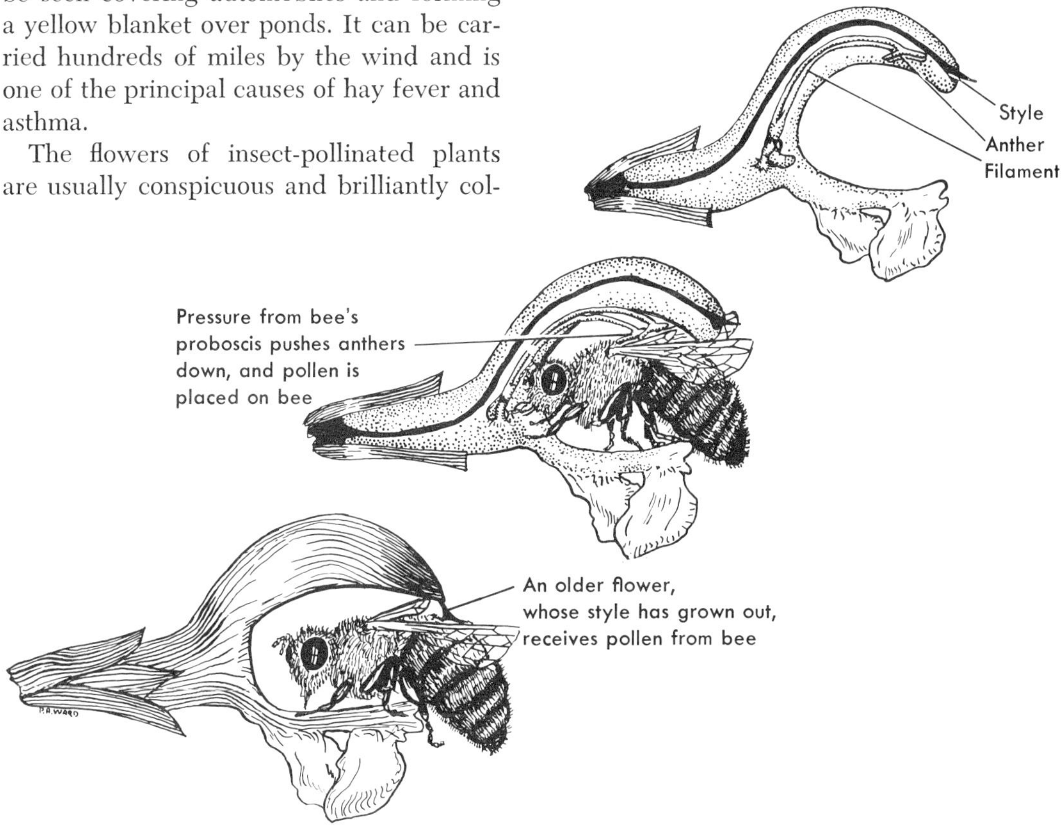

Fig. 13-32
Insect pollination (pollen carriers) in salvia. The bee travels from flower to flower carrying pollen from one to another. The anthers of young flowers ripen and release pollen, then dry. The older flowers have a mature style that receives the pollen.

ored and possess marked odors or nectaries (Fig. 13-31) which attract insects to them. The pollen is usually sticky and adheres readily to the body of an insect. In this way it is carried from flower to flower and deposited on the stigma (Fig. 13-32). Bees are more important than any other insect in pollination, although flies, butterflies, moths, and even beetles and bats account for some pollen transfer. Few flowers are pollinated by water, compared with those pollinated by insects and wind.

The subject of pollination and pollination ecology is much too involved for a further discussion in the space allowed.

FRUITS

Fruit is the usual result of the growth and development of a fertilized flower; therefore it is to be expected that great variation will be found in these structures, paralleling those found in the flowers. When parts other than the ovary adhere to or enclose it, the fruit is called an **accessory fruit.** A familiar example is the strawberry, in which individual fruits or achenes are borne upon a sweet and fleshy receptacle (Fig. 13-33). A strawberry is an accessory fruit that is also designated an **aggregate fruit.** The apple is another type of accessory fruit, but it is called a **pome** (Fig.

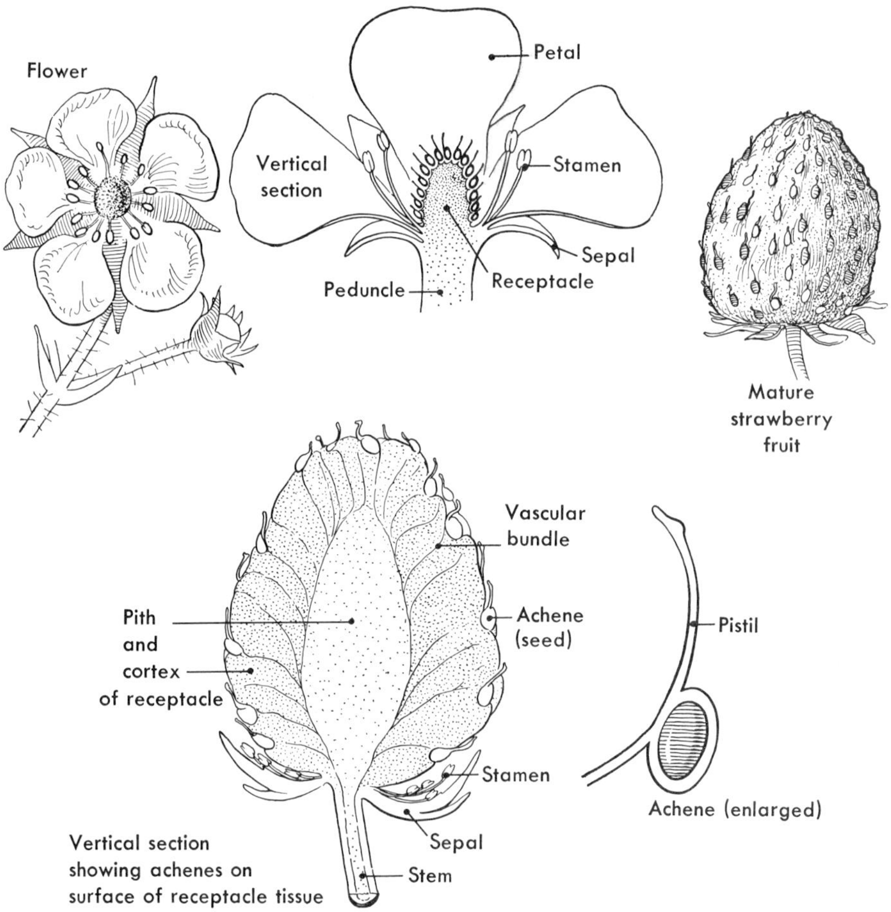

Fig. 13-33
Development of an aggregate fruit (strawberry) from the receptacle of the flower.

12-4). The ovaries of the apple become surrounded by the enlarged receptacle and the tissues of the floral tube. The true fruit of the apple is the **core,** which is not eaten. Therefore, in accessory fruits such as the strawberry, apple, pear, and quince, the edible portions are not mature ovaries, but instead are the stem and floral tubes in which the ovaries are embedded.

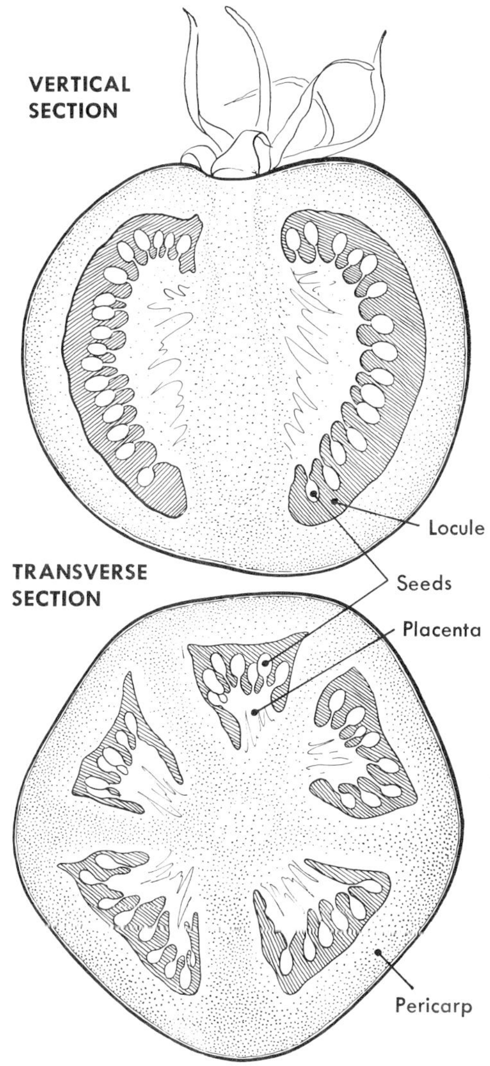

Fig. 13-34
Example of a berry (the fruit of the tomato plant).

Fruits are usually classified as **simple** aggregate, or **multiple.** A simple fruit is the result of the maturation of the ovary of a single pistil, whether it is simple or compound. The ovary of the pistil may be superior or inferior; thus the fruit may be composed of the ovary alone or of the ovary and any other part of the flower. Simple fruits may also be dry or fleshy. A **fleshy fruit** is one in which all or most of the pericarp is soft and fleshy. A fleshy fruit in which the entire pericarp is soft is called a **berry.** Good examples of berries are grapes, bananas, and tomatoes (Fig. 13-34). Watermelon, cucumber, and cantaloupe are berries having a hard outer covering, and citrus fruits are berries with leathery coverings. Another kind of fleshy fruit is called a **drupe.** Peaches, cherries, and plums are drupes (Fig. 13-35) in which the **exocarp** is thin, the **mesocarp** thick and fleshy, and the **endocarp** hard and stony. Drupes are fruits that have a stone or pit in the center and only one seed.

A second kind of simple fruit is called a **dry fruit.** In a dry fruit the **pericarp** becomes either leathery, papery, or woody. Dry fruits are further differentiated by the manner in which they split open to expel the seeds. If they split in a definite pattern, they are **dehiscent fruits.** If they do not split open along a definite seam they are **indehiscent fruits.** Legumes (pea and bean) are good examples of dehiscent fruits (Fig. 13-36). The samara seeds of the maples and the acorns of the oak are examples of indehiscent fruits.

An aggregate fruit is a cluster of from several to many ripened ovaries produced by a single flower (Fig. 13-33) and borne on a single receptacle. Each individual ripe ovary may be a drupe, for example, blackberries, or an **achene,** in which a single seed is inseparable from the ovary wall except where it is attached to the pericarp, as in the dandelion.

Finally, multiple fruits are a cluster of several ripened ovaries produced by many

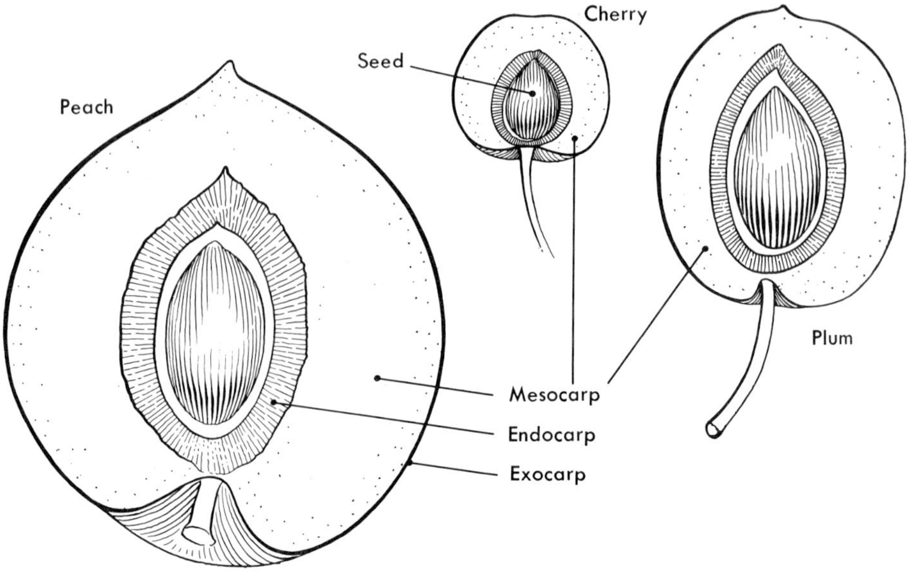

Fig. 13-35
Examples of drupes.

flowers on the same inflorescence. The mulberry, fig, and pineapple (Fig. 13-37) are multiple fruits. Although many fruits are popularly regarded as vegetables, few, if any, vegetables are considered fruits. However, there is one vegetable, rhubarb, that is used as a fruit. The edible portion of that plant is the petiole of the leaf and hence does not meet the botanist's definition of a fruit. The date and the grapefruit are among the many economically important fruits used by man.

SEEDS

Seeds vary in size, shape, and structure. In fact, many plants may be recognized by their shed seeds alone.

The structure and texture of the seed coat vary. In some it is thin and fragile; in others it is hard and stony or even fleshy. The attachment of the funiculus to the placenta is recognized externally as a scar or **hilum.** Leading away from the hilum may be a short ridge called the **raphe,** and on the opposite side of the hilum may be the micropyle opening.

Seed dispersal. Although the new generation of the flowering plant begins with the embryo in the seed, it is a limited beginning. The growth of the embryo is usually arrested until the seed becomes detached from the parent plant and is dispersed. It is after the seeds are disseminated that germination takes place. In the meantime the seeds may have survived extremes of heat, cold, and drying.

Many plants develop novel and valuable structural modifications for the dispersal of seeds over great distances. It becomes difficult to separate the methods of dissemination of seeds from those of fruits, since the same or similar modifications are noted in both. Furthermore, it is often difficult to distinguish the seeds from the fruit in some plants, since they are often both disseminated as a unit. Seeds are dispersed by agents such as wind, water, birds, insects, and mammals, including man.

Seeds that are small and light in weight are usually dispersed by the wind. Orchid seeds are so small that more than a million may be found in a single capsule. Thus they are easily airborne. Others (Fig. 13-38) may have plumes, modified floral parts that aid in their dispersal by the wind. Winged seeds, such as those of the maple,

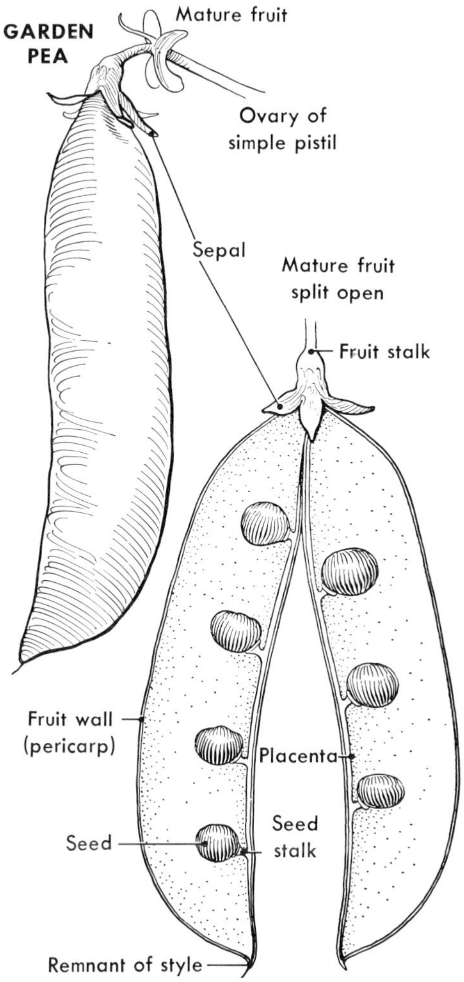

Fig. 13-36
Pea pods with peas (the fruit), an example of a legume.

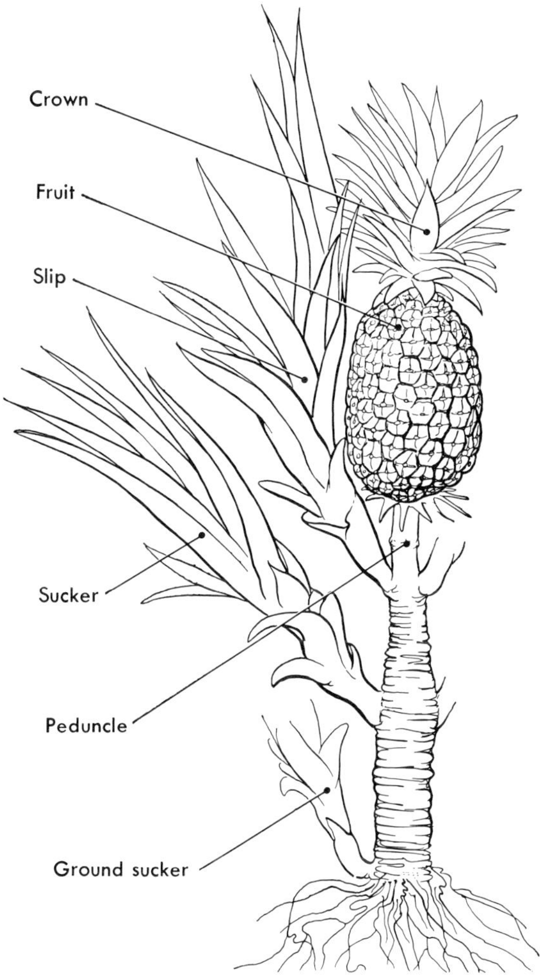

Fig. 13-37
Pineapple plant showing parts, including the fruit.

are a common sight in the spring of the year.

The action of water can be almost as important as that of wind in disseminating seeds. Large and small seeds and fruits are transported over long distances by the washing of rain, especially during flash floods, into streams and rivers, particularly alpine rivers. Seeds that have light, spongy coats or those relatively unaffected by water are modified water-dispersed seeds. Contrary to popular belief, coconuts are not fit for transportation by ocean currents and owe their widespread distribution instead to native transportation.

Animals are responsible for the dispersal of many seeds. Birds and other animals eat the fruits and seeds of plants, which pass through the digestive tract unharmed, to be dispersed during their travels. Other seeds have hooks, bristles, or spines, such as the cocklebur (Fig. 13-38), which catch in the fur of animals or the clothing of man, to be transported over a wide area. Through

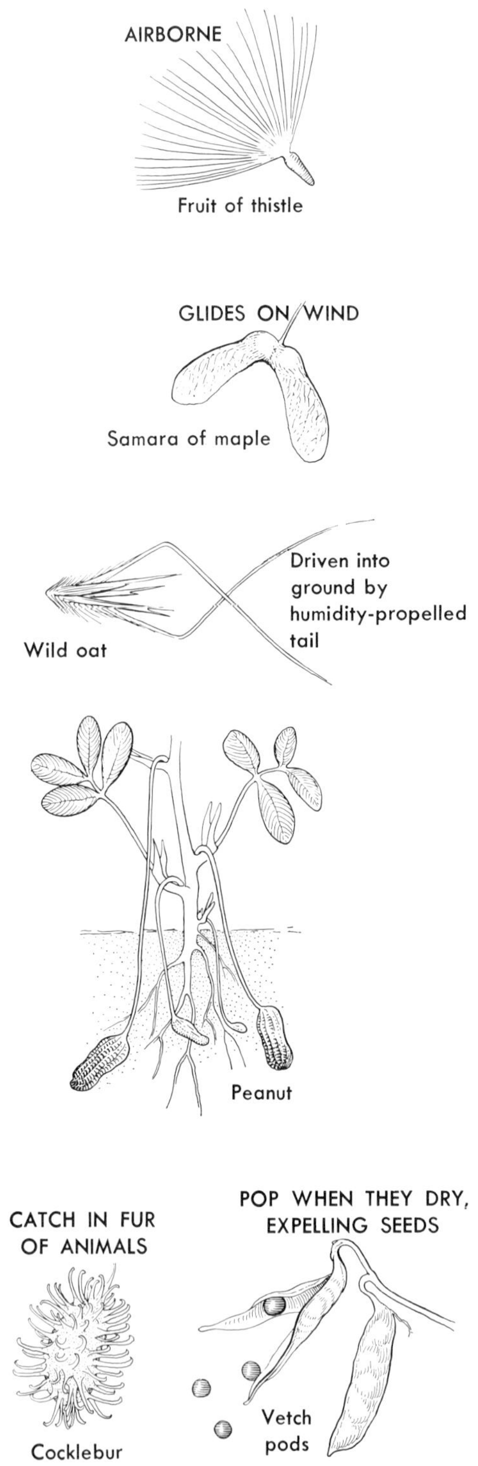

Fig. 13-38
Examples of seed and fruit dispersal.

commerce and travel, man accounts for the dissemination of a variety of seeds. Other adaptations for seed dispersal are the exploding of the dry pod of the vetch (Fig. 13-38). Many forms of seed-catapulting take place, but the most striking is the famous squirting cucumber of Italy, *Ecballium elaterium.* The internal turgor pressure extends the outer wall of the fruit, and at a certain point the stalk end is forced out exactly like the cork of a champagne bottle. The fruit wall contracts, and the entire juicy contents, including the seeds, are squirted out. In contrast, some seeds, such as the peanut, are prevented from being dispersed. The peanut fruit contains seeds that develop on a long "stem" that extends into the ground (Fig. 13-38).

DISTRIBUTION MECHANISMS OF PLANTS

All plants have some kind of dispersal mechanism, most of which are discussed elsewhere in this book. Therefore the following outline will serve merely as a summary. Although plants are usually thought of as stationary organisms, at some time during the life history of every species locomotion is possible and usually extensive.

1. Floating distribution
   a. Entire vegetative plant—unicellular and multicellular algae, bacteria, and some vascular plants are dispersed by both freshwater and saltwater currents.
   b. Vegetative parts—storm action or partial decay sometimes results in the dispersal of plant parts that will continue to live and grow into a new plant by rooting, once a new location is reached.
   c. Reproductive parts—spores and zygotes of many plants are dispersed by floating in water.
   d. Seeds—seeds are dispersed by water currents.
2. Swimming
   A few of the algae are flagellated in the vegetative stage and are dispersed by active swimming. Many algae have flagellated gametes and zygotes that allow this type of dispersal.
3. Ground distribution
   Some vegetative dispersal is possible by the

growth of runners, either as stems or roots. New plants become established in this manner, finally breaking away from the parent plant. A few species become established in new areas by the rooting of growing leaves after becoming detached from the parent.

4. Aerial distribution
   a. Vines—the growth of vines over large areas is the simplest means of aerial distribution.
   b. Spores—the terrestrial cryptogams depend almost entirely on the dispersal of spores formed vegetatively, rarely by sexual reproduction. These light, microscopic plants are distributed by wind currents and often have worldwide distribution. In past geological ages, before the evolution of seed plants, this means of dispersal was nearly universal among terrestrial plants.
   c. Seeds—many seeds have, as accessory parts, fine hairs or winglike projections that are caught by wind currents, to be carried long distances. Most are lost—but even with the establishment of one in a new locality, a new population is started. It is, of course, advantageous to a plant species to have the widest possible seed dispersal because the seed stage is precarious. Many things can happen to prevent the germination of seeds; therefore great numbers are produced by the plant. The greater the number of seeds, the better they are protected; and the more generally they are distributed, the better are their chances for germination.
5. Distribution by animals
   a. Fruit—many plants produce a fleshy fruit that may be eaten by animals who may carry it for some distance before it is consumed. If large, the seeds are removed and tossed aside to germinate. Small seeds are eaten with the fruit and passed through the digestive tract unharmed by chewing or digestive juices, finally to be deposited on the soil along with the fertile feces, providing an ideal situation for germination and growth. Many barren islands remote from continental land masses have been populated by plants brought to them in the intestines of birds.
   b. Seeds—nuts are gathered and stored by rodents and similar animals. A few roll away and may germinate in new ground. Other seeds may be equipped with spines that catch in the fur (or clothing) of animals and are thus transported. Sometimes complicated mechanisms are devised; for example, the seeds of parasitic mistletoe are transported by the missel thrush. This bird is especially fond of mistletoe berries but does not eat the seeds. The plant is dependent upon finding a suitable branch of a juniper tree to parasitize. The missel thrush rubs the seeds of the berries off against the branch on which it perches, and the seeds become fixed to the branch by the viscid internal flesh of the berry. A large number of woodland plants are dispersed by ants that pick out seeds made attractive to them by special oil bodies or **elaiosomes.** The desire of the ants to imbibe this oil causes them to gather the seeds and carry them off. They do not eat the seeds, and they are cast away after the oil is consumed, thereby dispersing them.
   c. Pollen—dispersal of pollen by animals (bees, etc.) serves as a means of mixing genetic traits throughout the range of the species and does not contribute directly to the extension of the range. However, when these genes are thus mixed, the mutation rate may be higher and the chance of adaptation to new areas somewhat greater.

This summary covers only the principal dispersal mechanisms. Other special features are discussed in books on ecology and plant distribution.

PLANT FORMS IN RELATION TO ENVIRONMENT

Plants, like animals, have definite shapes and forms according to the habitat in which they live. Animal form is usually geometrical, whereas plant form is randomly organized. Even so, it is not difficult to see habitus types in plant species and to correlate them with their habitat. Some of the more obvious of these types are listed and described.

Hydrophytes. Plants growing in water are characterized by lanceolate leaves if they are submerged, with stems reaching nearly to the surface and stem length depending upon the depth of the water, or with floating leaves having dorsal stomata for direct contact with the atmosphere. Hydrophytes may have air bladders to keep the plant afloat. All such plants are turgid since water is never a limiting factor, and the tissues are usually thick, with large alveolar spaces.

Freshwater habitats are often separate climax situations, with many cyclic activities going on independently of land plants and animals. One such situation was created at the Panama Canal when several river valleys were flooded for the canal. Large mats of floating vegetation (Fig. 13-39) were formed that not only are discrete ecological niches but also that provide vast areas for the breeding of larvae of anopheline mosquitoes, the adults of which are carriers of malaria.

Mesophytes. Most terrestrial species are mesophytes, the body form of which varies according to the closeness of an adequate water supply for the roots and the method of obtaining maximum exposure to light. This type of plant is characteristic of most of the earth's surface, and their particular variations and characteristics can only be described in a lengthy systematic floral catalog.

Xerophytes. Desert and arid land plants provide a series of limited habitus features familiar to any desert traveler. The spiny cactus is characterized by lack of leaves, and the chlorophyll is confined to thick, water-filled stems that form a storage system capable of holding the large amounts of water absorbed during the limited period of rainfall. The stomata of these plants are usually located within pits on the stem surface, and the body of the plant is covered by a waxy secretion, making the plant nearly impervious to loss of water by transpiration. Other desert plants may resemble mesophytes in many respects but are often leafless during much of the year, or the leaves are reduced to small leaflets. The former produces leaves during the rainy season only and remains nearly dormant throughout the rest of the year. The root system of xerophytes either extends very deep into the ground to reach deep, subsurface water or is very shallow but widespread, as in cactus, to absorb as much of the surface water as possible during the brief rains. This action is called **hydrotropism** (Fig. 13-4).

Fig. 13-39
Large mats of floating vegetation in Gatun Lake in Panama.

Aerial plants. A few plants live in aerial habitats completely independent of soil and are therefore without a root system. Spanish moss in the southeastern United States is an example. These plants are dependent upon a humid climate and absorb water through their stomata directly from the atmosphere. Other plants, such as certain bromeliads of the tropics, are epiphytes, some saprophytic and some not.

Other types. Saprophytes, commensals, and parasitic plants are likewise characterized by a particular body form, easily recognized in all cases. The Indian pipe is an example of a seed-producing saprophyte. Certain lichens form a type of Spanish moss on trees in California (not to be confused with the Spanish moss bromeliad of the southeast). Orchids are also examples of this, but are capable of extensive photosynthesis as well. Mosses and fungi may take on particular forms in special habitats. There are many other examples.

Sometimes plants have devised elaborate mechanisms to aid in obtaining food. There are many species of insectivorous plants (Fig. 13-23) with trapping devices for the capture of insects, the decaying bodies of which are a supplementary source of minerals and salts needed for the metabolism of the plant. Plants such as the Venus' flytrap and the pitcher plant mentioned earlier are, by their insect-capturing abilities, capable of living in areas deficient (for them) in nitrogen. These are usually boggy areas that are low in nutrients because of the downward movement of water that prevents the accumulation of minerals in the soil.

SPECIAL TERMS USED IN CHAPTER 13

abscission zone
accessory fruit
achene
adnate
adventitious bud
aerial stem
aggregate fruit
alternate leaves
annual
anthocyanin
apetalous
asepalous
axil
axile placentation
axillary bud
berry
biennial
blade
bud
bud scale
bud scar
bulb
central placenta
coalescent
compound leaf
compound pistil
core
corm
cross-pollination
deciduous
dehiscent fruit
determinate flower
dioecious
drupe
dry fruit
elaiosome
endocarp
epigynous
exocarp
fibrous root
fleshy fruit
funiculus
gamopetalous
gamosepalous
hilum
hydrotropism
hypogynous
herbaceous stem
imperfect flower
incomplete flower
indehiscent fruit
indeterminate flower
inflorescence
internode
irregular flower
leaf axil
leaf mosaic
leaf scar
lenticel
liana
mesocarp
monoecious
multiple fruit
naked flower
node
opposite leaves
palmate
parietal
perennial
perfect flower
pericarp
perigynous
petiole
pinnate
pinnately bicompound
pinnately compound
pistillate flower
placenta
pollination
pome
prickles
rachis
raphe
rhizome
self-pollination
shrub
simple fruit
simple pistil
staminate flower
stipule
stolon
taproot
tendrils
terminal bud
thorn
tree
trifoliolate
tuber
veins
vine
whorls
woody stem

QUESTIONS TO CONSIDER

1. Give the function for each of the following: root, stem, leaf, flower, fruit, and seed.
2. What type of adaptation is found in roots?
3. What type of adaptation is found in stems and leaves?
4. What type of adaptation is found in the flower?
5. What type of adaptation is found in the fruit?
6. How does each of these adaptations correlate with the effective environment?
7. A showy flower is likely to be an adaptation of what? Spines on the leaves and stem? An underground stem?

DISCUSSION FOR LEARNING

1. Many tropical plants have very large leaves, and most desert plants have very small leaves, yet cactus grows equally well in both habitats. How do you explain this? Include in your discussion the effect of watering desert plants.
2. Plants in arctic and alpine habitats grow very rapidly. At the same time, they require a very long photoperiod. What happens when such plants are grown in more southern, warmer climates? Consider in your discussion the factors that induce flowering.
3. Discuss the reasons why aquatic plants cannot grow on land and why land plants cannot grow in water. Are there exceptions?

Chapter 14

Plant life cycles

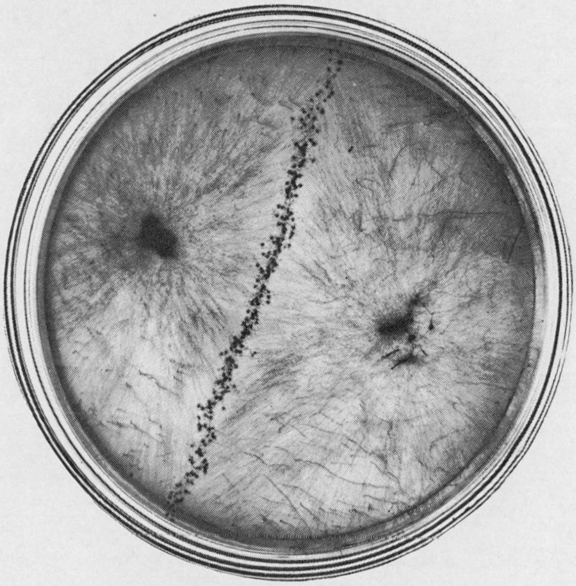

Courtesy Carolina Biological Supply Co., Burlington, N. C.

Life cycles of plants have long been fundamental to the understanding of plant evolution. At the same time, students have struggled to master these cycles. With this in mind, twenty-two life cycles are set apart in this section and presented in a comparative sequence.

The photograph above illustrates the fundamental aspect of sexual reproduction. It shows two sexually compatible strains of the black bread mold. Growing close to each other, they form gametes, and sexual union takes place. Such associations are responsible for the variety of life cycles.

Each one of the life cycles in this section is illustrated by a series of habitus drawings as well as by a diagram of the life cycle. The descriptions follow a fixed format, regardless of the variation in the cycle illustrated. The classification of the example chosen is given; then the vegetative characteristics of the plant are described. A discussion of asexual reproduction then follows, accounting for the vegetative plant, sporophyte, **sporangium,** and spore; next sexual reproduction is described, accounting for the gametophyte, antheridium, **archegonium** (or **oogonium** in lower plants), sperm, egg, zygote, and embryo in each case. The means of dispersal is also described because reproduction not only provides for the formation of new individuals but also for the dissemination of the species. If one or more of these stages is missing, this is so stated in the proper place by the word "none." It should be remembered, however, that in many cases the relationship between these stages from plant to plant is functionally analogous rather than structurally homologous. The student should never assume that, because a phase of the life cycle is described after one of these headings, it is equivalent homologously to that structure in other plants.

LIFE CYCLE 1—NOSTOC Sp.

DIVISION 1 Cyanophyta

CLASS Myxophyceae

Characteristics

The vegetative body of *Nostoc* sp. consists of short filaments of rounded cells embedded in a firm matrix of gelatinous material (Fig. 14-2). These filaments are referred to as trichomes, and each may have one or more enlarged cells, the **heterocysts.** The filaments are grouped together into colonies, which at first are microscopic but which grow into balls 30 to 60 mm. in diameter. These balls are jellylike, sometimes hollow, and may be irregularly shaped. The species of this genus are either aquatic or terrestrial. Many occur as floating colonies in sunny ponds and others in swift streams attached to stones. Some live as deep as a meter in the soil, and still others have a symbiotic association with certain fungi to form a lichen.

Asexual reproduction

Vegetative reproduction. The breaking apart of the colony may take place as a result of storms and similar disturbances, but otherwise there is no usual vegetative propagation.

Sporophyte. There is no true sporophyte; however, some of the cells are differentiated into **hormogones** forming a **hormogonium.** The hormogones are sporelike in function. They can germinate to produce new colonies.

Sporangium. None.

Spore. None. Thick-walled **akinetes** are formed from any vegetative cells in mature colonies that act as spores and that are capable of germinating to form a new filament in the colony.

Sexual reproduction

No form of sexual reproduction has been observed in *Nostoc* sp. or related genera.

Means of dispersal. The floating balls of filaments provide the only dispersal mechanism for these plants.

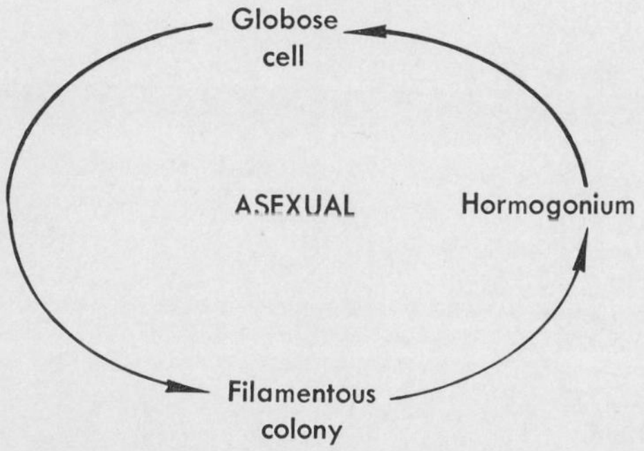

Fig. 14-1
Life cycle of *Nostoc* sp.

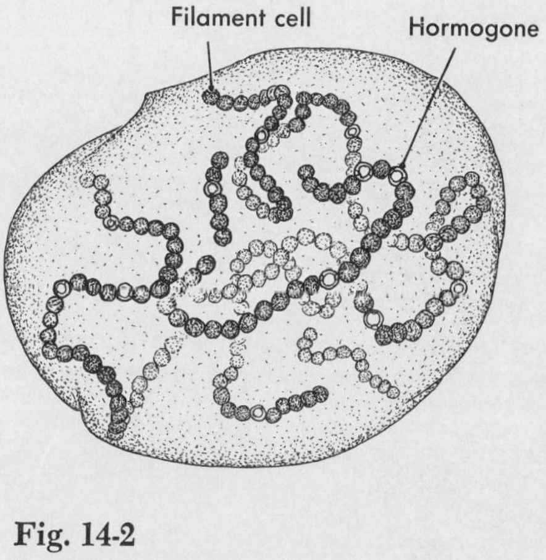

Fig. 14-2
Nostoc sp. Vegetative stage.

LIFE CYCLE 2—VAUCHERIA Sp.
DIVISION 7 Chrysophyta
CLASS Xanthophyceae

Characteristics

This genus is not really representative of the class because it is one of the most advanced forms. It occurs in both fresh-water ponds and moist or muddy land areas, where it may form a dark green, feltlike mat. It is filamentous, with very few cross-walls, and forms large, multinucleated cells (Fig. 14-4). The filaments may extend for a foot or more within the mat. The absence of crosswalls makes these filaments a syncytium. Thousands of small, disk-shaped chloroplasts occur in the cytoplasm. If a branch of the filament is injured, the opening may be sealed off by the generation of a crosswall. Food is stored only as oil droplets. This is a commonly available species and therefore is used as an example of the yellow-brown algae.

Asexual reproduction

Vegetative reproduction. There are no vegetative structures for asexual reproduction. However, the vegetative plant produces zoospores as well as gametangia.

Sporophyte. None.

Sporangium. Zoospores are produced by a terminal division of a filament, the zoosporangium. This area is separated from the remainder of the filament by a cell wall but is still multinucleated.

Spore. The spore is in the form of a multinucleated, flagellated zoospore of relatively large size. It is liberated under suitable conditions through a crack in the wall of the sporangium. It swims slowly away to a new site, where it settles down to form another vegetative thallus.

Sexual reproduction

Gametophyte. Sexual reproduction is effected by the formation of adjacent sex organs on the same filament. This area, then, is analogous to the gametophyte.

Antheridium. The antheridium is a special branch with a crosswall at the tip of the branch separating and isolating a small number of nuclei with the surrounding protoplasm. This antheridium branch is elongate and somewhat narrower than the main filament of the plant. The apex of the branch is tapered, and when the branch

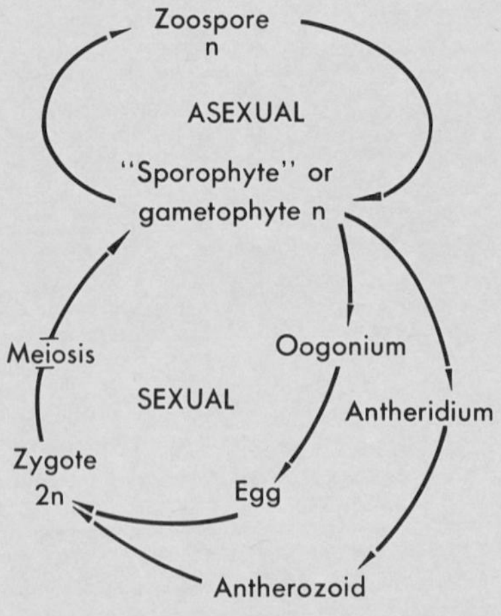

Fig. 14-3
Life cycle of *Vaucheria* sp.

is mature, a small pore opens for sperm release.

"Archegonium." The oogonium is formed from a branch near the antheridium. It is separated from the rest of the plant by a crosswall at the base. It swells to form a large rounded oogonium. When the branch is mature, a pore similar to that of the antheridium is formed.

Sperm. Uninucleated, biflagellated antherozoids are formed within the antheridium. These are released through the pore and are free swimming.

Egg. The single egg cell produced in the oogonium has only one nucleus and occupies the entire space within the oogonium.

Zygote. The antherozoids swarm to the pore of the oogonium. Even though several may enter, only one of them penetrates the egg. It does not fertilize the egg in the usual way, however; instead, a single sperm nucleus penetrates the ovum and enlarges considerably before it fuses with the nucleus of the ovum. This then becomes a central nucleus in the zygote. As soon as fertilization has taken place, a multilayered wall is formed, and the zygote remains dormant for several months.

Embryo. The germinating zygote develops directly into a new sporophyte. However, there is evidence that the nucleus divides meiotically before germination; therefore the sporophyte may be haploid.

Means of dispersal. The zoospores, by slow but free-swimming movement, provide the principal means of dispersal for this plant.

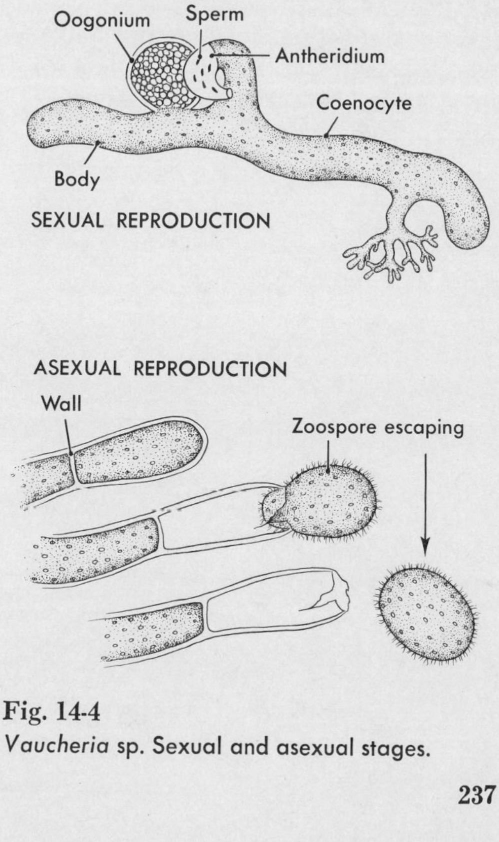

Fig. 14-4
Vaucheria sp. Sexual and asexual stages.

LIFE CYCLE 3—NEMALION Sp.
DIVISION 11 Rhodophyta
CLASS Rhodophyceae
Characteristics

Nemalion sp. produces a narrow, branching, wormlike thallus, reddish brown in color and composed of a single row of cylindrical cells, each with several disk-shaped chloroplasts (Fig. 14-6). The reproductive organs are inconspicuous at the apices of some of the branches. The plants are small, are several inches in length, and are attached to rocks. It is a marine species growing in the intertidal zone during the summer months and is exposed during low tide.

Asexual reproduction

Vegetative reproduction. None. If the plants become detached and floating, they will usually die.

Sporophyte. The sporophyte is represented by the **carposporophyte,** a small parasitic plant located on the gametophyte. This portion of the gametophyte is referred to as the **cystocarp,** and is developed from the **procarp.**

Sporangium. The terminal cell of the lateral filaments enlarges to become a carposporangium. The haploid carpospore is liberated from the carposporangium to become attached to some substratum.

Spore. The carpospore is liberated from the carposporangium as a naked spore. It attaches itself to some substratum and becomes enclosed within a thickened wall. Some time later it will germinate to produce a filament that becomes organized into the plant body.

Sexual reproduction

Gametophyte. The gametophyte is vegetative and is developed from the carpospore. It produces both antheridia and carpogonia, sincc it is homothallic.

Antheridium. The spermatangia are produced on the same plant as the carpogonia. Many spermatangia are produced on the terminal ends of the spermatangial branches.

"Archegonium." The oogonium is present on the carpogonial filaments. Each carpogonial filament consists of from three to five cells. The terminal cell has an elon-

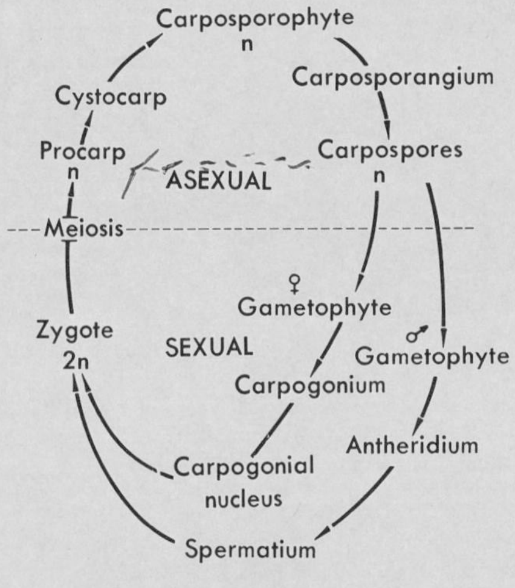

Fig. 14-5
Life cycle of *Nemalion* sp.

gated protuberance, the trichogyne, which functions during fertilization.

Sperm. The male gametes are known as spermatia. They are nonmotile, and a single spermatium is formed within each spermatangium. The single nucleus of the spermatium divides to form two male nuclei after contact with the trichogyne.

Egg. None. The single carpogonial nucleus serves as the female gamete and is found at the base of the carpogonium.

Zygote. The nonmotile spermatia rest against the trichogyne, and both male nuclei pass into it. They migrate toward the base of the carpogonium, and one of the male nuclei fuses with the female or carpogonial nucleus to form the zygote nucleus, which undergoes meiosis. Continued divisions of the daughter nuclei with the formation of cell walls produce lateral filaments.

Embryo. None as found in higher plants.

Means of dispersal. The liberation of the carpospores results in dispersal of the plants.

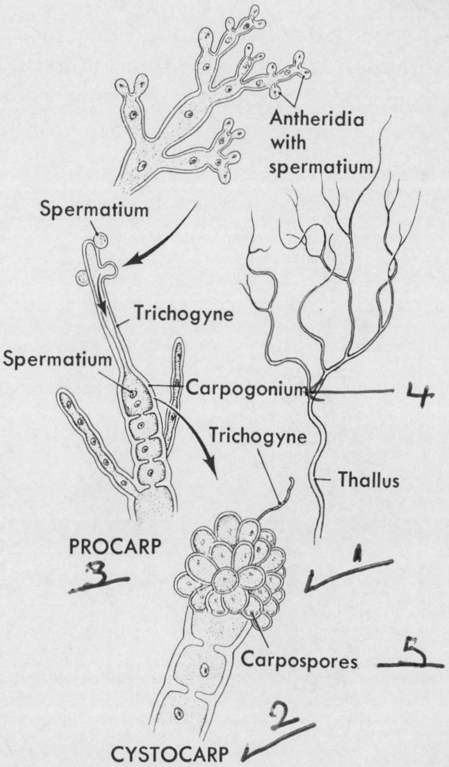

Fig. 14-6
Nemalion sp. Reproductive stages.

LIFE CYCLE 4—ECTOCARPUS Sp.

DIVISION 12 Phaeophyta

CLASS Phaeophyceae

Characteristics

The common brown alga *Ectocarpus* sp. is found in cold seawater around the world. It is attached to other kinds of algae as an epiphyte and consists of a main filament from which arise small branches (Fig. 14-8). The cells are uninucleated and contains many small plastids. The plants form small tufts several inches long on other algae, both in the tidal zone and below low tide. The life cycle of this species exhibits an alternation of generations that is characteristic of some of the higher forms of plant life. However, all the plants look similar but can be distinguished by their fruiting bodies. Hence, the plant is heterothallic.

Asexual reproduction

Vegetative reproduction. None.

Sporophyte. The sporophytes are indistinguishable from the filamentous gametophyte except as they produce zoosporangia (Fig. 14-8). The cells of the filament are diploid.

Sporangium. The sporangia are most typically **unilocular.** They are formed from a single uninucleate cell at the tip of a filament. The nucleus in this cell divides many times, producing, as a result of meiosis, anywhere from 32 to 128 small, biflagellated, and haploid zoospores.

The sporophyte of *Ectocarpus* sp. may produce on the same filament in addition to the unilocular sporangia, structures called **plurilocular** or neutral sporangia. They resemble the gametangia of the gametophyte filaments but produce zoospores instead of gametes. However, these zoospores are not produced by meiosis. Therefore they are diploid and upon germination give rise to other diploid filaments like the parent plant.

Spore. Both haploid and diploid zoospores are produced. The former produce gametophyte filaments, whereas the latter

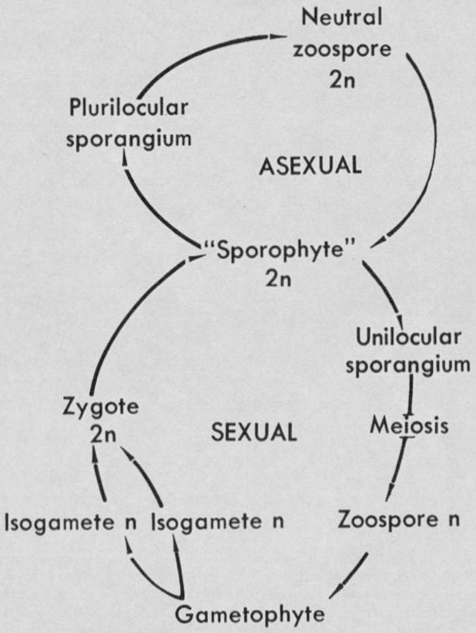

Fig. 14-7
Life cycle of *Ectocarpus* sp.

produce sporophytes. Both are biflagellated and free swimming.

Sexual reproduction

Gametophyte. The zoospores produced within the unilocular sporangia are haploid and develop into the haploid filament (gametophyte generation). These filaments are morphologically like those of the sporophyte filaments but are obviously different physiologically.

Antheridium. None.

Archegonium. None.

Sperm and egg. None. However, this alga does produce isogametes that have been demonstrated to be physiologically different. Those from a single gametangium will not fuse together. They must come from two different gametangia. Should they not fuse, a single gamete may develop parthenogenetically into a new gametophyte filament.

Zygote. Two haploid isogametes from different filaments fuse to proudce a diploid zygote. The zygote formed germinates and produces a new filament (the sporophyte filament) that bears zoosporangia. Thus there is an alternation of a sporophyte and a gametophyte generation.

Embryo. None.

Means of dispersal. Dispersal is almost entirely dependent upon the free-swimming neutral zoospores and the zygotes.

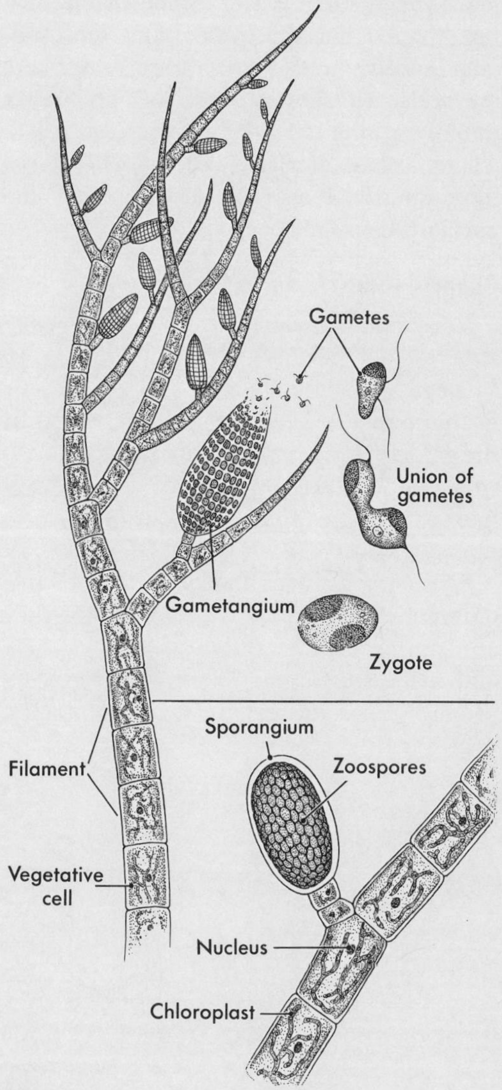

Fig. 14-8
Ectocarpus. sp. Vegetative and reproductive stages.

LIFE CYCLE 5—FUCUS Sp.
DIVISION 12 Phaeophyta
CLASS Phaeophyceae

Characteristics

Fucus sp. is so commonly available that it is used as an example of the brown algae even though it is not really representative of the type of life cycle of the class. The thallus is a dichotomously branching shoot arising from a flattened, disk-shaped holdfast (Fig. 14-10, *A*). The "branches" are ribbon shaped but short and blunt with evident midribs. Hollow, bladderlike air sacs are scattered along the thallus and serve to buoy up the plant when it is submerged. These plants are widely distributed throughout the cold coastal waters of the northern hemisphere.

Asexual reproduction

Vegetative reproduction. No special structures. The only known method is by fragmentation. Broken pieces may live as floating plants. This is especially true of the genus *Sargassum.*

Sporophyte. The *Fucus* sp. plant is diploid—hence the sporophyte or spore-producing plant.

Sporangium. Since the spores are of two different sizes, the sporangia wherein they are produced are called **microsporangia** and **macrosporangia.** These sporangia are produced in small cavities or **conceptacles,** which are found on the tips of branches in swollen, cone-shaped structures called receptacles (Fig. 14-10, *B*).

Spore. Meiosis of the nucleus occurs in the microsporangium as well as in the macrosporangium to produce four meiospores in the sporangia. Each of the four meiospores within the microsporangium undergoes four mitotic divisions to produce sixty-four haploid cells. These haploid cells eventually become microgametes or sperm. In the macrosporangium each meiospore undergoes a single mitosis to produce a total of eight nuclei. A small amount of cytoplasm eventually is massed around each to separate them into spherical, nonmotile macrogametes or eggs.

Sexual reproduction

Gametophyte. The sixteen haploid cells formed from each meiospore within the microsporangium may be regarded as the male gametophyte; similarly, the eight haploid nuclei formed from the meiospores within the macrosporangium may be regarded as the female gametophyte.

Antheridium. Each of the sixty-four haploid cells within the microsporangium develops into a microgamete or sperm with the haploid chromosome number.

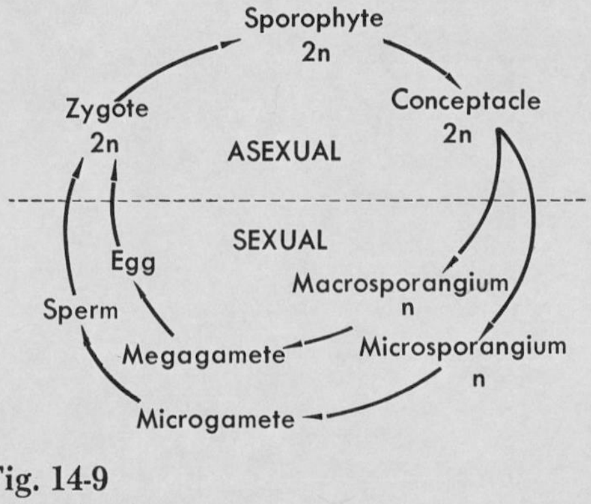

Fig. 14-9
Life cycle of *Fucus* sp.

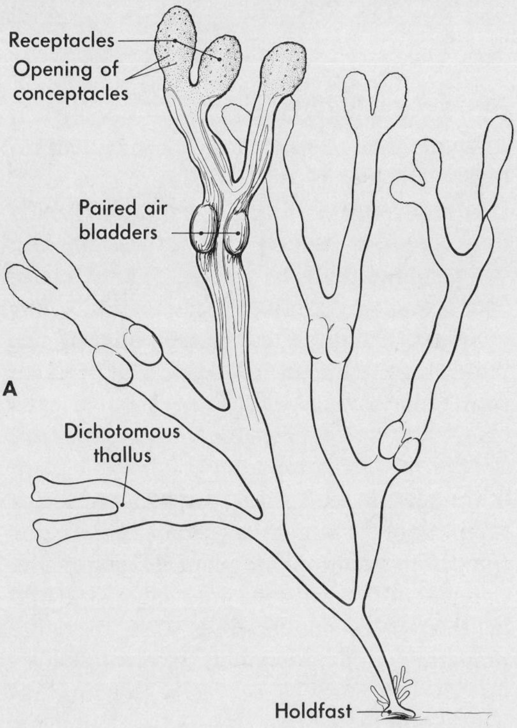

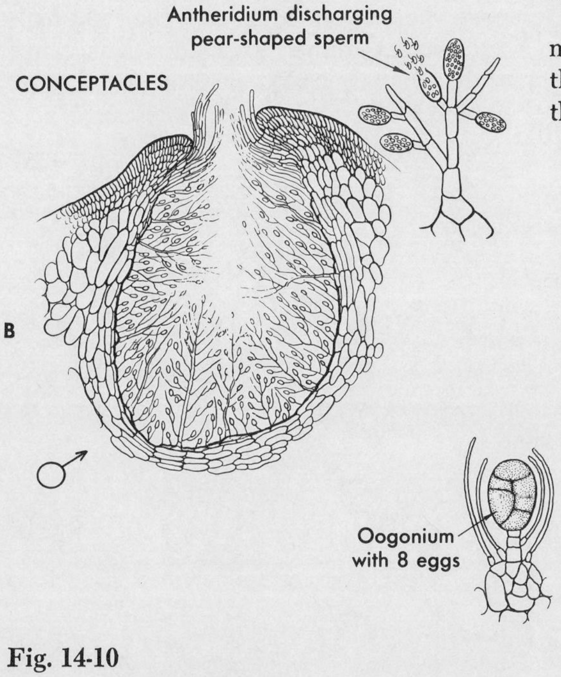

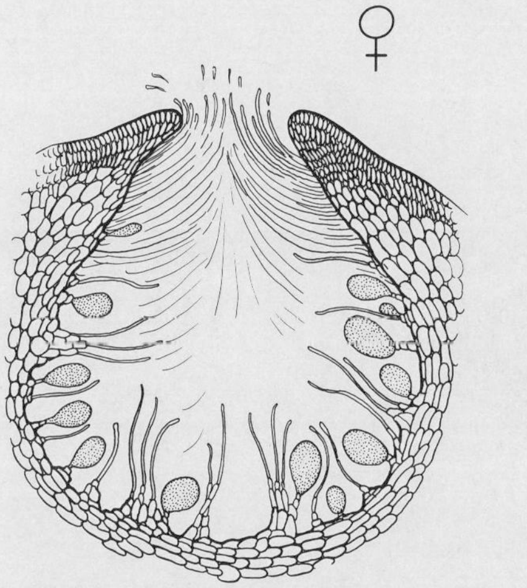

Fig. 14-10
Fucus sp. **A,** Vegetative stage. **B,** Reproductive stage.

"Archegonium." The oogonium arises from the megagamete already described.

Sperm. Each sperm has two lateral flagella and is motile. When mature they escape from the microsporangium and swim to the eggs.

Eggs. The eight nonmotile eggs develop within the megasporangium from the haploid cell masses. When mature, they are set free to float in the water.

Zygote. Fertilization of the egg by the sperm occurs in the water. Each floating egg becomes surrounded by many minute sperm. The sperm attach themselves to the egg by one **flagellum** and lash the water with the other, causing the egg to rotate. Finally, a sperm penetrates the egg, thus fertilizing it. This union of egg and sperm results in a diploid cell, the zygote.

Embryo. None as identified in the higher plants. However, the zygote immediately grows and develops into the sporophyte plant. It attaches itself to some suitable support and becomes the mature *Fucus* plant.

Means of dispersal. By means of fragmentation and at the time of fertilization the egg is dispersed by the ebb and flow of the tide.

LIFE CYCLE 6—CHLAMYDOMONAS Sp.
DIVISION 13 Chlorophyta
CLASS Chlorophyceae

Characteristics

The genus *Chlamydomonas* sp. is probably one of the most primitive of the green algae; it is found in standing water, in wet soil, and in swimming pools. The cells may become so numerous as to color the water green. It is a unicellular alga, the cell shape of which varies from spherical to ovoid. One large chloroplast almost completely fills up the cell. Within the chloroplast is a single, round pyrenoid that appears to be the center of starch formation. The cell also contains a centrally located nucleus, an eyespot sensitive to light, and two contractile vacuoles that probably function as organelles of excretion. At the anterior end of the cell are two flagella that pull the cell through the water. The entire cell is bounded by a conspicuous wall composed of cellulose.

Asexual reproduction

Vegetative reproduction. At certain times, two, four, or eight zoospores may be produced within the parent cells, which they resemble except that they are smaller.

Sporophyte. None.

Sporangium. None.

Spore. Daughter cells swim out of the parent cell through the dissolved wall to form new plants.

Sexual reproduction

Gametophyte. There is no structural differentiation of a gametophyte plant as distinct from a sporophyte plant. However, the parent cell may divide into eight, sixteen, or thirty-two smaller cells that resemble miniature *Chlamydomonas* sp. plants. These are the gametes or sex cells that are released into the water.

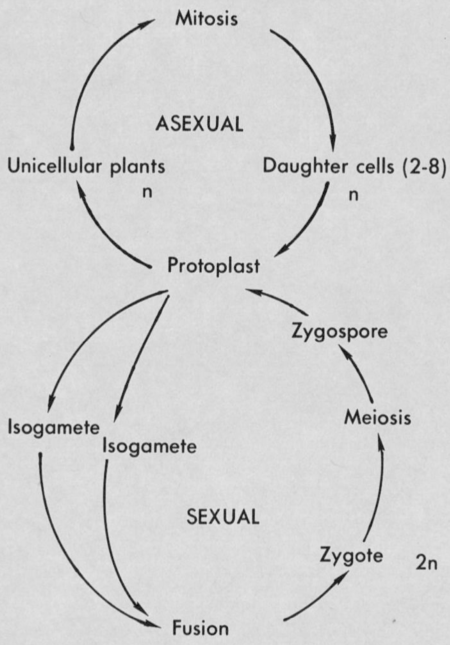

Fig. 14-11
Life cycle of *Chlamydomonas* sp.

Antheridium. None.

Archegonium. None.

Gamete. All are exactly alike; therefore, they may be called **isogametes.** They are not differentiated into egg and sperm. The gametes in some species look like the **zoospores** except that the former are smaller. Apparently in this life cycle there are sex cells and sex, but in a primitive type such as *Chlamydomonas* sp. the fusing gametes are isogametes; therefore no indication of maleness or femaleness is found. In other species there may be heterogametes.

Zygote. The gametes fuse in pairs (Fig. 14-12); ordinarily each fusing member of the pair comes from different parent cells. Since the gametes that fuse are alike, the resulting fusion process is called isogamy. However, the fusion cell formed is a zygote. A thick wall forms around the zygote and it becomes a resting cell that is adapted to withstand adverse conditions (such as drying).

Although there is no sporophyte, the zygote is a single cell containing one diploid nucleus that undergoes reduction division to produce four nuclei that eventually become incorporated into four zoospores. Each released zoospore develops into a *Chlamydomonas* sp. plant.

Embryo. None.

Means of dispersal. The flagella of the plant cells are the mechanism for the zoospores' dispersal; they allow the species to spread as far as physiologically suitable conditions permit.

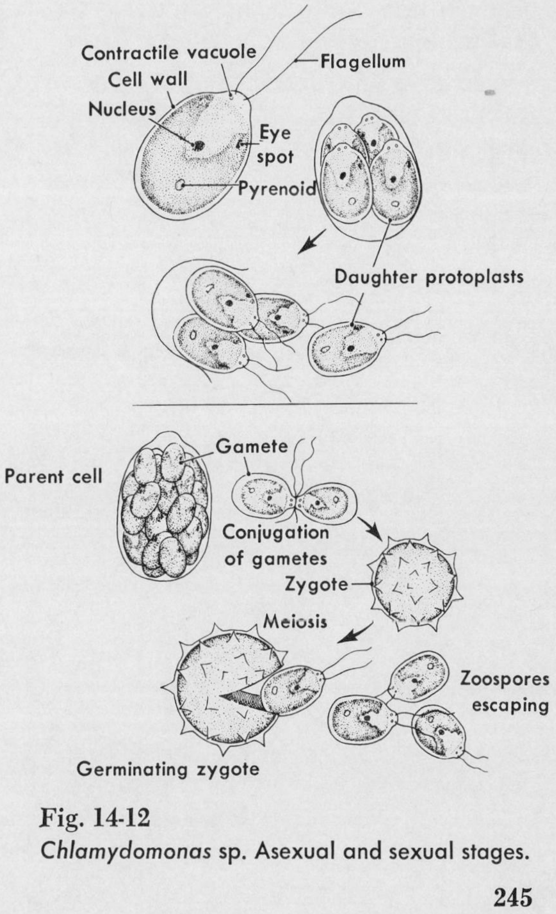

Fig. 14-12
Chlamydomonas sp. Asexual and sexual stages.

LIFE CYCLE 7—VOLVOX AUREUS

DIVISION 13 Chlorophyta

CLASS Chlorophyceae

Characteristics

The cells composing the vegetative body of *Volvox aureus* are individual organisms grouped together to form an organized colony (Fig. 14-14). Each cell is connected by a thin protoplasmic strand to the adjacent cells. The whole macroscopic colony slowly moves and rotates by the coordinated efforts of the projecting flagella. The colony is usually spherical or nearly so, is hollow within, and varies in size according to the number of cells. Mature colonies have from 500 to over 60,000 cells, depending upon the species. *Volvox aureus* occurs in freshwater pools, both temporary and permanent types, and often becomes abundant enough to give a greenish color to the water.

Asexual reproduction

Vegetative reproduction. The vegetative growth of the colony takes place until it reaches the size normal for the species. Further growth of the colony is very limited or stops entirely.

Sporophyte. None. The colony acts as the sporophyte. Daughter colonies are produced by special cells, the **gonidia,** which by rapid cell division produce hollow, spherical daughter colonies. These project into the center of the parent colony. When the inside is nearly filled with the smaller colonies, the parent breaks apart and liberates the new forms.

Sporangium. None. The gonidium is analogous to the sporangium of higher plants.

Spore. None.

Sexual reproduction

Gametophyte. Vegetative colonies may at times become gametophytes, producing either sperms or eggs or both. These develop from special cells.

Antheridium. Sperm-producing cells enlarge and undergo mitosis, producing hundreds of sperm that collect in a small pocket, the antheridium.

"Archegonium." A few of the cells in a colony enlarge, lose their flagella, and form

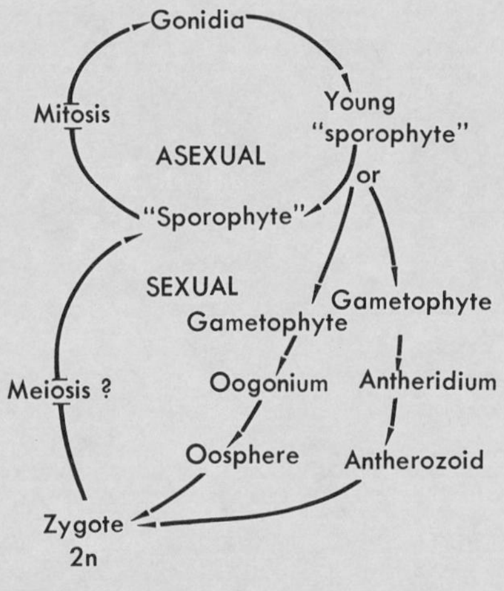

Fig. 14-13
Life cycle of *Volvox aureus.*

the egg cells, which are held in a small pocket, the oogonium.

Sperm. The sperm cell or **antherozoid** is smaller than the vegetative cells of the colony, spindle shaped, and biflagellated, but it has poorly developed chloroplasts.

Egg. The egg or **oosphere** is larger than the vegetative cells, is rounded, and lacks flagella. If this cell is not fertilized, it may develop into a new colony parthenogenetically. It is then known as a **parthenospore.**

When meiosis takes place is not clear. In some members of this order, alternate generations of haploid and diploid cells are known.

Zygote. The antherozoids are released from the antheridium as free-swimming cells. They swim to the oosphere and fuse with it. After fertilization the zygote produces a thick covering and becomes an **oospore.** This is released from the egg pocket and remains in the spore condition for some time before germination. During this resting period meiotic division of the nucleus may take place. Once the oospore wall is broken down, cell division proceeds and a new colony is formed.

Embryo. None.

Means of dispersal. Floating, combined with slow swimming of the colony, permits the dispersal of the species.

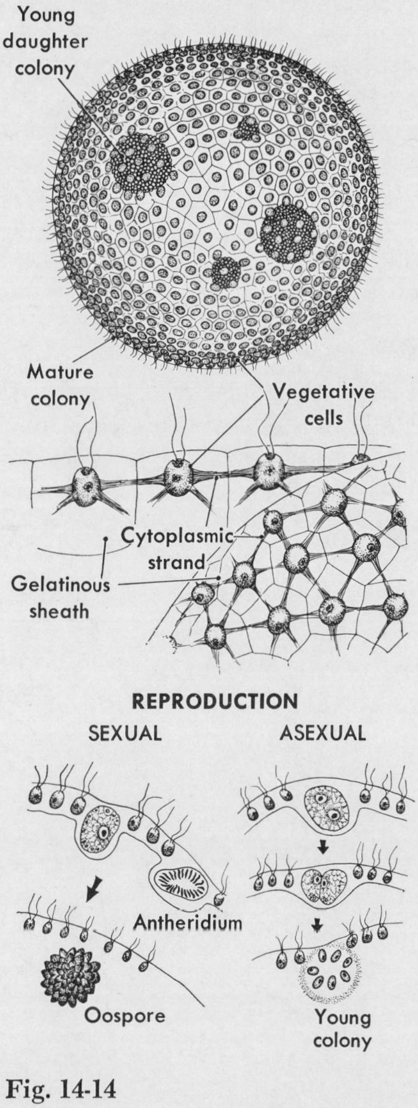

Fig. 14-14
Volvox aureus. Vegetative, sexual, and asexual stages.

LIFE CYCLE 8—ULOTHRIX Sp.
DIVISION 13 Chlorophyta
CLASS Chlorophyceae

Characteristics

Species of *Ulothrix* are common in fresh-water situations, where they form bright green, unbranched filaments having a small diameter (0.025 mm.). These filaments are composed of a single row of small cells, each surrounded by a cellulose wall (Fig. 14-16). At the base of the filament there is a well-developed **holdfast cell** that anchors the plant to some substrate. Although the holdfast lacks a chloroplast, it is otherwise not unlike the rest of the cells. In all the other cells the chloroplast forms a broad, transverse, ring-shaped band lying near the cell wall. This plant is truly multicellular; there is differentiation in both structure and function. The plant filaments increase in length by mitosis of the individual cells.

Asexual reproduction

Vegetative reproduction. None, other than fragmentation.

Sporophyte. None.

Sporangium. Zoosporangia.

Spore. Zoospores. Anywhere from two to four, eight, sixteen, or thirty-two zoospores are produced within unicellular zoosporangia. The zoosporangia are transformed vegetative cells. The zoospores each bear four flagella; they form new filaments by cell division after coming to rest in a suitable location.

Sexual reproduction

Gametophyte. The cells making up the filaments are the haploid generation. Gametes are produced in cells morphologically like those that produce zoospores.

Antheridium and "archegonium." None as such. The vegetative cells serve as unicellular gametangia that bear eight, sixteen, thirty-two, or sixty-four gametes. These gametes are similar to the zoospores, except smaller, but bear only two flagella.

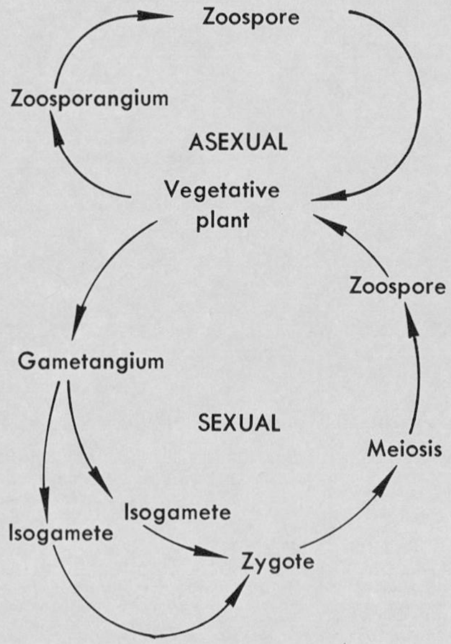

Fig. 14-15
Life cycle of *Ulothrix* sp.

Ulothrix sp. is considered to be **heterothallic,** the fusing gametes are produced by different filaments.

Sperm and egg. None. However, the isogametes act as sperm and egg and probably differ biochemically. Fertilization occurs between isogametes and is termed isogamy.

Zygote. A diploid cell is formed by the fusion of haploid gametes and then undergoes a period of rest prior to meiosis and the formation of four haploid zoospores. Once the zoospores are formed, they attach themselves and produce new filaments by cell division.

Embryo. None.

Means of dispersal. *Ulothrix* sp. is disseminated by the actively swimming zoospores.

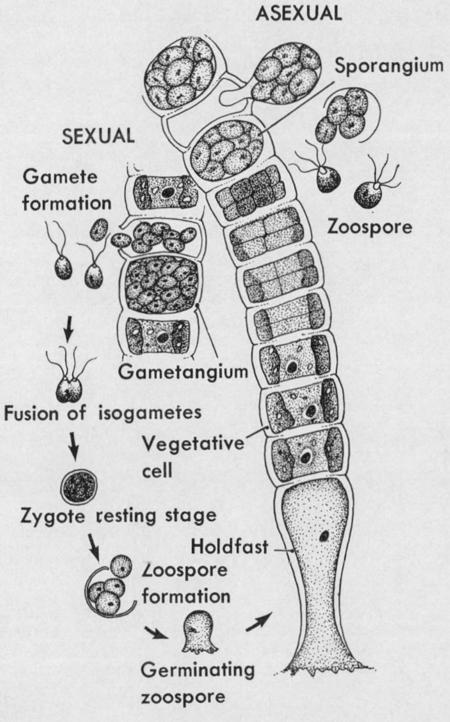

Fig. 14-16
Ulothrix sp. Asexual and sexual stages.

LIFE CYCLE 9—CHARA Sp.

DIVISION 14 Charophyta
CLASS Charophyceae

Characteristics

These plants are members of a single order, Charales. They are a group of green plants that grow submerged and are attached by rhizoids to the mud at the bottom of pools or slow-moving water. The chief feature of the class is a series of internodes with narrow, leaflike branches at the nodes (Fig. 14-18). Since plants of the genus *Chara* are inhabitants of waters especially high in calcium, they frequently become encrusted with lime—hence their common name stonewort. Their continued growth in water and their eventual accumulation result in a deposit of marl on the bottoms of ponds.

The cells of these multicellular plants contain more than one nucleus, and there are numerous small chloroplasts embedded in a peripheral layer of cytoplasm. The vegetative plants may have both male and female fructifications attached at the nodes (Fig. 14-18).

Asexual reproduction

Vegetative reproduction. The only type of reproduction known to occur is by the detachment of vegetative outgrowths that develop into new plants.

Sporophyte. None.

Sporangium. None.

Spore. None.

Sexual reproduction

Gametophyte. The vegetative plant is the haploid generation and bears separate sex organs at the nodes. Both the male **globule** and the female **nucule** consist of a gamete-producing part surrounded by a multicellular sheath. Most species are homothallic.

Antheridium. The globule in *Chara* sp. is borne below (Fig. 14-17) the female nucule. Motile sperm or antherozoids are produced by the globule.

"Archegonium." The oogonium of *Chara* sp. is developed above the globule and is

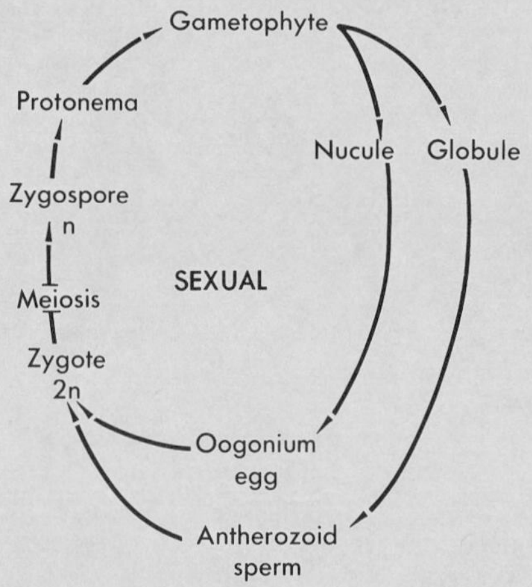

Fig. 14-17
Life cycle of *Chara* sp.

called a nucule. The initial cell of this structure divides to form a row of cells, the apical one of which is the oogonial mother cell. The rest of the cells grow around it while the apical cell enlarges to become the egg.

Sperm. Motile sperm or antherozoids are produced in large numbers within the globule to escape as free-swimming sperm.

Egg. The egg is nonmotile and remains within the nucule.

Zygote. Fertilization takes place within the nucule. The fusion of the nonmotile egg and motile sperm forms the zygote, around which is secreted a thick wall. Eventually, the thick-walled zygospore along with other parts of the nucule drop to the bottom of the pond or stream.

Embryo. None. Before germination the zygote nucleus divides by meiosis into four nuclei. At the time of germination the four nuclei present in the zygote contribute to the initial development of the multicellular plant. However, the three nuclei in the basal cell soon disintegrate, and the new plant is formed by the continued division of this unincleate cell to form the protonema and eventually the gametophyte.

Means of dispersal. Dispersal of the species is entirely due to the passive scattering of the zygospores. The movement of the water causes their scattering as well as that of the detached vegetative outgrowth.

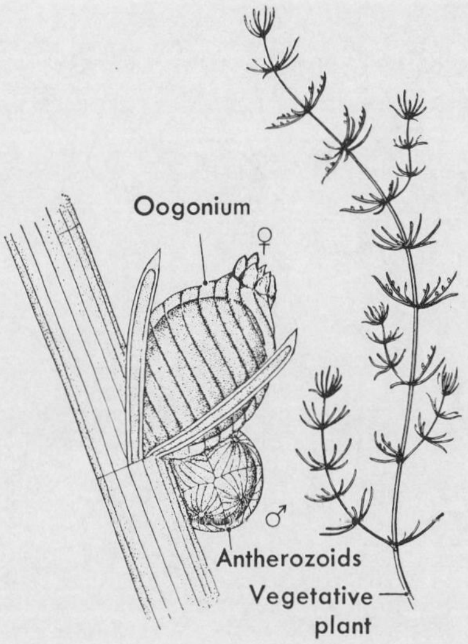

Fig. 14-18
Chara sp. (stonewort). Vegetative plant (with reproductive organs on right). Male and female fructifications in detail on left.

LIFE CYCLE 10—MARCHANTIA Sp.
DIVISION 15 Hepatophyta
CLASS Hepatopsida

Characteristics

The vegetative body is the gametophyte, composed of a prostrate dorsiventral thallus typical of the class (Fig. 14-20). Small unicellular rhizoids project into the substrate. The dorsal cells are green. Growth is radial from the point of origin of the thallus, which develops from the protonema. The surface of the thallus appears to have a protective, scalelike covering. Epidermal tissue surrounds the thallus, but only the upper surface has stomata. Beneath the epidermis the tissue differentiation is slightly more complex than that of the algae. The middle portion of the thallus is filled with chlorenchyma, interspersed with parenchyma storage cells. Two gametophytes of opposite sex are produced, and these can be readily distinguished by the nature of the **gametophores;** those of the male are stellate and those of the female have an umbrellalike shape (Fig. 14-20).

Asexual reproduction

Vegetative reproduction. When conditions are such that sexual reproduction is delayed, vegetative reproduction may take place by the production of structures called cupules located on the upper surface of the thallus. Within the **cupules** appear small buds, the gemmae. These break off from the cupules and are washed away. When they find a suitable growing site, they develop a new gametophyte thallus. If mechanically broken, the thallus will also regenerate into a new plant.

Sporophyte. The sporophyte is a small diploid plant growing from the undersurface of the archegoniophore of the female plant. It consists of a foot, a stalk or seta, and the sporangium. It is partially parasitic upon the gametophyte and is surrounded by a perianth produced by the gametophyte.

Sporangium. When the sporophyte is mature, the apical sporangium develops within the archegonium (see discussion of embryo that follows). The spore mother cells divide by meiosis and mitosis to produce simple spores. Within the sporangium are developed peculiar springlike elater cells that aid in spore dispersal.

Spores. As soon as the spores become

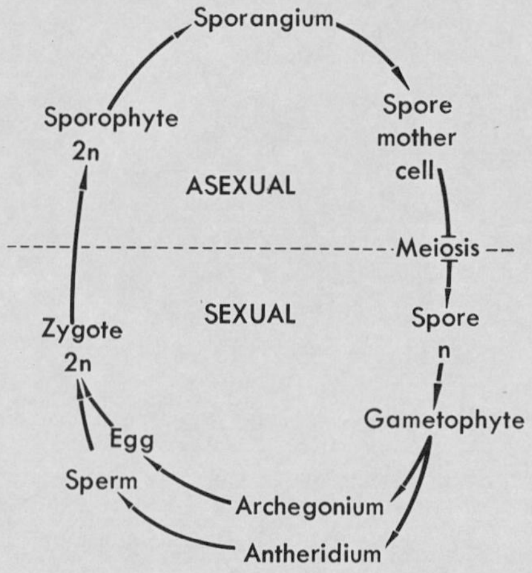

Fig. 14-19
Life cycle of *Marchantia* sp. (liverwort).

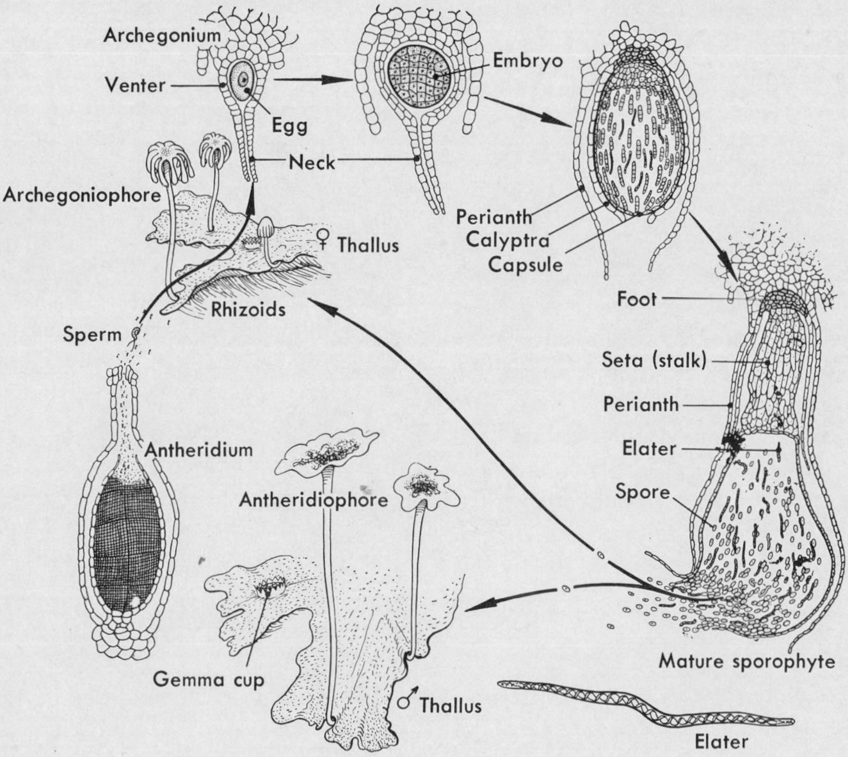

Fig. 14-20
Marchantia sp. (liverwort). Gametophyte and sporophyte generations.

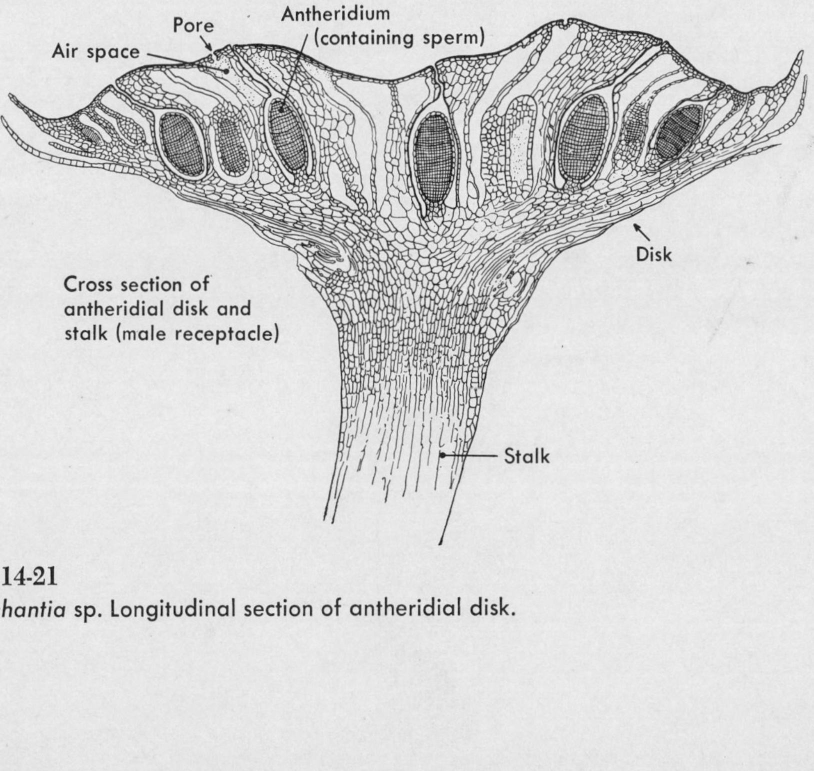

Fig. 14-21
Marchantia sp. Longitudinal section of antheridial disk.

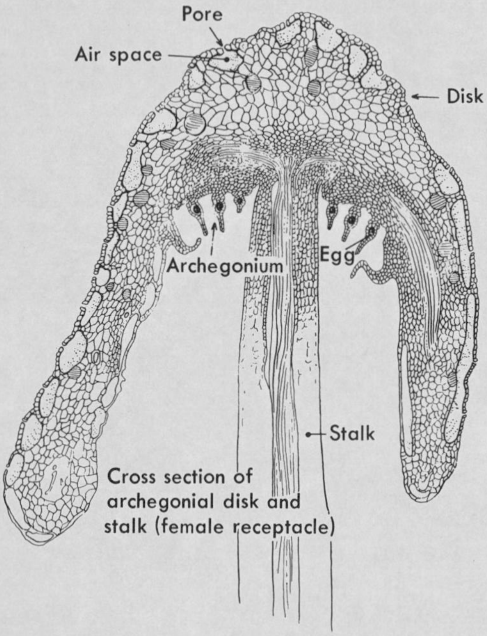

Fig. 14-22
Marchantia sp. Longitudinal section of archegonial disk.

mature, the covering or perianth of the sporangium and the archegonium ruptures, and the microscopic spores are ejected. Upon reaching a suitable habitat, they germinate to produce a small protonema, a filament-like, embryonic thallophyte.

Sexual reproduction

Gametophyte. The vegetative and dominant plant in this species is the gametophyte, which has just been described. It develops from a small protonema. Upon reaching maturity, it produces either **antheridiophores** or **archegoniophores,** depending upon the sex of the thallus. These stalks, several per plant, extend a few millimeters above the surface of the thallus.

Antheridium. The antheridia (Fig. 14-21) are developed in cavities in the expanded disk on the antheridiophore. They consist of small sacs of developing sperm.

Archegonium. The archegoniophore develops archegonia (Fig. 14-22) on the undersurface of its disk. These small structures are saclike, with an entrance channel or canal, the **neck canal.** When the egg is mature, this canal opens.

Sperm. True single-celled, flagellated sperm cells are produced in the antheridium. They are haploid, being produced from the haploid tissue of the gametophyte. Since they must swim to the egg, they are released only when the plant is flooded with water.

Egg. A true haploid egg, much larger than the sperm, is produced in the archegonium. This remains in the center at the base of the neck canal.

Zygote. A diploid zygote is formed when a sperm reaches the egg and fuses with it. The sperm enters the archegonium by way of the neck canal and swims to the egg. Both cytoplasm and nuclei fuse.

Embryo. The zygote grows by mitosis into an embryo, which eventually develops into the mature sporophyte described. Neither the egg, embryo, nor sporophyte leaves its original position within the center of the archegonium.

Means of dispersal. Spores and gemmae provide the distribution mechanism of these plants.

LIFE CYCLE 11—POLYTRICHUM Sp.

DIVISION 16 Bryophyta

CLASS Musci

Characteristics

This typical moss forms matlike colonies on soil or soil-covered rocks (Fig. 14-24). The gametophytes are fairly large, with numerous erect "stems" and small "leaves," giving them a bushy appearance.

Asexual reproduction

Vegetative reproduction. Several gametophytes develop from the protonema of a germinating spore. This structure persists and may be responsible for a small original colony of gametophytes. Gemmae and bulbils occur in mosses, and fractured plants are capable of regeneration.

Sporophyte. The embryo develops into the sporophyte, which is almost completely parasitic upon the female gametophyte. It extends dorsally above the archegonial branch of the thallophyte, however, and is readily seen in sporulating moss. It consists of the following parts: (1) a foot, which penetrates the tissue of the gametophyte from which nutrients are absorbed, (2) a stalk or seta, which arises from the foot, and (3) the sporangium, which is formed at the apex of the stalk.

Sporangium. The sporangium is a smaller, somewhat lanceolate **capsule** with a central **columella** of sterile tissue surrounded by the spore-bearing layer. The apex of the capsule is fitted with a small caplike deciduous cover or **operculum**, which opens to release the mature spores.

Spore. Spores are produced by meiosis of the spore mother cells. Each spore mother cell produces four haploid spores. The microscopic spores are stored within the capsule until conditions are dry enough for their release and dissemination. After the spores are released, they may remain dormant for long periods of time. When conditions of temperature and moisture are optimum, the spore absorbs water, and the cell within divides. It soon bursts open the spore case. Continued division of the spore forms a protonema typical of the group.

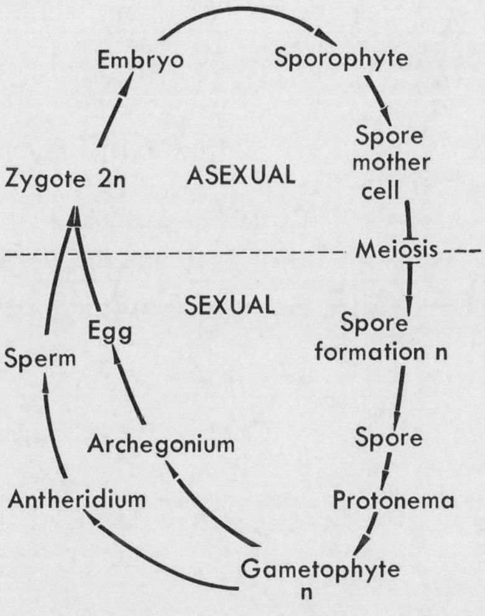

Fig. 14-23
Life cycle of *Polytrichum* sp. (moss).

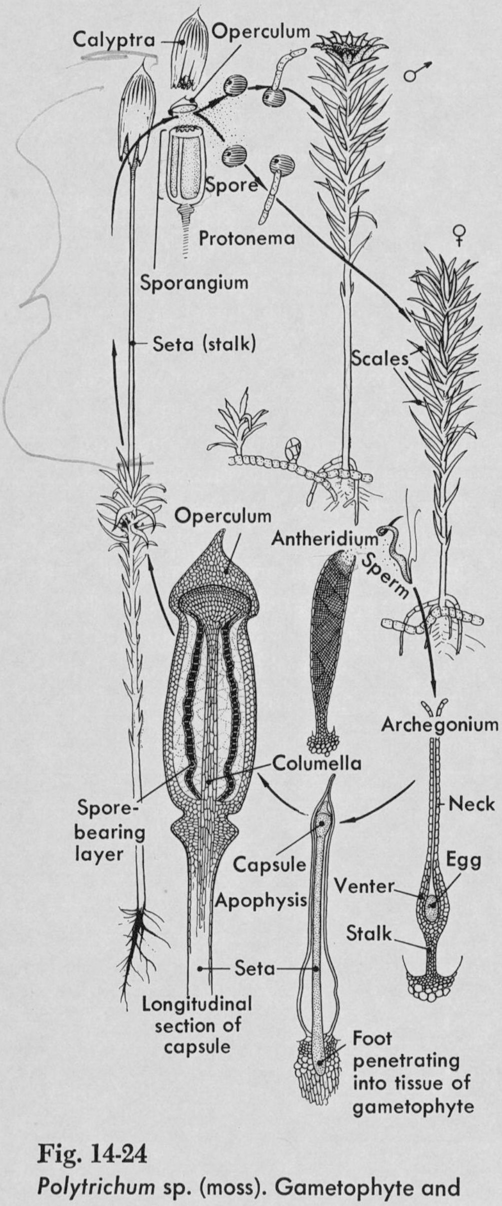

Fig. 14-24
Polytrichum sp. (moss). Gametophyte and sporophyte generations.

Sexual reproduction

Gametophyte. The **protonema,** which is somewhat flattened and resembles an alga, continues to grow, producing rhizoids that anchor the plant to the substrate. After a short growing period, "buds" composed of a compact group of cells are formed. These buds develop into "stems" and soon put forth green "leaves," thus forming the young, vegetative moss plant or gametophyte.

When this plant reaches maturity, sex organs are developed, usually at the apex of a thallus stem. In some species the male and female organs are borne on the same plant, whereas in others they develop on separate plants but in the same colony.

Antheridium. The sperm are produced by mitosis in a small capsule, the antheridium. When the sperm mature, a pore opens at the apex of the antheridium, allowing the sperm to leave.

Archegonium. The female structures are similar to those of *Marchantia* sp. and, as in the antheridia, are borne at the apex of the fertile shoots. A short stalk supports the **venter** in which the egg is formed. A neck is produced that hollows out to form a neck canal when the archegonium is mature.

Sperm. The male sex cells are biflagellated. They swim from the antheridia only when the plant is drenched in water.

Egg. The egg cell is large like that of *Marchantia* sp. and remains in the venter of the archegonium.

Embryo. The zygote develops shortly after fertilization and grows into a sporophyte.

Means of dispersal. Like *Marchantia* sp., these plants depend upon dissemination of spores for dispersal and formation of gemmae.

LIFE CYCLE 12—SELAGINELLA Sp.
DIVISION 18 Microphyllophyta
CLASS Glossopsida
Characteristics

Selaginella sp. is a small bushy plant resembling some mosses (Fig. 14-26). The vegetative plant is the sporophyte, producing loose and inconspicuous strobili at the apices of some of the branches. The leaves are of two types arranged in rows along the length of the stem, two rows of small leaves and two rows of large leaves.

Asexual reproduction

Vegetative reproduction. No special structures are present for vegetative reproduction; however, portions of the plant may run along the ground so that several clumps may be produced.

Sporophyte. The sporophyte has already been described.

Sporangium. The sporangia are of two types, both of which are found on the same **strobilus** (Fig. 14-27). The strobilus is characterized by having, in addition to the sporophyll leaves, a small tonguelike scale, the **ligule,** which fits into a pit at the base of the sporophyte. The basal portion of the strobilus produces **megaspores** and the apical portion **microspores.** Both types of sporangia are borne in the axil of the leaves, forming the strobilus. These are the **sporophylls** and are either **megasporophylls** or **microsporophylls,** according to the type of sporangia they subtend.

Spore. Each megasporangium produces four megaspores by meiosis of the diploid spore mother cell. The mature spore is covered by a rough coat, which protects the spore until germination. The microspores are produced in a similar fashion, except

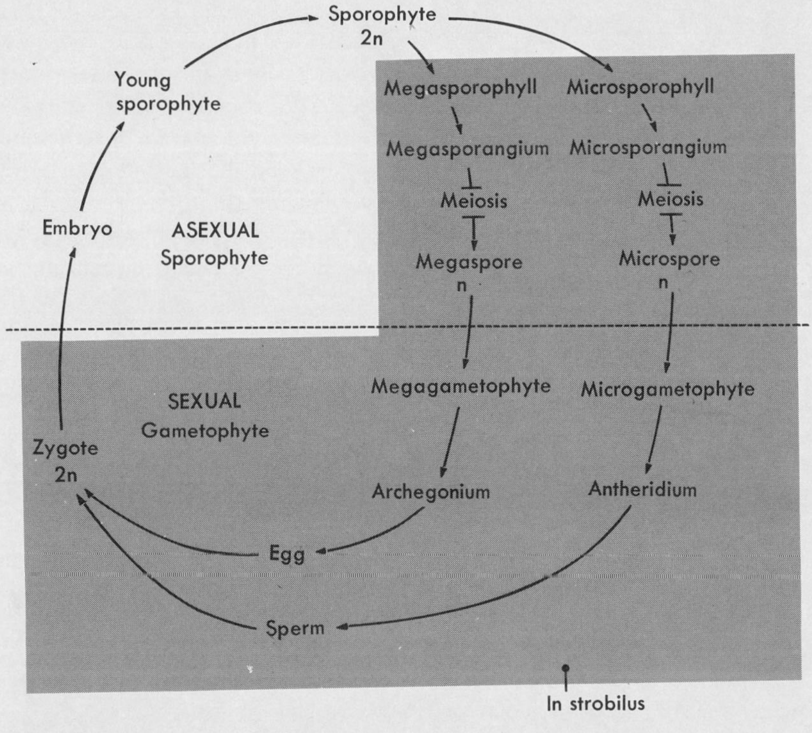

Fig. 14-25
Life cycle of *Selaginella* sp.

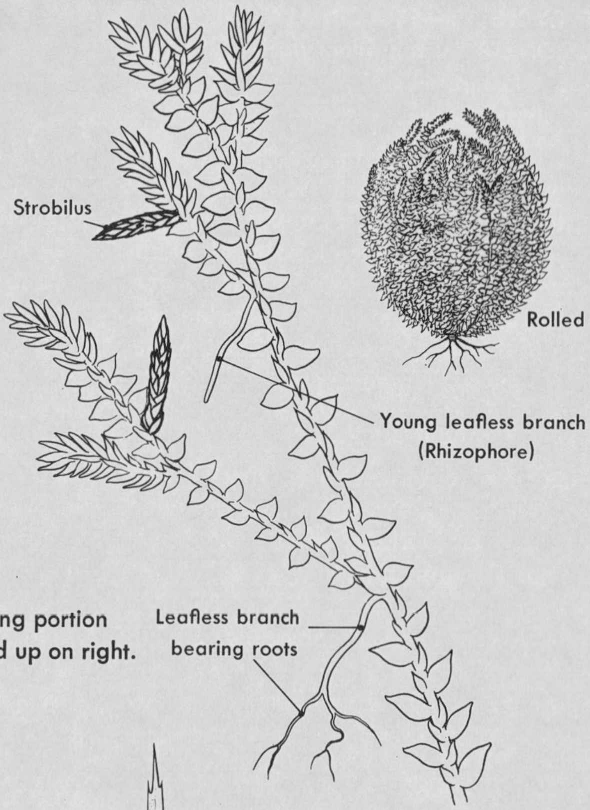

Fig. 14-26
Selaginella sp. (resurrection plant) showing portion of plant on left and an entire plant rolled up on right.

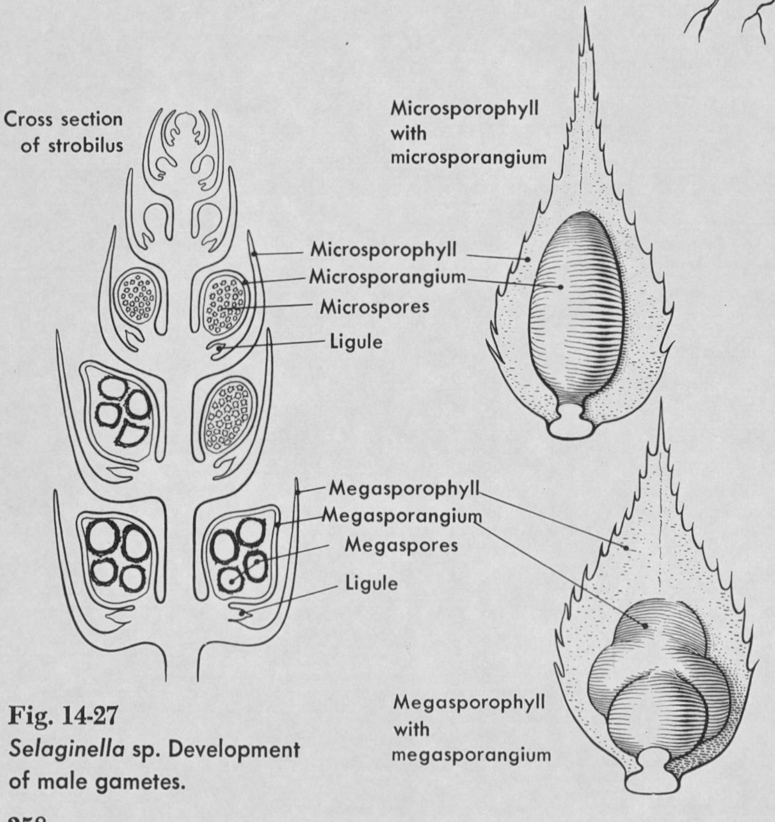

Fig. 14-27
Selaginella sp. Development of male gametes.

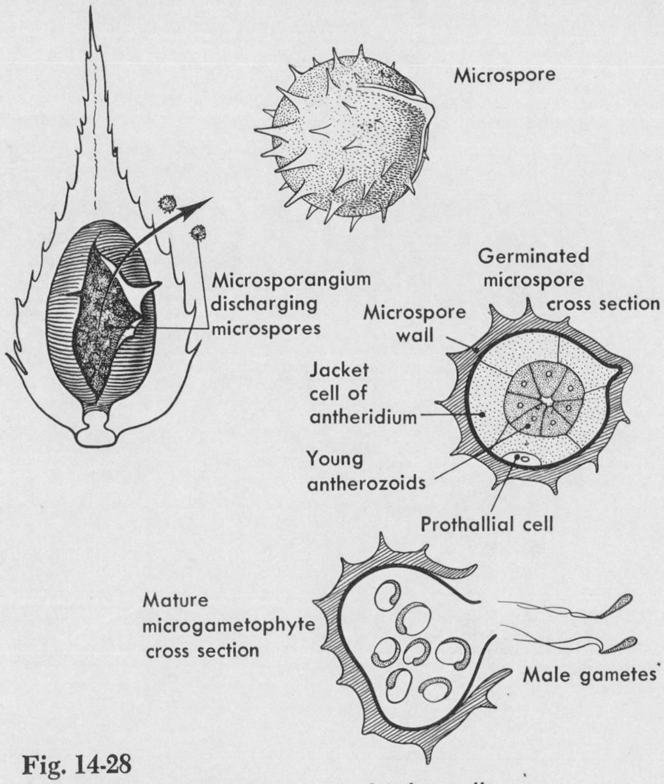

Fig. 14-28
Development of the microspore of *Selaginella* sp.

that the many microspore mother cells in the microsporangium produce hundreds of minute microspores.

Sexual reproduction

Gametophyte. The megaspores remain in the megasporangium where they develop into a megagametophyte. This small plant, while not parasitic on the sporophyte, foreshadows parasitism in the higher plants. As the megagametophyte develops, a layer of cells is produced to form a female **prothallus.** The rest of the spore serves as a reserve food supply. Several archegonia with a venter and a venter canal cell are produced on the prothallus. An egg develops in the venter, awaiting fertilization.

Meanwhile the microspore in the microsporangium develops into microscopic **microgametophytes** or prothallial cells (Fig. 14-28). While the antheridium is still in the spore stage, ciliated sperm are produced.

Antheridium. The antheridium is attached to the sporophyte.

Archegonium. The archegonium is attached to the sporophyte.

Sperm. The spores mature into the gametophyte and act as pollen grains. They are released from the microsporangium and are either washed from the strobilus or are wind borne. Some eventually land on the female portion of the strobilus. When moist, they burst open, and the ciliated sperm leave the old spore case.

Egg. The venter canal of the archegonium opens when the megagametophyte is mature. The egg is a small cell located at the base of this canal.

Zygote. Several of the sperm enter the ruptured spore case of the megameto-

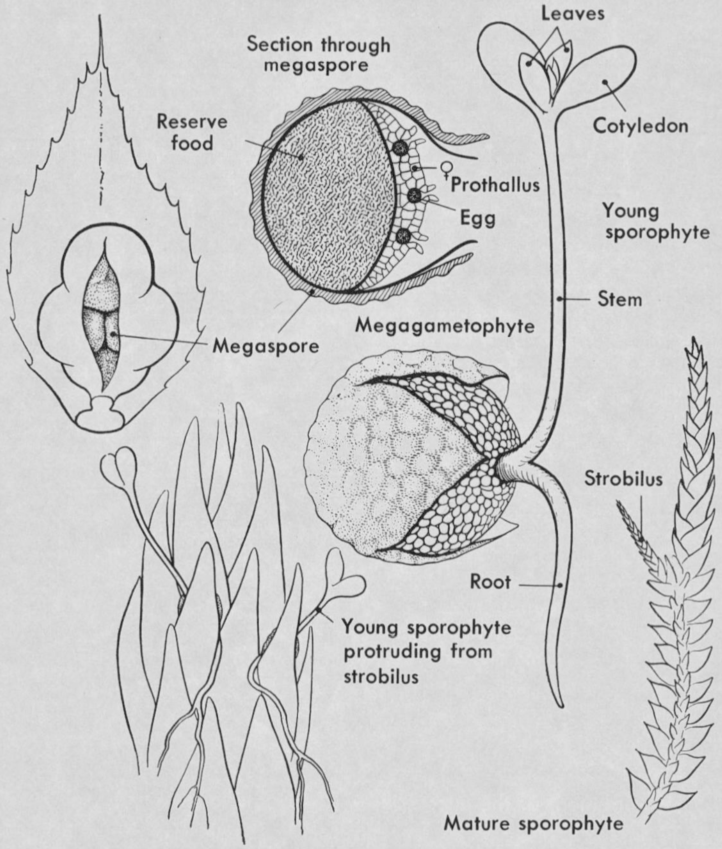

Fig. 14-29
Selaginella sp. Development of the sporophyte.

phyte and find a venter canal. The two cells fuse, restoring the nucleus to the diploid condition.

Embryo. As soon as fertilization takes place, an embryo develops that obtains its food from the reserve remaining in the spore. Soon a young sporophyte is developed (Fig. 14-29) that eventually drops from the strobilus of the parent sporophyte and becomes established in the soil.

Means of dispersal. The lack of spores producing independent gametophytes has prevented the wide dispersal of these plants so characteristic of other cryptogams. Most dispersal is the result of water washing the young sporophyte to new locations.

LIFE CYCLE 13—EQUISETUM Sp.

DIVISION 19 Arthrophyta
CLASS Sphenopsida

Characteristics

The vegetative and reproductive characteristics of the horsetails fit the characteristics of the class; they are described in Chapter 23.

Asexual reproduction

Vegetative reproduction. The rhizome or subterranean stem may run under the ground for several feet, putting forth aerial branches. If they become detached, the separate branches will continue to grow and produce new individuals.

Sporophyte. The sporophyte is the vegetative plant, many branches of which are sterile (Fig. 14-31). The spore-producing branches are separate and colorless and receive their nourishment from the underground rhizome connecting these branches with the vegetative ones.

Sporangium. A well-differentiated strobilus is borne at the apex of the fertile branch. This is a compact structure that darkens as it matures. It is a series of short stalks, the sporangiophores containing sporangia. These are probably modified sporophylls. The apex of the sporangiophore is six sided, giving a geometrical appearance to the strobilus before it ripens. A sporangium consists of six bracts grouped around the sporangiophore (Fig. 14-32). Inside, the spore mother cells divide by meiosis to produce small spores.

Spore. The spores are all alike in this plant. They are unique in that as accessory structures they have coiled appendages, termed **elaters.** As the sporangium dries, these elaters expand and force open the strobilus, thus aiding in release and dissemination of spores.

Sexual reproduction

Gametophyte. The spores are released and are blown about. After about 12 hours they will usually germinate. They will each grow into a small thalluslike green gametophyte partly submerged in the soil. When mature, the gametophyte produces gametangia. The thallus may persist for some time after fertilization.

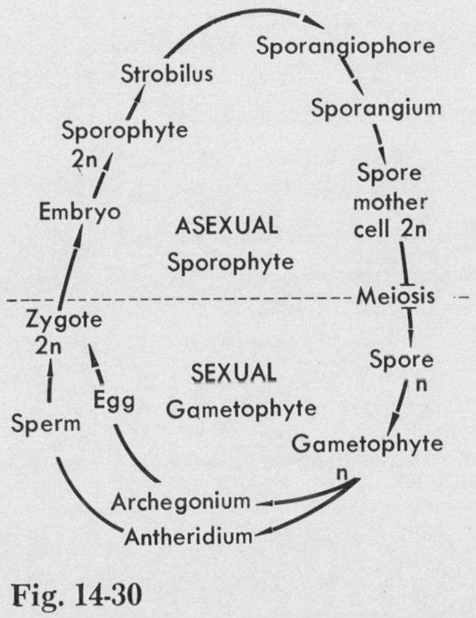

Fig. 14-30
Life cycle of the *Equisetum* sp. (horsetail).

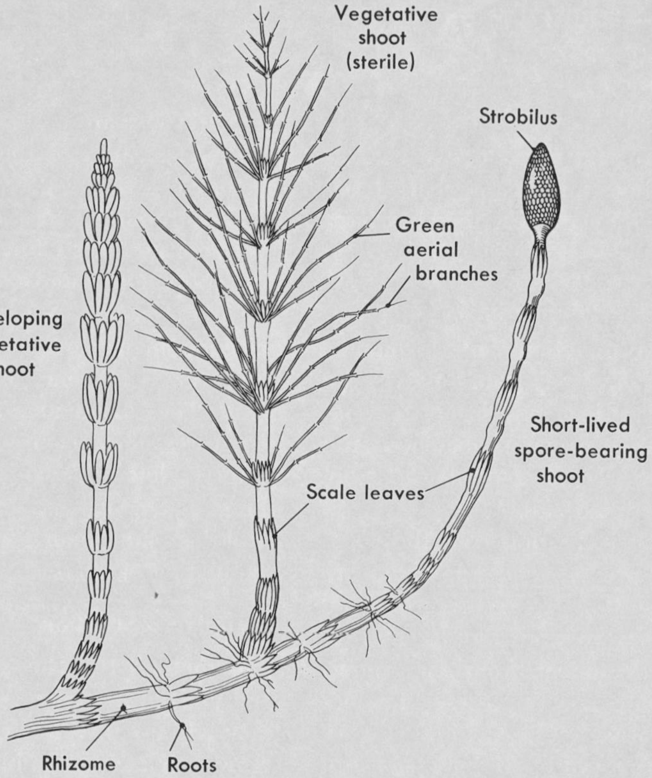

Fig. 14-31
Equisetum sp. (horsetail). Vegetative plant.

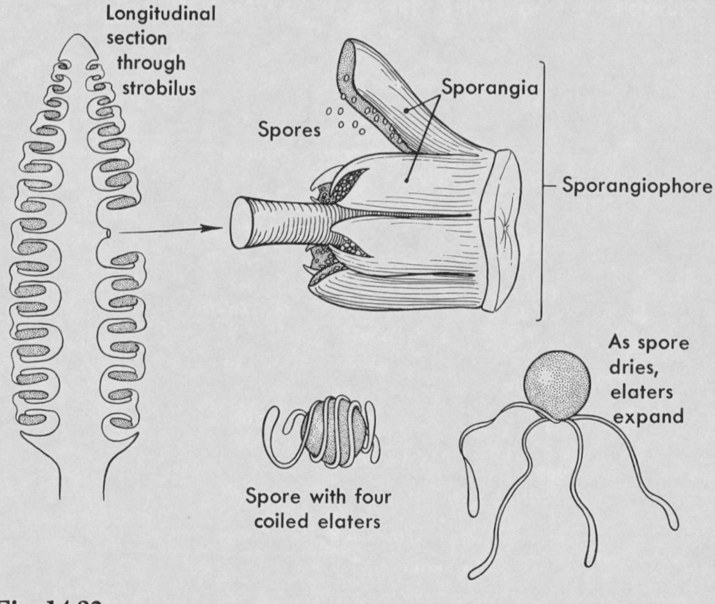

Fig. 14-32
Equisetum sp. (horsetail). Development of spores.

Antheridium. Antheridia are produced at the upper end of the prothallus, the gametophyte.

Archegonium. Archegonia are produced at the base of the short branches of the prothallus.

Sperm. The sperm are coiled and multicilated and are capable of swimming to the archegonium.

Egg. The eggs are located in a venter at the base of a fertilization canal.

Zygote. Fertilization takes place during rainy weather, the sperm swimming from one gametophyte to another to fertilize the ripened eggs.

Embryo. A young sporophyte develops within the fertilized archegonium. It has primary roots, stem, and leaves in addition to a small "foot" by means of which it obtains its first food from the gametophyte. Eventually it grows into a sporophyte and replaces the parent gametophyte, which gradually dies.

Means of dispersal. Spore dissemination by the sporophyte provides the means for the distribution of the species.

LIFE CYCLE 14—POLYPODIUM Sp.
DIVISION 20 Pterophyta
CLASS Pteropsida
Characteristics

These ferns are fairly large sporophytes with long pinnate **fronds** (Fig. 14-34). They are common in wooded areas throughout North America and are often grown as ornamentals. The sporangia are arranged in **sori** along the margin of the **pinnae** of the vegetative fronds. The plants of this genus sporulate in the summer and fall and remain green throughout the year.

Asexual reproduction

Vegetative reproduction. Growth of the rhizomes may produce several clusters of fronds interdependent upon each other. If these become separated, however, the plants persist and develop new clones, which grow up in the spring by uncoiling (**circinate** vernation).

Sporophyte. The spore-producing plant is vegetative. The frond is composed of many pinnae arranged laterally along the petiole or rachis. Each pinna is also pinnate.

Sporangium. The sporangia are located in small clusters along the margin of the undersurface of the pinnae of the fronds. These clusters are called sori and are partly covered by a leaf fold, the **indusium.** When the spores are ripe, this area appears dark brown. The sporangium is suspended from the frond by a short petiole. Each sporangium has a remarkable device for the release of the spores. Along one side is a row of cells with thick inner walls and thin outer walls. This structure is called the **annulus.** Along the other edge is a row of thickened cells, two of which are called **lip cells.** The cells of the annulus are normally filled with water so that turgor pressure keeps the cell inflated and in position. Upon drying, the thick inner walls spring together and rupture the sporangium at the area of the lip cells, releasing the spores. If moist conditions return, the sporangium will again close because the cells of the annulus become reinflated, forcing the walls apart. Eventually the sporangium ruptures beyond closing. This mechanism ensures the release of spores during dry weather when they are most likely to be widely distributed. The spores are produced by meiosis of the spore mother cells.

Spore. The spores are microscopic in size,

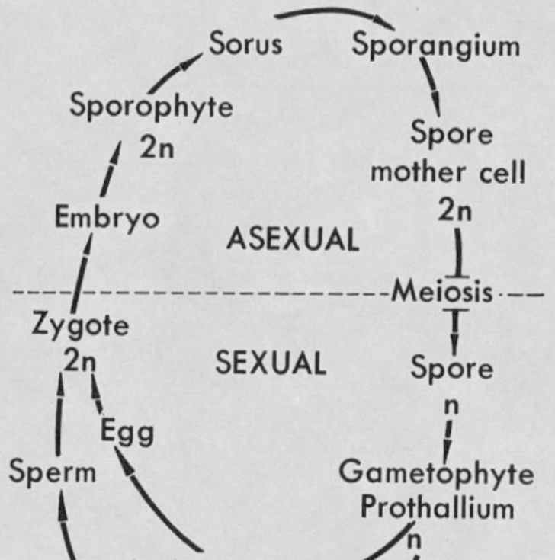

Fig. 14-33
Life cycle of *Polypodium* sp. (fern).

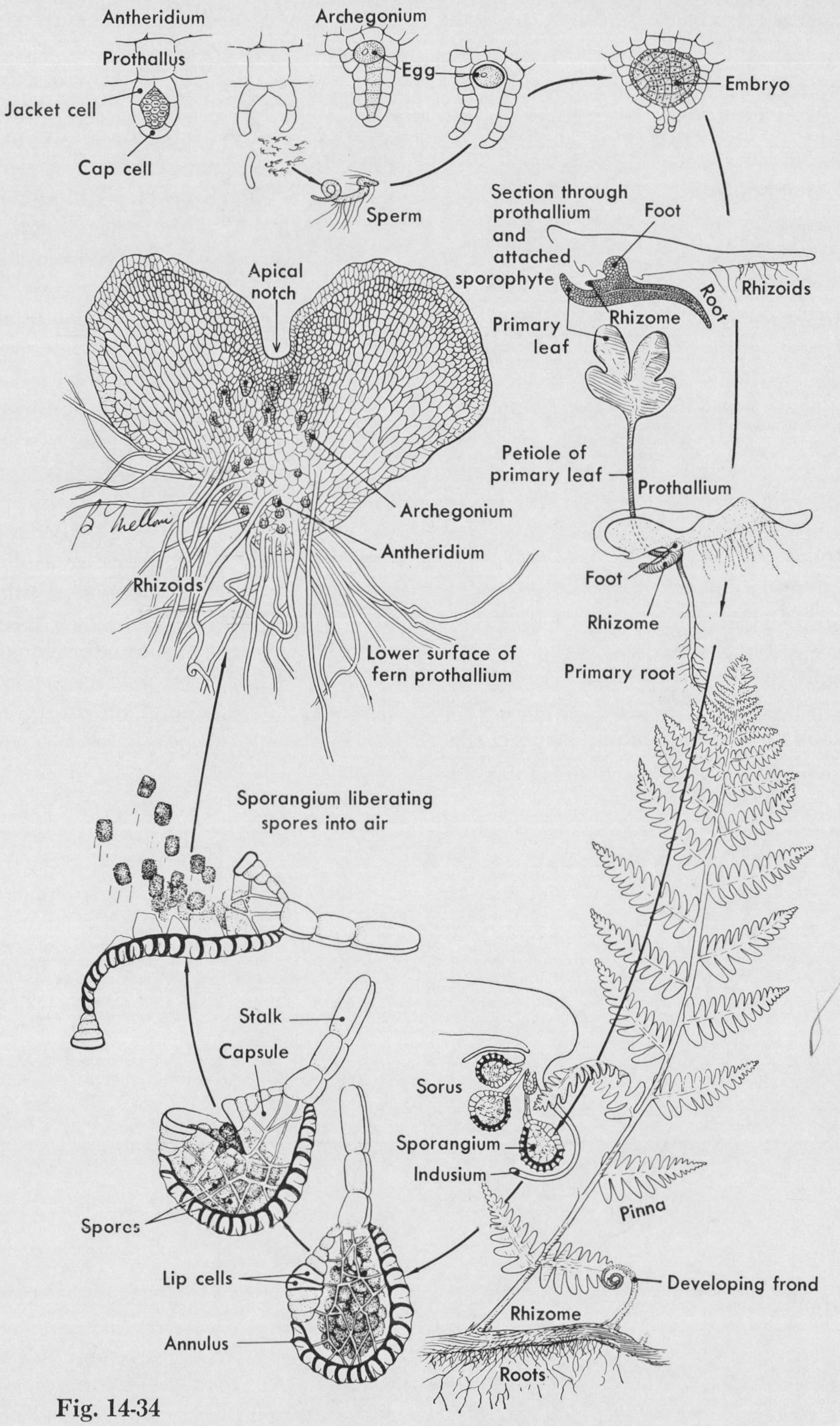

Fig. 14-34
Polypodium sp. (fern). Sporohpyte and gametophyte generations.

are haploid, are covered with a resistant coating, but are without elaters or other structures for dissemination. They are small and light enough to be carried great distances by air currents.

Sexual reproduction

Gametophyte. Upon reaching a suitable habitat, the spores germinate, usually during the following spring, and grow into a rhizoid-bearing green filament. However, the filament is not a true protonema. This filament eventually develops into a small heart-shaped **prothallium** about ½ to 1 inch in diameter, and this prothallium, when mature, develops the gametangia. The prothallium, which is green in color, is capable of limited photosynthesis. It is composed of a thin layer of cells but lacks a vascular system and is therefore devoid of true roots, stems, and leaves.

Antheridium. The male or sperm-producing structures develop as small lobes on the lower surface of the prothallium among the rhizoids.

Archegonium. The female structures are also located on the lower surface of the prothallium, usually near the apical notch some distance from the antheridia. The archegonium consists of a venter harboring a single egg cell and the venter canal.

Sperm. The sperm are irregularly shaped cells with many cilia, enabling them to swim freely.

Egg. The egg is nonmotile and remains in the archegonium.

Zygote. During rain the sperm are released from the antherida, swim to the archegonium, enter the venter canal, and fertilize the egg. The zygote thus formed remains in the archegonium.

Embryo. An embryo soon develops at the expense of the gametophyte. It first puts forth a primary root and a primary leaf. A small foot remains in the archegonium to absorb nutrients. As growth continues, a rhizome develops, and the new sporophyte appears. As the sporophyte grows, the prothallium is destroyed and disappears.

Means of dispersal. The species, like all ferns, depends upon spores for wide dispersal.

LIFE CYCLE 15—PINUS Sp.
DIVISION 23 Coniferophyta
CLASS Gymnospermae

Characteristics

Members of the genus *Pinus* are well known as valuable trees generally distributed in the northern hemisphere of both the Old World and the New World. The trees have a central stem or trunk with lateral branches and branchlets. The leaves are in the form of needles, usually in groups of two to five, and are uniform in length, the length depending upon the species. Pine trees are evergreen and monoecious, producing both male and female pendulous **cones.** The female cones have **scales** and **bracts** in spirals and are held tightly closed, except for a slight opening at the time the megagametophyte matures to permit pollination. Until the seeds mature, they remain closed.

Asexual reproduction

Vegetative reproduction. None.

Sporophyte. Like all seed plants, the sporophyte is the vegetative plant, and it produces microsporangia and megasporangia on the same plant.

Sporangium. The reproductive parts are in cones and represent a strobilus formed from modified leaves. Each cone is a cluster of sporophylls with a short central stem. The two types of cones are distinguished according to the sex of the gametophyte they produce. Microsporophylls are the staminate cones and are the smaller of the two; megasporophylls are the carpellate cones.

Microsporophyll. Each microsporophyll is a scale attached to the axis of the cone

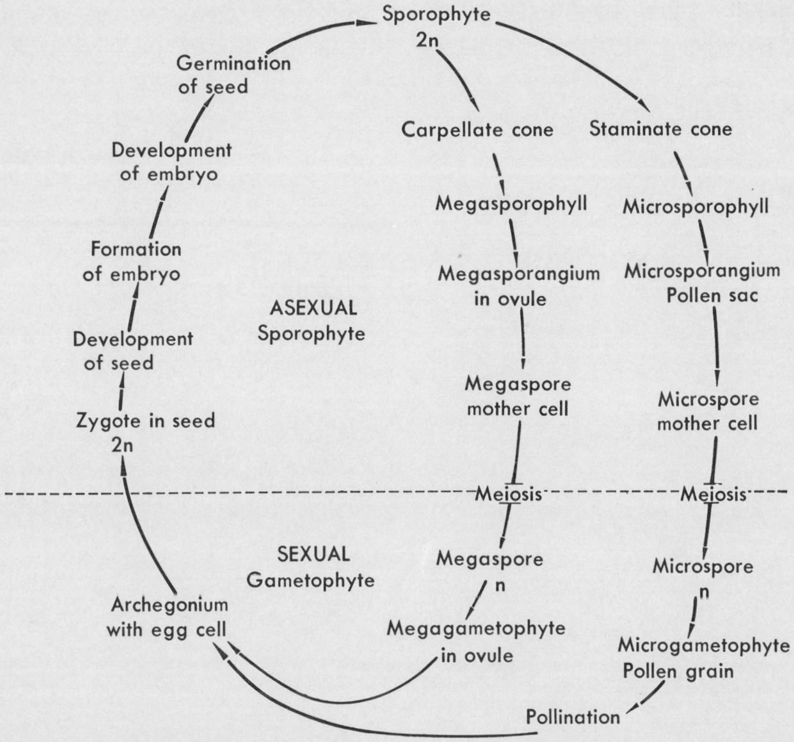

Fig. 14-35
Life cycle of the *Pinus* sp. (pine).

by a short stalk and produces two microsporangia on its lower surface.

Megasporophyll. Each megasporophyll is also a scale subtended by a bract much larger than that of the **staminate cone.** It bears two ovules near its base, each containing a megasporangium. The bracts are conspicuous only in young cones, but in some conifers the bracts may exceed the scales in the mature cone.

Microsporangium. The tissue of the microsporangia gives rise to numerous diploid cells, microspore mother cells, which produce a tetrad of haploid cells by meiosis. They separate into four microspores.

Megasporangium. The megasporangia, like the microsporangia, produce diploid generative cells, the megaspore mother cells, which divide by meiosis to form, in this case, 4 haploid megaspores, three of which eventually degenerate. Unlike the cryptogams, these megaspores are produced in megasporangial tissue modified to form an ovule or nucellus, a structure similar to the venter of the cryptogam gametophyte. A section through one of these ovules shows that it consists of a single integument with an opening, the micropyle, directed toward the axis of the cone. This structure is equivalent in function to the venter canal of the archegonium of the cryptogams but is not homologous in structure because it is formed by the sporophyte and not by the gametophyte.

Microspore. As soon as microspores (Fig. 14-36) are produced, they develop lateral projections or wings. They remain in the microsporangia, however, for further development into a microgametophyte.

Megaspore. The megaspore remains in place in the ovule for further development into the megagametophyte.

Sexual reproduction

Gametophyte. The gametophytes of all seed plants are greatly reduced to microscopic parasites on the sporophyte. The megagametophyte never leaves the parent,

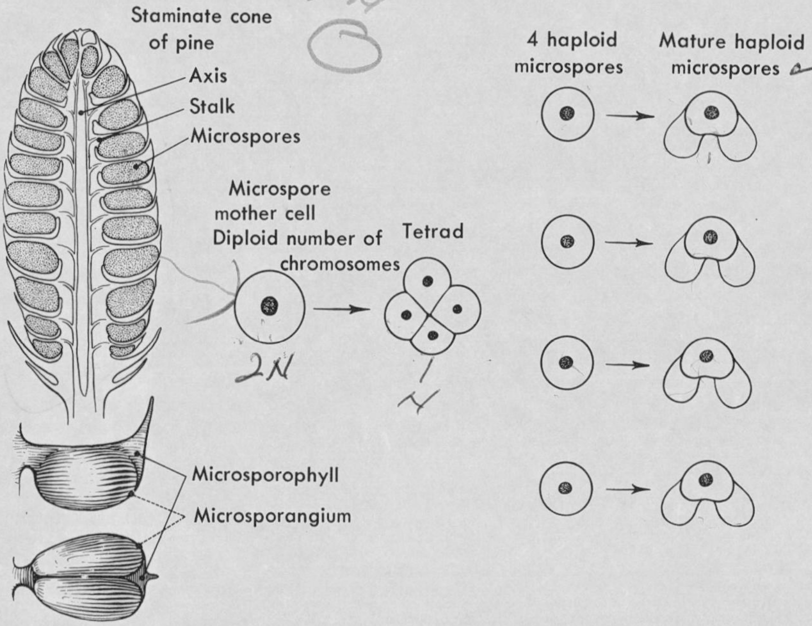

Fig. 14-36
Pinus sp. (pine). Development of the microspore.

but the entire microgametophyte is in the form of a **pollen** grain, which is usually wind borne. One of the sperm or male nuclei fertilizes the egg of an archegonium.

Microgametophyte. The haploid microspore cell divides by mitosis to form a large central cell and a smaller vegetative cell (Fig. 14-37). The vegetative cell, which is now equivalent to the prothallium of the cryptogams, again divides to form a first and a second vegetative cell. Then both cells degenerate and remain as remnants throughout the further development of the pollen grain. This is the extent of the vegetative part of the microgametophyte and the end of what was, in the lower plants, the dominant vegetative plant.

Megagametophyte. Further development of the megaspore does not take place until a few pollen grains enter the micropyle, after which the latter closes, preventing entrance of more pollen (Fig. 14-38). However, before actual pollination, the megaspore mother cell divides meiotically to produce four megaspores, three of which degenerate. The remaining functional megaspore enters the free-nuclear stage; that is, the haploid nucleus divides several times by mitosis to form the megagametophyte.

Antheridium. The pollen grain does not develop a true cellular antheridium. However, the central cell divides to produce a generative cell and a tube nucleus. It is in

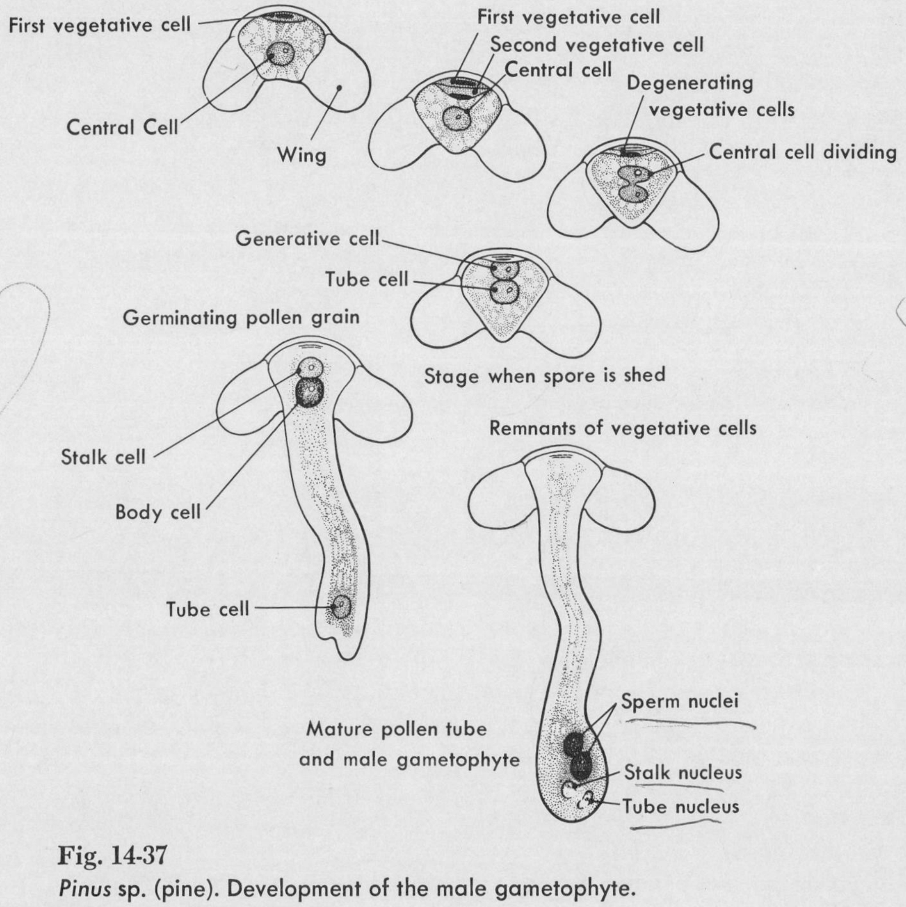

Fig. 14-37
Pinus sp. (pine). Development of the male gametophyte.

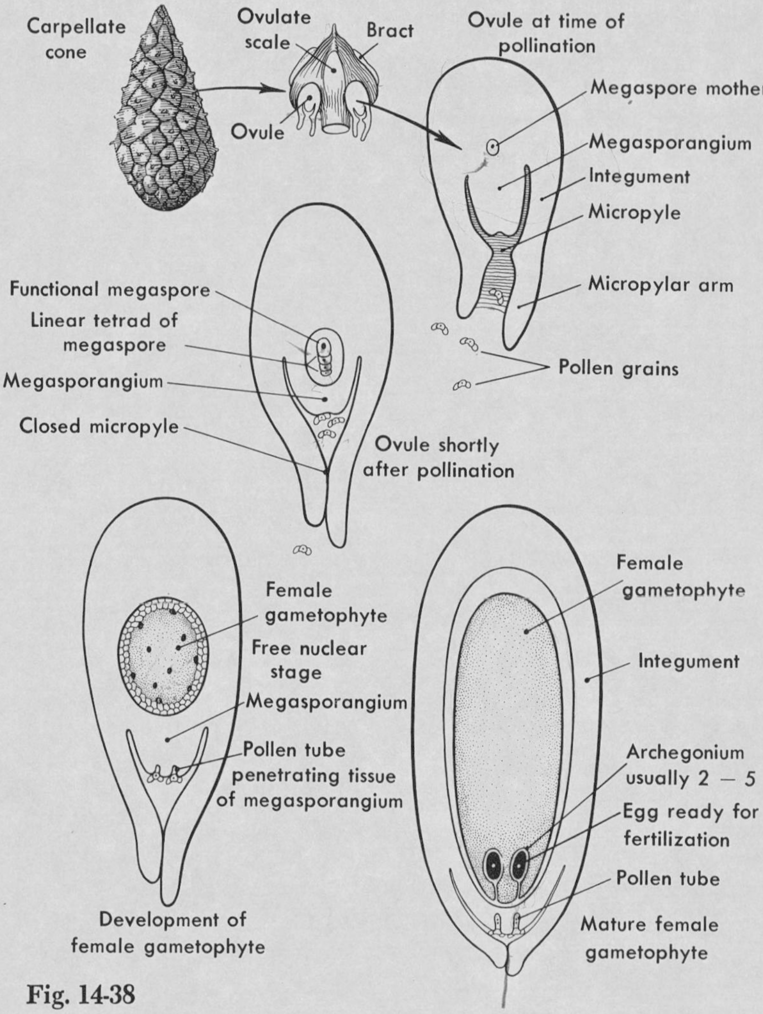

Fig. 14-38
Pinus sp. (pine). Development of the female gametophyte.

this stage that it is shed from the microsporophyll and then carried by the wind to the **carpellate cone.**

Archegonium. The megagametophytes within the two ovules of every megasporophyll develop archegonia, each with an egg, an archegonial or neck canal, and an endosperm, which is the vegetative body of the gametophyte and which will become the food supply of the seed. At the same time the sporangium is enlarging to form the seed coat and wings of the seed.

Sperm. Sperm nuclei in pollen tube.

Egg. Egg nucleus within the megasporangium.

Zygote. Once it is inside the megasporangium, the pollen grain is attached near the micropyle by a sticky fluid exuded at this stage by it. The pollen grain now produces a pollen tube. The generative cell divides to produce a stalk cell and a body cell. These cells then migrate down the pollen tube, where the body cell again divides to form two sperm nuclei.

Fertilization (Fig. 14-39) takes place

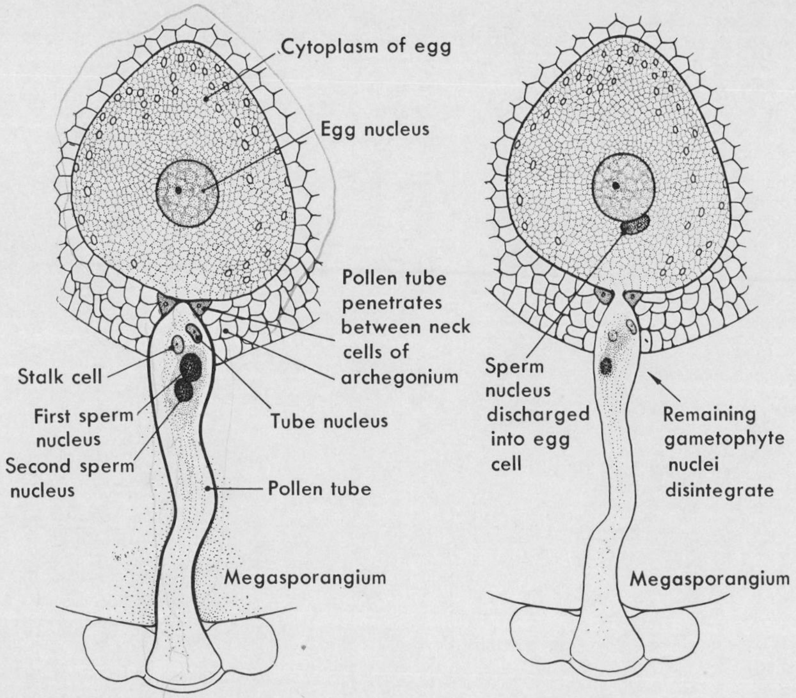

Fig. 14-39
Pinus sp. (pine). Fertilization of the egg by the sperm nucleus.

about a year after pollination. The pollen tube penetrates between the neck cells of the archegonium, and one of the sperm nuclei enters the cytoplasm of the egg to fuse with the egg nucleus to form a dippoid zygote. The remaining nuclei of the microgametophyte degenerate along with the rest of the structure.

Embryo. As soon as fertilization has occurred by **syngamy** (Fig. 14-40, *A*), a proembryo is formed. The egg nucleus divides into four separate nuclei geometrically arranged (Fig. 14-40, *B*) within the zygote. These in turn divide twice, walls form between them, and four tiers of four cells each are formed (Fig. 14-40, *C*). Each cell of the lower tier forms an embryo, which now begins to elongate (Fig. 14-40, *D*) so that there is a bottom rosette, a suspensor in the middle, and an embryonic tube with an apical growing region. The suspensor cells elongate, and the apical region develops (Fig. 14-40, *E*) until the immature embryos or **proembryos** penetrate into the **endosperm** of the seed (Fig. 14-40, *F*). Shortly thereafter all but one of the proembryos disintegrate, and the true embryo survives embedded in the endosperm (Fig. 14-40, *G*). By continued growth making use of the nutrients of the endosperm, this embryo differentiates into a hypocotyl, an epicotyl, and cotyledons. The embryo is surrounded by the endosperm and this in turn by the seed coat (Fig. 14-41). Thus the ovule has become transformed into a seed which develops wings that aid in its dispersal when it is released from the pine cone.

By absorption of water from the soil, the embryo enlarges and splits open the seed coat (Fig. 14-42). A radicle or primary root grows down into the soil for the absorption of more water. The remainder of the endosperm is used for further

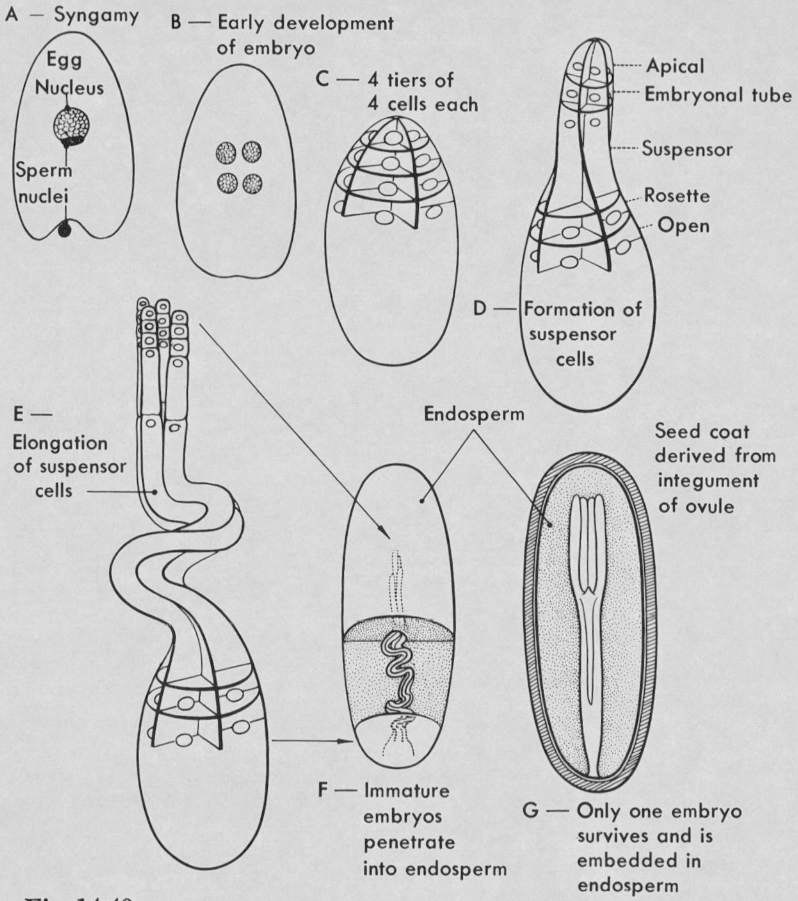

Fig. 14-40
Pinus sp. (pine). Development of the embryo in the ovule.

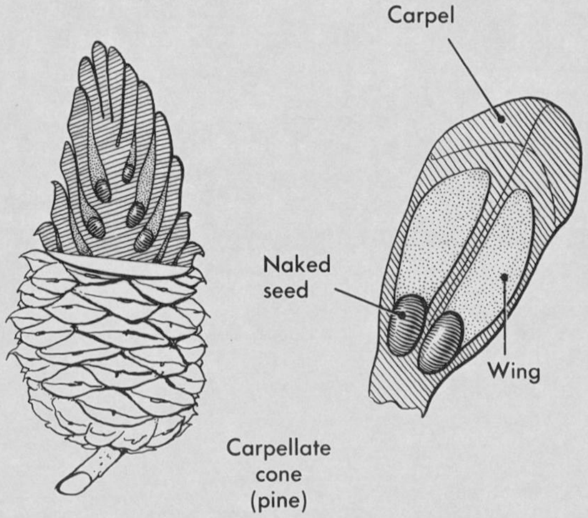

Fig. 14-41
Arrangement of seeds in gymnosperms showing how they are exposed in the axil of the bract.

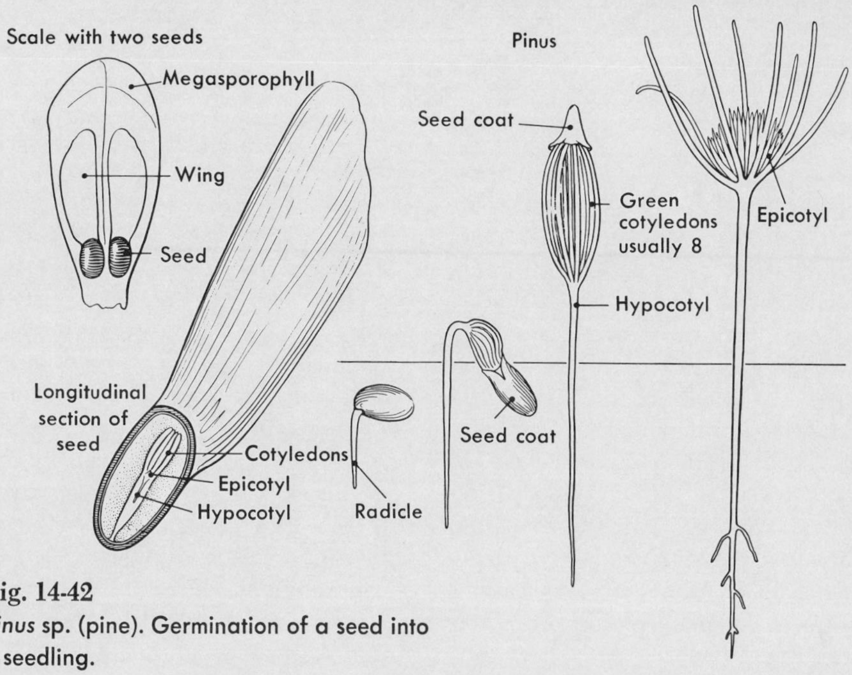

Fig. 14-42
Pinus sp. (pine). Germination of a seed into a seedling.

growth of the embryo, which soon appears above ground. The hypocotyl elongates to form a stem. Cotyledons develop chlorophyll, push off the remainder of the seed coat, and spread out. The epicotyl develops into the true leaves, which by continued growth through photosynthesis produce a pine tree.

Means of dispersal. When ripe seeds are formed in the carpellate cone, it loosens and opens to release them. By means of the wings that are attached, these seeds are blown about by the wind until they reach suitable soil, usually not far from the parent sporophyte.

LIFE CYCLE 16—LILIUM Sp.
DIVISION 25 Anthophyta
CLASS Angiospermae
SUBCLASS Monocotyledoneae

Characteristics

The genus *Lilium* has a very large number of species. The tiger lily is a common example. The lily family, Liliaceae, is easily identified by its flower, which has a perianth composed of six parts, and a superior ovary with three carpels and six stamens (Fig. 12-2). The lily flower is widely used to illustrate the flower of angiosperms, the dominant plant life of the curren geological era. The life cycle of these plants developed after a long period of evolutionary change, which has resulted in a highly specialized reproductive structure known as the flower. The steps leading to this are seen in many of the preceding life cycles (Fig. 14-44).

Asexual reproduction

Vegetative reproduction. Various means of vegetative reproduction can be found in the flowering plants, all of which have been discussed in Chapters 11 and 12. The lily is usually propagated by means of bulbs instead of seeds.

Sporophyte. A very large and independent (except for a brief and early period) sporophyte is present. It consists of roots, stems, leaves, and flowers—the plant body previously discussed. Angiosperms differ from gymnosperms in that flowers instead of cones are present; the ovules and seeds are covered, that is, enclosed by the carpel (megasporophyll) rather than naked as in the pine; the carpel or carpels constitutes the fruit when mature.

Sporangium. A microsporangium (pollen sac) in the anther and a megasporangium in the ovary form a part of the flower. A cross section of a very young anther reveals four pollen sacs or microsporangia in an anther (Fig. 14-45). Two of these lie on either side of the supporting tissue through which runs a single vascular bundle (Fig.

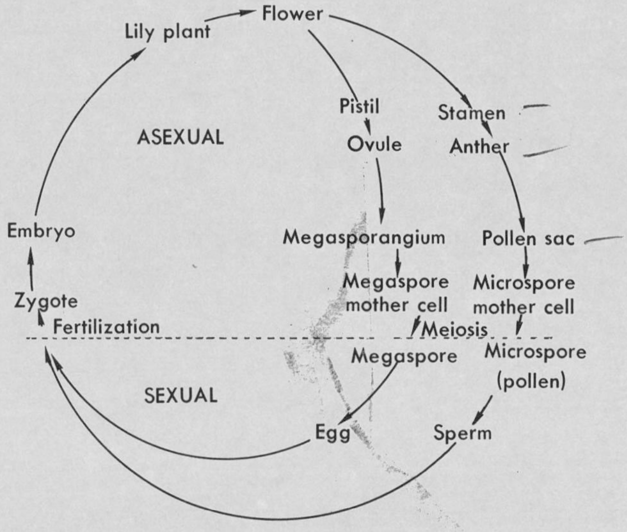

Fig. 14-43
Lilium sp. Life cycle of the lily.

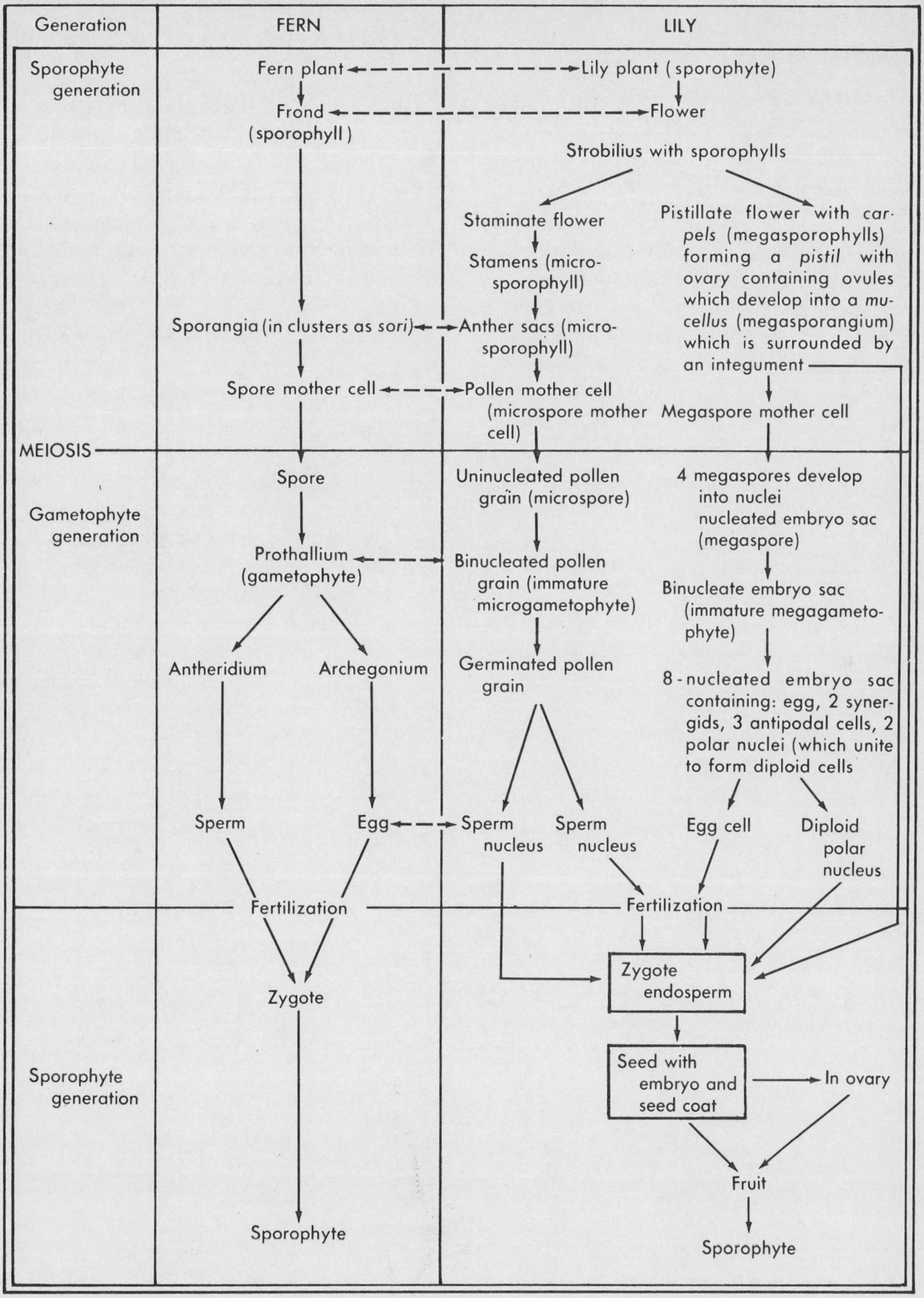

Fig. 14-44

Comparison of the life cycle of two plants, a fern and a lily. The fern has two independent generations, each living as a separate plant, whereas the lily has comparable stages, but the gametophyte is parasitic on the spermatophyte. This diagram shows that the two cycles are essentially the same, but that the angiosperm cycle is a refinement of the primitive cryptogamic cycle of the fern.

14-45). In each are many microspores or pollen mother cells. The anther is an elongate structure, and pollen mother cells are found throughout its length. When the pollen mother cells are first formed, they are closely packed together. Later, however, they separate and become spherical. As the growth of the anther progresses, the nucleus of each pollen mother cell divides by a special process termed sporogenesis (Fig. 14-46). The chromosomes line up during metaphase in pairs (synapsis), and, after each has duplicated itself, individual members of the pair will separate and move

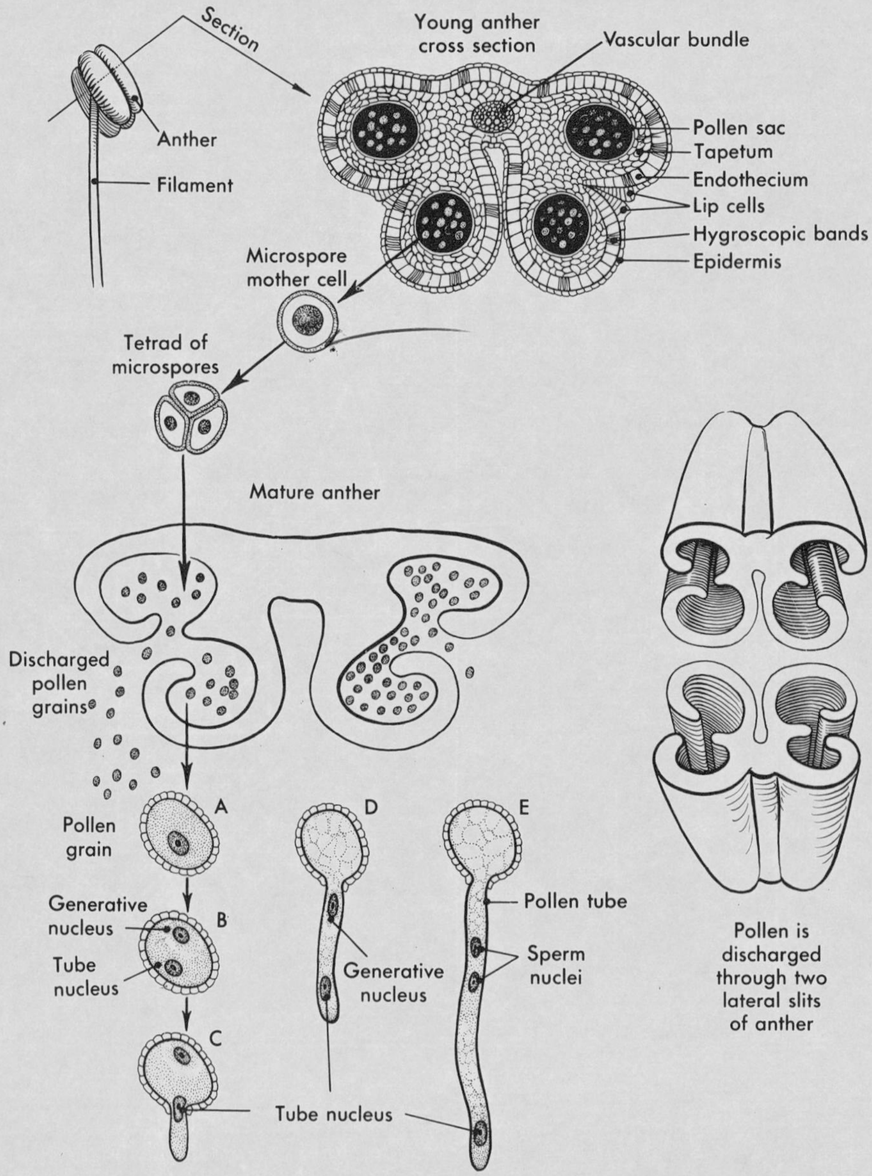

Fig. 14-45
Development of microspore and pollen tube of the lily.

to opposite poles, eventually producing two daughter nuclei that have one half the number of chromosomes of the mother cell. The chromosomes of each daughter nucleus undergo a second and equatorial division, and the half chromosomes separate

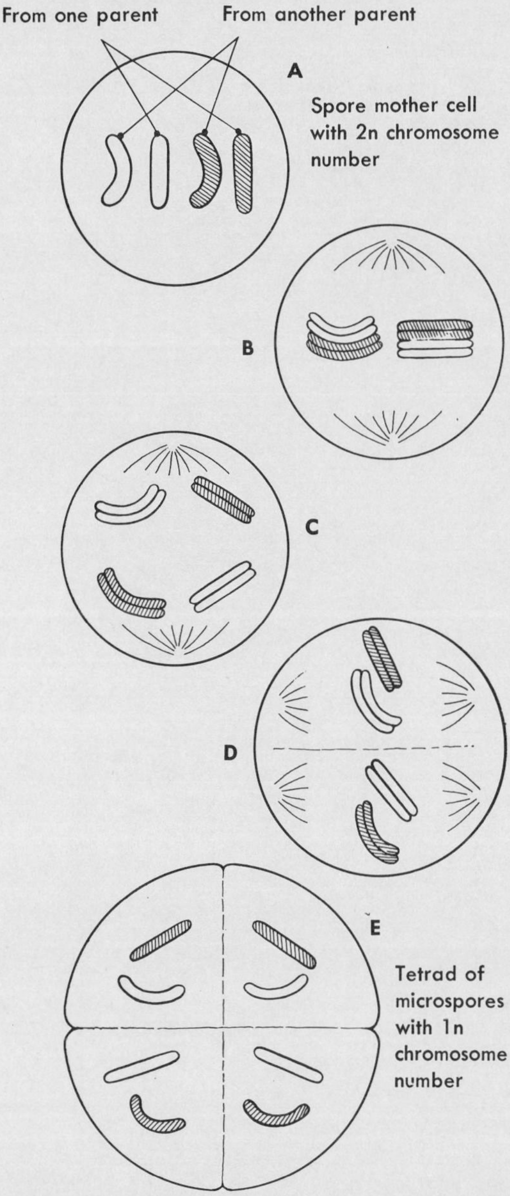

Fig. 14-46
Diagram of sporogenesis.

into four nuclei, each with the haploid or reduced number of chromosomes. Cell walls form, giving rise to a group of four cells known as microspores. This cluster of four spores is designated a tetrad of microspores. These spores may remain together in a tetrad arrangement. Usually they separate and develop into pollen grains. This is the only male gametophyte.

The development of the megasporangium usually occurs concurrently with that of the pollen. The carpels of the angiosperm pistil are spore-bearing leaves that have ovules and bear the megasporangia. Within the megasporangium is a megaspore mother cell that undergoes two nuclear divisions, resulting in four haploid nuclei. The production of these nuclei is essentially like that of the microspores inasmuch as both have the haploid number of chromosomes. At first they are arranged at the micropylar end of the cell, but later, three migrate to the opposite (**chalazal**) end (Fig. 14-47, *D*). Next the three chalazal nuclei divide by mitosis to form two large triploid nuclei. Meanwhile the haploid nucleus at the micropyle divides by mitosis, forming two haploid nuclei. Finally, a fourth mitosis takes place in all four nuclei, resulting in eight nuclei, four triploid and four haploid (Fig. 14-47, *E*). This is a functional female gametophyte.

Spore. Both megaspores (Fig. 14-47) and microspores (pollen) are produced by the floral parts as a result of meiosis.

Microspore. The conversion of a microspore into a pollen grain is the result of the division of the microspore nucleus into two nuclei, the **generative nucleus** and the **tube nucleus.** The wall of the microspore thickens, and its outer surface may become covered with ridges, spines, or other characteristic markings of the species producing it. Thus the wall of the microspore becomes the wall (**exine**) of the pollen grain and is one of the most resistent natural substances known. Each pollen sac is eventually filled with a powdery mass of mature

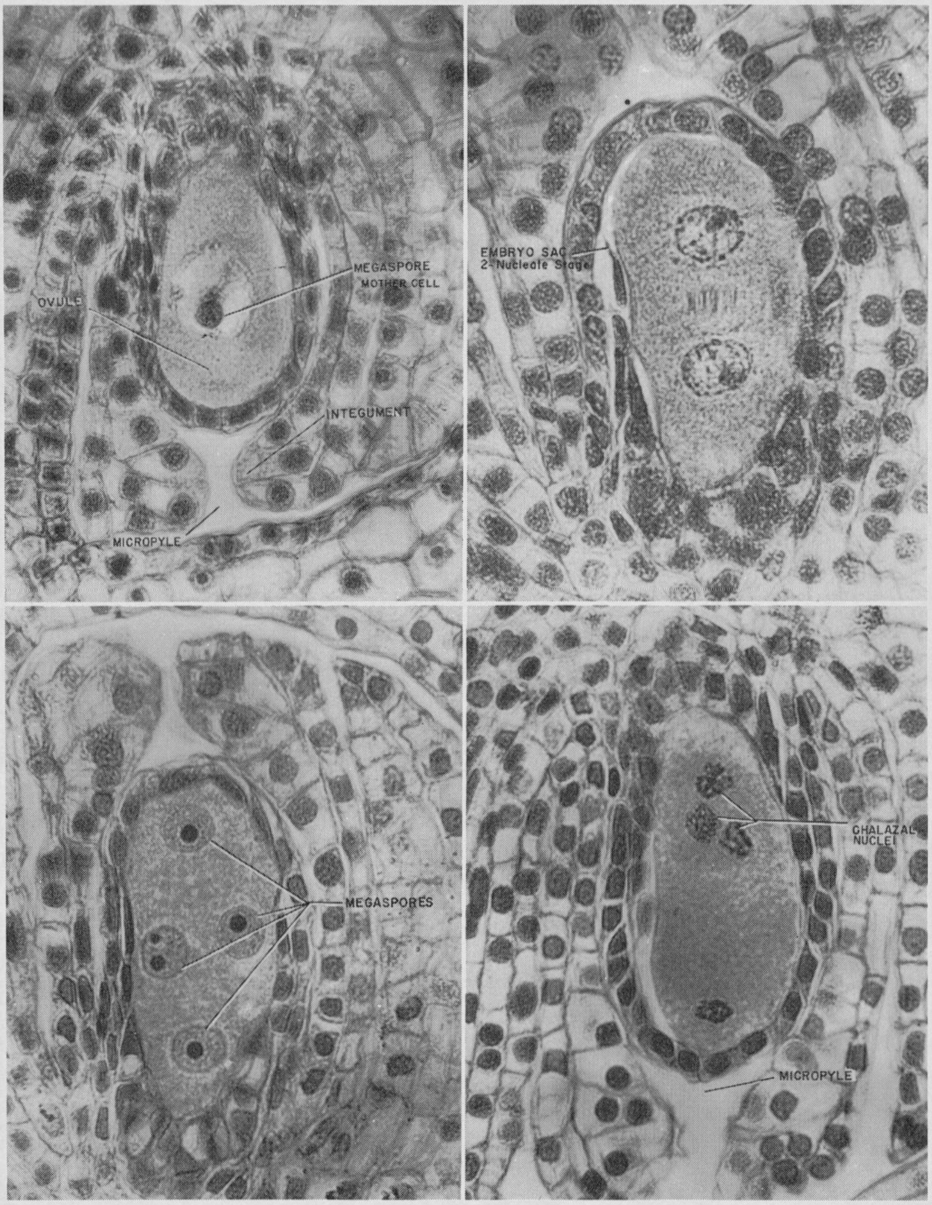

Fig. 14-47

A-H, Development of megaspore and female gametophyte in an angiosperm. **A,** Cross section of a lily ovary showing an embryo sac with a megasporocyte about to undergo division (×500). **B,** Lily embryo sac showing the 2-nucleate stage; remains of spindle fibers can be seen between the nuclei (secondary megasporocytes) (×500). **C,** First 4-nucleate stage with the nuclei (megaspores) arranged in the diamond-shaped pattern, typically preceding migration of the three chalazal nuclei (×500). **D,** Same lily embryo sac showing a later migration (×500). (Courtesy George H. Conant, Triarch Products, Ripon, Wis.)

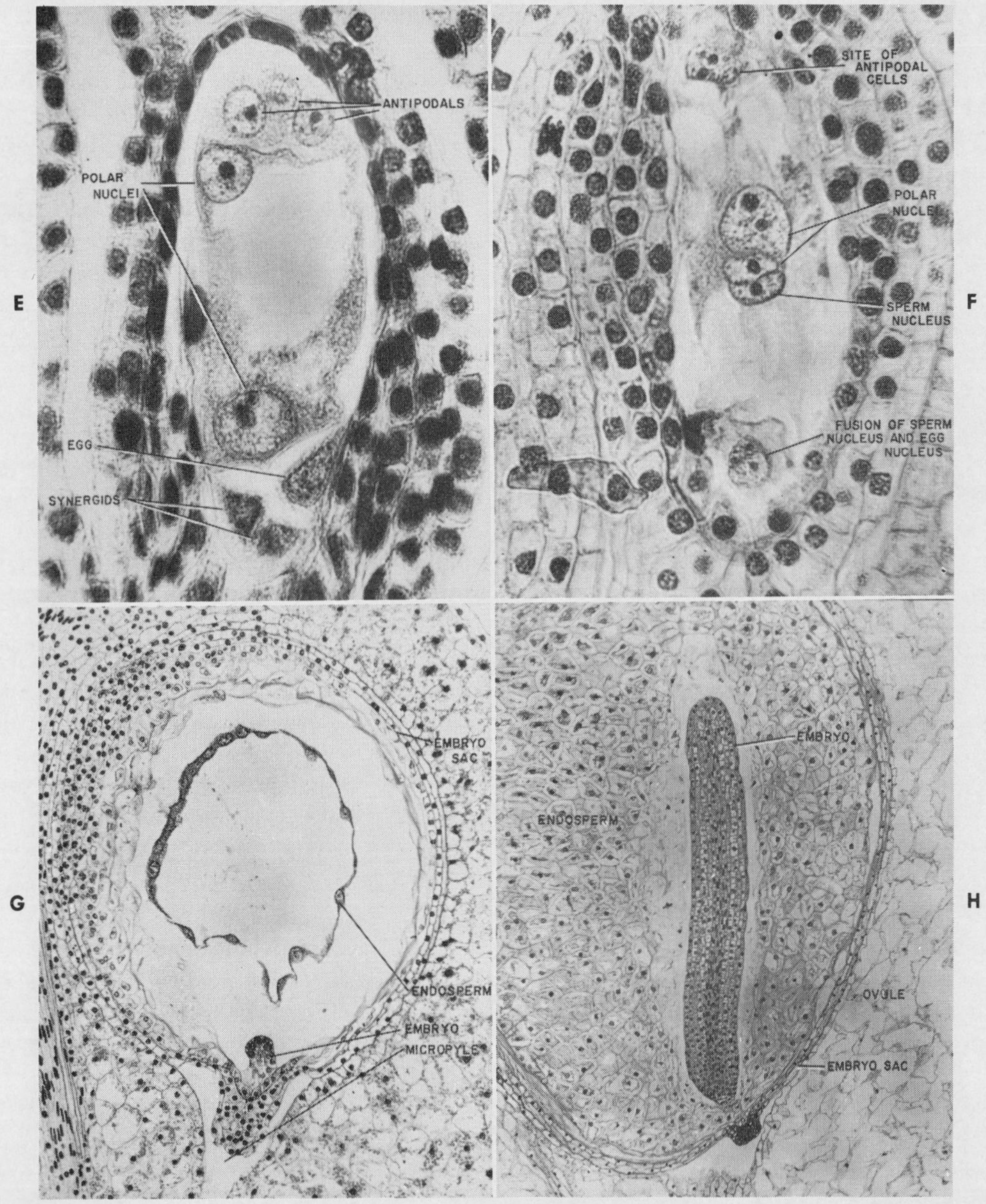

Fig. 14-47, cont'd.
E, Rarely occurring true 8-nucleate stage, with all nuclei clear and distinct (×500). **F,** "Double fertilization"; two innermost nuclei of the 8-nucleate sac migrate to the center and are referred to as "polar nuclei"; these fuse together with one of the male nuclei from the pollen tube; the other male nucleus fuses with one of the micropylar nuclei (egg), while the rest of the micropylar nuclei (with synergids) disintegrate or remain nonfunctional (×500). **G,** Cross section of a lily fruit showing a young embryo sac and the free-nucleate endosperm (×80). **H,** Cross section of a lily fruit showing an older embryo (×35). (Courtesy George H. Conant, Triarch Products, Ripon, Wis.)

pollen grains. As the anther matures during the development of pollen, the pollen sac splits longitudinally, and the pollen is free to be dispersed.

Megaspore. The eight nuclei are embedded in a common mass of cytoplasm, without separation into individual cells. This is called the embryo sac in the flowering plants. The number of nuclei in the embryo sac varies according to the species, and although eight is the common number, there may be as few as four.

Sexual reproduction

Gametophyte. There are two gametophytes—male and female—as in the pine. But both gametophytes are much reduced in size, the pollen grain and the embryo sac; both are completely dependent upon the sporophyte generation, since neither gametophyte has chlorophyll.

Antheridium. Only the two sperm nuclei of the pollen grain represent the male sex organ.

Archegonium. The only reminders of the female sex organ are the **egg nuclei** (Fig. 14-47, *E*). One nucleus of each group of four migrates to the center of the embryo sac to become **polar nuclei.** Each ovule is often bent back upon itself and enveloped by one or more **integuments,** except for a small opening at the tip called the micropyle. The integuments form as the female gametophyte develops. The entire ovule is attached to the placenta by a basal stalk or funiculus.

Egg. After the migration of the polar nuclei, three haploid nuclei remain at the micropylar end of the embryo sac. Two are called **synergids,** and the other is the egg. The three nuclei remaining at the end of the embryo sac opposite, chalazal, to the micropyle are known as the **antipodals** (Fig. 14-47, *E* and *F*).

Sperm. Two male gametes or sperm are produced as a result of the division of the generative nucleus of the pollen grain (Fig. 14-47, *F*).

Pollination. The pollen grain settles on the stigma. The tube nucleus forms the pollen tube that must grow through the tissues of the stigma, style, and ovary to the ovule instead of merely landing in or near the **micropyle** of the ovule (Fig. 14-47), as in the gymnosperms.

The development of the egg sac within the ovule, the presence of the sperm in the pollen, and the transfer of pollen to the stigma are all events leading up to fertilization in angiosperms (Fig. 12-3). The stigma of the pistil is covered with a sticky mixture of water, sugar, and other substances. Once the pollen falls on this stigma, it is held there, and in this environment the pollen immediately starts to grow or germinate. It is possible to induce many kinds of pollen grains to germinate in a sugar solution on a slide so that the development of the pollen tube can be observed.

The pollen grain bulges out to produce a **pollen tube** that grows down through the style. As the protoplasm flows out into the tube, the pollen grain is left empty on the stigma. The pollen tube produces digestive enzymes that decompose the tissues of the style as it makes its way toward the ovule. Usually the tube nucleus is near the tip of the elongating tube, and the generative nucleus is somewhat behind it (Fig. 14-45). The generative nucleus divides so that three nuclei may be present in the tube. Eventually the pollen tube arrives at and penetrates the ovule, either through the funiculus or through the micropyle, as indicated in Fig. 12-3. Once the pollen tube reaches the embryo sac, the two sperm nuclei, which have been derived from the division of the generative nucleus, are discharged into the sac. The tube nucleus and pollen tube are left to die and degenerate. One of the sperm nuclei fuses with the egg and the other with the fused polar nuclei. The fused egg and sperm have the diploid (2n) number of chromosomes. The fused polar nuclei result in a 5n chromosome number. The fertilized egg marks the beginning of the new sporophyte or diploid generation and will develop into the embryo of the seed by a series of mitotic divisions.

Zygote. In the life cycle of the angiosperms water is not necessary for fertilization. Unlike the pine, the endosperm of the lily is formed by the fusion of two polar nuclei and one of the sperm nuclei (Fig. 14-47, *E* and *G*). Hence the endosperm is 5n, which, by a series of mitotic divisions becomes the food storage tissue of the seed. In the pine the endosperm was formed from the female gametophyte; hence it was haploid (n). The final union of egg and sperm (nuclei) results in a fertilized egg, or zygote. This fusion of egg and sperm nuclei is fertilization.

Embryo. Following fertilization, the zygote undergoes a series of divisions leading to the development of the embryo. The female gametophyte becomes the functioning embryo sac (Fig. 14-47, *H*), within which the embryo is found. The embryo is attached to the embryo sac by a suspensor at its basal end. The apical cell produces the embryo proper. In many plants, such as the bean, the endosperm degenerates before the seed is mature (Fig. 11-1). When this occurs, food is stored in the embryo seed coat, or the **nucellus,** which is the megasporangium of the ovule enclosing the megagametophyte. The union of one sperm nucleus with the egg nucleus and the other with the fused polar nuclei is frequently called **double fertilization** and is characteristic only of angiosperms. The presence of a polyploid tissue for food storage has no apparent advantage over a haploid or diploid one.

Although fertilization in most angiosperms is required for the production of seeds, there are some in which this is not true. The production of seeds without fertilization is known as **apomixis.** In apomictic plants it is not unusual for more than one embryo to be present inside a seed, although normally fertilized plants have only one.

Means of dispersal. During the growth of the embryo and the endosperm there is also rapid growth of the other tissues of the ovule. The ovule increases in size, the integument may harden, the embryo becomes dormant, and the ovule has thus ripened into a seed. The ovary usually increases in size, and its tissues become differentiated into those parts involved in the production and dissemination of the seeds. The mature ovary containing the seeds, together with any accessory parts, is now called a fruit.

LIFE CYCLE 17—SAPROLEGNIA Sp.
DIVISION 29 Oomycota
CLASS Oomycetes

Characteristics

Members of this genus are common water molds, and some are known to occur also in soil. Most species are saprophytic on plant and animal debris in fresh water, but a few are parasitic on algae and one on fish. The plant body is small, forming a mass of white hyphae upon the food source (Fig. 14-49). The hyphae are coenocytic and branching. Note the similarity between this life cycle and that of *Vaucheria* sp. (life cycle 2).

Asexual reproduction

Vegetative reproduction. Growth by continuous hyphae permits extensive growth, and fracture of parts of the plant allows for some additional spreading and the formation of new plants.

Sporophyte. None.

Sporangium. Special branches of the hyphae project outward from the substrate. The apical portion of this branch develops a crosswall that separates the sporangium from the remainder of the plant. The cytoplasm of this area darkens as zoospores are formed. When the branch matures, an apical pore opens from which the zoospores may escape. Spore production results after a cell wall appears around the nuclei in the sporangium and by the mitotic division of the already haploid nuclei.

Spore. The spores are biflagellated. Upon maturity, they swim from the sporangium, swarm for some time and then withdraw their flagella and secrete a cyst wall. After remaining dormant for some time, they germinate by producing another zoospore of a somewhat different shape. These secondary zoospores again swim about until they find a suitable location, after which they settle down and germinate into new sporophytes. The phenomenon of having two periods of motility of zoospores is known as **diplanetism**, but the effect of this can only be assumed to be a means of dispersal and at the same time a means of providing a mechanism for passing through adverse conditions.

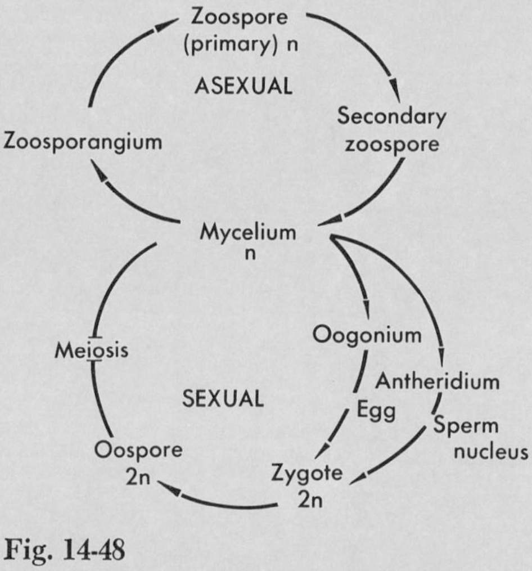

Fig. 14-48
Life cycle of *Saprolegnia* sp. (water molds).

Sexual reproduction

Gametophyte. The vegetative plant produces well-developed oogonia and antheridia. This plant is **heterogamous.**

Antheridium. The antheridia are produced on hyphae located around the oogonia or from an entirely different hypha. They contain several nuclei, and the antheridia grow until they come into contact with an oogonium. On contact, the antheridium produces a tube that pierces the wall of the oogonium.

"Archegonium." There are produced unicellular oogonia that arise as swellings on terminal hyphae separated by a crosswall. Most of the nuclei of the oogonium degenerate, but one remains to form the egg nuclei of each egg.

Sperm. None. Only "males" present are the sperm nuclei of the antheridium.

Egg. One or more eggs is present in each oogonium.

Zygote. The fertilized eggs develop into thick-walled oospores. Several months later they will germinate to develop hyphae that terminate in zoosporangia, in which are produced zoospores.

Embryo. None.

Means of dispersal. Dissemination is most effective by means of the motile zoospores. The oospores are also means for dispersal.

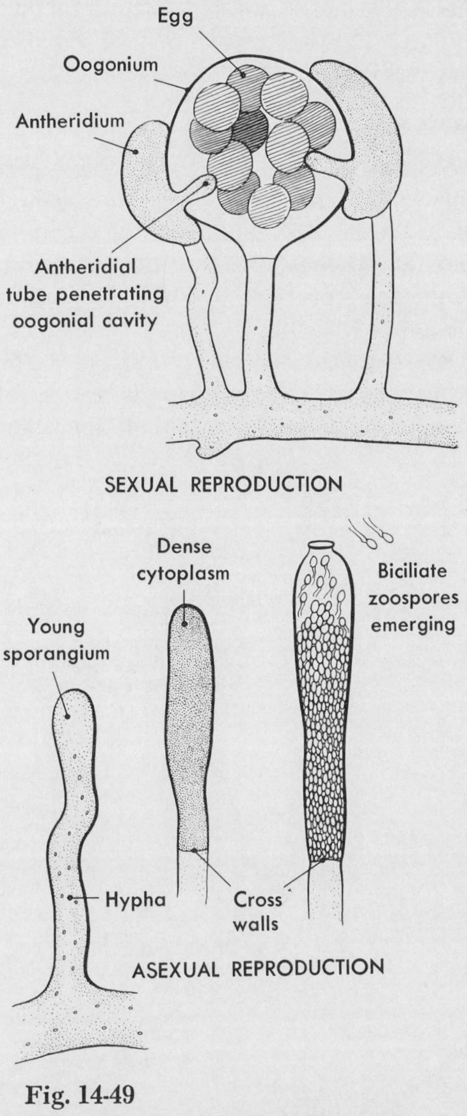

Fig. 14-49
Saprolegnia sp. Sexual and asexual reproduction.

LIFE CYCLE 18—RHIZOPUS STOLONIFER
DIVISION 31 Zygomycota
CLASS Phycomycetae

Characteristics

The common black bread mold, *Rhizopus stolonifer,* has long been a familiar example of the Phycomycetae. Members of this group are of minor economic importance only because of their damage to stored fruits and vegetables. Bread mold is characterized by white mycelia, prominent black zygospores, and black fruiting bodies (Fig. 14-51). The growth may be extensive and readily visible to the unaided eye.

Asexual reproduction

Vegetative reproduction. Stolons grow out from a group of rhizoids to penetrate the substrate. This is a means of vegetative growth and enlargement of the plant body. If these stolons become broken, separate plants are formed.

Sporophyte. None.

Sporangium. The sporangia are visible as small, black globules on the mycelium. They develop as terminal swellings on upright hyphae (Fig. 14-51). These swellings become filled with the multinucleated protoplasm, which then is separated from the rest of the hypha by a dome-shaped crosswall called the columella.

Spore. The many nuclei in the sporangium collect a small mass of cytoplasm around them and become transformed into uninucleated spores. The sporangium wall ruptures when the spores are mature. On reaching a suitable medium and under conditions of proper temperature and moisture they will germinate and give rise to a new mass of hyphae.

Sexual reproduction

Gametophyte. Under certain circumstances the formation of gametes takes place. This occurs when two separate strains of *Rhizopus stolonifer* are growing near each other (Fig. 14-52). These have

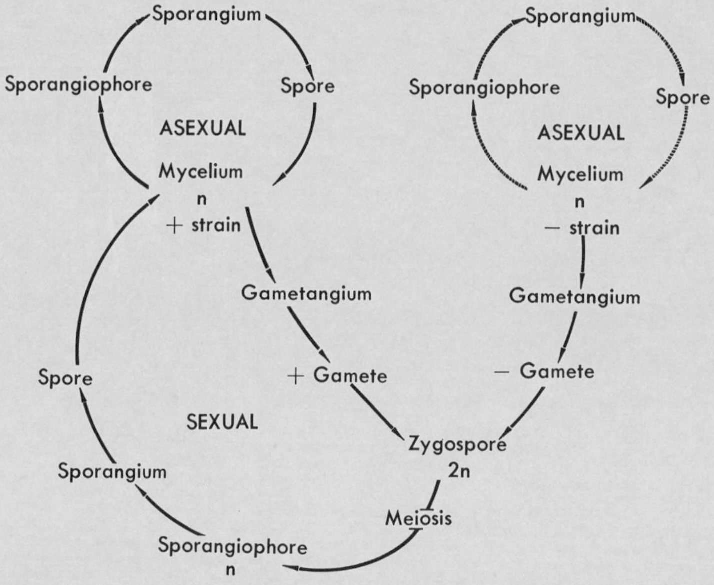

Fig. 14-50
Life cycle of *Rhizopus stolonifer.*

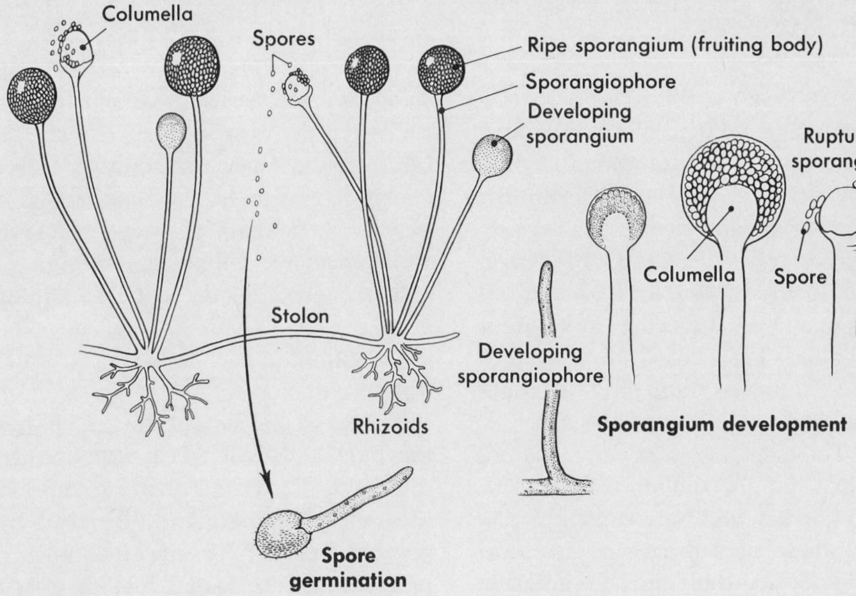

Fig. 14-51
Rhizopus stolonifer (black bread mold). Vegetative and asexual stages.

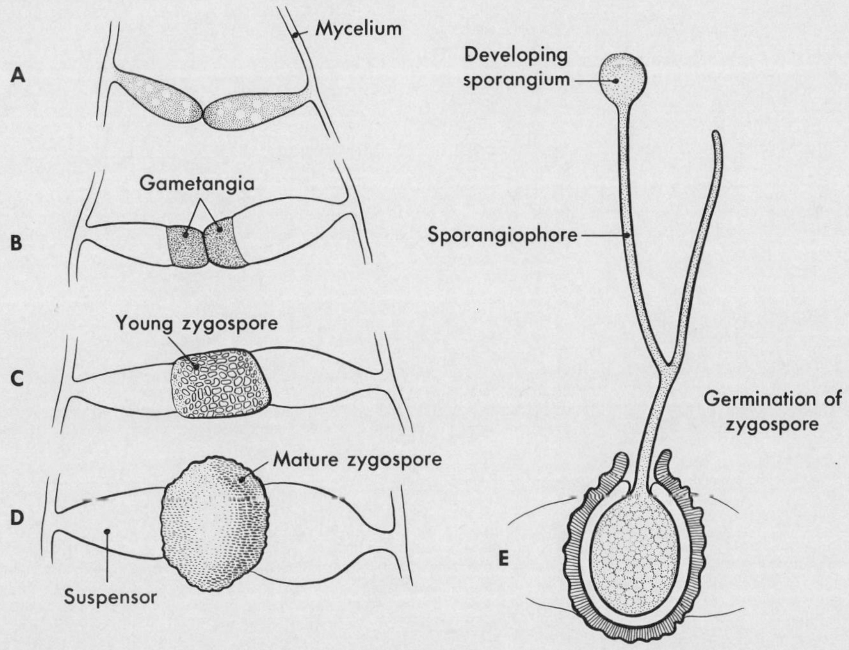

Fig. 14-52
Rhizopus stolonifer. Sexual reproduction.

been termed plus and minus strains and are equivalent to the separate sexes of other plants.

Gametangium. When the mycelia of two plants of opposite strains come near each other, lateral lobes are developed. These are the progametes, and this portion is darker than the mycelium proper because of thickening of the wall of the gametangia in the vicinity of their union. The apical portion is separated by a crosswall to form the gametangia. These are equivalent in function to the antheridium and archegonium of the higher plants.

Gamete. The gametes are multinucleated. When the gametangia come in contact with each other, the cell wall between them dissolves, and their nuclei pair off to form diploid nuclei. Unfused nuclei probably disintegrate.

Zygote. The multinucleated zygote formed at the time of copulation of the gametangia enlarges to form a zygospore. This is held in place by the original lobes of the respective mycelia. The zygospore remains protected by a heavy covering for some time. Many such zygospores give the growth a black appearance along the line of fusion (see illustration at the beginning of this chapter). If conditions remain favorable, the **zygospore** will germinate and produce new mycelia and develop **sporangiophores** and sporangia while remaining in its original location.

Embryo. None.

Means of dispersal. The spores from the asexual stage are the principal means of dispersal. That these spores are present everywhere can be readily proved by leaving homemade* bread in a moist, warm place, where it soon develops a luxurious growth of mold.

*Bakery bread has preservatives added that prevent molding for several days.

LIFE CYCLE 19—SCHIZOSACCHAROMYCES Sp.

DIVISION 32 Ascomycota

CLASS Ascomyceteae

Characteristics

The yeasts are typically unicellular and microscopic in size (Fig. 14-54), with oval cells, large vacuoles, and scattered inclusions of reserve food. The nucleus is obscure except when the plant is budding. The nucleus divides under favorable conditions, and when rapid budding is taking place, small chains or clusters of cells may be seen.

Asexual reproduction

Vegetative reproduction. The most common kind of asexual reproduction is by budding. A small protuberance forms on the side of the yeast cell. The bud grows for a time, after which it may become detached from the parent cell. The cells may separate, although in some instances they cohere and continue to bud, producing a long chain.

Sporophyte. None.

Sporangium. A yeast plant may become a kind of spore-bearing structure called an **ascus.**

Spore. By parthenogenesis eight ascospores may develop within the ascus, which in turn develops into new yeast cells.

Sexual reproduction

Gametophyte. The yeast plant (each single cell) is the haploid generation.

Gametangium. None.

Gamete. The haploid yeast cells form an ascus and behave like sex cells inasmuch as two of the cells are of one strain and two of another. They may pair and fuse to form a diploid cell.

Zygote. After fusion of two yeast cells the fused cell can be considered a zygote, since it is diploid. The fused cell nucleus undergoes meiosis, producing eight haploid **ascospores** within the ascus. The ascospore, as a spore, develops into the new yeast cells.

Embryo. None.

Means of dispersal. Ascospores are produced as a result of both sexual and asexual reproduction. Dissemination of yeast is accomplished upon their release from the ascus.

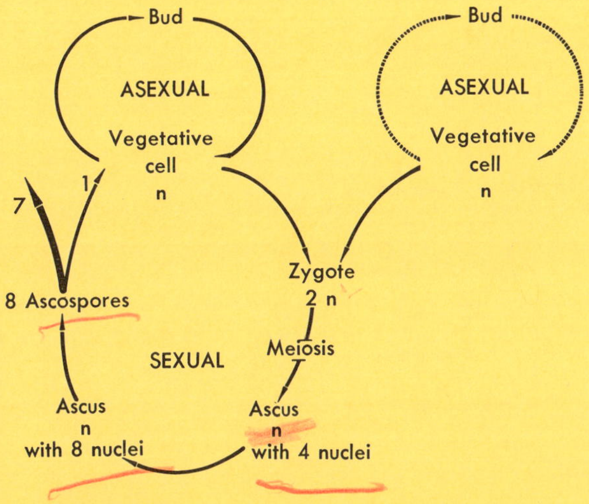

Fig. 14-53
Life cycle of *Schizosaccharomycetes* sp. (yeast).

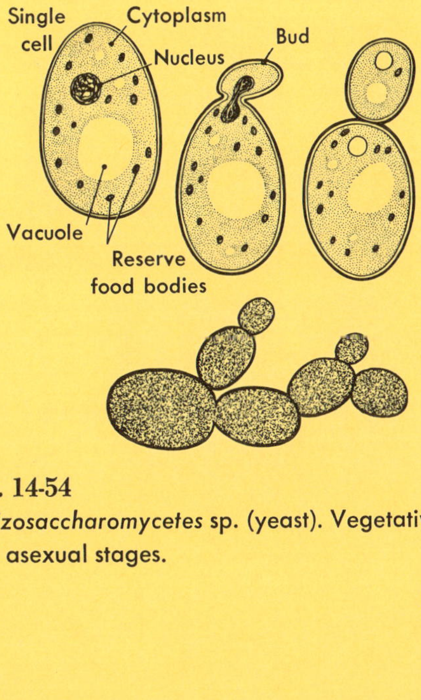

Fig. 14-54
Schizosaccharomycetes sp. (yeast). Vegetative and asexual stages.

LIFE CYCLE 20—MICROSPHAERA Sp.
DIVISION 32 Ascomycota
CLASS Ascomycetae

Characteristics

The powdery mildews represent a group of plants that produce diseases in other plants. They get their common name from the **conidia** that are produced and that give the surface of the host plant a powdery appearance (Fig. 14-56). The mycelium of *Microsphaera* sp. is not extensive, since it infects only the epidermis of the leaf. The **cleistothecium** containing the asci is a small body attached to the epidermis. This changes color as it matures—it is white at first and then becomes darker until it is finally black.

Asexual reproduction

Vegetative reproduction. The growth of the mycelium as it covers a greater area of the leaf is the only form of vegetative reproduction. The vegetative plant consists only of a short **haustorium,** a special hypha that penetrates the cell for absorption of nutrients and that is rhizoid in nature, the interconnecting stolons, and the **conidiophores,** short, erect hyphae producing the conidia.

Sporophyte. None.

Sporangium. The conidiophore is the equivalent of the sporangium, producing the conidia in rows.

Spore. The conidia, therefore, are the equivalent of the spores of other fungi. These small bodies are released from the apex of the conidiophore and are capable of germinating into new plants, thus completing the asexual cycle.

Sexual reproduction

Gametophyte. The mycelium, composed of septate hyphae, represents the gametophyte generation.

Antheridium. Late in the season the pro-

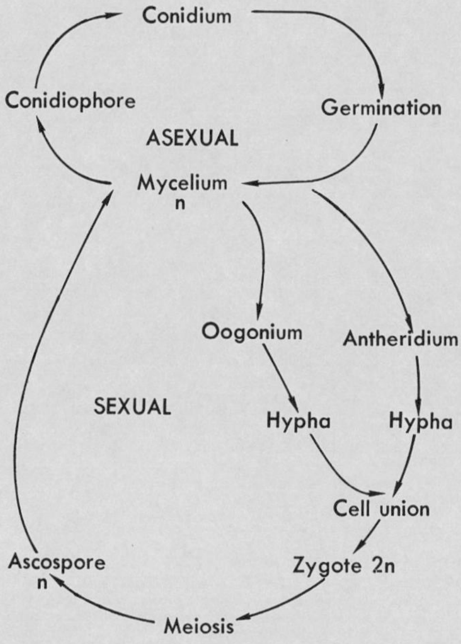

Fig. 14-55
Life cycle of *Microsphaera* sp.

duction of conidia ceases, and the mycelium develops poorly differentiated sex organs. From adjacent hyphae two very short branches arise side by side. One of these bears the single-celled antheridium. The antheridial cell becomes closely applied to the female reproductive organ or oogonium.

"Archegonium." No true archegonium, however, an oogonium termed **ascogonium** is produced adjacent to the antheridial cell-bearing hypha. After an opening forms between the antheridium and the ascogonium, the nucleus of the antheridial cell moves into the ascogonium, but nuclear fusion is delayed.

Sperm. None as such, but the oogonial nucleus acts as an egg.

Zygote. The union of the antheridial and oogonial nuclei produces a diploid cell or zygote. After fertilization, a vegetative cover of haploid hyphae, the **perithecium,** grows around the developing ascospore to enclose it completely. The **ascus mother cell** forms within the cleistothecium and then goes through three successive nuclear divisions to form eight haploid ascospore nuclei, and eventually ascospores. These develop new mycelia.

Embryo. None as found in the vascular plants.

Means of dispersal. Dispersal is by the conidia shed from the conidiophore and also by the ascospores produced within the ascus.

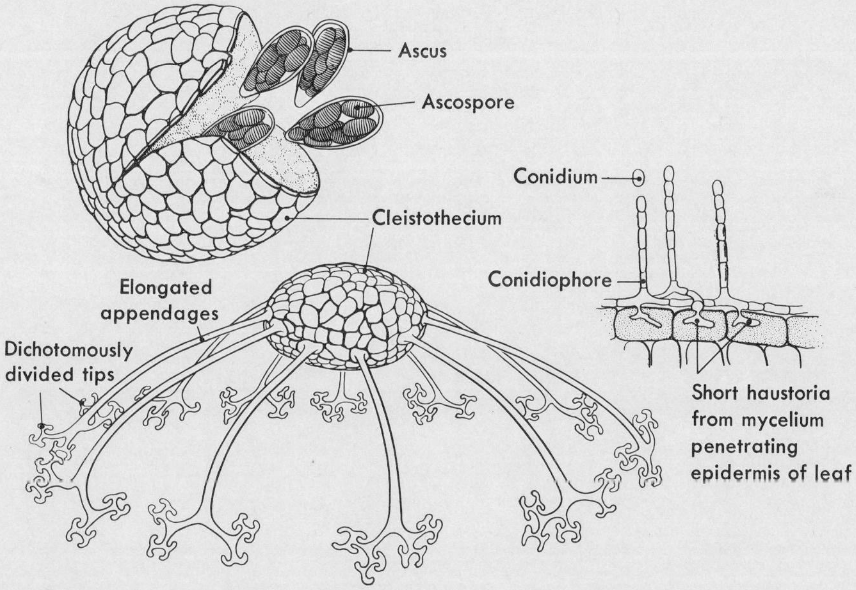

Fig. 14-56
Microsphaera sp. (powdery mildew). Spores are produced as a result of both asexual (conidium) and sexual (ascospore) reproduction.

LIFE CYCLE 21—PUCCINIA GRAMINIS
DIVISION 33 Basidiomycota
CLASS Basidiomyceteae

Characteristics

For the most part, the wheat rust fungus is a noxious parasite living on its host, the wheat leaves and stems. However, only part of its life cycle (Fig. 14-58) is spent on the wheat plant. To complete its life cycle, a part of its growth period must occur on the American barberry bush. Therefore it alternates between the wheat and barberry bush but does the most serious damage to the wheat. The most effective control of the ravages of the wheat rust has been accomplished by the elimination and destruction of the barberry bush in areas where wheat is grown. The only source of infection of wheat plants each spring is a spore (aeciospore, Fig. 14-59) produced by the fungus on the barberry leaf.

The wheat rust fungus produces a number of different kinds of spores, which develop into filaments or hyphae that are found between the cells of the leaves and stems. This is called a rust because of the reddish brown color of the spores that are found chiefly upon the surface of the host leaves and stems.

Asexual reproduction

Vegetative reproduction. The mycelium (mass of hyphae) grows between the cells of the host plant, frequently rupturing through the surface to produce the rust spots characteristic of this parasitic fungus.

Sporophyte. None as described for the higher plants. The binucleated mycelium growing between the cells of the barberry

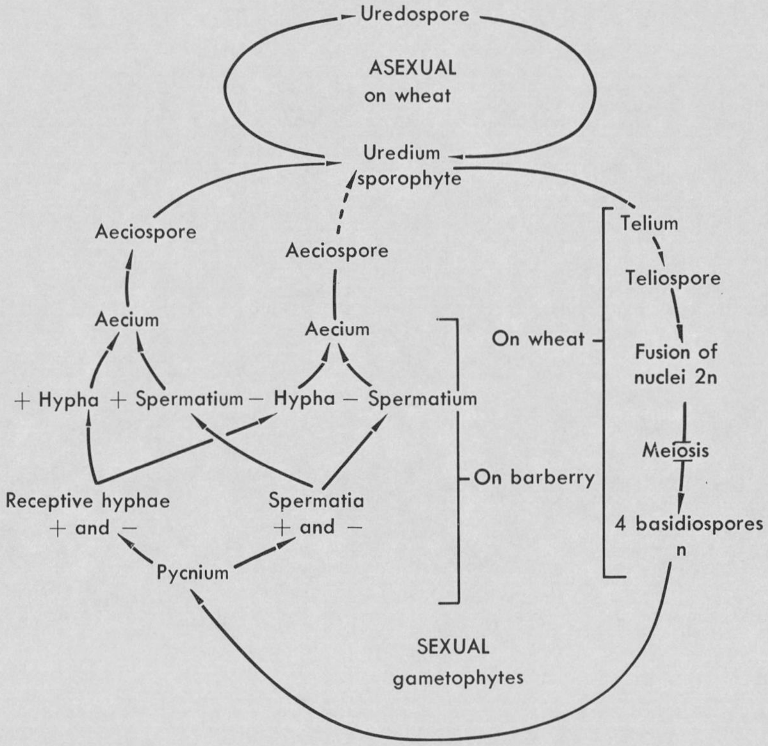

Fig. 14-57
Life cycle of *Puccinia graminis* (wheat rust).

leaf leads to the production of the binucleated cells of the aecidia. Aecidia are the cuplike formations found on the barberry leaf (Fig. 14-59), which finally rupture through the leaf epidermis to release the binucleated **aeciospores** (Fig. 14-59). The aeciospores are shed in the spring and are wind borne to the young wheat plants that they infect. Upon germination of these spores a mycelium is produced between the cells of the wheat leaf and stem, which finally ruptures through the epidermal surfaces as "red rust" spots. The mycelium on the wheat produces **uredospores,** one-

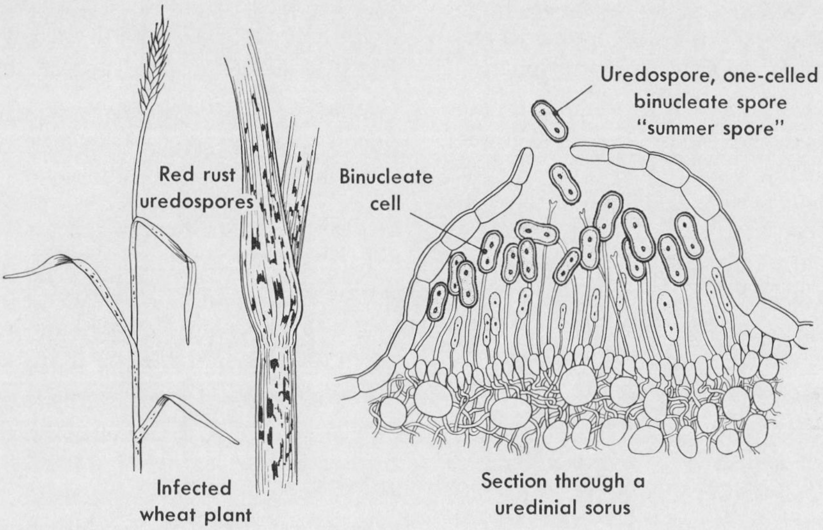

Fig. 14-58
Puccinia graminis (wheat rust). Uredial stage (asexual).

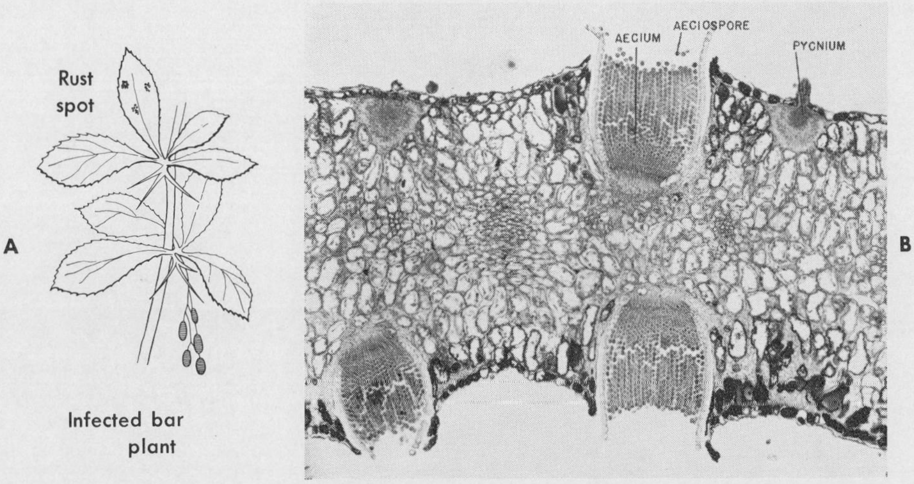

Fig. 14-59
Puccinia graminis. Sexual stage. **A,** Infected barberry plant. **B,** *Puccinia graminis* on barberry leaf showing at least one median pycnium and one median aecium. (×100.) (**B** courtesy George H. Conant, Triarch Products, Ripon, Wis.)

celled, orange-colored spores (the "summer spores") within a sporangium called a **uredium.** Again, the uredospores are wind borne and are capable of infecting other wheat plants. In this way the disease may be spread throughout an entire field. In late summer the mycelium on the wheat will form a **telium** and produce two-celled "winter spores" or **teliospores** (Fig. 14-60). Each of the two cells contains two nuclei, which initially fuse as the teliospore matures. Since the teliospore is a dark brown spore producing what appears as black rust spots on the surface of stems and leaves of the wheat, this stage is called the "black rust." It is the most destructive stage because it weakens the stems and causes them to break, thus preventing the harvesting of the wheat.

The teliospore has a thick, protective wall and is found lying in the wheat stubble or on the soil surface, where it remains all winter. Teliospores remain dormant until spring, when they germinate on the ground. Each teliospore produces a short hypha that develops a **basidium** (the basidial stage, the only stage having no host). Four **basidiospores** (Fig. 14-60, *A*) are produced on the basidium after meiosis. They form two strains. These are carried by the wind; if they fall on a barberry leaf, they will germinate and give rise to an intercellular mycelium.

Sporangium. The aecidium, basidium, and uredium are all sporangia producing their respective spores.

Spore. There are binucleated aeciospores, uredospores, and teliospores, which are produced on their respective host plants, as well as the haploid basidiospores.

Sexual reproduction

Gametophyte. None as such. However, the haploid basidiospores develop into haploid hyphae of two strains mixed together on the leaf surface of the barberry leaves. Eventually small, flask-shaped structures, pycnia, are formed, primarily on the upper surface of the leaf (Fig. 14-60, *B*). These produce hyphae called spermagonia. Small cells called **spermatia** are produced at the ends of the hyphae. The spermatia ooze out on the leaf surface in a sweet liquid that is attractive to insects.

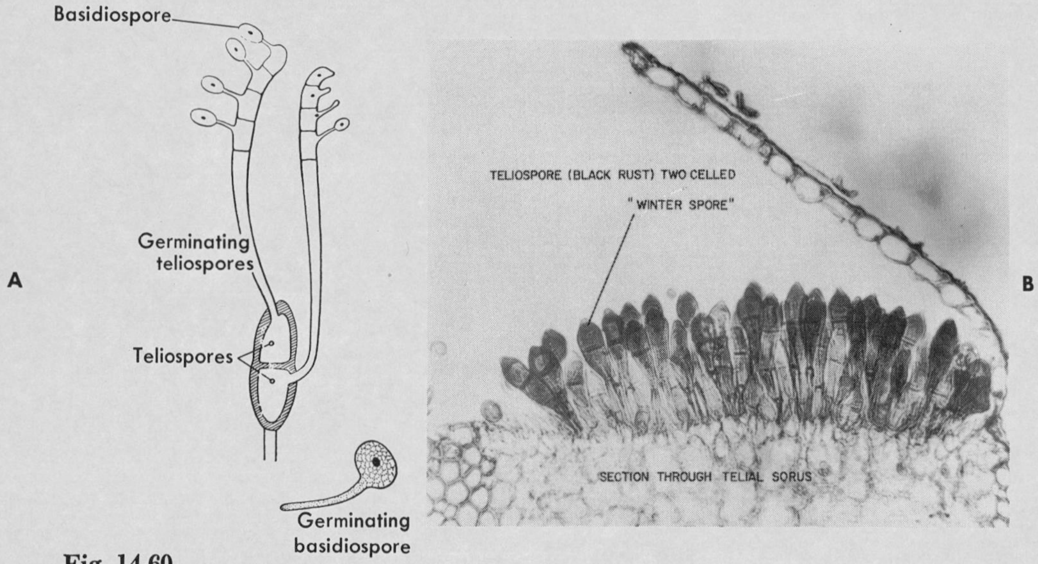

Fig. 14-60

Puccinia graminis. **A,** Telial stage showing teliospores on wheat stem (×300). **B,** Basidial stage. (**B** courtesy George H. Conant, Triarch Products, Ripon, Wis.)

Gametangium. None as such. However, the pycnium also produces either plus or minus receptive hyphae depending upon the basidiospore that produced the **pycnium.**

Gamete. Sperms are not produced, but plus and minus spermatia serve the same purpose. The receptive hyphae act as eggs.

Zygote. None as such. When a plus spermatium is carried, by an insect visiting the sweet spermatia mass, to a minus hypha, or vice versa, the plus and minus cells fuse. Following this fusion a binucleated mycelium will develop on the underside of the barberry leaf; this eventually leads to the production of aecidia. Aeciospores are formed after nuclear meiosis to form binucleated hyphae cells. This completes the life cycle of the wheat rust fungus.

Embryo. None.

Means of dispersal. This dispersal of the fungus is accomplished by the windborne basidiospores transported to the barberry leaf, from the barberry leaf to the wheat plant by the wind-borne aeciospores, and from one wheat plant to others by the wind-borne uredospores.

LIFE CYCLE 22—AGARICUS CAMPESTRIS

DIVISION 33 Basidiomycota

CLASS Basidiomyceteae

Characteristics

Mushrooms are well known to everyone. The portion that is eaten, however, is merely the fruiting body, which is composed of a **stipe** or stem and a **pileus** or cap (Fig. 14-62). When the fruiting body is mature, the underside of the pileus exposes the **gills.** Each surface of the gill bears spores. The annulus is part of the covering of the pileus remaining on the stalk after it has broken to expose the gills so that the spores may be shed. The vegetative plant is more extensive than the fruiting body, since the former consists of mats or masses of subterranean hyphae that feed on organic matter. Mushrooms grow best around decayed trees or in fertilized soil such as a pasture. They form the "fairy rings" of literature because they grow in a circle around the area of the roots of trees long since decayed.

Asexual reproduction

Vegetative reproduction. The extensive growth of the hyphae forms interconnecting plants that send up fruiting bodies.

Sporophyte. The sporophyte is composed of uninucleate cells forming extensive underground growths. Their interwoven and matted mycelial strands are called **rhizomorphs.** At maturity the rhizomorphs form buttons that are pushed above the ground to form the fruiting body.

Sporangium. The gill region of the fruiting body is the sporangium. The gill proper is composed of interlaced hyphae. Two kinds of structures develop along the sides of each gill, the paraphyses or sterile filaments and the basidia or fertile filaments. The basidium is binucleated.

Spore. As the basidium develops, the two nuclei, one plus and the other minus, fuse to form a delayed diploid cell. Meiosis im-

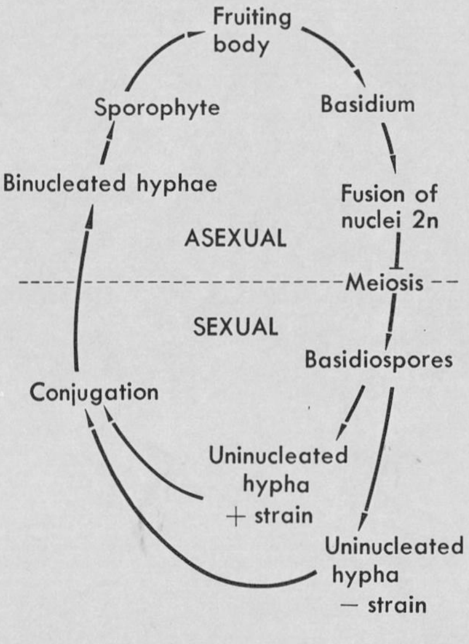

Fig. 14-61
Life cycle of *Agaricus* sp. (mushroom).

mediately follows, producing four nuclei. Meanwhile the basidium develops four lobes into which the new haploid nuclei migrate. In these lobes they develop basidiospores subtended by a short stalk or **sterigma.** Basidiospores are shed when they are ripe.

Sexual reproduction

Gametophyte. The reproductive cycle of this common plant takes place in the soil. The basidiospores germinate to produce small uninucleated hyphae. Hyphae formed from basidiospores are either plus or minus.

Gamete. Hyphae from germinating basidiospores of opposite strains conjugate to form binucleated cells. It is these cells that develop into the sporophyte.

Zygote. None as such.

Embryo. None.

Means of dispersal. Spores shed from the fruiting body are blown about by the wind and germinate when they come in contact with a suitable substrate. These spores are gathered by commercial growers and used as seed in mushroom farms in caves.

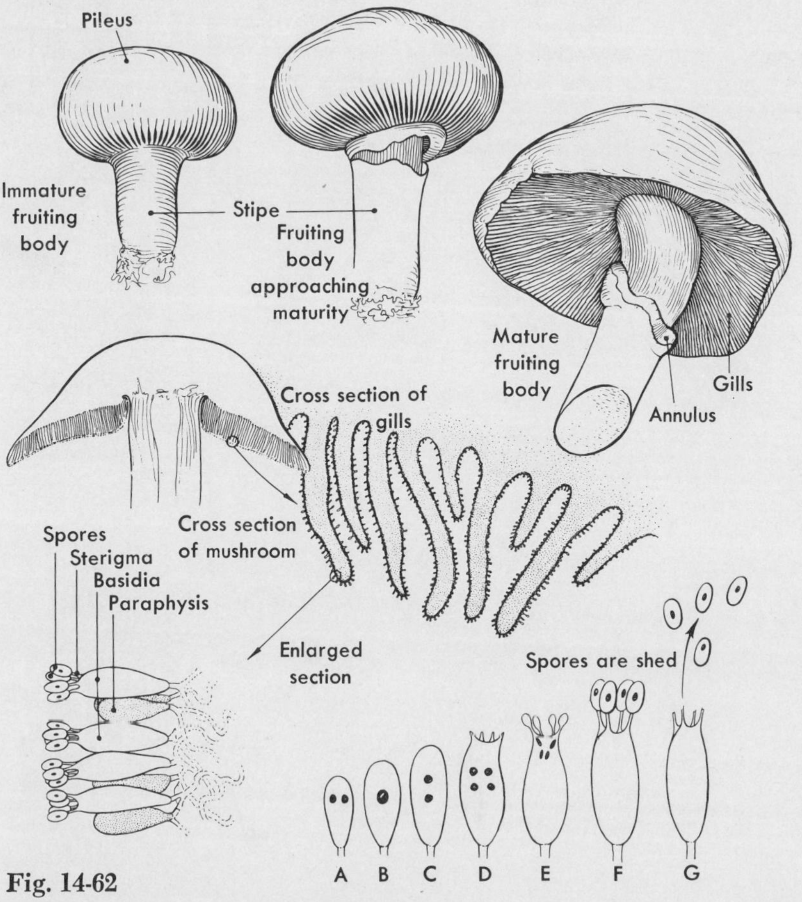

Fig. 14-62
Agaricus sp. (mushroom). Fruiting body and spore formation.

SPECIAL TERMS USED IN CHAPTER 14

aeciospore
aecium
akinete
annulus
antheridiophore
antherozoid
antipodal
apomixis
archegoniophore
archegonium
ascogonium
ascospore
ascus
ascus mother cell
basidiospore
basidium
bract
capsule
carpellate cone
carposporophyte
chalazal
circinate
cleistothecium
columbella
conceptacle
cone
conidiophore
conidium
cupule
cystocarp
diplanetism
double fertilization
egg nucleus
elater
endosperm
exine
flagellum
frond
gametophore
generative nucleus
gill
globule
gonidium
haustorium
heterocyst
heterogamous
heterothallic
holdfast cell
hormogone
hormogonium
indusium
integument
isogamete
lip cell
ligule
macrosporangium
megaspore
megasporophyll
microgametophyte
micropyle
microsporangium
microspore
microsporophyll
neck canal
nucule
oogonium
oosphere
oospore
operculum
parthenospore
perithecium
pileus
pinna
plurilocular
polar nucleus
pollen
pollen tube
procarp
proembryo
prothallium
prothallus
protonema
pycnium
rhizomorphs
scales
sorus
spermatium
sporangiophore
sporangium
sporophyll
staminate cone
sterigma
stipe
strobilus
synergid
syngamy
teliospore
telium
tube nucleus
unilocular
uredium
uredospore
venter
zoospore
zygospore

Chapter 15

Plant communities and distribution

The environment of plants, combined with their adaptive structures and physiological latitude, determines where they live. This exceedingly complex web of functions programmed in the genes of all organisms results in a changing world of great variety. The few plants pictured above are in a portion of the Chihuahuan desert near Big Bend National Park, Texas, where there is a struggle for existence in a changing habitat. A closer look at habitats, however, reveals certain patterns of distribution that make it possible to describe large areas of the earth's surface in terms of the flora and fauna found there. Each species, no matter how abundant or how rare it may be, has a range that may be circumscribed in geographical and environmental terms.

The ecology of both plants and animals is difficult to consider in stages or subdivisions because each area investigated is dependent upon almost all other branches of biology. It is customary for ecological studies to be made from one of two approaches. One deals with the study of the individual organism, species, or group of related species. This organismic approach may be termed **autecology.** For example, a single plant may be studied under varying growth conditions, or, the ecology, that is, the environment and its relationships, may be described for a single species. From data obtained for each of a group of species, generalized statements may be made about the ecology of a genus, a family, or even an order. The larger the taxon, the less precise any of these statements may be. More recent studies have been concentrated on local breeding populations of a single species, and interest in the comparative autecology of separate populations leads to information about speciation pathways.

The second subdivision of ecology is termed **synecology** and deals with the study of **communities**, or associations, of different kinds of organisms functioning more or less as a unit. Communities may be described according to the dominant (most obvious) plants in the association, as for example, an oak-hickory-magnolia association. Such studies are often called population ecology, community ecology, or ecosystem ecology.

The tendency of several fields or branches of biology to concentrate on populations, that is, several different biological approaches, has lead to some confusion. At the present time the term population is used by three separate biological disciplines in three different ways. The systematist uses the term to describe an interbreeding group of individuals that form one part of the total living members of a species. The geneticist is concerned with populations according to their genotypes and studies them only as a sample of the species. The ecologist treats a population as a segment of a community of species with interrelated functions.

In Chapter 8 we described the various aspects of autecology concerned with the environment. Further autecological information on the various plant groups is discussed in Part VI. This chapter describes the synecological data available for the regions of the world. The energy cycles in ecosystems is beyond the scope of this text. The field of ecology is vast, and many important topics cannot be mentioned in an introductory text, for should we attempt to do so, we would become involved in a complex terminology that would tend to magnify this part of botany out of due proportion. All the basic principles of

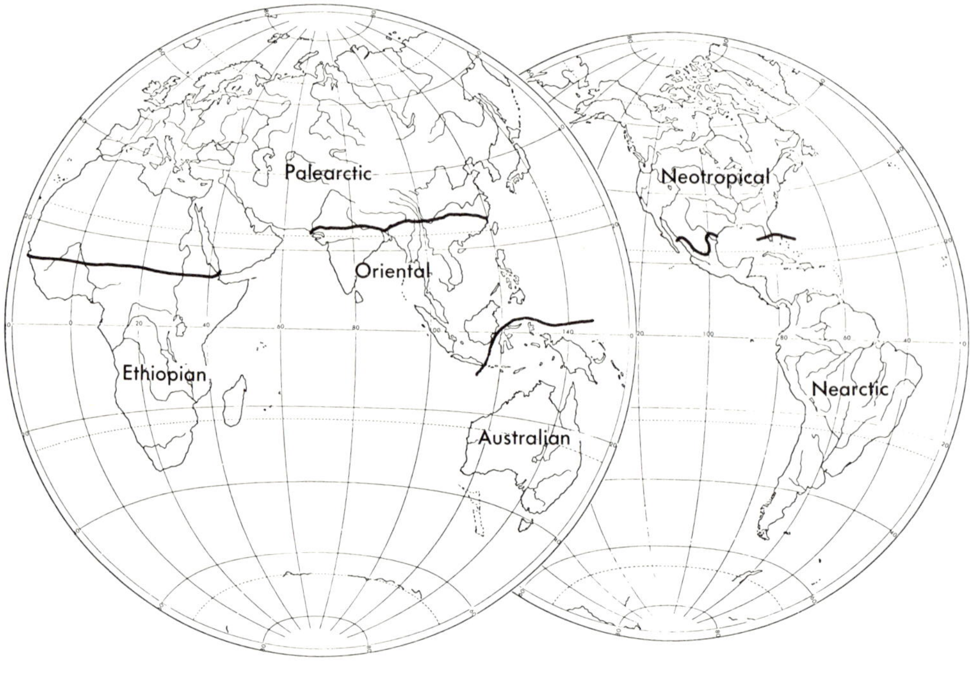

Fig. 15-1
Zoogeographical realms. These are the classical regions used by zoologists, but serve no more than a convenient reference for the discussion of the distribution of most organisms.

botany are involved in the study of ecology so that the student is assured of the necessary foundation for more specialized studies if desired.

One aspect of ecology of great concern to modern society is the ecology of human life. As mentioned in Chapter 4, land management, air and water pollution, and conservation of wildlife are major problems of our times. Each of these are of direct concern to ecologists, and applied ecology is a rapidly developing branch of biology.

CLASSIFICATION OF BIOTIC COMMUNITIES

No entirely satisfactory system of divisions of the World's flora can be devised except as a purely arbitrary descriptive device. Many widely different systems are used, but certain ones are commonly encountered in the literature. The three means of classification are based on (1) major or dominant species; (2) the physical habitat, for example, aquatic, terrestrial, and marine; (3) function, for example, the energy systems, or ecosystems.

The term **biota** refers to the entire assemblage of plants and animals, in all their stages, living on the earth at any given period of time. Thus it is the biota that we classify into various levels of communities based on their kind and their distribution.

Realm

Realms deal with the gross distribution patterns. Several systems are available to describe the division of the world into floristic regions or realms. Each of these is rather complicated and is based both on climate and major or dominant plants. They have not been as widely accepted or as useful as the one used by zoogeographers. These regions are given here exactly as used by zoologists (Fig. 15-1). Generally botanists do not use this system of realms but prefer to use geographical regions instead. Thus one would study the flora of North America rather than the flora of the Nearctic region. There are many floras of states or geographical regions.

GEOGRAPHICAL REALMS

Holarctic division

- Arctic—northern Europe, Asia, and North America above the tree line, the tundra
- Palearctic—Europe south of the tundra, including the Mediterranean area and the Siberian steppes, eastern China north of Viet Nam, Japan, southern Russia, and central Asia to eastern China
- Nearctic—North America south of the tundra (except for southern Florida and isolated spots near Brownsville, Texas) and the Mexican Plateau

Ethiopian—Africa south of the Sahara

Oriental—India south of the desert region and south of Tibet, southern China, Burma, Viet Nam, the Malayan Peninsula, and the large islands north of Australia

Australian—the Hawaiian Islands, all the remaining islands of the Pacific Ocean, Australia, Tasmania, and New Zealand and neighboring islands

Neotropical—southern Florida, islands of the West Indies, Central America, the southern half of Baja California, Mexico except the plateau region, certain isolated areas near Brownsville, Texas, and South America.

Biomes

It is primarily the physical environment that determines the distribution of plants. Rainfall, sunlight, and temperature variations of the earth's surface are reflected in the vegetation of each area. These conditions bring about rather well-marked areas known as **biomes**, the major vegetation zones of the continents. Each biome may appear in several areas in the world. These regions will be composed of a distinctive set of plants so that the separate biomes of each kind, although superficially resembling each other, differ greatly in their species communities.

Several classifications of biomes are possible, but there is general agreement on these areas, except some are occasionally lumped together for simplicity of treatment. The ten biomes treated here are found in many or most of the realms.

PLANT BIOMES
Ocean—Reef—Shore
Tundra—Grassland and savanna—Chaparral—Desert
Northern (taiga) coniferous forest—Deciduous forest—Tropical forest

Ocean. Only recently biologists have turned their interest to studies of the ocean as a biome. Its complexities form almost a separate subdivision of ecology. Certainly little is known yet about the distribution of plants and their interrelationships in the ocean environment. This vast area, 70% of the earth's surface, compared to only 3% occupied by freshwater, has been referred to as the last frontier. It is not a single barren biome, but rather, one characterized by a very diversified fauna and flora. Most of its surface is occupied by microscopic plants able to withstand the constant motion of the waves. Such environmental factors as light, temperature, salinity, and pressure, in harmony and disharmony with many animals, form an important site of biological energy exchange. As one approaches the shore, the environment becomes more severe. It is well known that the interface (**ecotone**) of a biome or a more restricted habitat is an area of tension, and so, one biome may abruptly change into another.

Reef. One such oceanic habitat is the reef. Many tropical islands and offshore areas near continents in tropical regions have formations called barrier reefs. Algae as well as coral (Coelenterata) form intricate habitats, the complexities of which are the subject of a great amount of literature.

Shore. True shore areas, especially in the intertidal region, are subjected to a wide range of environmental changes, and here

Fig. 15-2
Tundra. Note particularly the absence of trees except at the interface, where they are dwarfed (left edge of scene). (U.S. Forest Service.)

one finds various uniquely adapted plants such as fucus (Fig. 21-2) and other typical intertidal plants.

Tundra. Tundra areas are located on the fringe of the northern and southern limits of the continent and on the tops of tall mountains, usually at elevations over 12,000 feet (Fig. 15-2). Such areas (Fig. 15-3) are characterized by long cold winters and short rapid growing seasons, which produce stunted plants. Characteristic plants are mosses, lichens, grasses, and sedges. True tundra has permanently frozen soil (permafrost) beneath a shallow surface.

Fig. 15-3
Oat field in Alaska where the growing season is short, but the sunlight hours each day are long. (USDA photograph.)

Fig. 15-4
Western grassland area with trees only along permanent stream beds, and foothill mountains at the edge of the grassland. (USDA photograph.)

This biome merges into the extreme northern and southern eternal snow fields, which are entirely devoid of all plant life except for occasional patches of unicellular algae and sometimes on exposed rocky surfaces, a few lichens.

Grasslands. The great central plains of the United States, Canada, Asia, and limited areas of Africa and Argentina, form the grasslands (Fig. 15-4). These regions are treeless or nearly so, except for small areas along the banks of streams. Characteristic plants are grasses and composites such as asters, goldenrod, and sunflowers. Rainfall is sufficient to support grass but not trees, so it is the controlling factor. Unfortunately much of the world's grassland has been destroyed by overgrazing and is now virtually a desert. In tropical regions with abundant rainfall, but long dry seasons, a special type of grassland termed savanna is found. Scattered clumps of trees may occur in these areas. Extensive areas of central and southern Africa, parts of Australia, and scattered areas in South America have this kind of grassland. It is here that much of the big game herbivores are found. Such areas are subject to widespread fires that serve to remove the dead grass stems and make room for the new growth.

Chaparral. Areas with heavy winter rains, and long, hot, dry summers, especially on mountain slopes, may develop a climax of trees and shrubs with thick, hard, evergreen leaves. These areas are called chaparrals, probably the most poorly de-

Fig. 15-5
Typical desert scene in western Texas. This area is nearly bare of vegetation. Similar areas with wind-eroded rock finely ground into sand may form dunes as in parts of the Sahara desert.

fined biome. Such areas are widely scattered but are best known from California, Mexico, along the shores of the Mediterranean Sea, and along the southern coast of Australia. Again, fire plays an important part in the growth and decline of this climax.

Deserts. Areas that receive an average of less than 10 inches of rainfall per year are considered deserts. Associated with these are semiarid areas that receive more rainfall but that remain arid because of the rapid runoff of rain. The deserts of North America occur in eastern Oregon, western Idaho, Utah, and Nevada, extreme western Colorado, New Mexico, the eastern part of southern California, Arizona, western Texas (Fig. 15-5), and much of northern Mexico. Cactus and similar xerophytes, yucca, mesquite, creosote bush, and grasses of various sorts characterize these areas. The desert regions of South America are found west of the Andes Mountains along the coast of Ecuador, Peru, Chile, and central Argentina. Extensive Old World deserts are found in Africa, Asia, and Australia. The barren areas of the Sahara desert may have wind-blown sand dunes of fine sand, but much of the area is rocky or gravelly, with an occasional oasis surrounding an artesian well. Southwest Africa has a large area of arid land somewhat similar to the desert of southwestern United States. The Arabian peninsula, eastern Pakistan, parts of Siberia, and much of central Australia complete the list of deserts of the world. Each of the

Fig. 15-6
Sonoran desert. This type of desert maintains a variety of distinctive plants and offers interesting and rather simple associations of plants and animals, each with special adaptations for survival in an area having less than 10 inches of rainfall annually. Note the giant cactus. (U.S. Forest Service.)

major deserts is subdivided into characteristic regions, and each has a unique flora. Most deserts lie in the leeward shadow of high mountain ranges that receive most of the water dropped by the clouds as they pass over the mountains.

True desert in North America is typified by the Sonoran desert in Arizona (Fig. 15-6). The presence of creosote bush (*Larrea* sp.) and saguaro characterize much of this region. Cactus may be an indicator of deserts, but many cacti are found outside of desert regions, and some deserts have cactuslike plants of the family Euphorbiaceae.

Coniferous forests. Coniferous trees, especially pine, spruce, and juniper, often grow in almost pure stands that create a distinctive biome. However, different regions have different compositions. Those of North America are described separately. Coniferous forests also occur across northern Europe and Siberia as well as at higher altitudes of most mountain ranges.

Northern evergreen forests. The area south of the tundra or below the tundra on mountains is composed mainly of conifers and forms the northern evergreen forests (also known as taiga). This area extends across Canada, where the climate is less severe than that of the tundra region but where also the winters are cold and the summers cool. Pine and similar conifers predominate, and there are a few deciduous and herbaceous plants.

Southern evergreen forests. The coastal plains from Virginia through Florida to eastern Texas form another type of evergreen forest (Fig. 15-7). These areas are often swampy, and much of the region is covered with longleaf or pitch pine. The wet areas grow live oak, water oak, bald cypress, gum, and magnolia. Spanish moss,

Fig. 15-7
Southern evergreen forest of pure stands of pine. Note the characteristic ground cover of palmetto. (U.S. Forest Service.)

an **epiphyte,** grows on the live oaks throughout most of the area.

Western evergreen forests. The Rocky Mountains and Pacific Coast areas at altitudes above 7,000 feet are covered with evergreens of one kind or another, even when deserts are found at lower altitudes in the same area. There is an abundance of rainfall at the higher altitudes. The plants vary according to the altitude, with ponderosa pine at the lower altitudes and Douglas fir (Fig. 15-8), white fir, quaking aspen (Fig. 15-9), alpine fir, and spruce in ascending order to the higher levels.

Deciduous forests. Forests with trees that drop their leaves in winter are found in many temperate regions throughout the world (Fig. 15-10) south of the northern evergreen forests. These deciduous forests begin in North America in southern Ontario and western and central New York State and extend southwest along the Appalachian Mountains to Louisiana and eastern Texas and west to Oklahoma and southern Minnesota. Oak, hickory, maple, chesnut, elm, ash, birch, rhododendron, mountain laurel, and ferns of various species are typical. Higher rainfall is characteristic of these regions, much of which has been severely lumbered. Almost no natural deciduous forests are left due to land clearing, lumbering, the growth of cities, and the building of roads.

Tropical forests. True tropical rain forests occur only in Central and South America. However, parts of Mexico, the lower Rio Grande region of Texas, the southern tip of Florida, the Florida Keys, and the West Indies (Fig. 15-11) have many of the elements characteristic of tropical forests. Palms, lianas, epiphytes, and orchids are characteristic plants of tropical rain forests, many of which are found in the less dense stands in North America. Other tropical rain forests occur in South America, in the Amazon Basin, in Central and South Africa,

Fig. 15-8
Mature Douglas fir in high-rainfall forest in the Pacific Northwest. Note density of undergrowth compared with that of a pure pine forest.

Fig. 15-9
Ponderosa pine in Kaibab Forest, Arizona. High rainfall occurs in the high mountains, which creates islands of forest surrounded by the lower desert. (U.S. Forest Service.)

Fig. 15-10
Eastern deciduous forest. A great variety of trees occur in these forests, creating variable habitats. (U.S. Forest Service.)

Fig. 15-11
Tropical rain forest. Note the large tree ferns and the epiphytes not found outside these forests. (U.S. Forest Service.)

India, Burma, and the vast tropical regions of Indonesia, New Guinea, and northern Australia.

Community

Although both realms and biomes are in the broad sense communities, they are far too vast to be studied and described in enough detail to demonstrate the variety of ecosystems established in these areas. Therefore further division of the biome is necessary, even though such divisions are not always clearly defined. The term community, as previously stated, is often applied to these aggregations of plants and animals. So many communities have been described for the terrestrial habitats of the earth that no attempt will be made to describe them here.

Communities are ever changing, so that even when they may appear at first to be uniform, close examination will reveal stages or **sere** through which portions pass in the development of a **climax** situation. A climax is the last sere of community development, and the most stable. As portions of a climax community die or are destroyed, a new series of **successions** take place, usually following an established order until the climax is again reached. Therefore certain plants, such as aspen or fire birch, are indicators of a sere, whereas others, for example, the ponderosa pine, indicate a climax. A recent treatment of the vegetation types of the United States (Küchler, 1964) lists and maps well over 100 distinctive communities. A similar treatment by Shelford (1963) describes these and others in detail.

DISTRIBUTION

As we have shown, each species of plant has not only a range but also a relationship with the physical and the biological environment. It is possible, therefore, not only to map the geographical range of the species but also to give a habitat description of each. Obviously, no attempt will be made here to do this for even the most dominant of the flowering plants. Certain general statements are made about the distribution of the various divisions of the plants in Part VI, which deals with diversity.

A close study of the range of any species shows that it is not continuous throughout the geographical limits of its distribution. Instead it is divided into **populations,** each with a range that is circumscribed by the limits of interbreeding. The populations fall into roughly three major types. The first is characteristic of a common widespread species. Here there are broad areas of overlap in the range of a population so that the gene flow is limited only under unusually adverse conditions, for example, long periods of foul weather, cold, or dry condition in part of the range of the species, or similar restrictions to free crossbreeding. The second situation is intermediate and characterizes a great many plants that occupy somewhat specialized habitats. Here the populations are more or less isolated, but there is considerable gene flow through wind or insect pollination. Again, during severe weather periods, portions of the range may become temporarily isolated from the rest of the species. The third condition is characteristic of those species often described as rare. They are composed of a series of isolated populations, usually with specialized habitat requirements. Each population may show sufficient variation from the others to exhibit a succession of characters following a geographical pattern. This is known as a cline. The gene flow between these populations is very limited and miscegenation may not take place except after a lapse of several years.

With this concept of a population in mind, combined with our knowledge of gene action, it is now possible to understand the mechanisms of isolation and evolution described in the next chapter. It is for this reason that the systematist is very much interested in the distribution and ecology of plants.

SPECIAL TERMS USED IN CHAPTER 15

autecology	epiphyte
biome	population
biota	realm
climax	sere
community	succession
ecotone	synecology

QUESTIONS TO CONSIDER

1. Where would you expect to find the following plants: maple, ponderosa pine, bald cypress, cactus, spruce, grass, willow?
2. Describe the dominant plant community of your particular area.
3. What are the three different meanings to population? What is meant, biologically speaking, by overpopulation?
4. What are the controlling factors in plant distribution, and how do these factors operate?

DISCUSSION FOR LEARNING

1. Discuss the differences between the flora of a region and the vegetation of the area.
2. Describe and discuss some obvious succession of plants in an area known to you.
3. What is the relationship between ecology and the evolution process? Discuss this in terms of your study so far in this textbook.

Chapter 16

Plant evolution

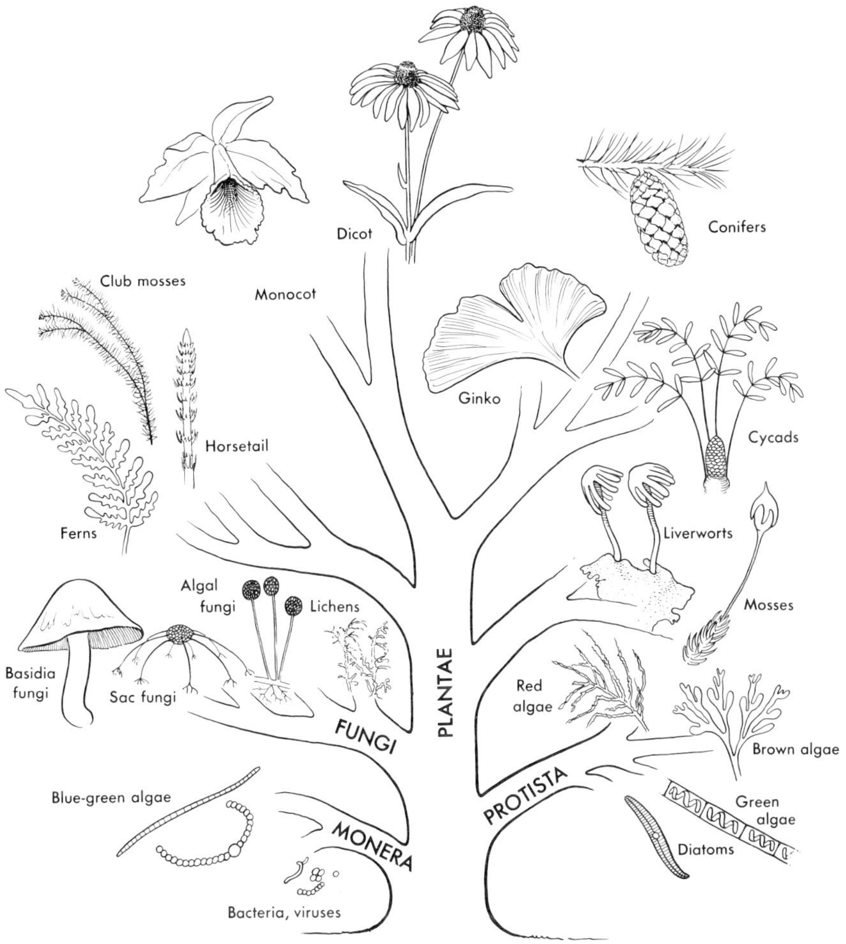

The history of the development and diversification of the Monera and most of the Protista is unknown. Some of the fungi mycelia and especially the spores are preserved as fossils, but little in the way of phylogenetic conclusions may be inferred from these scanty records. The Plantae, especially the vascular plants, have a substantial, but far from complete, fossil record.

In Chapter 10 we discussed the genetics of evolution at the cellular level and the population level, and in Chapter 15 we pointed out the possibilities of evolution at the population level as affected by the changing environment. Here we bring these facts together to describe the actual evolution, or the history, of plant life. The phylogenetic "tree" at the top of this page, however, gives some idea of the relationships of the kingdoms (except for Animalia) as they probably evolved through time. In this case, time is the distance from the base of the tree to the extremity of each branch.

Age of the earth

The origin of the earth in the solar system is of concern to astronomers and geologists; it is currently believed to have separated from the sun much more than two billion years ago because it is estimated that sedimentation started at that time. The fact that sediment could form indicates that the earth had cooled enough to permit stratified or layered rock to form. Prior to that time volcanic action was so widespread that all rock was molten and continually changing. Water was present only as steam or vapor. More information about the origin of the earth and the solar system will soon become available as a result of recent space exploration.

Origin of life (see also p. 141)

There is little doubt that life originated in steps as conditions changed on the earth and that these steps took place following the laws of chance. In other words, a succession of events resulted in the formation of primitive life forms as a part of many other chemical formations. These one-time events persisted through natural selection of chemicals and, in the same way that modern species develop, as conditions became suitable. Life as we know it developed under situations no longer present.

The great interest in solving the problem of the origin of life has resulted in the testing of several clever, interesting, and simple systems to show how life might have developed. It seems apparent that the early atmosphere of the earth was reducing in character due to the lack of oxygen. However, the presence of a dense cloud of water vapor produced oxygen at the upper levels through the reduction of the water either by electrical energy or perhaps by cosmic energy. The resulting hydrogen escaped into space, but the heavier oxygen remained and gradually increased. One of the interesting experiments performed to show a possible step involved a "spark-discharge" apparatus into which was placed a mixture of methane, water vapor, ammonia, and hydrogen, the primordial atmospheric gases. When these were exposed to a series of electrical discharges, the gases circulated past the charge by boiling water on one side and condensing it on the other; small quantities of more complex chemicals were formed, including amino, hydroxy, and aliphatic acids, the three basic types of organic molecules needed for protein structure. In fact, glycine, alanine, aspartic acid, and glutamic acid appeared, all necessary for protein formation. It is not difficult for a biochemist to imagine further steps that might take place, for example, ultraviolet light as an energy source, to continue the building process toward the formation of life particles. As we have pointed out in previous chapters, once the replication of molecules took place, life was present, since no known, naturally occurring system of replication occurs outside of life forms as we know them.

Such steps as these are not the only possibilities for life formation. Very likely many processes actually took place, only one of which was successful in surviving. All the elements of the scientific method are present in the current approach to these studies; therefore it seems logical to believe that those currently discussed may well have been the ones that took place.

FOSSIL FORMATION

Of the myriad of individuals that have inhabited the earth during its two billion year history of life, the chance of even one being preserved as a fossil is so remote that it is almost beyond belief. It seems, therefore, that the hundreds of thousands of fossil specimens available and still being discovered are some indication of the abundance of life during past ages.

Fossils are formed in a variety of ways. One of the more spectacular is petrification. The logs scattered over hundreds of acres in and near the Painted Desert in Arizona are an example of this. Petrification takes place when living forests are covered by soil and rock. The weight of the cover presses the plants. Then by a slow process, mineral-containing water seeps through the covering rock so that, molecule by molecule, the carbon compounds composing the wood are replaced by mineral deposits, and the wood is turned to stone. Every detail of the grain is preserved in an array of colorful rock. Other forms of petrification occur, especially in limestone areas. Sometimes rocks are found that contain perfectly preserved plants or animals, showing every detail of structure. These concretions are often inside round stones, which may be cracked open to reveal the fossil contents within. Still another form of petrification, complete preservation, occurs. Tar and resin deposits reveal nearly perfect plants and animals, the actual body of the organism having been kept intact for thousands of years. Tar pits contain plants that have fallen into this material. The chemical composition prevents decay so that the original structure is preserved. Resinous secretions from trees are sometimes so abundant that they form fossil deposits of amber. Insects, small animals, leaves, and needles are often trapped in this material when it is soft. These are kept in a preserved state comparable to the mounted slides used in laboratory studies.

Most fossils are compressions formed by a covering of rock. The usual process is for silt to cover plants and animals that have either fallen into or live in water. The steady deposits of silt cover these dead bodies, gradually building a rock cover hundreds of feet thick. After millions of years these rocks are raised up and form land that is eroded away to expose their contents. Coal is formed in a similar way. In this case the vegetation was growing in marshy or swampy areas. After long periods of time a thick mat of vegetation was formed, much like present-day peat deposits. If the land sank and became covered by silt, veins of coal eventually formed because of the great weight of the rock that compressed and turned the vegetation into various types of coal. Under certain circumstances the shape and structure of plants are preserved in such a way that the original plants may be studied.

Finally, impressions of various kinds are found. Leaves that fell into mud left impressions that were later covered by fine silt deposits. Because the two materials, mud and silt, were different, they later separated as stones so that the leaf impressions may be seen in the form of casts.

GEOLOGICAL RECORD

The fossil record of rocks is very incomplete. Obviously, if all the individual organisms that had lived on earth were preserved, this planet would soon have been filled with dead bodies. The fact that only certain kinds of organisms are suitable for fossil preservation and the fact that there is no orderly process of fossil formation prevent any hope that the history of the evolution of all organisms can be obtained from fossils. Because of cellulose in their cell walls, woody plants in particular, as well as many others, are more likely to be fossilized than animals. The fossil record of plant evolution in general is more complete than that of animals. Even so, there are many gaps in this record so that there is never an unbroken chain of specimens from one stratum to the next.

Geological time

Geologists have worked out a scale, the geological time scale, which names the various periods of time since the beginning of the earth. This scale is based primarily on the layers of stratified rock. Volcanic rock is not dated in this manner because it cannot be measured in time units. It is

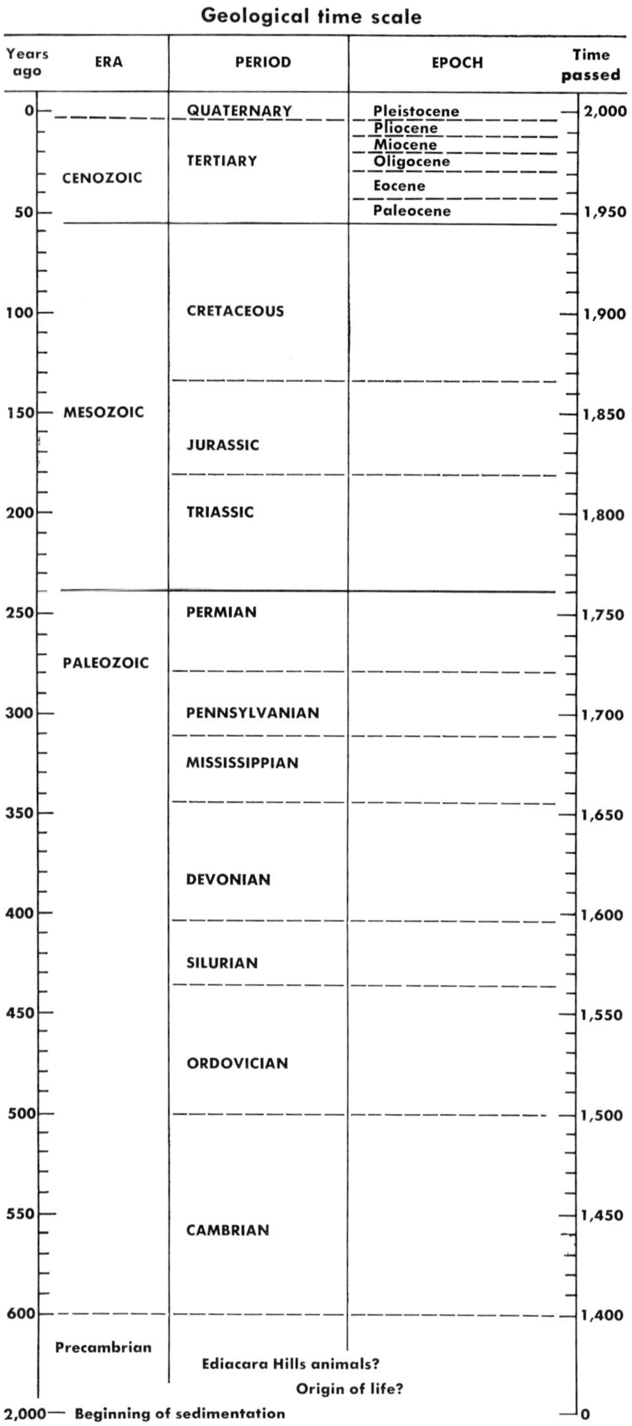

Fig. 16-1
For legend see opposite page.

possible to date both types of rock by radioisotope studies. The original time scales were dated by measuring the thickness of each layer, and an estimate was then made of the thickness of known silt deposits for a few years. By careful computations, and estimate of the time it took for a given deposit to accumulate could be made. After long study, the various deposits were coordinated and arranged into a time scale. Actually, the geologists used fossils to help date the rocks, and this practice resulted in a paradox because the paleontologist used the rocks to date his fossils! However, when it became possible to date organic matter by C^{14} tests, the entire time scale was in need of only minor adjustments. These tests showed that the amount of time for the periods was somewhat underestimated and was actually longer than previously thought.

A revised geological time scale (Fig. 16-1) shows that approximately two billion years of sedimentation have occurred on the earth. During the first 1,400 million years very few fossils were preserved, although the geological record is well known. The geologists divide this long period into various eras, but the biologist generally refers to it as the Precambrian era. Good deposits of Precambrian animals have been found at Ediacara Hills in South Australia, but their exact date has not been determined. The time for known fossils is divided into eras, periods, and epochs. These divisions usually indicate a major geological change, some of which will be mentioned.

Continental drift

Nearly half a century ago the theory of continental drift was formally proposed, maintaining that the continents were once together (Fig. 16-2) and that they separated slowly by drifting apart to their present position. But the match of the configuration of the shores on both sides of the Atlantic Ocean had not escaped the notice of geologists for many decades previously. Biogeographers have long postulated **land bridges,** narrow connections of land between continents, as a means of explaining the distribution of plants and animals. The theory of continental drift as originally proposed lacked general appeal, mainly because there was little evidence that such a movement could or had taken place, and although it was hard to believe that the fit of the continents could be entirely coincidental, it still explained very little of the biogeographical problems.

For many years the proposal remained as a controversial, speculative subject. During the past few years very significant new data have suddenly given the theory not only renewed attention but very widespread acceptance. The first new evidence comes from an increased knowledge of deep-sea geography and geology. Originally it was difficult to imagine an energy source great enough to move entire continents. It is now known that there are midocean ridges in both the Atlantic and the Pacific and that in these areas there is considerable volcanic action. This has occurred in cycles throughout the known history of the earth.

Fig. 16-1

Geological time scale showing the periods during which fossil remains are known. All time is in millions of years. The date 2,000 million years ago, at which time this scale starts, represents the beginning of sedimentation and the formation of the first known stratified rock. This scale is based upon new methods of dating that use the radioactive materials in the rock as indicators of the time that has passed since the sediment was deposited. (Modified from Kulp, J. L. 1961. Science **133**:1111.)

Fig. 16-2
Map showing how continents fit. Placing of present-day land masses in this position is only to show how they once formed a continuous land mass and later separated at approximately the same shores as present continents. This "restoration" omits the very important land mass Antarctica.

Through studies that include such items as the precise measurement of minute changes in the distance between two points in Europe and North America, the magnetic polarity of the layers of rock in the ocean floor, and other data, the theory of continental drift is now amplified to the sea floor spreading theory. The latter proposes that new rock is coming up in the middle of the oceans at the present time, causing the spreading of the ocean floor at the rate of about 1 inch each year.

Continental drift theory now proposes that the continents have moved apart and together over long periods of time. The movement takes place not by a rise and fall of land mass but by the actual floating of portions of the earth's crust on its hot, molten interior, much the same as a cut piece of pie crust might float on a fresh baked pie. The upwelling of rock in the ocean ridge will separate the continents, and the down flow of rock in another area will allow the movement of the opposite coast. It is likely that the rise of mountains occurs at the same time due to wrinkling of the surface.

Evidence for this theory now includes, in addition to that already mentioned regarding continental fit and **paleomagnetism,** data such as the increased age of islands the greater their distance from the midocean ridges, earthquake data, and paleoclimatic information. It is now estimated that the current separation of the

Table 16-1

Geological history and origin of four kingdoms and major plant groups

Era	Period	Kingdoms containing "plants"				
		Monera and Protista	Meta-phyta	Anthophyta		Fungi
				Gymno-sperms	Angio-sperms	
Cenozoic		Present	Present	Present	Present	Present
Mesozoic	Cretaceous	Present	Present	Present	Present	?
	Jurassic	Present	Present	Present	Origin	
	Triassic	Present	Present	Present		
Paleozoic	Permian	Present	Present	Present		
	Pennsylvanian	Present	Present	Present		?
	Mississippian	Present	Present	Present		
	Devonian	Present	Present	Origin		?
	Silurian	Present	Origin (Psilophyta)			
	Ordovician	Present				
	Cambrian	Present				
Precambrian		Origin				

continents bordering the Atlantic Ocean began 150 to 175 million years ago, or during the Mesozoic era (discussed later). This is recent enough to have considerable effect on the distribution of plants, and biogeographic studies support this.

Paleozoic era

The evolution of the major groups of plants is fairly well known. The Paleozoic era saw the development of all these groups, except the flowering plants and perhaps the Fungi, as shown in Table 16-1.

The Paleozoic era was the longest period in the development of plants. At the beginning of the era there was no life on land. The climate was very hot, and there was almost constant rain. Fresh water probably had some simple algae living in it, but the entire evolution of major groups of Protista is generally thought to have taken place in the sea, which at that time still had a relatively low-salt content.

Cambrian. The Cambrian period is characterized by times of flooding of the land forming shallow seas, which resulted in a much different geography from that of today. Continents little resembled the shapes we know them to be today. Undoubtedly the Protista were well developed by the end of the Cambrian period.

Ordovician. The Ordovician period is introduced by the Taconic revolution, a period of land elevation that resulted in the cooling of the land and the diminishing of rain so that various kinds of algae emerged from the sea. In the still very wet condition they were able to gain some footing. Their growth on land caused the first soils so that by the next period, the Silurian, land plants were beginning to develop rapidly.

Silurian. Among the first of the Silurian plants were woody Psilophyta, distant relatives of the Pterophyta. Many other Silurian plants are known, but their relationships have not been studied very well as yet. These first plants were, of course, all spore producers and were probably generally distributed across this rather barren earth. The climate of the time was uniform, with no seasons.

Devonian. The late Silurian and Devonian periods were marked by an increasingly arid climate, but certainly the dryness was relative because during the Mississippian and Pennsylvanian periods, known together as the Carboniferous period, the climate was still much wetter than that of present day tropical rain forests and probably much hotter. There are few known survivors from the Silurian to the Devonian, yet Psilophyta of various kinds remained as the most characteristic plants of the period. Other major groups are present in the Devonian as well, including members of the order Cladoxylales, an early fernlike plant that may have been the start of the Pterophyta (ferns). Plants of the order Hyeniales, also present in this period, may be the ancestors of the Arthrophyta (horsetails). A group of gymnosperms, the Pityeae, formed the first seed plants, but these were short lived. The dominant plants of the landscape of the period were members of the Psilophyta, Microphyllophyta (club-mosses), and the Arthrophyta. Although these plants developed early, they were temporary and were replaced by other plants in the later half of the Devonian. Much of the Psilophyta flora was replaced by the primitive ferns that became the dominant plants of the Mississippian period. The Devonian period ended, nevertheless, with a well-developed land flora.

Mississippian and Pennsylvanian. The Carboniferous period (Mississippian and Pennsylvanian combined) was characterized by a mild, moist climate all over the world. A heavy cloud cover kept carbon dioxide close to the earth and caused the phenomenon known as the "greenhouse" effect, producing high heat and humidity. This and the dense clouds resulted in optimum conditions for luxuriant plant growth, probably the most extensive swamp flora (Fig. 16-3) the earth has ever known if one can judge from the coal deposits that remain. It was during this period that the seed plants were perfected, although throughout these periods the ferns seem to have been well represented. Many of the fronds once thought to belong to ferns are now known to be early Cycadophyta (cycads), the seed ferns. Even though we consider them the most primitive division of the Metaphyta (higher plants) it is not until the Pennsylvanian period that the Bryophyta appear. Their late appearance makes it rather certain that the group is a sideline development and not a contributor to future plant evolution. At the end of the Pennsylvanian period the first conifers appeared. These were the first pollen producers. All the rest of the seed plants of that time were dependent upon motile sperm.

Fig. 16-3
Carboniferous swamp showing reconstruction of ancient horsetails. (Courtesy Field Museum of Natural History, Chicago, Ill.)

Permian. The Permian period was ushered in by a general rise of land masses, which was followed by the building of mountains, the cooling of the climate, the draining of large swamp areas, and the extinction of many plants. South Africa was covered with a glacier. Thus the first great changes in land flora took place. The spore plants, so dependent upon an abundance of water for fertilization, were less well fitted in the new climate. The higher ferns, conifers, and cycads developed to become the dominant features of this flora. Because of the climatic diversity, floral regions became evident for the first time. The early conifers were still fertilized by ciliated sperm partly dependent upon rain for their dispersal, as are the cycads today. By the end of the Permian period practically all the Carboniferous plants were replaced by modern groups, many of which have survived to the present time. It was also at this time that the continents began the splitting apart and drifting that became much more evident in the Mesozoic era.

Mesozoic era

The Mesozoic era began with the Appalachian revolution, the formation of the Appalachian Mountains in eastern North America, as well as of mountains in other parts of the world, and is characterized by a cooler climate. At this time many of the Paleozoic plants were extinct.

Triassic. Taking the place of the Poleozoic plants were more lycopods, horsetails, and ferns, the latter rapidly becoming the dominant plants of the Triassic period. It was during this period that the Araucariaceae developed. Plants of this family have been studied for the distributional evidence they show in support of the theory of continental drift.

Jurassic. Cycadlike plants, the true cycads, and the ginkgos were the dominant plants of this period, and there was a rapid rise in growth of conifers. The angiosperms had their origin in the Jurassic period and developed rapidly. The flora of the Mesozoic era would be recognizable and somewhat familiar today. The whole era is characterized by a climate that became gradually more dry and progressively more severe.

Cretaceous. The Cretaceous period saw the extinction of several groups of plants and the decline of the cycads, horsetails, and ginkgos to minor groups. The flowering plants rose to become the dominant plants, along with the conifers. Unfortunately very little of the early history of the flowering plants has been found in the fossil record; the earliest known examples are fully evolved and can be placed in modern families, and even in genera. The end of the period is marked by the Laramide revolution, another period of mountain building. It was during this period that high mountains such as the Rocky Mountains were formed and that all ancient forms of plants became extinct.

Cenozoic era

The Cenozoic era is, of course, among the best known. Its vegetation was very modern. All the surviving plant groups now on the earth were present at the beginning of the period so that no new plant divisions or major groups are recorded as originating at this time. There was a gradual evolution of nearly all the land masses that form our present-day geography. All large areas of swamp ground disappeared, except those that remain today. Deserts were extended on the leeward side of high mountain ranges, and extensive grasslands were formed. The various epochs of the Cenozoic era are not marked by distinctive changes in vegetation types, except for the shifting of biomes. The climate was characterized by well-marked seasons, but because of alternate periods of cold and heat, tropical vegetation gave way to temperate and subarctic vegetation, only to have tropical conditions return again. The periods of the era are only the Tertiary and the Quaternary.

Tertiary. The early part of this period is

characterized by the spread of tropical vegetation far north due to a general warming trend. This caused the melting of ice, the rise of the oceans, and this in turn, provided water for an abundance of more snow in the north. Therefore a reverse trend took place toward the end of the Tertiary, eventually resulting in the series of glaciations of the last part of the period.

Paleocene. The first epoch of the Tertiary, the Paleocene, is better known from animal fossils than from plants. However, those present show a continuation of a warm climate with a subtropical vegetation extending far into the northern part of North America and Europe.

Eocene. By the Eocene a marked tropical flora extended as far north as England, and, strangely, this flora had a strong Indo-Malayan element. These later gave way to forms that may have come from the less tropical Eocene flora of Asia.

Oligocene. The end of the northern tropical flora was reached by the middle of the Oligocene epoch. However, palms, sequoia, and ginkgo reached as far north as Greenland where abundant fossils of deciduous trees indicate that forests once existed. In South America the modern plants of the genera *Nothofagus*, and *Araucaria* developed. These plants are shared with Australia, a phenomenon still not clearly explained by continental drift. The genus *Nothofagus* probably existed on Antarctica from the late Cretaceous to the Miocene and was distributed from this region to its current circumpolar sites. The epoch ended with new mountain building, including such mountains as the Alps.

Miocene. Mountain building reached its climax during this epoch, accompanied by considerable volcanic activity. The Himalayas in Asia and the Cascade range in western North America were formed, and they represent our newest mountains. The flora of the time lost its Indo-Malayan forms and became temperate in character. Tropical plants slowly retreated until today the only remains of Miocene flora in the United States are found in Florida. The epoch closed with the beginning of glaciation in the far North, and Miocene ice still remains there in a few areas.

Pliocene. At the beginning of the Pliocene epoch the temperate regions still retained certain tropical elements in its flora, but by the end, the Ice Age had begun. Most of the flora of the period is known from flowering plants. The trend was certainly toward herbaceous and away from woody plants as shown by the preserved seeds and fruits. This is the final period of the Tertiary of the Cenozoic.

Quaternary. The Quaternary period contains only the Pleistocene epoch, which is characterized by successive ice ages, ages of widespread and successive glaciations. Since this epoch is the most recent, it is not surprising that it is the best known. It is also the period during which man evolved; thus more intensive study has been given to it. Four glacial periods existed, with only about 10,000 years having passed since the last glacial cover of North America. The fascinating history of the Quaternary cannot be dealt with in this introduction. However, it was a period of wholesale extinction of both plant and animal species and the almost explosive development of modern groups. Some northern areas either were not covered by ice, or were only lightly covered so that some of the earlier Cenozoic plants were presevered in these **refugia.** These areas also provide an isolated and interesting flora in that the study of these plants gives us an insight into preglacial conditions. Many acid and alkali bogs of the northern part of North America are such refugia.

Recent. Our present time period began at the end of the last ice age, and no one knows whether a new ice age is on its way; therefore we cannot determine its exact geological state in relation to other periods and epochs. The flowering plants are now the dominant plants, with conifers as a

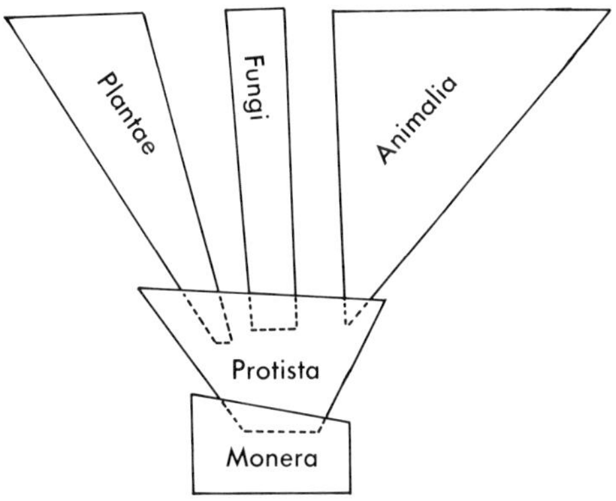

Fig. 16-4
A five kingdom system as proposed by Whittaker (1969), based on cell structure and the means of food getting. Note the three levels of organization: bottom, procaryotic; middle, eucaryotic unicellular; top, eucaryotic multicellular.

prominent relic of past ages. The ferns, mosses, bryophytes, and similar spore-producing plants of the Paleozoic and Mesozoic remain on the earth only as a minor part of the flora both in appearance and as a part of the energy exchange cycles. They are a silent reminder of the past.

COMMENTS ON PHYLOGENY

The study of evolutionary mechanisms (Chapter 10) and of the history of the evolution of plants leads botanists to speculate on the phylogeny of plant taxa. They are interested, in other words, in tracing the historical relationships of the plants, and therefore, their evolutionary lives. As this brief discussion has shown, our fossil records are far from complete. Therefore any phylogeny proposed is subject to change as new evidence becomes available. All data are used when creating a phylogeny or family tree of plants. Since it is only the higher plants, and particularly the vascular plants, that have left a fossil record, the arrangement of other groups is based primarily on structural similarity. The current phylogeny of the kingdoms (Fig. 16-4) is based on cellular structure and function. The arrangement of the higher divisions of Plantae as given in this text is according to structure and fossil record. Unfortunately there is little paleontological data available to show the arrangement of the orders of the flowering plants, and this group still remains in a state of continued change as research taxonomists continue their attempts to show relationships that are as close to the actual history of the groups as is possible through logical deduction.

SPECIAL TERMS USED IN CHAPTER 16

continental drift	paleomagnetism
land bridge	refugium

QUESTIONS TO CONSIDER

1. How would you describe the environment of the earth at the time of origin of life? Of terresterial life? Of man?
2. How do these primeval conditions relate to the fermentation process described previously?
3. What is meant by "mountain building"? What are the major periods of mountain building? What effect did this have on climate?
4. What evidence of the earth's history might one expect from the moon? Do you think there is evidence of life on the moon?
5. Why have the major changes in the development of a plant's anatomy taken place on the earth's surface rather than in the sea?
6. Name the geological eras, periods, and epochs in order and give the approximate dates.

DISCUSSION FOR LEARNING

1. Discuss the theory of continental drift in relation to the dispersal mechanisms of plants. In other words, how much passive riding on these land masses have plants done?
2. Discuss spore dispersal and pollen dispersal in terms of the known fossil history. What are the advantages of one over the other?
3. Discuss the conditions under which the flowering plants developed and those under which they became the dominant plants of the earth's surface. Why are they apparently better fit for present day conditions than are other kinds of plants?

Part VI

Diversity

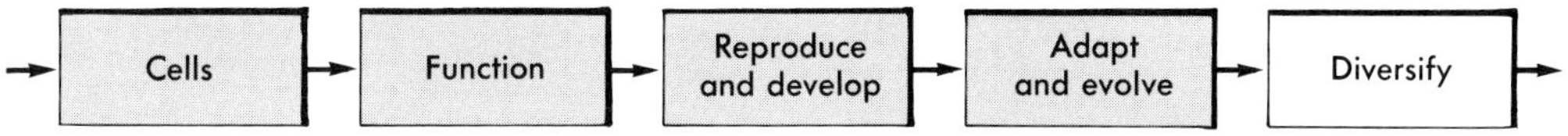

- Variation within well-circumscribed limits is permitted by the simple codification of the macromolecules—variation that potentially exceeds numerically $10^{10^{10}}$.
- Levels of variation progress from the individual, the sexes, to the species, signified by distinctive gaps at any time period.
- Variation or diversification of the species is a natural process permitting the orderly operation of diverse life systems.
- The recognition of this diversity and its classification is the turnstile of information processing.

Chapter 17

Classification of plants

The classical problem of how to separate plants and animals is no longer a real scientific problem. Plants and animals are not separate groups of organisms existing in nature. The two groups are separated only as a means of referring to common life forms. As can be seen in the previous chapters, all living organisms have common fundamental features, but not all have the same life plan. In this chapter a classification scheme is outlined, based as much as possible on our knowledge of the evolution that has taken place. The remainder of the text describes many of these groups to complete our survey of those organisms known as plants.

The great diversity of plant life is due primarily to changes in body form as a result of adaptation to different environments. Tropical rain forests such as the dense forest on the slope of the Andes pictured above is an example of this. Fog and rain cover the area much of the time, yet so great is the diversity of form that hardly any of the species of plants in this picture are duplicated in any single scene.

Many different body types within the overall limit of a particular group are possible.

Some of these forms appear to be much simpler than others. These are referred to as primitive. It is difficult to be sure whether a particular anatomical type is primitive or whether it merely represents a reduced condition in a certain stage due to some specialized function or habit. Modern methods have been developed by students of evolution so that in well-known groups very logical and statistically reliable data can be pieced together in order that the phylogenetic lines are fairly clear. Often knowledge of the complete life cycle and also of allied species helps to determine whether a particular species is simple or represents a reduced form of a more complex group.

The anatomical structure forms the basis for assigning species to higher groups. It has been shown in some representative forms of these groups that anatomical structure is an indicator of physiological functions and habits. Therefore it is assumed that organisms having similar anatomy also have other features in common, even if this is not known by direct observation. By the use of this principle all the plants and animals are classified.

PLANTS VERSUS ANIMALS

Many attempts have been made in the past to make clear-cut distinctions between plants and animals. All have failed; therefore modern systematists no longer try to place all organisms in one of two kingdoms, plant and animal, but instead they recognize at least three. Once the step was made in the direction of grouping organisms on the basis of major body plans, the two-kingdom, traditional system seemed obviously wrong. The most logical scheme at the present time calls for five kingdoms, and these are recognized in this text. Four of the five kingdoms are treated here, three in their entirety: Monera, Plantae, and Fungi, and the fourth, Protista, in part. Only the kingdom Animalia is omitted entirely.

Each of the kingdoms is further divided into either divisions or phyla. The International Code of Botanical Nomenclature requires that the first category of a "plant" kingdom be called a division, whereas the International Code of Zoological Nomenclature requires that the term phylum be used for the same category. Neither code recognizes, as yet, the five-kingdom system. Some organisms are included in both systems, but those interested in these groups seem little concerned with the problem of nomenclature. Therefore the names applied depend upon the view of the author and his audience.

Certain features of each of the five kingdoms should be pointed out here, and these will be repeated in more detail in the respective chapters to follow.

The kingdom **Monera** contains those organisms that lack, essentially, distinct nuclei. Because the bacteria fall here, the group is of major importance. It may be debated whether the viruses and similar agents should be included here, or whether a sixth kingdom should be erected for them, or even whether they should be considered as organisms at all. The strong tendency is to place them in the Monera.

The kingdom **Protista** is the group that contains organisms formerly thought to be either plants or animals, that is, algae or protozoans. By placing all single-celled organisms with distinct nuclei in this kingdom, there is no longer a taxonomic problem. Some obviously closely related multinucleated or multicellular organisms are placed here too, but this causes no real taxonomic problem. The presence or absence of chlorophyll long ago was shown to be of no taxonomic importance in the recognition of this kingdom.

The kingdom **Plantae** provides for the remainder of the photosynthetic organisms, all of which are multicellular in their vegetative stages. A few forms, through secondary modification, no longer have photosynthetic pigments and exist as saprophytes, but again, these cause no taxonomic problems.

The final kingdom, **Fungi,** is now recognized as a distinctly separate group of early origin related to neither the higher multicellular plants or the higher animals, although they share some characters of each. Their mode of life and cytological organization is so different that there is no problem now in discussing them at the end of the text. This does not imply, however, that they are believed to be more advanced than the Plantae.

Since most of the species to be dealt with in these kingdoms are either in the Plantae, or the Animalia, much of this discussion is academic. The common features of each are listed here so that it can be seen that most biological processes are fundamental to both kingdoms:

1. Both are composed of protoplasm and similar included substances such as water, fats, and starch particles. In general, all protoplasm is the same. A few substances, such as cellulose, are characteristic of plants but are rarely found in animals.

2. The protoplasm in both is organized into individually operating units called cells. Although these are structurally distinct and may be highly specialized, they are basically similar in all organisms. Their fundamental physiological processes are so alike that it takes highly refined testing methods to detect any differences, and, even then, cells cannot be grouped in such a way as to say that one is typically animal and the other typically plant.

3. Both are capable of reacting to changes in the environment. These responses include reactions to temperature change, increase or decrease in moisture, pressure changes, gravity, and light. Both also react in varying degrees to chemicals, especially those that are ionized or dissociated. Variations in wavelength other than light and heat may also result in responses. Some of these wavelengths fall within the range of audible sound. Other wavelengths, for example, X rays and gamma rays, cause a response, but usually plants and animals are adversely affected by such wavelengths. So far as is known, neither will respond to unconverted radio or magnetic wavelengths.

4. Both have diverse but characteristic shapes. Plants tend to be asymmetrical and superficially unorganized in shape, whereas animals tend toward definite body plans; however, both also share similar shapes. For example, many marine animal colonies are often mistaken for plants by laymen.

5. Reproduction is usually sexual, and new organisms are formed by the union of two cells, the egg and the sperm. This holds true for all but a few organisms in both groups. Even many of the exceptions in the plant kingdom have been observed to exchange genetic material or to have a plus (+) and a minus (−) arrangement for the formation of new organisms at some stage in the life cycle of the species.

6. All animals and plants maintain a distinctness of kind; that is, there are species that remain genetically distinct for definite periods of time, under definite conditions, and in a definite geographical or physiological range.

From this brief statement of similarities, it seems apparent that living organisms are a single manifestation of our physical world. It is when life is examined in this light that we can easily see how the evolutionary facts now know indicate a single origin and how these facts result in a great unifying principle.

SOME OLD CLASSIFICATIONS

The classification of plants has had a long history, part of which was discussed in Chapter 2. Several of these classifications are still used; therefore it is felt that these should be compared, briefly, as is done in Table 17-1, so that when the group names are found in collateral reading, their use will be understood (see also the glossary).

The following classification of plants used in this book was adapted from Whittaker (1969).

Kingdom MONERA (procaryotic plants)
- Branch Myxomonera (Monera without flagella)
  - Division 1. Cyanophyta (blue-green algae—life cycle 1)
  - Division 2. Myxobacteriae (gliding bacteria)
- Branch Mastigomonera (Monera with flagella, and related nonmotile forms)
  - Division 3. Eubacterae (true bacteria)
  - Division 4. Actinomycota (mycelial bacteria)
  - Division 5. Spirochaetae (spirochetes)

Kingdom PROTISTA (unicellular or colonial-unicellular organisms)
- Division 6. Euglenophyta (euglenoid organisms)
- Division 7. Chrysophyta (golden algae—life cycle 2)
- Division 8. Pyrrophyta (dinoflagellates and cryptonomads)
- Division 9. Hyphochytridiomycota (hypochytrids)
- Division 10. Plasmodiophoromycota (endoparasitic slime molds)

Kingdom PLANTAE (multicellular organisms usually with photosynthetic pigments)
- Subkingdom Rhodophycophyta
  - Division 11. Rhodophyta (red algae—life cycle 3)
- Subkingdom Phaeophycophyta
  - Division 12. Phaeophyta (brown algae—life cycles 4 and 5)
- Subkingdom Euchlorophyta
  - Branch Chlorophycophyta
    - Division 13. Chlorophyta (green algae—life cycles 6, 7, and 8)
    - Division 14. Charophyta (stoneworts—life cycle 9)
  - Branch Metaphyta
    - Division 15. Hepatophyta (liverworts, hornworts—life cycle 10)
    - Division 16. Bryophyta (mosses—life cycle 11)
    - Division 17. Psilophyta (whisk fern and allies)
    - Division 18. Microphyllophyta (club mosses and allies—life cycle 12)
    - Division 19. Arthrophyta (horsetail and allies—life cycle 13)
    - Division 20. Pterophyta (ferns—life cycle 14)
    - Division 21. Cycadophyta (cycads)
    - Division 22. Ginkgophyta (ginkgos)
    - Division 23. Coniferophyta (conifers—life cycle 15)
    - Division 24. Gnetophyta (gnetales)
    - Division 25. Anthophyta (flowering plants—life cycle 16)

Kingdom FUNGI
- Subkingdom Gymnomycota
  - Division 26. Myxomycota (plasmodial slime molds)
  - Division 27. Acrasiomycota (cellular slime molds)
  - Division 28. Labyrinthulomycota (cell-net slime molds)
- Subkingdom Dimastigomycota
  - Division 29. Oomycota (oosphere fungi—life cycle 17)
- Subkingdom Eumycota
  - Branch Opisthomastigomycota
    - Division 30. Chytridiomycota (true chytrids and related fungi)
  - Branch Amastigomycota
    - Division 31. Zygomycota (conjugation fungi—life cycle 18)
    - Division 32. Ascomycota (sac fungi—life cycles 19 and 20)
    - Division 33. Basidiomycota (club fungi—life cycle 21 and 22)

Table 17-1
Some former systems of classification of plants

Traditional	Recent	Reproduction*	Morphology
1. Algae	Thallophyta	**Cryptogams**	Nonvascular
2. Fungi			
3. Bryophytes	Bryophyta		
4. Pteridophyta Classes Psilophyta Lepidophyta Calamophyta	Tracheophyta		Vascular
5. Spermatophyta Classes Gymnospermae Angiospermae		**Phanerogams**	

*This terminology is based on the location of the reproductive organs, that is, whether they are hidden or exposed.

Now that the similarities have been pointed out, the great diversity of the thirty-three divisions will be shown. Each of the divisions is listed in the preceding outline (p. 326).

SPECIAL TERMS USED IN CHAPTER 17

cryptogam
phanerogam

QUESTIONS TO CONSIDER

1. What is the International Code of Botanical Nomenclature? Who uses it?
2. What is the relationship, if any, between classification and evolution?
3. What is a division, a category, or a taxon? Give an example of each.
4. What is the difference between a division and a phylum?

DISCUSSION FOR LEARNING

1. Discuss the advantages of a system of classification that uses more than the "standard" two kingdoms; also, the disadvantages.
2. Discuss the pros and cons of a classification based on convenience of handling information and one that attempts to be phylogenetic.
3. Discuss the theory of classification in relation to the theory of information storage and retrieval discussed in Chapter 1.

Chapter 18

Blue-green algae and bacteria

Courtesy Chas. Pfizer & Co., Inc., New York, N. Y.

Many fundamental biological phenomena are illustrated by the bacteria and their relatives, not the least of which are the principles of pathogenicity set down by Pasteur and Koch about a hundred years ago. As the photograph above illustrates, the search for new cures for disease goes on. In this case the scientists are looking for antipathogens that may be found in the soil of the Sahara desert region.

The organisms discussed in this chapter are an extremely diverse group, sharing two principal features: they are unicellular, and they lack a nucleus surrounded by a nuclear membrane. Such cells, as pointed out previously, are called **procaryotic.** One group, the Actinomycota, although unicellular, form a **mycelium.** The five divisions grouped here are placed in the kingdom Monera, and they probably represent the most primitive living organisms. It is for this reason that the viruses are considered with them, if they are included as living organisms, although they really share none of the characters of this kingdom. They may represent degenerated Monera that, as parasites, have lost their cytoplasm much the same as the fungi have lost their locomotion, if indeed, either ever had these additional attributes. In this text, however, we have treated the viruses and their allies separately, and we make no attempt to place them in an evolutionary sequence.

The kingdom Monera may be defined as solitary, unicellular, colonial-unicellular, or filamentous organisms, rarely forming a mycelium, without nuclear membranes, usually without plastids or mitochondria, and sometimes with flagella. They are saprophytes or parasites, a few with chlorophyll (blue-green algae) are photosynthetic, and some bacteria are chemosynthetic. They reproduce either asexually by fission, budding, or by a simple sexual process without differentiated sex cells. These organisms are usually nonmotile, but some have flagella, and others move relatively fast, compared to their size, by gliding.

DIVISION 1. CYANOPHYTA—BLUE-GREEN ALGAE

Blue-green algae are unicellular or filamentous organisms with cells having no well-defined nucleus. They occur mainly in the sea as a part of the plankton, but many species occur in freshwater pools, ditches, and stagnant water, and a few species are found in lakes or water reservoirs. Some are located in damp situations such as stream banks and moist soil or in hot springs. The species found in hot springs can withstand temperatures as high as 185° F. These algae cause the trass color around the famous geyser in Yellowstone National Park, Old Faithful.

In addition to these characteristics the cells, and in filamentous types the thallus, are always enclosed in a gelatinous sheath. Besides chlorophyll, the cells contain **phycocyanin** (blue) and carotene (orange) pigments, which often give them a blue-green color, from which they get their common name. All species do not show this feature, however; many species vary in color from blue to brown, or even red. These photosynthetic pigments are not located in definite plastids. The cell wall is very thin cellulose, but the surrounding gelatinous sheath may be very thick. Several cells or filaments may be surrounded by this sheath, binding the whole mass of cells into a single colony-like organism (Fig. 18-1). No species of this group forms flagellated cells or gametes.

Their cell structure is very simple. There are no vacuoles or organoids of any type, except for granules of chromatin. Usually these are grouped near the center of the cell in an area of colorless protoplasm. The granules divide prior to cell division and represent the genetic material of the cell.

Fig. 18-1
Colonies of *Nostoc* sp. form floating, jellylike balls as shown here. (Courtesy General Biological Supply House, Inc., Chicago, Ill.)

Mitosis is atypical. The pigment material is distributed in the outer portion of the cell. Also present are oil droplets, small granules of glycogen, and other substances.

Reproduction. Reproduction is by simple cell division in the unicellular species. The filamentous species break into small segments of cells periodically. These are set free from the sheath. The new organism is capable of slow movement, but it soon settles down to the substrate to form a new filament. During adverse conditions hormogones (Fig. 18-2) are formed (life cycle 1, *Nostoc* sp.).

There are about 150 genera and 2,500 species.

Many genera are commonly encountered in the field. These all exhibit similar structures and may be readily recognized as members of this group. In the more primitive genera such as *Gloeocapsa* and *Chroococcus*, both nonfilamentous, the resemblance to bacteria is apparent. The more advanced genera such as *Lyngbya* and *Oscillatoria* (Fig. 18-3) show their relationship to the green, filamentous algae of the Plantae. *Nostoc* sp. forms complex colonies and has a primitive but definite reproductive cycle (life cycle 1). Such genera as *Hyella* and *Chamaesiphon* resemble some of the members of the kingdom Fungi. *Hyella* sp. grows within the shells of mussels, and *Chamaesiphon* sp. occurs on old filamentous *Oedogonium* sp., a green alga. The *Oedogonium* sp. occurs as a tangled branching mass of filaments; the *Chamaesiphon* sp. occurs as epiphytic cells producing endospores.

Economic importance. It is not infrequent that some species of this group of plants grow so abundantly in water reservoirs that they must be eradicated because the decaying plants cause diagreeable tastes and odors. Occasionally, lakes "blooming" with these algae cause the death of fish or render the water injurious to the livestock that drink from it, but this is not common.

Fig. 18-2
Nostoc sp. filament showing hormogones.

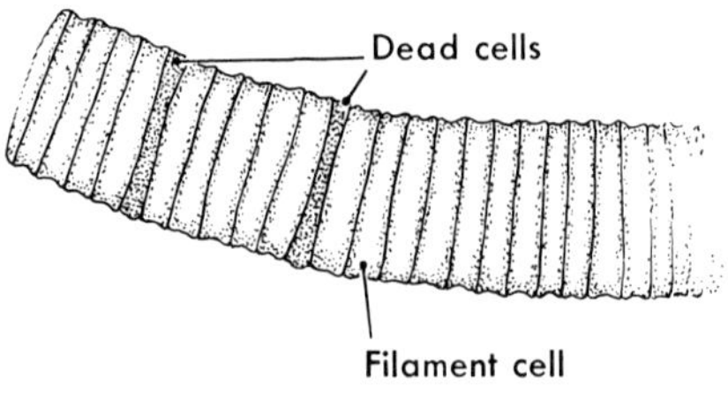

Fig. 18-3
The vegetative cells in this filament of *Oscillatoria* sp. include several dead cells that are responsible for breaking it up into smaller segments (diagrammatic).

DIVISION 2. MYXOBACTERIAE—GLIDING BACTERIA

This small group, sometimes called the slim bacteria, are separated from the true bacteria. The cells are rod-shaped, flexible, and grouped into variously shaped fruiting bodies that attain macroscopic size. Most species occur in the soil; some are aquatic; and a few are parasitic. Because they move by gliding, they are classified as a distinct division resembling the Cyanophyta.

The two divisions, the Cyanophyta and the Myxobacteriae, are grouped together to form the branch Myxomonera of the kingdom Monera.

DIVISION 3. EUBACTERIAE—BACTERIA

The organisms assigned to the Eubacteriae are all microscopic or submicroscopic in size. They vary from 0.1 to 20 μ in diameter. Most are very simple in structure (Fig. 18-4) without readily demonstrable or distinctive nuclei. They feed on organic

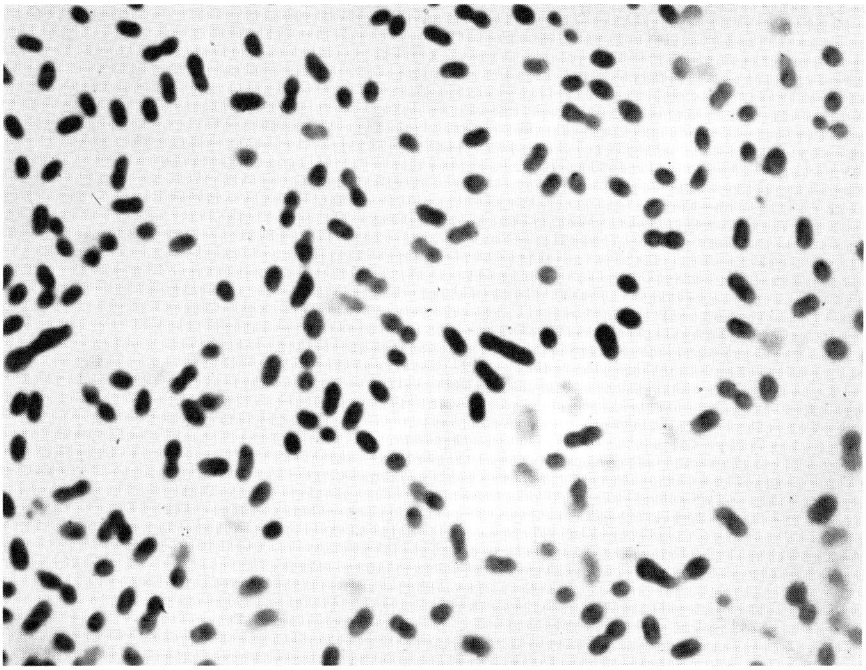

Fig. 18-4
Although the pneumococcus *(Diplococcus pneumoniae)* shown in this photograph is the common cause of pneumonia, it is only one of several bacterial agents responsible for the disease. (Courtesy Chas. Pfizer & Co., Inc., New York, N. Y.)

materials and are usually incapable of synthesis of organic materials from inorganic substances. Usually they are not dependent upon other cells for their life processes.

Because of the small size of bacteria, the structure of the cell appears to be very simple, without specialized organelles. The cell wall is a rigid layer composed of fatty substances, proteins, and carbohydrates chiefly polysaccharide in nature. Other combinations of these substances are used in the cell walls of the group. Sometimes the outer wall forms a thick, mucilaginous capsule that protects the cell from drying and makes aerial dispersal possible. The chromatin material is distributed throughout the cell in the cytoplasm. This material divides when the cell divides. In this manner it acts as a nucleus. When special preparations are made, some bacteria can be demonstrated to have a central mass of chromatin. An exchange of chromatin material by conjugation has been observed in a few species. This is the genetic material, and in this respect it seems likely that some form of sexual reproduction takes place perhaps in all or most species. Other details of cytoplasmic structure show that obscure vacuoles are sometimes present, and a few inclusions have been observed. Many bacteria have a strong affinity for basic stains, which further indicates the presence of nuclear materials.

The external anatomy differs in such a way as to be characteristic of groups of species. Some species possess flagella and are therefore motile. Others are motile only by the effect of Brownian movement on their small size. On the basis of their general shape, bacteria are classified as **bacillus** if rod shaped, **coccus** if a sphere or

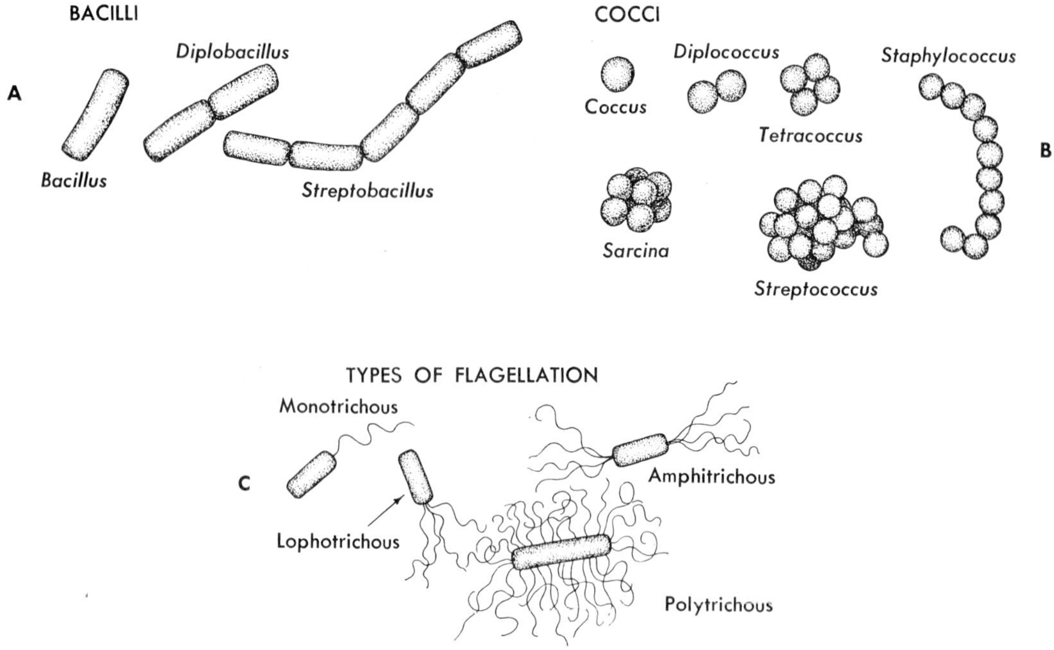

Fig. 18-5
Bacteria are differentiated by their shapes, colony characteristics, and appendages, as well as by their physiological reactions. Some shapes are diagrammed here.

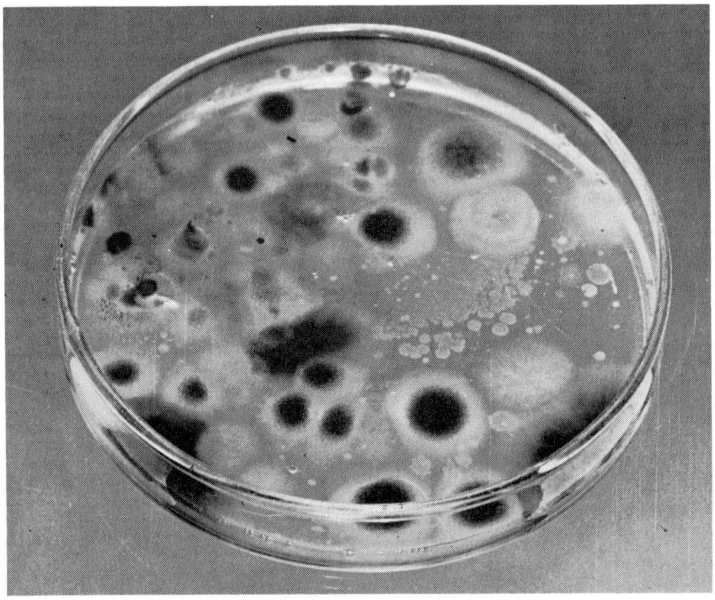

Fig. 18-6
Microorganisms of many kinds are grown in pure or mixed culture (as above) on agar plates. (Courtesy Chas. Pfizer & Co., Inc., New York, N. Y.)

spirillum if a spirally twisted rod (Fig. 18-5). Other details of structure vary within these groups. Sometimes cocci form chains or cubes in small colonies.

Reproduction. The usual means of reproduction in the bacteria is simple cell division or binary fission. This division is rapid in a suitable medium providing optimum conditions (Figs. 18-6 and 18-7). In twenty-four hours under optimum conditions a single individual would produce 281,472,976,710,656 descendants. Fortunately, conditions are never perfect, and this number is never reached. As just alluded to, certain forms have been observed to carry on conjugation. To this extent, they show sexual reproduction. When proper conditions are present, some species of bacteria are capable of forming spores in the following manner. The cell loses water and the protoplasm becomes separated from the wall, causing the entire cell to shrink. This is followed by the secretion of a very impervious cell wall, and often a change in cell shape occurs. Except in a few cases, spore formation is confined to the family Bacillaceae. These spores are resistant to great extremes of temperature, to many

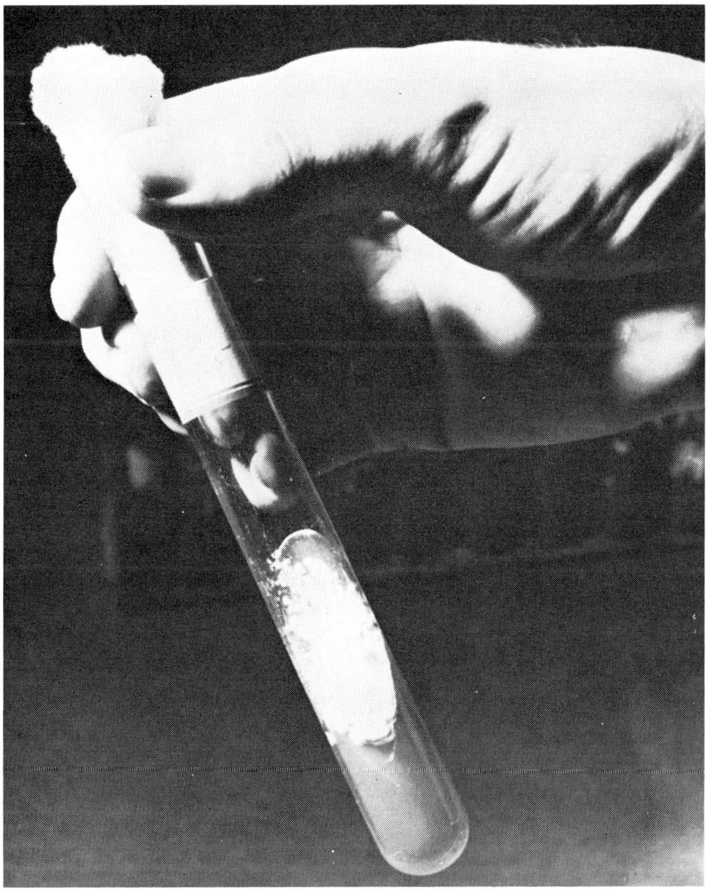

Fig. 18-7
Bacteria may be isolated from an agar plate (Fig. 18-6) and grown in pure culture on agar slants as is the case here. (Courtesy Chas. Pfizer & Co., Inc., New York, N. Y.)

strong chemicals, and to drying. For example, anthrax bacillus can withstand boiling water for over an hour. Others will germinate after being immersed in liquid helium (−269° C.) for several hours. Spores in the dry state have been known to remain viable for thirty years and more, awaiting the return of or the chance contact with a suitable environment. Under certain conditions bacteria are known to conjugate (Fig. 18-8), a process permitting the exchange of genetic material.

Colony formation. After division a daughter cell may remain in contact with the parent to form colonies of characteristic shapes. Bacilli often remain attached in pairs or form long chains. Cocci form pairs, chains, four-celled plates, cubes of eight, or large masses resembling a cluster of grapes (the most common type).

Physiology. Bacteria may be saprophytic, chemosynthetic, or photosynthetic, but most forms are either saprophytes or parasites. Those types feeding on nonliving organic material (**saprophytes**) are responsible for the process known as decay. This may take place by means of a succession of species that are responsible for various stages of decay. Sometimes the result of this process yields useful by-products such as alcohol and cheese or nitrogen in a form available to higher plants. The parasitic species live on or within living plants and animals, where they may cause little noticeable damage or, in varying degree, several pathogenic conditions known as disease. Not a few species are symbiotic with other organisms; that is, they contribute something useful to the host organism while receiving protection and food material from the host. Examples of this include the normal flora of bacteria inhabiting the human digestive tract. They aid in the digestion and processing of food material by breaking down cell walls through the secretion of various enzymes. In return, the bacteria thrive on the otherwise unusable food material.

Some species are **chemosynthetic,** for example, the iron and sulfur bacteria that are

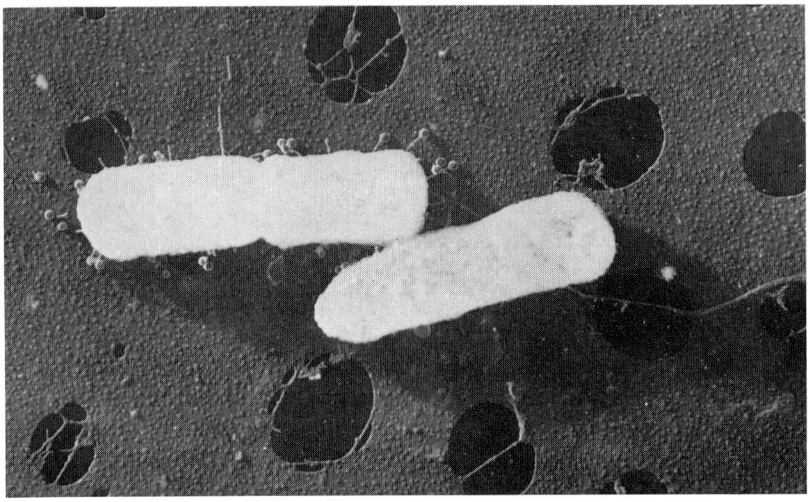

Fig. 18-8
Electron photomicrograph of two bacteria in conjugation. Note also the bacteriophages attached. (Courtesy Thomas F. Anderson, Philadelphia, Pa.; from Wollman, E. L., and F. Jacob. 1956. Scient. Amer. **195**:109, July.)

capable of reducing compounds of these elements into useful materials for their metabolism. Thus they are independent of organic substances and probably represent primitive forms capable of living on the earth in early geological ages when sunlight was obscured by dense cloud layers. The sulfur bacteria play a prominent part in the purification of sewage and are found in all natural waters. A few species of bacteria contain **photosynthetic** pigments and carry on photosynthesis like higher plants.

Several families comprising the suborder Rhodobacteriineae carry on primitive photosynthesis, making use of a variety of pigments.

Bacteria in general have remarkable powers of food synthesis, which means that few substances are indispensable to them. In a general way their metabolic requirements may be classified as follows:

1. Bacteria that can use ammonia as a source of nitrogen for protein synthesis, and simple carboxylic acids as a source of carbon for proteins, carbohydrates, and fats, the nitrogen-fixing bacteria (Fig. 18-9). These include common soil species.
2. Bacteria that require amino acids as a source of nitrogen. A high percentage of the species fall in this classification. Tryptophan seems to be the only amino acid indispensable to them. They can convert all other amino acids from higher plants and animals into those suitable for their own needs.
3. Bacteria dependent upon a complex of nutritional needs, including many vitamins (vital amino acids). These are represented by a few specialized parasitic species.

Besides these special features of metabolism, bacteria have the same requirements as other forms of life, that is, water, oxygen, and carbon compounds. Not all bacteria require free oxygen, however. Some species known as **anaerobes** are capable of reducing other materials to obtain oxygen necessary for their metabolism. In fact, such bacteria cannot live in the presence of free oxygen and are found only in the spore stage when not deep within tissues or other dense media. Other bacteria, dependent upon free oxygen for their energy source (for respiration), are known as **aerobes.**

Waste products from metabolic activities are produced by the bacteria, as in all living organisms. Often these substances are poisonous to the host and are therefore known as **toxins.** Waste material is always poisonous in varying degrees to bacteria. If these products are not removed somehow, the bacteria form spores until the conditions change.

Fig. 18-9
Nodules on red clover caused by nitrogen-fixing bacteria. (USDA photograph.)

All physiological processes of bacteria are possible by diffusion within the cell and by osmosis through the cell wall, as in other organisms.

Ecology. Bacteria are universally distributed. Two features have permitted them to inhabit every place occupied by other forms of life: (1) extremely simple structure requiring exactly the same nutritional materials as other organisms and (2) spore formation for the preservation of the individual organism. They are found in soil as deep as 16 feet, in the upper atmosphere, and on and in all other organisms, even though movements by their own efforts are very restricted. Movement measurements record their speed at 7 inches per hour if these movements were sustained. Transposed to the size of man, these speeds would be magnified to the astounding rate of 90 miles per hour.

Classification. A natural classification of bacteria is extremely difficult to formulate because of their lack of morphological diversity and their apparently poorly organized hereditary mechanisms. Lack of stability of species to the degree commonly observed in higher organisms is perhaps the main reason for classification difficulties. Biologists are used to a reasonable amount of readily observable stability of species. Therefore attempts to define bacteria species in the same terms used for higher organisms is virtually impossible. This is not to imply that species do not exist in the bacteria. It simply means that they have a wider range of adaptability and therefore a more rapid evolution. Mutations occur frequently, and because of the simple nature of the bacteria, they are more likely to survive than are higher plants. However, it is well to think of the bacteria as a group with a fluid "gene pool"; that is, temporal species may be short lived, constantly changing entities. There is little doubt that they have a genealogical lineage like that of other organisms. It is equally certain that any possible attempts to demonstrate this phylogeny will be highly subjective because of the lack of any direct evidence.

It is no surprise, therefore, that the current classification systems are rapidly changing and do not meet with universal acceptance. The classification proposed in various editions of *Bergey's Manual of Determinative Bacteriology* is in general use today but is frequently changed. This manual describes and classifies all known species of bacteria, rickettsias, phages, and viruses. The rickettsias are small bacteria and, like viruses, are dependent upon living cells for their growth. Most rickettsias are transmitted by arthropods. Several diseases, including typhus and Rocky Mountain spotted fever, are caused by them.

A discussion of the various classes, orders, and other subdivisions of the bacteria would be of little service here, since it would involve details beyond the scope of this text. However, in order that relationships between the various forms of common bacteria do not remain limited to a miscellaneous assemblage of disease organisms, we have made an abridged and annotated classification that should enable one to organize the bacteria in his mind somewhat (see opposite page).

Some principles illustrated by bacteria. Several important biological principles are illustrated by the bacteria, most of which are found in other groups as well. One reason these principles are illustrated by bacteria in formal course work is the relative ease of culturing and experimenting with them. Standard methods have been developed so that testing can be done easily. When all or nearly all environmental factors can be controlled, it is then relatively simple to alter one of these factors and observe the results. Of all the living organisms studied today, this empirical method is best illustrated by the bacteria. In this respect, then, bacteriology is one of the few branches of biology close to becoming an exact science. However, this is relative because there are many perplexing problems

Family 1. **Nitrobacteraceae** (nitrogen-fixing bacteria and similar chemosynthetic species found in this family)
Family 2. **Pseudomonadaceae** (common bacteria of soil and water belong here; some species cause plant disease; others cause fermentation)
Family 3. **Azotobacteraceae** (soil forms, nonsymbiotic nitrogen fixers; example: *Azotobacter* sp.)
Family 4. **Rhizobiaceae** (nitrogen-fixing bacteria, water and soil bacteria, and a few plant pathogens included in this family; example: *Rhizobium* sp.)
Family 5. **Micrococcaceae** (parasites and saprophytes; parasitic species live on skin, in skin glands, or in skin gland secretions; principal genus, *Micrococcus,* also known as *Staphylococcus*)
Family 6. **Neisseriaceae** (all known species parasites of man and animals; among species are those that cause gonorrhea and meningitis)
Family 7. **Enterobacteriaceae** (many species pathogenic; *Escherichia* sp. causes colitis; others cause galls, wilts, and rots in plants; *Salmonella* sp. causes severe diseases of several types, particularily enteric fever, septicemia, and gastroenteritis)
Family 8. **Brucellaceae** (many parasitic and pathogenic species of man and animals, including *Pasteurella* sp. and *Brucella* sp. that cause plague and undulant fever, respectively, are members of this family)
Family 9. **Lactobacillaceae** (two principal types of bacteria in this family, one causing infections, such as *Streptococcus* sp., and the other causing fermentation of dairy products and vegetable juices; many harmless residents of mouth and intestinal tract of man and other animals)
Family 10. **Bacillaceae** (a few important parasites or pathogens of insects; example: *Clostridium* sp.)

in the field, not the least of which is the rapid mutation of species immune to antibiotics. Incidentally, as a result of these new drugs, a flourishing industry has been developed.

Principle of pathogenicity. Disease-producing bacteria are certainly in the minority as far as the total number of bacteria is concerned. When a particular species of bacteria lives within another organism and causes sufficient damage to impair its normal function, the bacteria species is a **pathogen.** The damage to the host is usually the result of noncompatible waste material, toxin, given off by the bacteria. It may be either caustic in effect or a protein to which the body reacts. Various degrees of pathogenicity are expressed, dependent upon several factors. The bacteria multiply and spread from organism to organism by various means.

Principle of disease. Two causes of disease are known: (1) metabolic or physiological disease, a malfunction of the body not directly due to living organisms and (2) pathological disease due to the invasion of the body by some living organism, which in turn does damage to the cells and tissues of the body and otherwise affects its proper function. Most disease is caused by the toxins of bacteria, that is, their metabolic wastes. These substances may be extremely poisonous to the body. The toxin of tetanus, for instance, is 200 times as poisonous as arsenic, 150 times as poisonous as strychnine, and 40 times as poisonous as the venom of rattlesnakes.

Principle of immunity. As a result of bacterial infection or the invasion of the body by a foreign protein (an **antigen**), an innate defense mechanism (an **antibody**) against the invading organism is brought into play. For each antigen introduced a specific antibody is developed that, by chemical union with the antigen, renders it ineffective. It is in this manner that immunity to pathogens is either present or built up in the animal body. These antibodies are called **antitoxins.** With the use of this principle, antitoxins can be induced to develop in certain animals, and the substance can be extracted and injected into human beings to provide an artificially built-up immunity. This is done in the prevention of diphtheria by growing an antitoxin in horses for later use in man. Vaccination is much the same, except that small doses of the organisms induce the manufacture of antibodies without the severe clinical symptoms of the disease itself, as is done in smallpox (a virus) vaccination.

Principle of parasitism. A parasite and a pathogen are the same. A distinction is made on the basis of the type of organism and its effect. The organism invading the body is the parasite, and it is parasitic. The result of this parasitism is pathogenicity when the effects become symptomatic. The process whereby the parasite enters into the relationship with the host is infection. A parasite causing pathogenicity is, therefore, a pathogen.

There are other characteristics illustrated by this group of organisms, for example, asexual reproduction and spore formation. However, the ones just described are those that are best illustrated by the bacteria.

Bacteria are frequently distributed in the spore stage and are able to multiply very rapidly once they find suitable conditions. This characterizes both the nonpathogenic and a few pathogenic species.

Disease causing bacteria, fortunately, because of various bodily defense mechanisms and sanitary measures, often cannot find the proper conditions in which to multiply in large enough numbers to overcome the natural resistance of the host. Once they do, however, and the disease condition is established, they are dispersed from the host in large quantities either orally or by defecation; unless proper measures are taken, an epidemic is in the making.

Beneficial bacteria. Most bacteria species are beneficial to man either directly or indirectly. Their use in industry, particularly in the production of fermentation products, is well known. They also have a part in the making of cheese and butter. Curing and ripening of tobacco, tanning of skins into leather, and disposal of sewage are also dependent upon bacteria.

From a biologist's point of view, the bacteria that cause decay play an important and necessary role in the cyclic changes of organisms. Nearly all decay is caused by bacteria or fungi or both. Dead plants and animals are immediately attacked by bacteria, since there are no longer any resistant elements to oppose their action. These bacteria live as saprophytes on the dead bodies. By a succession of species, eventually the entire body, plant or animal, is decomposed into the simplest inorganic and organic compounds. The effect of weather as well as scavengers is minor compared to the action of bacteria. If it were not for the effect of bacteria, dead plant and animal material would accumulate, making a shortage of the raw materials necessary for new growth. Thus the concept of "from dust to dust" is not only a fact but is also necessary for the continuance of life.

The well-known effect of nitrogen fixation by some of the primitive bacteria has a very beneficial effect in agricultural practices by building up the available useful nitrogen in the soil, one of the most necessary raw materials for growth. Thus the planting of legumes, which harbor nitrogen-fixing bacteria in a symbiotic relationship, is a routine procedure for the rejuvenation of the soil by nitrification. Bacteria also add to the fertility of the soil by their decay action. Old roots and similar organic materials are broken down into useful plant food.

DIVISION 4. ACTINOMYCOTA—MYCELIAL BACTERIA

The members of the division Actinomycota form filaments and some are stalked or branched. Some are pathogenic, but most are soil forms. However, this group includes the pathogens causing tuberculosis and leprosy.

DIVISION 5. SPIROCHAETAE—SPIROCHETES

The spirochetes (Fig. 18-10) are thin, usually flexible, spiral, filaments without flagella, but with an undulating membrane attached along the body. All species are motile. One group of spirochetes is found in stagnant fresh or salt water and in the intestinal tract of mollusks. The other group are mostly parasitic. This group includes *Treponema pallidum,* the causative agent of syphilis. *Borrelia* spp. are the causitive agents of relapsing fever and other diseases.

The three divisions of bacteria form the branch Mastigomonera of the kingdom Monera.

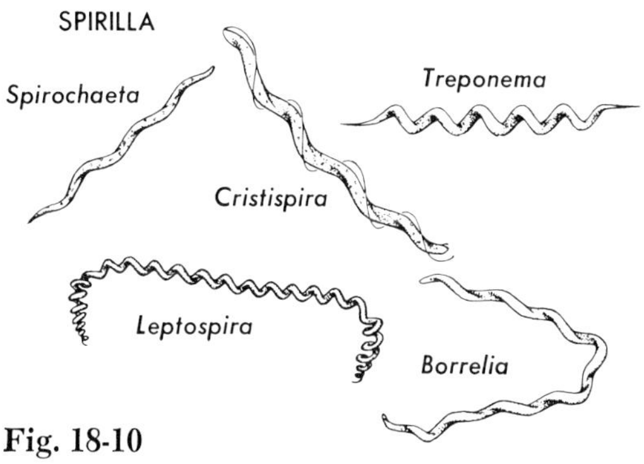

Fig. 18-10
Examples of spirilla.

SPECIAL TERMS USED IN CHAPTER 18

aerobe	mycelium
anaerobic organism	pathogen
antibody	photosynthetic
antigen	phycocyanin
antitoxin	procaryotic
bacillus	saprophyte
chemosynthesis	spirillum
coccus	toxin

QUESTIONS TO CONSIDER

1. Name some common diseases caused by bacteria. Name at least one that is not.
2. Why are bacterial diseases, in general, not as serious as they used to be?
3. How does a disease-producing bacterium cause a disease? Describe the disease cycle.
4. List at least two features of the blue-green algae that are similar to the bacteria, and two that are different.
5. What great evolutionary principle was established by Pasteur using bacteria as his study organisms?
6. Why can we not consider bacteria plants?

DISCUSSION FOR LEARNING

1. Some bacteria have become resistent to ordinary disinfectants. Discuss possible ways that this might have come about.
2. Bacteria are classified principally by means of their physiological reactions, and the plants are classified by means of morphological characters. Discuss the reasons why both means of classification are in reality the same.
3. Discuss some ordinary health practices in relation to the previous discussion of these microorganisms, the bacteria. Consider also, food preservatives in relation to the bacterial flora of the digestive tract.

Chapter 19

Viruses and viruslike agents

Certain submicroscopic agents are termed viruses and are treated here because they may possibly be degenerate parasites related to the bacteria. They are very primitive life-forms, but about the only feature they have in common with other life-forms is the presence of replicating DNA or RNA molecules, but never both.

The agents assigned to the group are very simple, hardly much more than complex macromolecules. They vary from 10 to 200 $m\mu$ in diameter ($m\mu = 1/1,000\ \mu$); they are usually so small that they readily pass through a very fine porcelain filter. It is for this reason they were once known as "filterable" viruses. They are now well described from numerous electron micrographs such as the one at the top of this page, but for a long time they were known only from there effect on tissues or bacterial cultures. There are two more or less distinctive groups now recognized as being viruses: (1) viruses containing DNA (Fig. 19-1) and (2) viruses containing RNA. They consist of three parts, a head, a tail, and tail fibers (Fig. 19-1). The head is surrounded by a protein coat, the **capsid,** and the center is a core of DNA or RNA. The tail portion consists of a sheath, a core, and tail fibers.

The mature virus outside the cell is termed the **virion,** and it is not infective by itself.

Several forms are no longer considered to be viruses. Those removed and returned to the Eubacteriae are the psittacosis agents (causitive agents of parrot fever), Eaton agents (a **pleuropneumonia-like organism,**

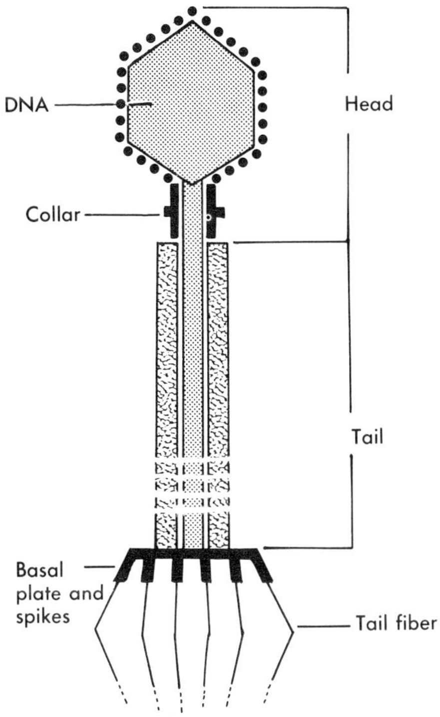

PPLO), the trachoma agents, and the rickettsias. These are all extremely small bacteria and therefore are confused with the viruses.

The viruses are divided into classes, orders, and families according to a new classification proposed in 1965 by the Provisional Committee for the Nomenclature of Viruses.

Viruses and bacteriophages deserve some further special mention because of the adverse effects of some and the indirectly beneficial effects of others. All known viruses are pathogenic because they require

Fig. 19-1
Diagram of a phage drawn from one shown in Fig. 19-3. The head measures about 950 Å, the tail about 3,000 Å, and the tail fibers 1,500 Å. Note the protein coat, represented by the large dots, surrounding the head.

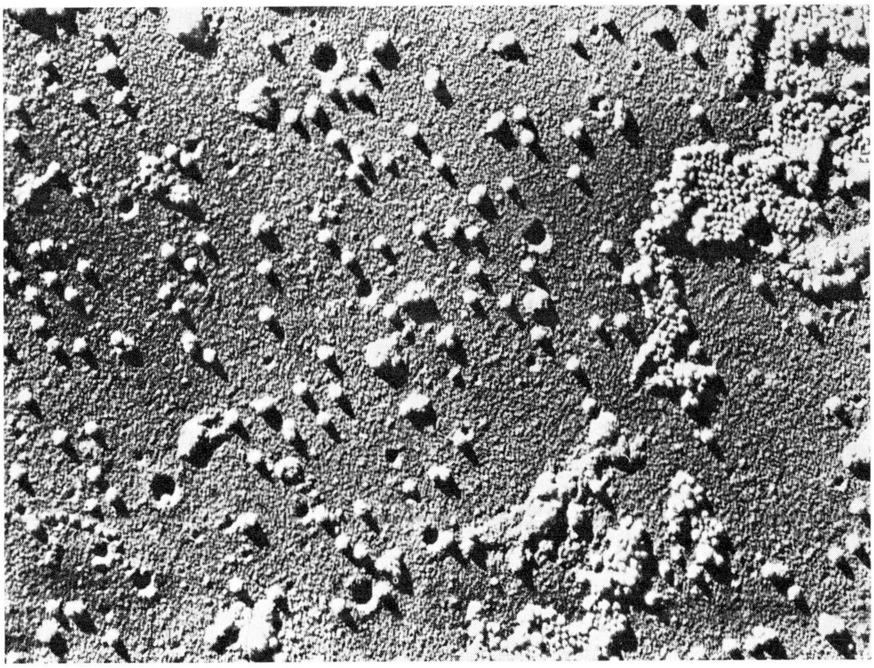

Fig. 19-2
Poliomyelitis virus. (×77,000.)

a living cell in which to live. Such well-known diseases as mumps, poliomyelitis (Fig. 19-2), and smallpox are caused by viruses. In this respect they are unique. However, both chemically and physiologically they resemble nuclei, which gives room for some interesting speculations. It has stimulated researchers in cancer to investigate the role of viruses in the abnormal cell division of malignancies. Viruses causing disease in both plants and animals are barely more than complex protein molecules. Some are even capable of crystallization and can be stored in jars indefinitely. Upon contact with a suitable cell, they will "come alive" and multiply.

Those viruses that infect bacteria, and there are many, are termed **bacteriophages,** or **phages.** Some viruses need to be "helped" by other viruses before they can replicate within the cell. These **defective viruses** seem to arise as mutants, and each has its satellite or **helper virus.** Tobacco necrosis virus is an example of this. Two size particles are present, one, the larger, is the virus that lacks sufficient information to code all the proteins necessary for their replication. The helper virus, the smaller particle, seems to supply the code needed to complete the life cycle.

Bacteriophages were first discovered in 1896 when it was found that the water of the Ganges River contained organisms that would check the growth of bacteria. The electron microscope shows that there are minute structures resembling a coccus with flagella. Phages infect only bacteria, within which they reproduce and which they eventually destroy. When first discovered, it seemed likely that phages could be used

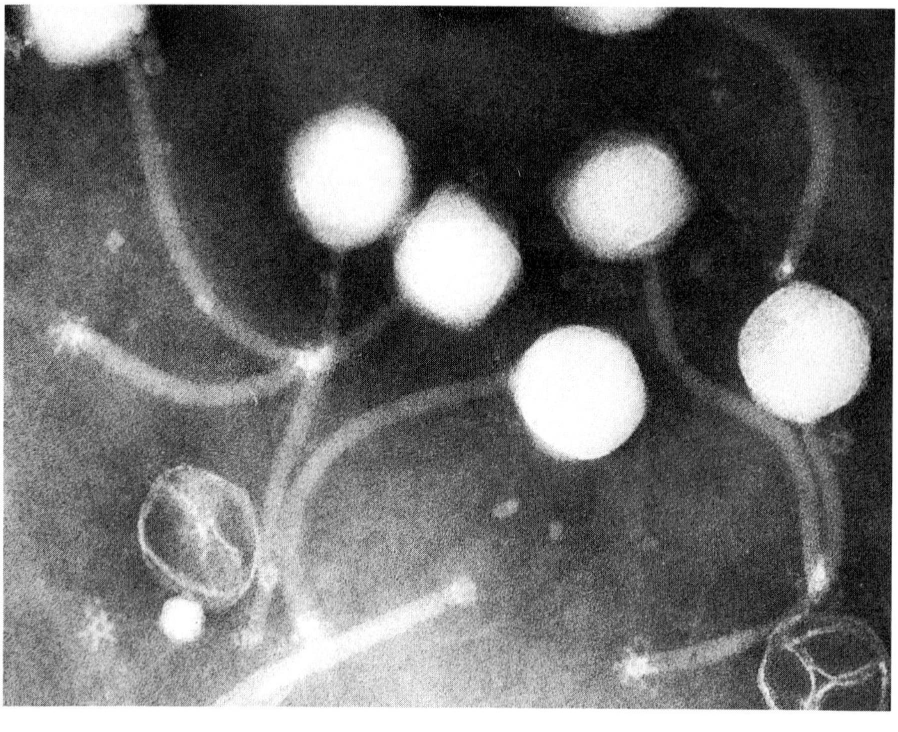

Fig. 19-3

D 32 Mycobacteriophages. (×400,000.) The head, tail, and tail fibers are easily seen in this photograph. (Courtesy Jeolco, Inc., Medford, Mass.)

to combat pathogens. Except in the treatment of cholera, all attempts in this direction have been unsuccessful.

A bacterium host cell for a phage is the cell whose metabolism is used for the growth and reproduction of a phage. One condition, termed a **lysogenic bacterium,** contains a prophage, a latent virus whose DNA is actually combined with that of the bacterium so that its genes are transferred with those of the bacterium at each cell division. These "strains" of bacteria eventually cause infection in others, and lysis takes place in the culture.

The multiplication of bacteriophages (Fig. 19-3) leads to the lysis of the host cells, eventually causing **plaques,** round clear areas in the colony of bacteria. The lysis kills the contiguous cells through several cycles of virus growth, and the plaques are formed. Unfortunately, little practical application of this effect has been discovered. Bacterial disease treatment still depends upon other antibiotics.

The life cycle of a phage (Fig. 19-4) is a good indication that these very small organisms are not as simple as was once thought. Two types of infections may take place,

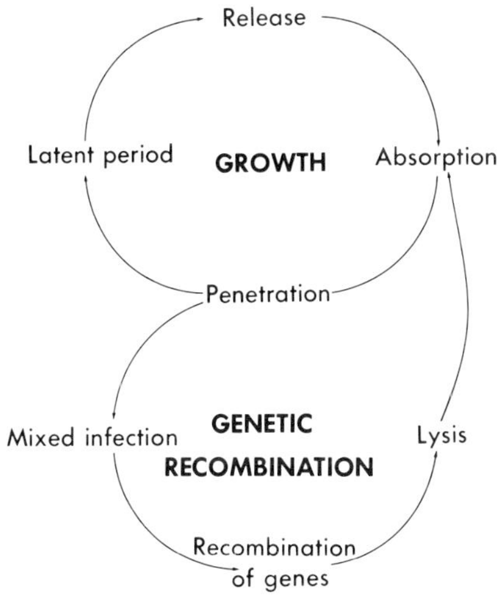

Fig. 19-4
Life cycle of a T-even phage.

single or mixed. If a single viron attaches to a bacterium (Fig. 18-8) a simple growth cycle takes place. If more than one attaches, under certain circumstances a "sexual" or genetic recombination cycle may occur.

The simple growth cycle of a T-even phage (under the old classification, a T2, T4, T6 phage, now members of the family Phagoviridae) will serve as a generalized example. The virion attaches to the surface of the bacterium. (In an animal virus the viron is engulfed or absorbed by the host cell.) The process is termed absorption, and the attachment is irreversible. Shortly thereafter, penetration takes place, a process consisting of the injection of the DNA contents of the virion into the host cell. The capsid remains outside and is no longer needed. Next is the latent period, so-called only because nothing is observed in the culture. However, inside the bacterium two processes take place. First, there is a period of protein synthesis known as the eclipse period. Enzymes, capsid coats, and DNA are produced from raw material provided by the bacterial cell. A second part of the latent period is termed the maturation period, during which time condensation of the DNA takes place and the capsids are added. About 200 to 1,000 virions are released, ending the cycle with the final, or release, phase.

During the recombination phase there is an exchange of protein material. This is dependent upon mixed infections that require the absorption of two or more virions by the cell. There are three types of such infections: (1) phenotypic mixing, which is caused by the combining of capsids of two different forms of virions during the growth process, resulting in a different external appearance of the newly formed virions; (2) complementation, which involves two different defective viruses that help each other by supplying each others missing genes (differs from helper viruses); (3) phage recombination, during which time portions of the DNA molecule recombine with others so that there is an actual exchange of genetic material that follows the same laws of heredity found in the higher plants.

The recombination phase of mixed infections has been likened to sexual reproduction, and the growth phase to asexual reproduction. We are therefore left with the question: are viruses with these attributes similar to life-forms, living organisms?

SPECIAL TERMS USED IN CHAPTER 19

- bacteriophage
- capsid
- defective virus
- helper virus
- lysogenic bacterium
- phage
- plaques
- pleuropneumonia-like organism (PPLO)
- virion

QUESTIONS TO CONSIDER

1. Why do we hesitate to call viruses living organisms? What was our definition of life in previous chapters, and do the viruses meet the requirements?
2. What two recent discoveries have enabled researchers to learn more about viruses?
3. How does the virus, like the "common" cold virus, cause disease?
4. How are viruses transmitted?

DISCUSSION FOR LEARNING

1. Discuss the reasons why you believe a virus is or is not a primitive organism. Include in your discussion reasons why you do or do not believe that a virus is a degenerate bacterium and therefore a living organism.

Chapter 20

Protista

U.S. Forest Service.

The kingdom Protista is a group of algae and protozoans, living in freshwater ponds like the one photographed above, in the sea, or as endoparasites. For convenience we will treat here only the groups of the Protista generally considered to be "plants," and leave the remainder for the zoologists. The "plant" divisions are never parasitic on animals, but many are parasitic on plants or fungi. The part of the kingdom treated here is grouped into five divisions. They differ from the Monera by the presence of distinct nuclei in their cells. The nucleus is surrounded by a nuclear membrane. They are generally unicellular, but some are multicellular, although obviously closely related to the unicellular forms. They reproduce in various ways, but typically by asexual division. The vegetative cells with the haploid number of chromosomes divide. This is followed by a sexual stage, involving the fusion of two cells to form a diploid zygote, followed immediately by meiosis as the usual sequence. Since the kingdom includes species with photosynthetic pigments and those without, both conditions characterize the group. Those without the pigments are generally treated as protozoans, but there are cases where the same species appears in two forms, one with pigment and one without.

DIVISION 6. EUGLENOPHYTA—EUGLENOID ORGANISMS

The euglenoid algae are often treated as protozoa. The zoologist classifies them as the family Euglenidae of the class Mastigophora. Some are colorless and therefore dependent on organic food for energy and growth. Others, like *Euglena* sp. (Fig. 20-1), are equipped with chlorophyll and can therefore manufacture their own food. All members of this division are flagellated, and most of the species are motile. Some, however, form sessile colonies and thus are nonmotile. They are all unicellular and simple in structure. It is impossible to distinguish such organisms as either plants or animals. Undoubtedly this division and the animal phylum Zoomastigina have common ancestors. If all members of the group lacked chlorophyll, they would be treated as protozoa, but they are more advanced forms of algae and are therefore treated in this position.

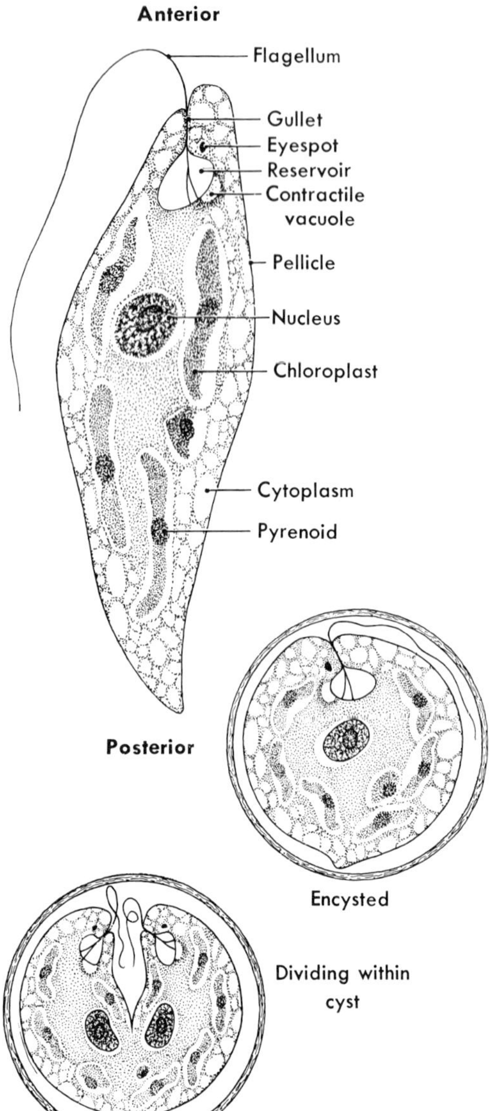

Fig. 20-1
Euglena sp. Above, the motile green alga stage; below, the encysted stage. Note the cell division for asexual reproduction.

Sexual reproduction so far is unknown in the group, although dubious cases of gametic union of vegetative cells have been reported. Asexual reproduction takes place by simple cell division; cyst stages are formed. Meiosis occurs apparently after the asexual fussion of nuclei in a process termed **autogamy**, but this is rare.

These organisms are very common in small, stagnant, freshwater pools rich in organic matter. Their common use in biological laboratories testifies to their abundance and rearing ease. Because of their simple growth cycle, no life cycle is shown in Chapter 14.

DIVISION 7. CHRYSOPHYTA—YELLOW-GREEN AND GOLDEN BROWN ALGAE AND DIATOMS

The plastids of this division contain carotenes and xanthophylls, giving them a variety of colors from yellow-green to brown. There are about 300 genera and 6,150 species divided among three classes, each of which are often treated separately. They are Xanthophyceae, the yellow-green algae; Chrysophyceae, the golden brown algae; and Bacillariophyceae, the diatoms.

Class Xanthophyceae—yellow-green algae

The plant body of the yellow-green algae is either unicellular or multicellular. Unicellular forms are flagellated, whereas the multicellular forms are filamentous and often multinucleated. All species have yellowish green chromatophores, which give them their characteristic yellow-green color.

The class is small, with only about 400 species, most of which are aquatic, occurring in freshwater. However, some are aerial, growing on moist tree trunks, damp cliffs, or with mosses, and a few occur in the soil.

In many species, structure of the cell wall is similar to that in the diatoms, that is, overlapping halves that fit together like a Petri dish. Asexual reproduction is usually by means of zoospores. Sexual reproduction is known only in a few genera in which it is effected by means of free zoogametes of similar form (life cycle 2, *Vaucheria* sp.).

Class Chrysophyceae—golden brown algae

The plastids of the cells of the golden brown algae contain carotenes and xanthophylls, giving the plants a golden brown color. The cells may be uniflagellate or biflagellate but rarely triflagellate. Most species are unicellular, but some form small, loosely held together colonies or loose filaments. They rarely appear as compact, branching filaments. Sexual reproduction has not been conclusively demonstrated.

The class contains about 70 genera and 250 species divided into several orders and families. Most of the species occur in cool, freshwater that is low in calcium. The motile species often form a large part of the

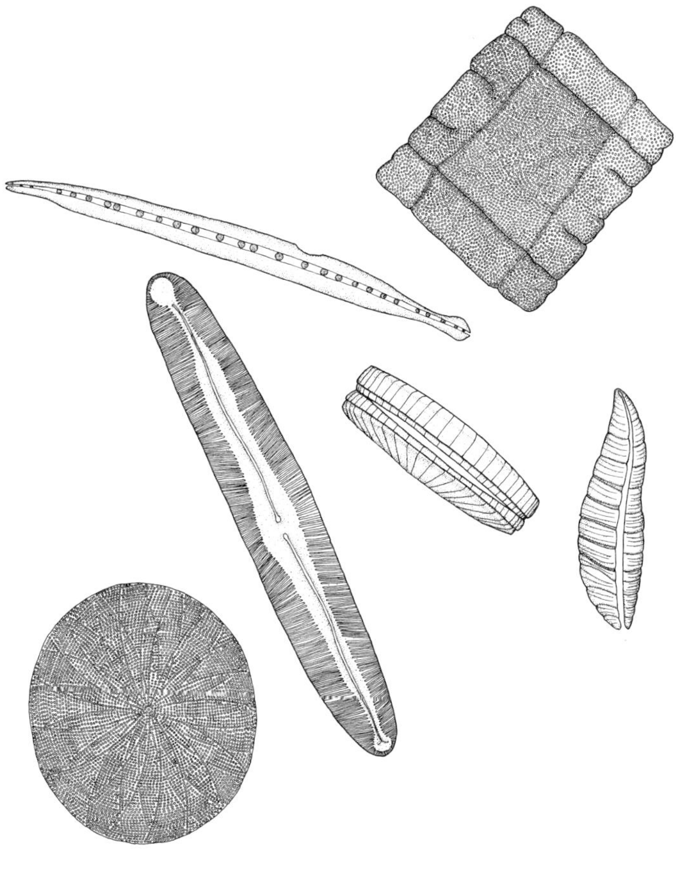

Fig. 20-2
A variety of diatoms from a sample of diatomaceous earth.

plankton of lakes. Others occur in cold springs and brooks.

Reproduction is asexual by either simple cell division of the unicellular species or by fragmentation of the motile colonies. Nonmotile genera produces zoospores formed singly within a cell. **Statospores** similar to those formed by the diatoms may be produced in some genera. These unique spores have silicified walls that form two overlapping halves and a pore closed by a conspicuous gelantinous plug, which enable the species to survive unfavorable conditions. Because of this type of spore the class is often placed in the same group as the diatoms.

Class Bacillariophyceae—diatoms

The diatoms are an ancient group of plants that form important deposits, called **diatomaceous earth** (Fig. 20-2), used in the manufacture of abrasive products, filters, and insulation.

The Bacillariophyceae are unicellular, rarely colonial, and the cells are surrounded by a two-part wall resembling a Petri dish (Fig. 20-3). They are microscopic in size. The cell contents are difficult to see, but their geometric shapes make them easily recognizable. The cell wall is composed of silicon, which accounts for their presence in great quantities in fossil formations. The chromatophores are yellow or brown. Food is stored as fat or as a proteinlike substance, not as starch. There are about 5,500 species, most of which are both living and fossil. A few are known only as fossils.

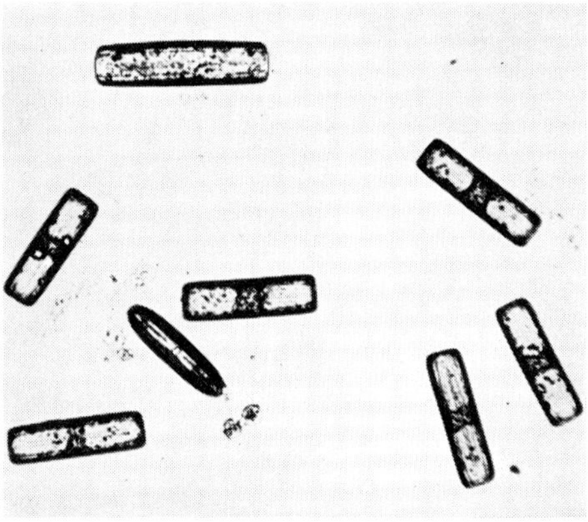

Fig. 20-3
Girdle and valve views of living diatoms. (Courtesy Carolina Biological Supply Co., Burlington, N. C.)

Asexual reproduction is by a cell division complicated by the necessity to secrete a new cell wall to match the one lost to the separated half of the cell. The new cell wall is secreted inside the old one so that with each successive generation the diatom grows slightly smaller. This small size is counteracted by the formation of sex cells.

Sexual reproduction is affected by a casting off of the cell wall. The cells then divide and form haploid ameba-like gametes. When they fuse, the zygote forms new cell wall halves, preserving the symmetry of the parent cell; thus they regain their original size. Sexual reproduction also occurs in other ways, such as by the formation of auxospores, reproductive cells formed by the fusion of two diatom cells. These may be produced by processes similar to conjugation in *Paramecium* sp., a protozoan.

DIVISION 8. PYRROPHYTA—DINOFLAGELLATES AND CRYPTOMONADS

Members of this division are mostly motile, flagellated forms. A few are filamentous. They are characterized by a transverse groove or girdle within which one of the flagella is placed in repose. The other flagellum extends backward from the girdle, as in *Peridinium* sp. (Fig. 20-4). The chromatophores contain a brownish pigment in most species, but some lack photosynthetic pigments and are saprophytic or holozoic like the Protozoa.

There are only about 300 species. They occur in both marine and freshwater habitats but are rarely abundant. Occasionally *Gymnodinium* sp. becomes exceedingly abundant, causing a "red tide" and a severe disease of fish.

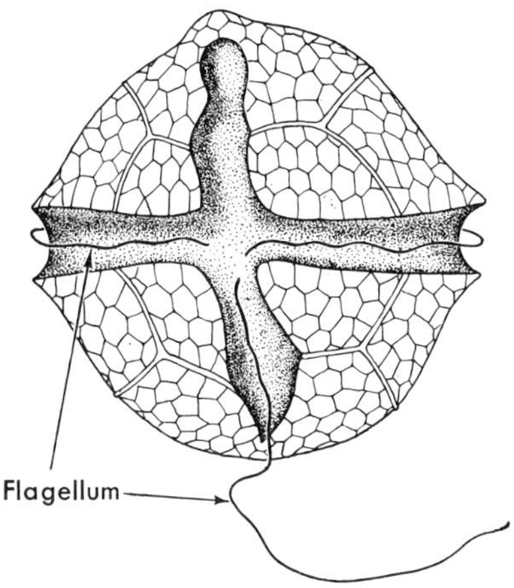

Fig. 20-4
Peridinium sp. is a chlorophyll-bearing unicellular alga having a platelike covering.

The cryptomonads are small, oval flagellates generally classed as protozoans and are found in foul water.

DIVISION 9. HYPHOCHYTRIDIOMYCOTA—HYPOCHYTRIDS

Members of this division are freshwater or marine algae often classed with fungi. The presence of motile cells with an anteriorly placed flagellum is the reason they have been removed from the kingdom Fungi and placed near the dinoflagellates.

They are parasitic on algae and fungi, or some live on plant and insect debris in the water in which they are found. There are only about fifteen species.

DIVISION 10. PLASMODIOPHOROMYCOTA—ENDOPARASITIC SLIME MOLDS

This group is removed from the kingdom Fungi. They have a naked multinucleate, plasmodial type of plant body, and they are entirely parasitic, causing various plant diseases of vascular plants, algae, and fungi.

SPECIAL TERMS USED IN CHAPTER 20

autogamy
statospore
diatomaceous earth

QUESTIONS TO CONSIDER

1. What are the features this group of organisms have in common?
2. Where are most of the species found?
3. What pigments are present in the plastids of the Protista?

DISCUSSION FOR LEARNING

1. Discuss a logical evolutionary sequence for the Protista. What do you think happened during the Precambrian evolution of this group?

Chapter 21

Algae

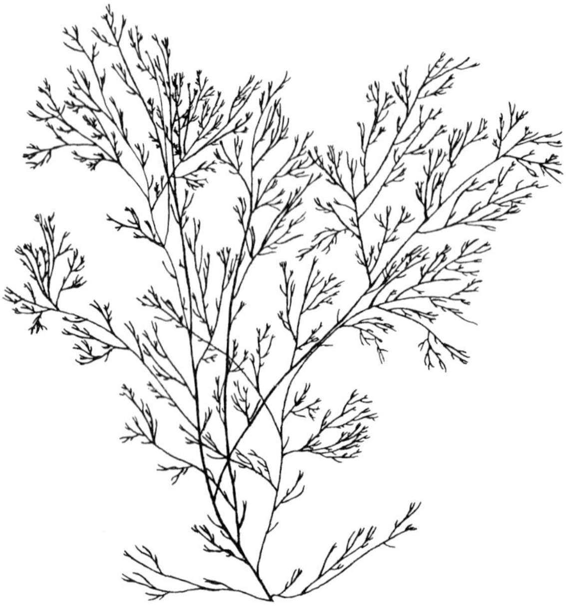

True plants, in the modern sense, include the remainder of the divisions, except for the Fungi, considered in this text. The kingdom Plantae is defined as a group of organisms, usually multicellular, with walled and frequently vacuolate **eucaryotic** cells, and almost always with photosynthetic pigments in plastids of at least some of the surface cells. The higher members of the kingdom are differentiated into specialized tissues and organs. Reproduction is by means of a sexual cycle of haploid cells and an asexual cycle of diploid cells. The four remaining divisions of algae are discussed in this chapter. Many of these resemble the higher plants as does the diagram of the browning alga, *Dictyosiphon* sp. at the top of the page.

DIVISION 11. RHODOPHYTA—RED ALGAE

The red algae are chiefly marine forms characterized by the dominance of the red pigment **phycoerythrin,** which masks the other pigments that are present. However, because of varying amounts of the blue pigment phycocyanin, many shades of red, pink, brown, purple, violet, and olive green occur.

Definition. The most characteristic feature of this group is their passive, nonflagellated sperm cells. Most species are multicellular and form a branching filament or a thallus of moderate size. Most of them are small, but a few reach a meter in length. The cells are usually uninucleated, with a large central vacuole. The cell wall is composed of cellulose and pectic compounds, from which is obtained commercial agar used medicinally or as culture media for bacteria. There are about 2,500 species of red algae.

Vegetative reproduction is rare. Asexual reproduction is effected by the formation of various kinds of spores. Sexual reproduction is unique in this group, and special terminology must be used. The life cycles of many red algae are incompletely known. The life history may be rather complicated, involving the alternation of as many as three separate generations. The gametophytes are the large vegetative plants. These produce, on the same plant, **carpogonia,** and **antheridia,** which in turn produce the egg and the spermatium. The latter are released into the water to eventually float into contact with a carpogonium. The carpogonium has an elongate terminal projection, the **trichogyne.** When the spermatium makes contact with this structure, a sperm nucleus penetrates the cell wall and migrates to the egg nucleus. The resulting zygote divides to form a parasitic diploid carposporophyte, which buds **carpospores** on tiny branches. They are liberated, float away, and form a third type of thallus, the **tetrasporophyte.** When mature, these plants resemble the gametophytes. They produce, in special internal cells in groups of four, spores known as **tetraspores,** which are formed by meiosis and are therefore haploid. When these are released, they germinate and develop into gametophyte plants.

Several other types of life cycles are known among the species of the red algae. Variation is mainly evidenced by the elimination of the tetrasporophyte and carposporophyte stages.

Some of the genera, such as *Porphyra* sp. have no marked alteration of generations. The vegetative plant or gametophyte produces gametes, which fuse to form the zygote. Meiosis takes place in the zygote stage, in which the zygote matures into a new gametophyte, so that the sporophyte is represented only by the zygote.

Nemalion sp. (life cycle 3) lacks the tetrasporophyte generation. The sporophyte is represented by the carposporophyte, parasitic on the gametophyte. However, it is haploid because of the formation of this generation after a series of divisions, one of which is meiotic, in the zygote. Therefore the diploid stage is represented again by the next zygote.

The life cycle of *Polysiphonia* sp. involves the three generations just mentioned. All three plant stages, gametophyte, carposporophyte, and tetrasporophyte, are macroscopic. The gametangia of the gametophyte are located on separate plants. In these the zygote divides to form the carposporophyte, but these divisions do not involve meiosis; therefore, the sporophyte is diploid. This plant, like *Nemalion* sp., develops from a cystocarp, a seedlike structure containing diploid carpospores. These eventually break away from the parent thallus and form new plants upon germination. The diploid tetra-

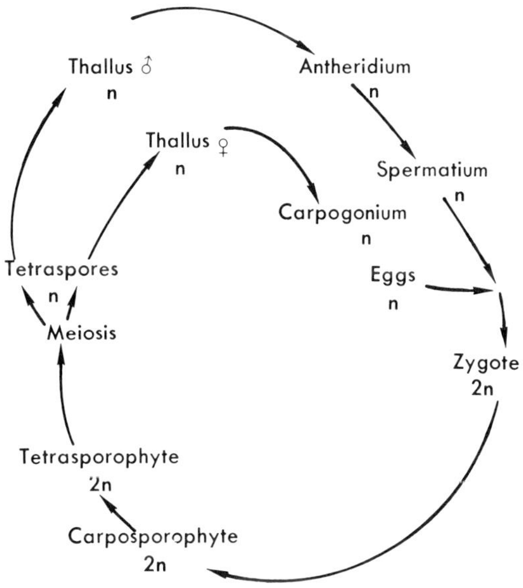

Fig. 21-1
Three alternate generations of the complex life cycle of the red algae.

sporophytes resemble the haploid gametophytes. When mature, these plants produce tetraspores after reduction division has taken place. These, then, are haploid, and they in turn produce new haploid gametophytes. This type of life cycle is common in this phylum.

The three alternate generations of a complex life cycle of the red algae are shown in Fig. 21-1. Each generation represents a separate macroscopic plant: (1) the gametophyte generation, with monoecious plants (male and female plants are not separate), (2) the carposporophyte, the sporophyte that is parasitic on the carpogonium of the gametophyte, and (3) the tetrasporophyte, which is a free-living plant resembling the gametophytes.

The economic importance of the red algac is increasing because some species, for example, species of *Porphyra*, are used as food. The Japanese are consumers of millions of tons of this plant, grown in bamboo bundles in shallow water near the shore. The plants are used either dried or fresh and are cooked or made into soup. Others, such as species of *Chondrus* and *Rhodymenia*, are similarly eaten. *Gelidium* sp. is a source of agar, which is used in the manufacture of jellies and in a variety of desserts, as well as in the familiar agar plates used by bacteriologists.

DIVISION 12. PHAEOPHYTA—BROWN ALGAE

The brown algae contain, in addition to chlorophyll, a brown pigment, fucoxanthin, which almost completely masks their green color. Unicellular, filamentous, and thalliform species occur. The great majority of species form the complex, relatively thick plant body commonly referred to as seaweed. This thallus is the most elaborate and complex of any group of algae. All but 3 of the 1,000 or so known species are marine. The freshwater forms are very rare.

The common marine *Fucus* sp. (Fig. 21-2) or rockweed can be seen at low tide along the coast of the United States. The kelps (*Laminaria* spp.) are familiar forms on bathing beaches. *Sargassum* sp. played a prominent part in the discovery of America. This seaweed forms such an extensive mat in the middle Atlantic ocean that it is

Fig. 21-2

Fucus sp. sporophyte. Note air bladders and the terminal receptacles. (From Russell, N. H. 1958. An introduction to the plant kingdom. The C. V. Mosby Co., St. Louis.)

known as the Sargassum Sea. Passing through this mat on his first voyage to the New World, Columbus convinced his sailors that they were nearing land, which gave them courage to continue the voyage (life cycle 5, *Fucus* sp.).

Classification. The relationship of the various orders of this division is well known except for the order Fucales, which contains the common rockweed *Fucus* sp. The life cycle of this group does not include the free-living haploid stage common in the other members of the division. There are three orders, with over 200 genera and several hundred species.

Laminaria sp. is a still more advanced type of brown algae, commonly known as a **kelp.** These large algae are confined to cold marine waters where they often grow to giant sizes, as long as 30 to 50 meters. These large plants are very highly developed and have some indication of a vascular system in the form of **trumpet hyphas,** filaments with large cells inflated at the ends. There is a well-defined epidermis one or two cells thick, with numerous small cuboidal chromatophores.

Unilocular sporangia are borne in sori, covering the leaf surface at certain times of the year. Haploid zoospores are formed and discharged during high tide. The zoospores, after a period of swarming, develop into small thalluslike gametophytes. Half of the zoospores develop into male gametophytes and the remainder into female ones. Under proper conditions they form oogonia or antheridia. Each oogonium produces a single egg. Fertilization takes place in the oogonium, and the result is a zygote. The plant develops without leaving the oogonium, and the cycle continues (also life cycle 4, *Ectocarpus* sp.).

The fucus type of life cycle seems to represent the end of the algal line of development because of the near elimination of the gametophyte. This condition is not present again on the evolutionary scale until the flowering plants are reached.

DIVISION 13. CHLOROPHYTA—GREEN ALGAE

The green algae have plastids containing primarily chlorophyll, with small amounts of carotenes and xanthophylls giving a bright green color. They produce motile reproductive cells with flagella. Reproductive cells are always borne in one-celled organs. The plant body may be one of many cells, or if many celled, it forms filaments.

Distribution. The green algae are aquatic, usually living in freshwater, but a few are marine. They are most abundant in freshwater ponds, lakes, and streams. A few species are attached to roots or debris, but some float and form so much plankton that they color the water. They are also found in brine lakes and as green snow in snow fields.

Structure and reproduction. Sexual reproduction occurs in most species. The gametes are formed in antheridia (male) and in oogonia (female) cells. Like all algae, these structures are always single cells, producing either a single female cell or sometimes many male cells. These may be re-

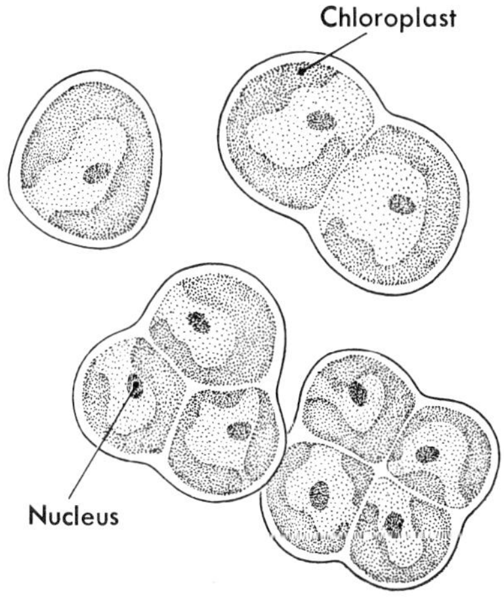

Fig. 21-3
Protococcus viridis is a green alga found growing usually on the north side of trees. The figure illustrates colonies of one, two, three, and four cells.

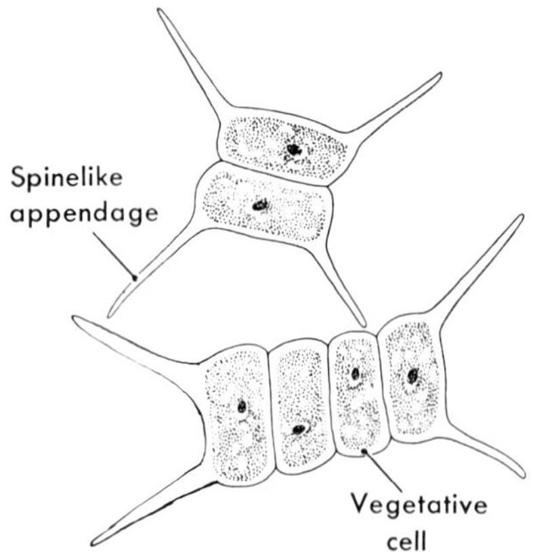

Fig. 21-4
Colonies of *Scenedesmus* sp., a green alga, are common in almost all standing water; this alga often appears in practically pure culture in aquariums and jars of water standing in a laboratory for some time.

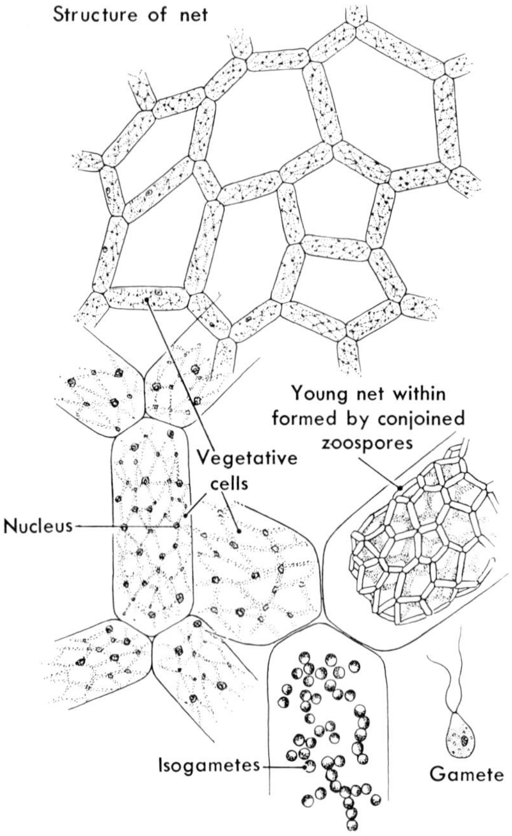

leased into the water where fertilization takes place, or the male cells, being motile, may go to the oogonium where fertilization takes place. In the asexual process small motile bodies termed zoospores are formed. They are released into the water, after which they swim around for a short time and then settle down for vegetative growth.

Classification. These algae are the most common forms encountered in freshwater habitats. More than 6,900 species are known. The following list gives the classification of the common forms and some examples, several of which are described in more detail in Chapter 14 on life cycles.

Classification	Some genera
Order Volvocales	*Chlamydomonas, Pandorina, Eudorina,* and *Volvox*
Order Tetrasporales	*Tetraspora*
Order Ulotrichales	*Ulothrix* and *Protococcus* (Fig. 21-3)
Order Ulvales	*Ulva*
Order Schizogoniales	*Prasiola*
Order Cladophorales	*Cladophora*
Order Oedogoniales	*Oedogonium*
Order Zygnematales	*Spirogyra* and desmids
Order Chlorococcales	*Scenedesmus* (Fig. 21-4), *Hydrodictyon* (Fig. 21-5), *Chlorella* (Fig. 21-6), and *Pediastrum* (Fig. 21-7)
Order Siphonales	*Vaucheria*
Order Siphonocladiales	*Valonia*
Order Dasycladales	*Acetabularia*

Progressive development. Progressive development of the Chlorophyta is shown by primitive forms consisting of single-celled motile organisms that progress to single nonmotile cells. Others are united end to end to form a platelike mass, and the most advanced forms are filaments.

Fig. 21-5
The structure of *Hydrodictyon* sp. is netlike in appearance and often traps small fish in an aquarium. It reproduces by zoospores and isogametes.

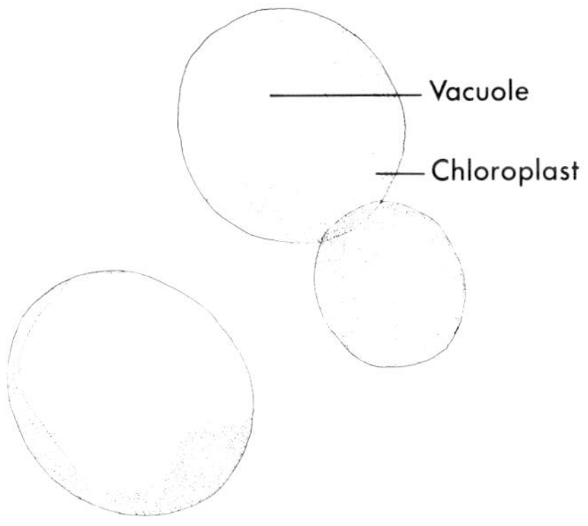

Fig. 21-6
Chlorella sp. A colonial alga widely cultivated for food as well as experimentation.

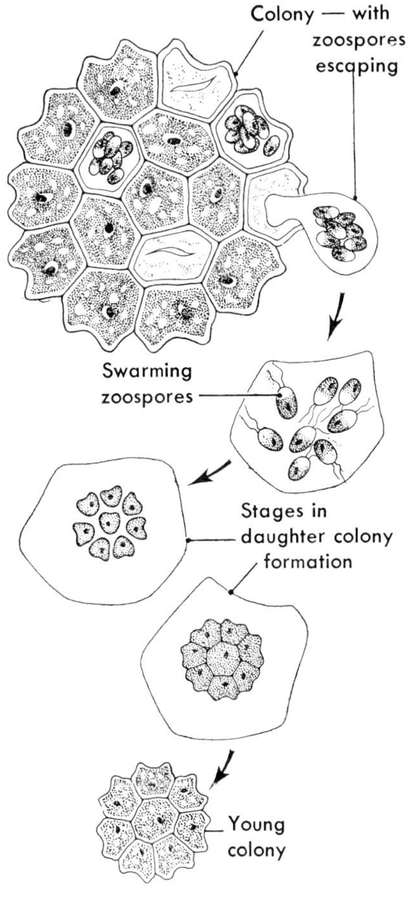

Fig. 21-7
Pediastrum sp. is a widely distributed colonial alga that grows free floating in ditches and in the plankton of lakes.

Examples of single-celled green form. *Chlamydomonas* sp., which is described in Chapter 14 (life cycle 6), is a good example of this type of single-celled organism. These plants are very similar to the Protozoa of the animal kingdom, with which they may be classified as the family Chlamydomonadidae of the class Mastigophora. A colonial alga, *Chlorella* sp. (Fig. 21-3), is extensively used in the study of physiology of photosynthesis and respiration. These plants are nonmotile and have a very simple organization. They may be cultured like bacteria, and attempts have been made, particularly by the Japanese, to grow them commercially as a cheap source of food. One of the most interesting features of *Chlorella* sp. is that these plants may be induced to live in the bodies of the protozoa *Paramecium* sp. and *Stentor* sp. and of the coelenterate *Hydra* sp. By so doing, they live symbiotically and by virtue of the ability to carry on photosynthesis are able to manufacture their own food. Thereafter, in *Paramecium* sp. at least, these plants are transmitted from generation to generation by cell division as a part of the animal.

Examples of motile colonies. Two common pond forms, *Pandorina* sp. (Fig. 21-8) and *Volvox* sp. (Fig. 21-9 and life cycle 7), illustrate motile colonies. These genera consist of species with groups of sixteen cells (*Pandorina* sp.) or many cells (up to 20,000 in *Volvox* sp.). They differ primarily in their complexity. *Volvox aureus* shows some cell specialization (Fig. 21-10) and a certain amount of food exchange, although in both of these exchanges, the individual cells morphologically resemble *Chlamydomonas* sp. and generally carry on independent physiological functions.

The cells in *Pandorina* sp. are somewhat pyramidal in shape because they are tightly packed in the colony so that their inwardly projecting portions are thus compressed. The whole colony is surrounded by a mucous covering, out of which project the two long flagella borne by each cell. Even in

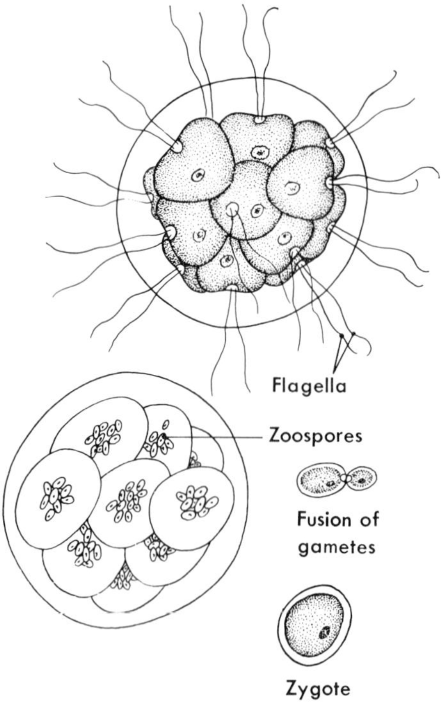

Fig. 21-8
Pandorina sp. is an example of a simple motile colony.

this simple association there is slight cell specialization. The colony has a definite anterior end, and the eyespots of the cells in this area are larger and better developed.

These genera belong to the plant family Volvocaceae, but they are also assigned to the Protozoa family Volvocidae of the class Mastigophora.

Examples of filamentous types. Many familiar pond species of this class are filamentous and are best represented by the order Ulotrichales. The filaments may be simple or branched. The individual cells are without flagella, except as reproductive forms.

Branching filaments and other chlorophytes. Many marine species of the chlorophytes grow rather complicated thalli. These plants (for example, *Canlerpa* sp. and *Bryopsis* sp.) are **coenocytes**—single, large multinucleated cells. The cell is differentiated into a main axis from which arise branches or pinnae. The lower portion of the axis forms a basal rhizome of branches that serves to anchor the plant. The plant body is one continuous vacuole joined by cytoplasm containing numerous, round chloroplasts and nuclei. Vegetative reproduction may take place by the break-

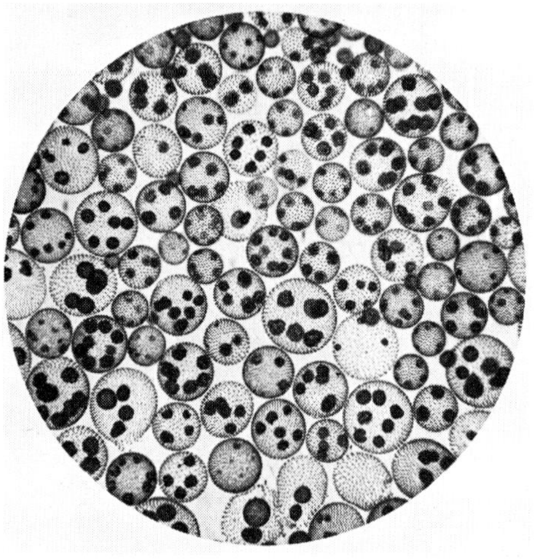

Fig. 21-9
Volvox sp. is an example of a complex motile colony. (Courtesy General Biological Supply House, Inc., Chicago, Ill.)

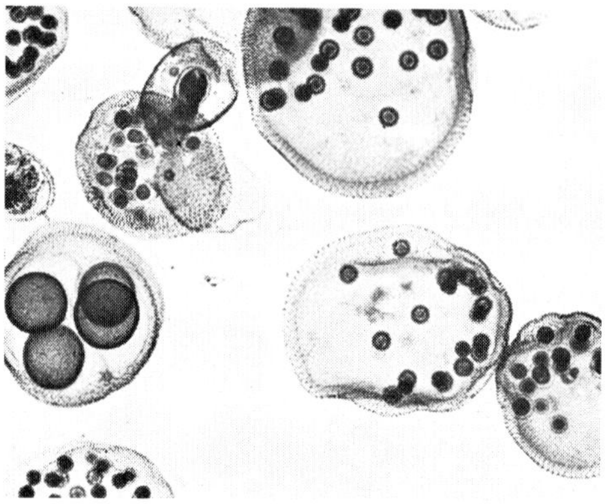

Fig. 21-10
The colonies of *Volvox aureus* pictured show specialization of cell types. (Courtesy Carolina Biological Supply Co., Burlington, N. C.)

ing off of small sections of the plant. Sexual reproduction is effected by flagellated gamete formation in a gametangium formed from one of the small branches.

The familiar *Oedogonium* sp. and *Spirogyra* sp. are similar to *Ulothrix* sp. (life cycle 8) but are somewhat more highly developed. *Oedogonium* sp. produces heterogametes and, after formation of the zygote (Fig. 21-11), **"swarm spores."** The process consists of rapid cell division of the zygote in which are produced flagellated cells that swim some distance from the parent filaments and serve to disperse the species. *Spirogyra* sp. (Fig. 21-12) carries on sexual reproduction by a process known as **conjugation** (Fig. 21-13). This consists of the passing of the contents of one cell through a conjugation canal formed between two cells in juxtaposition to unite with the neighboring cell. Both cells are identical, but the cell contents that move are considered to be the male. Movement is effected by the pumping action of many contractile vacuoles. A closely related genus, *Zygnema* sp. (Fig. 21-14), reproduces like *Spirogyra* sp., by **scalariform conjugation,** which is conjugation between cells of different filaments. Conjugation between adjacent cells of the same filament is called **lateral conjugation.**

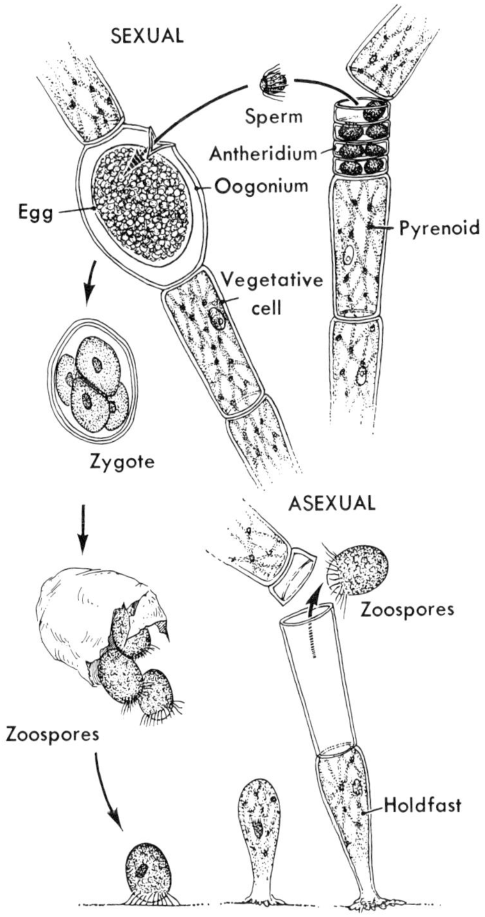

Fig. 21-11
The life cycle of *Oedogonium* sp. illustrates the initial development of heterogametes and "swarm spores."

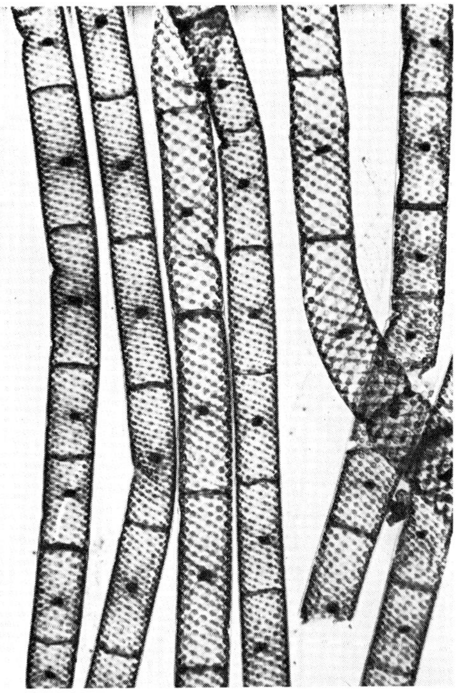

Fig. 21-12
Spirogyra sp., a familiar pond and laboratory plant, is characterized by spiral chloroplasts. (Courtesy General Biological Supply House, Inc., Chicago, Ill.)

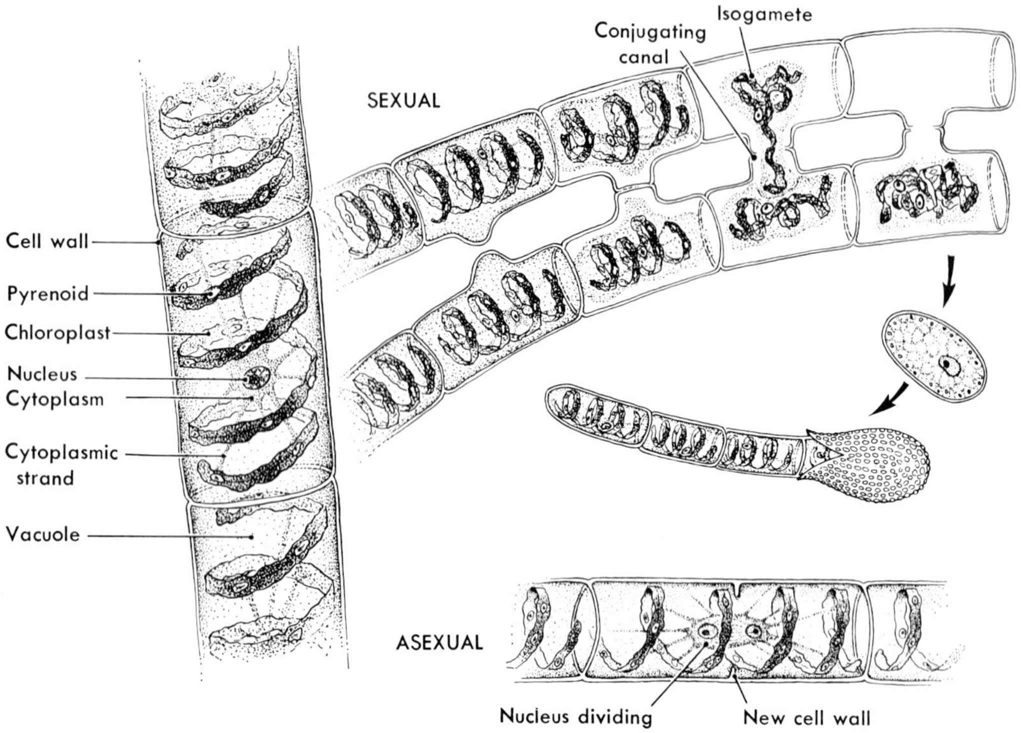

Fig. 21-13
Sexual and asexual reproduction in *Spirogyra* sp. diagrammatically shown.

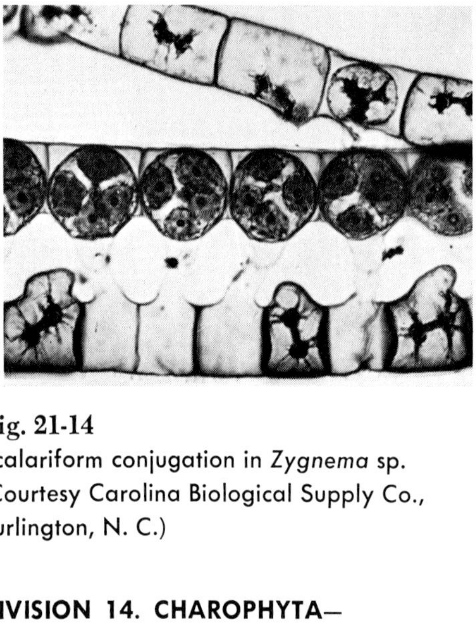

Fig. 21-14
Scalariform conjugation in *Zygnema* sp. (Courtesy Carolina Biological Supply Co., Burlington, N. C.)

DIVISION 14. CHAROPHYTA—STONEWORTS

These plants are commonly called stoneworts because of the heavy calcium deposits in the cells. The habitus of these plants is so similar to certain aquatic vascular plants that they are often mistaken for the latter by the layman. The thallus consists of an erect, branched body differentiated into a regular succession of nodes and internodes. The latter branches at the nodes resemble small needlelike leaves. The cells contain chlorophyll and carry on independent metabolic process, with only slight food exchange. The leaves or branches are capable of limited growth, but the stem has unlimited growth and may also give rise to branches with unlimited growth. Each stem is organized in similar fashion, with nodes and internodes. The plants may be 1 to 3 feet or more in length and are anchored to the bottom of the pond or lake in which they are growing. The holdfast cells are colorless and appear to be rhizoids. They may form extensive subsurface meadows in favorable areas.

Distribution. Stoneworts grow in freshwater pools, ponds, and small lakes. Some species grow in brackish water.

Reproduction. Asexual reproduction is affected by the production of vegetative structures of several types but not by zoospores. These structures are called **amylum stars, bulbils,** and protonema-like outgrowths. Amylum stars are star-shaped aggregates of cells developed from lower nodes filled with starch. These break away and generate into a new vegatative plant. Bulbils are small nodules developed on the rootlike basal extremities of the plant, which eventually may break away and form new plants. Protonema-like outgrowths sometimes appear at a node and resemble small filaments of algae. When they are freed, they will mature into new vegetative plants.

Sexual reproduction is by the formation of gametes in multicellular antheridia and oogonia of fairly complex structure. Fertilization takes place in the oogonium. The resulting zygote forms an oospore, which is shed. This eventually germinates to form a small protonema and finally the young plant (life cycle 9).

Classification. The division is a small one with about 6 genera and 250 living species. Many of these plants are also found as fossils, having occurred as long ago as the Devonian period, as evidenced by the calcium content in their cells. All extant species are assigned to the family Characeae, but three fossil families are also described in the botanical literature.

BIOLOGICAL SIGNIFICANCE OF ALGAE

Although the algae are of significant economic importance, they are not currently an important part of the economy of man. They could conceivably become an important food source in the event that the land becomes overcrowded and not suitable for cultivation. They are extremely important as food for aquatic animals and are a part of many food chains, involving all kinds of marine and freshwater animals used as food by man. It goes without saying that they are likewise a part of the natural balance of all aquatic communities.

From the viewpoint of the biologist they also occupy an important place because of diversity of form and the functions they exhibit. This single group illustrates so many lines of development that recent systematists have divided them into many divisions, and included them in two separate kingdoms. The ranking of the groups is not important to students of general botany.

The biological significance of these divisions may be summarized as follows:

1. The unicellular algae and the several animal phyla comprising the protozoa cannot be distinguished in a natural way. It is currently believed that the latter were an offshoot of the algae, and that these in turn gave rise to the remainder of the animal kingdom. The vascular plants are a direct line of development from the algae, with the fungi and bryophytes as sideline developments.

2. The various life cycles beginning with simple asexual growth and ending in the complex alteration of three generations suggest all of the reproductive processes found in the higher plants, plus many special forms unique to the algae. The essential features of these reproductive processes are (a) the formation of gametes having haploid nuclei, (b) the union of gametes by various mechanisms to ensure their fusion, (c) the development of spores for rapid vegetative multiplication and dispersal, and (d) motile stages for gamete dispersal to ensure a mixing of genes throughout the breeding population.

3. The development of the plant body from a simple single-celled plant to a large plant with specialized cells and tissues occurs in this kingdom and leads to the development of a vascular system. A larger size is possible by the secretion of gelatin-like or cuticle-like materials that can occur

only in aquatic situations. The thickness of the thallus is never great, since it lacks an efficient vascular system.

4. Unlike most species in the plant kingdom, the gametophyte is dominant.

SPECIAL TERMS USED IN CHAPTER 21

amylum stars
antheridium
bulbil
carpogonium
carpospores
coenocyte
conjugation
eucaryotic cell
kelp
lateral conjugation
phycoerythrin
scalariform conjugation
swarm spores
tetraspore
tetrasporophyte
trichogyne
trumpet hypha

QUESTIONS TO CONSIDER

1. Why does the term "algae" no longer have any classification significance?
2. What are the major characteristics used to separate the divisions of organisms formerly known as algae?
3. What is the role of these various divisions (as discussed in this chapter) in their biota?
4. What is seaweed?
5. Of what economic importance are algae?

DISCUSSION FOR LEARNING

1. Review the various types of life cycles found in the "algae" and discuss which ones are the most successful and which ones are the least likely to give rise to new groups. Why?
2. Discuss the pros and cons of man's ancestors being "seaweeds."

Chapter 22

Mosses, liverworts, and hornworts

Courtesy Carolina Biological Supply Co., Burlington, N. C.

The two divisions included in this chapter are sometimes collectively referred to as the bryophytes, the amphibia of the plants. They probably evolved from some group of algae as the first true land plants. However, their members are still dependent upon moisture for sexual reproduction, and it is for this reason that they are likened to the animal amphibia. The first clear-cut example of alternation of generations by land plants is found in the groups included here, the other land plants being certain algae that formed terrestrial mats but reproduced while submerged in water. There are distinct sporophytes and gametophytes with sexual reproduction taking place on separate plants. The photograph at the top of this page shows a typical male gametophyte of a liverwort.

There are about 23,000 species in these two divisions, but most of them are mosses. The two divisions are Hepatophyta, with the Hepatopsida (liverworts) and the Anthoceropsida (hornworts), and the Bryophyta, with the single class, Musci (mosses). The Hepatophyta produce gametophytes that are dorsiventral and differentiated into leaves and stems. The sex organs of the Hepatopsida of the Hepatophyta are arranged on stalks on the dorsal side of the plant and are separate, frequently on different plants. The sporophyte is small and retained in the archegonium. The Anthoceropsida of the Hepatophyta have their sex organs embedded in the gametophyte. The sporophyte is prominent.

The bryophyte gametophyte is erect and has stems and leaves, the apices of which develop the sex organs. The sporophyte, which grows from the archegonium as in the other groups, may or may not be prominent.

DIVISION 15. HEPATOPHYTA—LIVERWORTS AND HORNWORTS

The two classes that belong to this division differ considerably and are treated separately in the following discussions.

Class Hepatopsida

Most members of this class grow in bogs or on the soil near running water. A few are found floating on water. The life cycle of *Marchantia* sp. (female, Figs. 22-1, *A*, and 22-2, and male, Fig. 22-1, *B*), a typical representative of the group (life cycle 10), involves an alternation of generation typical of the division.

The liverworts are characterized by thalluslike or straplike plant bodies without prominent stems. These are prostrate, having rhizoids that anchor the thallus to the soil. The rhizoids are unicellular and unbranched. The cells of the thallus are several layers thick; the upper epidermal layer is usually covered with a sturdy cuticle. Stomata open into the plant body and allow the transfer of gases in the spongy layer of chlorophyllic cells. The plastids of these cells are small and numerous. Photosynthesis is carried on by all species. The protonema is very small and transitory or absent. Elaters are present in the capsules and serve to disseminate spores.

Classification. Liverworts are found as fossils, and they were probably more abundant in past geological periods than they are today. Extant species are assigned to four orders and several families. There are about 8,500 species.

Class Anthoceropsida

The hornworts have a similar structure, but appear somewhat different because of a more prominent sporophyte that extends above the thallus in stemlike fashion. The plastids in the cells of the hornwort are

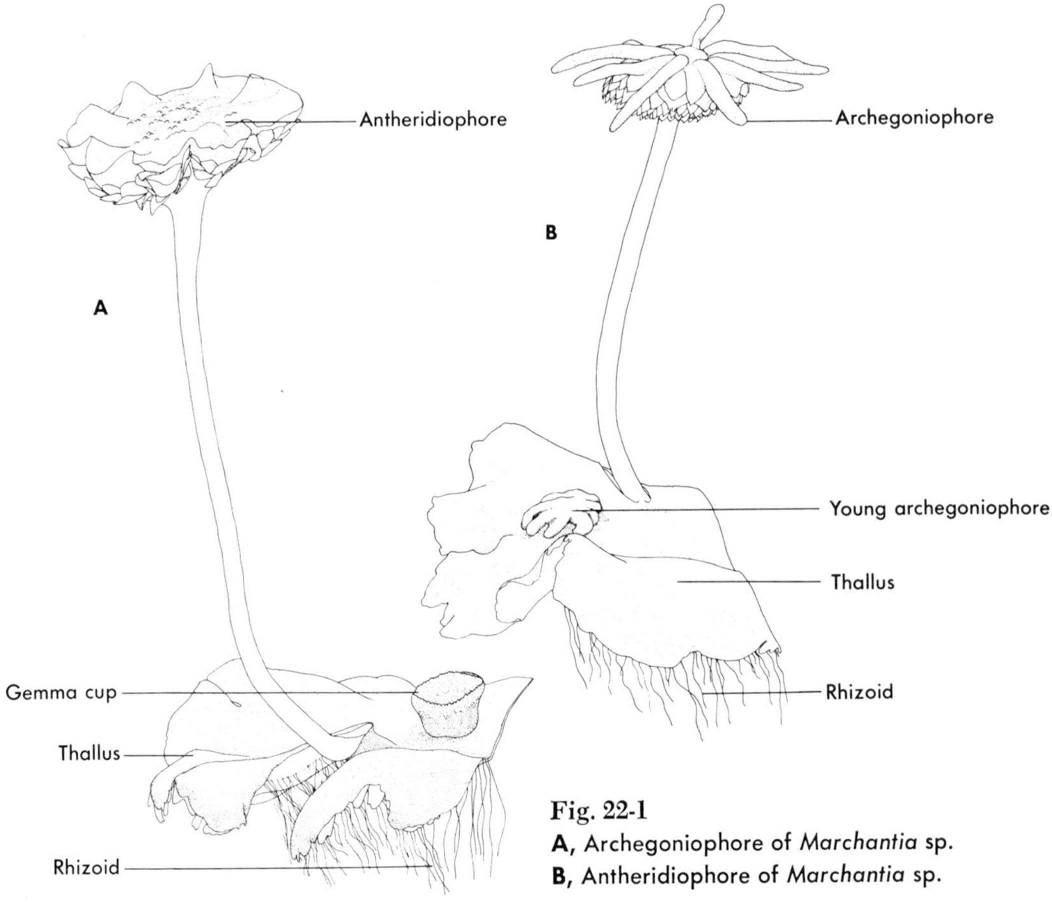

Fig. 22-1
A, Archegoniophore of *Marchantia* sp.
B, Antheridiophore of *Marchantia* sp.

Fig. 22-2
Marchantia sp. with archegoniophore. (Courtesy Carolina Biological Supply Co., Burlington, N. C.)

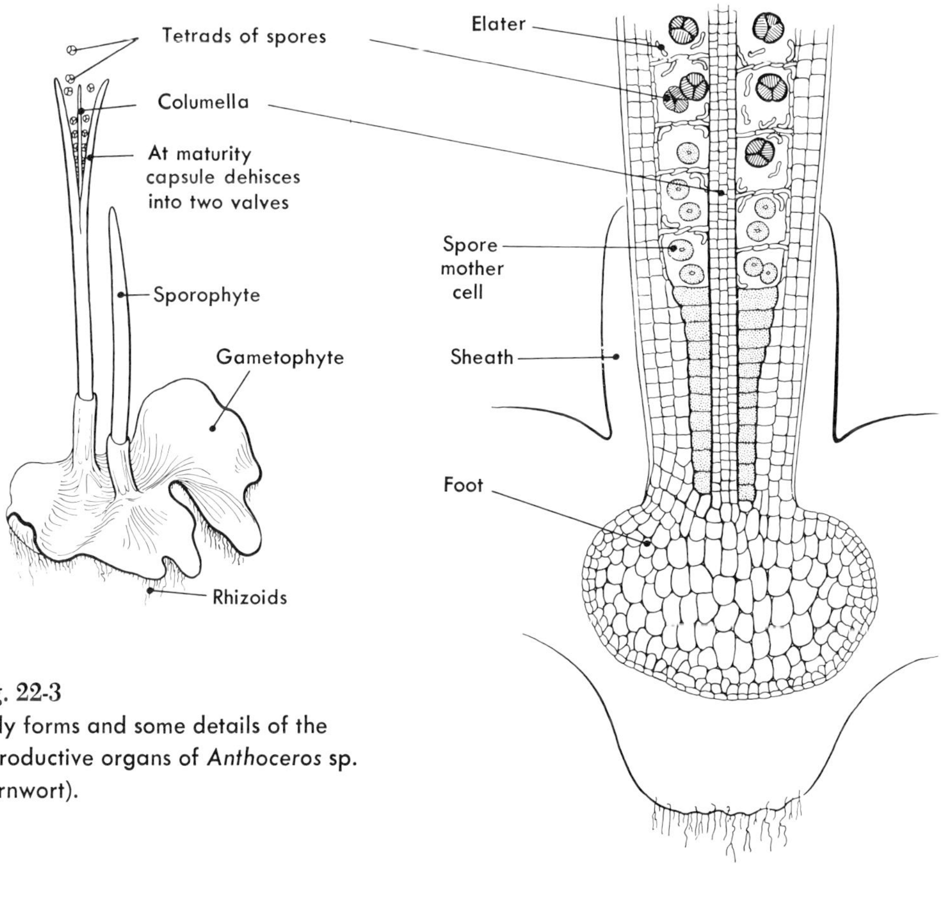

Fig. 22-3
Body forms and some details of the reproductive organs of *Anthoceros* sp. (hornwort).

larger, and pyrenoids are also present. The body form and some details of the reproductive cycles are illustrated in Fig. 22-3. The sporophyte differs from that of the liverworts by having a central column of sterile tissue, the columella, in the capsule. Growth in the capsule is continued over a long period of time.

The spore capsules contain elater cells, which are coiled in a springlike fashion for spore dispersal. The spores themselves form a tetrad of cells.

The gametophyte tissues are not so well differentiated as those of the liverworts; however, like them, they develop antheridia and archegonia on the thallus, but without stalks. This thallus often contains colonies of the alga *Nostoc* sp. growing inside cavities on its surface.

Classification. Only one order and one family belong to this class. The hornworts are a small group of less than 500 species that live on soil, rocks, and, in the tropics, on the trunks of trees. Of course, they are found always in moist places.

DIVISION 16. BRYOPHYTA—MOSSES

The mosses represent the most complex development of the two groups. They have a greater degree of tissue differentiation than do the others. The thallus develops false stems and leaves, with a thickened portion resembling a midrib to give it support. They all, of course, lack vascular tissue. The rhizoids are septate, branched, and more extensive than those of the other division. Their rhizoids serve as anchoring roots, and the thallus is stemlike and erect. Small leaves branch from the stems. Both the stem and the leaves contain chlorophyll and carry on photosynthesis. Because there is no vascular tissue and therefore no efficient supporting tissue, the size of mosses is limited. Exchange of food and raw materials is the result of diffusion through the cells.

Instead of single cell specialization for reproduction and other processes, groups of cells in the form of tissues carry on these functions. However, bryophytes lack true organs, which decidedly limits their development in other habitats. These plants are restricted to moist regions.

The sex organs are multicellular, with the outer layer of sterile cells, in direct contrast to the algae studied. This means that a true archegonium is developed in place of the oogonium, a distinct advance over the previous divisions.

A conspicuous protonema is formed by the germination of spores, and this persists throughout the life of the plant. Their sporangia lack elater cells, and spore dispersal is effected entirely by air currents. However, the spores are smaller and lighter than those of the other division and float away easily from the capsule. In fact, spores are often seen leaving the capsule like smoke so that a patch of moss looks as if it were on fire. A greater portion of the sporophyte is devoted to the manufacture of food so that, although it is still attached to the gametophyte, it is not so completely dependent as are sporophytes of the liverworts and hornworts.

This division is much more abundant than the Hepatophyta, with approximately 14,000 species widely distributed over the surface of the earth and sometimes in large enough patches to form a conspicuous feature of the vegetation of some areas. Although they are dependent upon moist conditions for sexual reproduction, they can be found in relatively arid areas. Arctic regions also abound in mosses.

The reproductive cycle is described in Chapter 14 (life cycle 11) and is illustrated by *Polytrichum* sp., a common woodland group. This cycle is typical of all mosses, the only variation being that of structure.

Occurrence. All species are terrestrial; they are found on the soil, on rocks, and on decaying wood or bark of trees. Some bryophytes are of commercial importance, but none are disease producing, and none are edible (reindeer moss is not a moss but

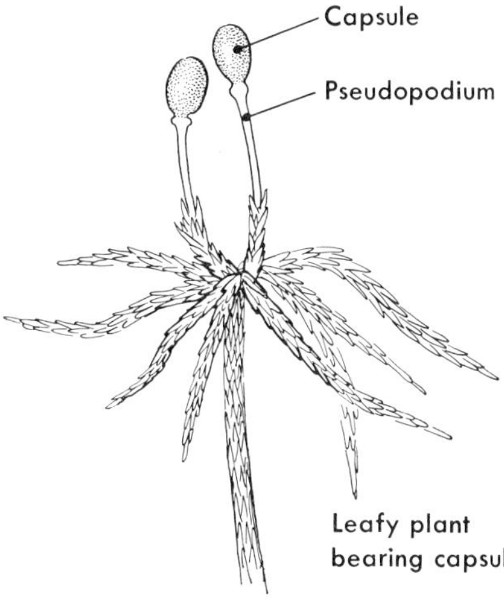

Fig. 22-4
Bog moss, *Sphagnum* sp., showing the sporophyte.

a lichen). However, *Sphagnum*, a genus of mosses that forms in bogs, is commercially valuable as peat moss. It is gathered and packed into bales for use by gardeners as a mulch. A unique property that prevents it from growing mold adds to its usefulness. Eventually bogs left undisturbed for many years will form peat, which in turn after many thousands of years, will form coal.

Classification. The division is divided into three groups. One, Sphagnopsida, is exemplified by the genus *Sphagnum* just mentioned, which is characterized by a single gametophyte produced from the protonema (Fig. 22-4). The sporophyte capsule (Fig. 22-5) is borne on a stalk, the pseudopodium, which is developed by the gametophyte. There is only one order, one

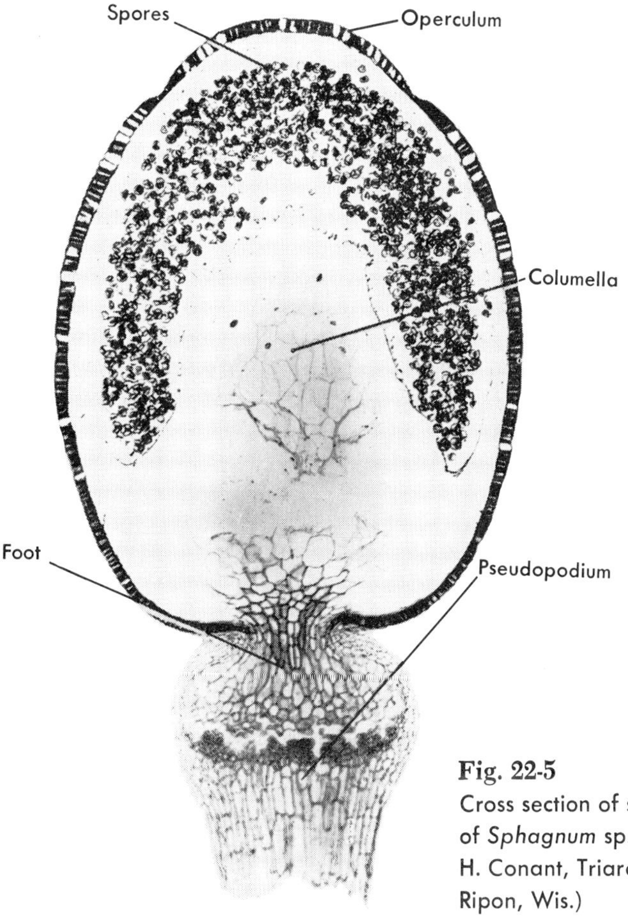

Fig. 22-5
Cross section of sporophyte capsule of *Sphagnum* sp. (Courtesy George H. Conant, Triarch Products, Ripon, Wis.)

family, and the single genus, *Sphagnum*, with over 300 species. They all grow in bogs in acid water.

The granite mosses, Andreaeopsida, are a small group with a single order, one family, and two genera. These mosses grow only on rocks in cold arctic climates or in the alpine zone of high mountains.

The true mosses belong to the Mnionopsida. The protonemata produce several plants with leaves having midribs. The pseudopodium is absent. *Polytrichum* is a typical genus. The vegetative plants, the gametophytes, are often sexually dimorphic. The male plant is devoid of parasitic sporophytes. The antheridial disks are crown shaped. The female plants (Fig. 22-6) with bristly archegonia are the hosts for the parasitic sporophytes (Fig. 22-6), which appear to be the fruiting bodies of the plant (life cycle 11).

The group is divided into about fourteen orders and eighty families, which are widely distributed in all regions of the world. As inhabitants of moist places, they seem to grow best in shaded areas. The familiar rule that moss grows on the north side of a tree (at least in the northern hemisphere) is a result of this habit. However, these tree mosses should not be confused with the algae that also grow on trees. Actually, the avoidance of intense light is not due to any physiological aversion to light, but rather it prevents desiccation. In general, these plants prefer acid soil.

Interesting ecological relationships have been found in the mosses. Moist but bare soil is soon covered by a feltlike mat of

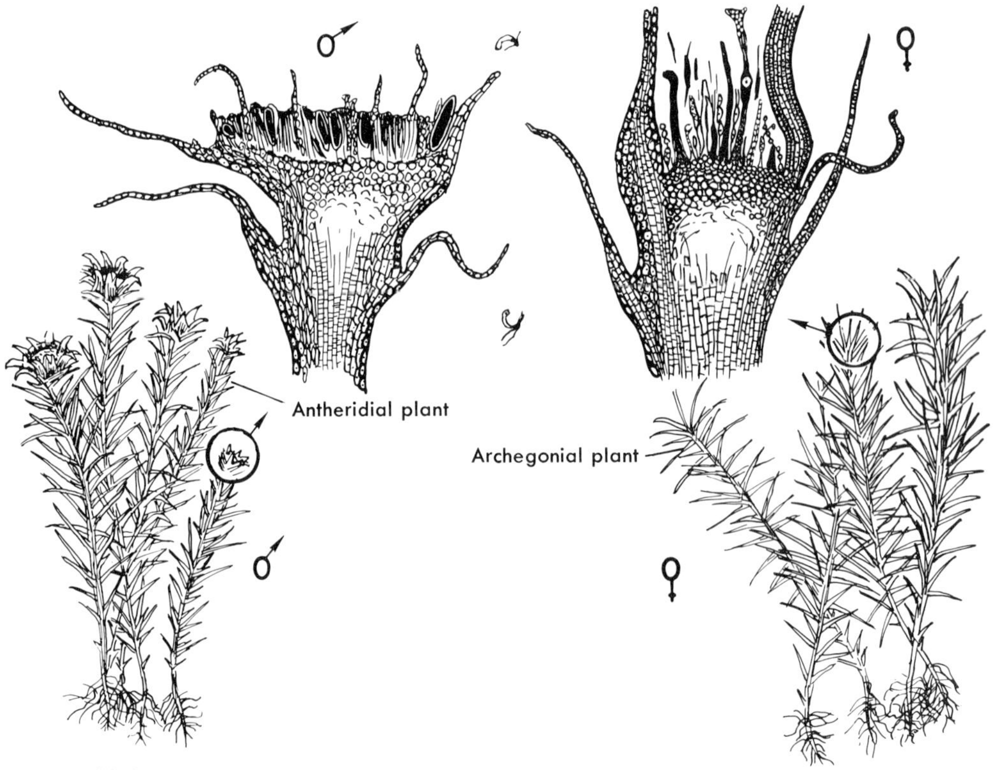

Fig. 22-6
Diagrammatic drawing of male and female gametophyte plants of moss; also enlargements to show longitudinal sections of antheridial and archegonial heads.

protonemata that develop into gametophytes. A new area may be rapidly colonized by a succession of species; the early species are replaced by others as the habitat becomes established. The effect of this is important under natural conditions because of the water-holding qualities of the mosses. In tropical areas where the soil cover is thin, this plays a major role in the prevention of complete erosion and the exposure of bare rock. Even bare rock may be colonized by mosses, but this is a much slower process.

QUESTIONS TO CONSIDER

1. What features are lacking by the divisions considered in this chapter to make them completely independent of water for reproduction?
2. How are mosses able to colonize, that is, what feature makes them available to occupy a new habitat?
3. Why are mosses rarely a dominant feature of the biota?
4. What is meant by true alternation of generations?

DISCUSSION FOR LEARNING

1. Mosses, liverworts, and hornworts are referred to as a sideline development, as are the animal amphibia. Discuss why this is so in terms of a more probable origin of vascular plants.

Chapter 23

Primitive vascular plants

From Russell, N. H. 1958. An introduction to the plant kingdom. The C. V. Mosby Co., St. Louis.

Several divisions of plants produce spores instead of seeds, but unlike those discussed so far, they have vascular tissue and are differentiated into roots, stems, and leaves. Because of the development of true organs and the conduction of water from the soil, they are truly terrestrial plants. We have included in this chapter three divisions, which probably represent the only extant forms of much larger groups. In fact, more fossil than extant species are known.

Table 23-1 showing the features and classification of the vascular plants summarizes our knowledge of these groups. The plant body of the surviving species, unlike those of past ages, is an obscure part of our landscape, as can be seen from the above photograph of *Lycopodium clavatum,* one of the living representatives of this group of primitive plants.

DIVISION 17. PSILOPHYTA—WHISK FERNS

The sporophytes of these plants are rootless but have a rhizoidlike underground stem and an aerial portion forming the vegetative plant. This portion is branched, sometimes profusely in fossil species, and

Table 23-1
Classification of the vascular plants

	Division Psilophyta	Division Microphyllophyta	Division Arthrophyta	Division Pterophyta	Divisions Coniferophyta and Gnetophyta	Division Anthophyta
Common name	Whisk ferns	Lycopods, club mosses, and ground pine	Horsetails	Ferns	Conifers	Flowering plants
Number of living species	3	780	25	8,000	675	270,000
Leaves	Scalelike	Minute	Whorled	Foliaceous	Needlelike or foliaceous	Usually foliaceous
Strobilus	None	Large, apical, or lacking	Large, apical	Usually none	Cones	Flowers
Distinctive feature	Rootless	Leaf with single, central, unbranched vein	Jointed stems	Spores usually in sori	Cones (usually)	Flowers
Life cycle	Sporophyte dominant; gametophyte subterranean	Sporophyte dominant; gametophyte subterranean or remaining in strobilus	Sporophyte dominant; gametophyte small green plant	Sporophyte dominant; gametophyte above ground, small	Sporophyte dominant; gametophyte parasitic	Sporophyte dominant; gametophyte parasitic
Fossil record	Abundant in Silurian and Devonian periods; now nearly extinct	Abundant during Carboniferous and Permian periods; now only a few species	Abundant during Carboniferous and Permian periods; now nearly extinct	First appeared in Carboniferous period; abundant since then, some more abundant in Cenozoic period	First appeared in Permian period; dominant in Mesozoic period; now diminishing	First known in Jurassic period and rapidly became dominant plants

either it is without leaves or the leaves are small, veinless, scalelike structures. In fossil species the sporangia are borne on the ends of branches or as lateral appendages, the latter condition obtaining in the extant species.

Most of the species are extinct, and about twenty-five genera are known as fossils, which first appeared during the Middle Silurian period but rapidly disappeared after the Devonian period. As far as the fossil record shows, these were the first vascular plants. Only 3 rather rare species have survived and are placed in two genera, of which *Psilotum* sp. (Fig. 23-1) is the more common. Their present-day distribution is

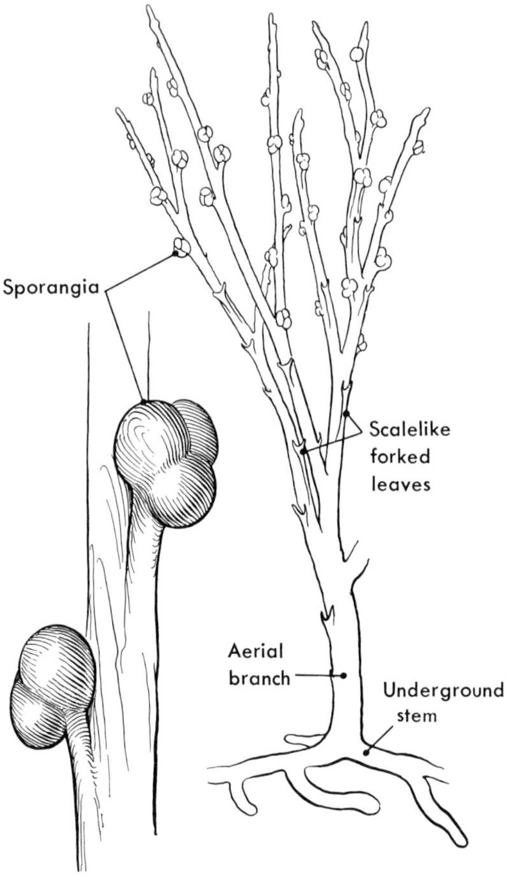

Fig. 23-1
Psilotum sp., an example of a rare living fossil.

confined to tropical and subtropical and southern regions. In the United States one species can be found along the coast from Florida through South Carolina, where it is confined to sandy areas. Outside the United States the other species are epiphytes of tropical trees.

DIVISION 18. MICROPHYLLOPHYTA—LYCOPODS AND ALLIES

The sporophyte of these plants is differentiated into true stems, leaves, and roots. The leaves, with a single, central, unbranched vein, are usually arranged spirally on the stem, but some may be opposite or in whorls. The sporangia of these and all remaining tracheophytes are located at the base of a special leaf, the sporophyll. In most cases the sporophylls are arranged in special groups, forming a strobilus.

Most species are known as fossils. They were once the dominant plants, even reaching treelike proportions. Their dead bodies contributed to our modern coal deposits. Today, no member of the group is a very familiar part of our flora.

Three distinctive orders are placed in the division.

Order Lycopodiales. The sporophyte is herbaceous, with short stems that often run along the ground. The leaves are small and evergreen and give the plant the appearance of a small pine—hence its common name ground pine. The sporophylls are borne in strobili, and in several species are on the end of stalks resembling clubs, which has given rise to the common name club moss. The gametophytes are either subterranean, or nearly so, and usually saprophytic or they may live symbiotically with a fungus. They are seldom seen.

Members of this order differ from those of the other two orders in having only one kind of spore instead of two. The spores are shed and may be dormant for many years, which eventually assures a new crop of plants even if unfavorable conditions temporarily exterminate the entire local population.

The order contains only two extant genera. One, *Lycopodium* (Fig. 23-2), is worldwide in distribution, including arctic regions. However, most species are tropical or subtropical. In the temperate regions it is fairly common in mountainous woodland habitats. The other genus is restricted to the Australian region.

Order Selaginellales. This order comprises the most abundant group in the class and is often confused with mosses. The sporophyte has leaves that are usually borne in four rows, two rows of small ones and two rows of large ones. At the apices of some branches are poorly developed strobili. Two types of spores are produced in the sporangia of these plants. The upper portion of the strobilus produces microspores, whereas the lower portion produces megaspores. Further details of the life cycle are described in Chapter 14 (life cycle 12). The separate male and female gametophytes foreshadow the condition found in the seed plants.

There is only one genus with 700 extant species, worldwide in distribution. Most species are tropical, and many of them grow in damp forests; however, several are well adapted for life in extremely arid areas. The resurrection plant is an example

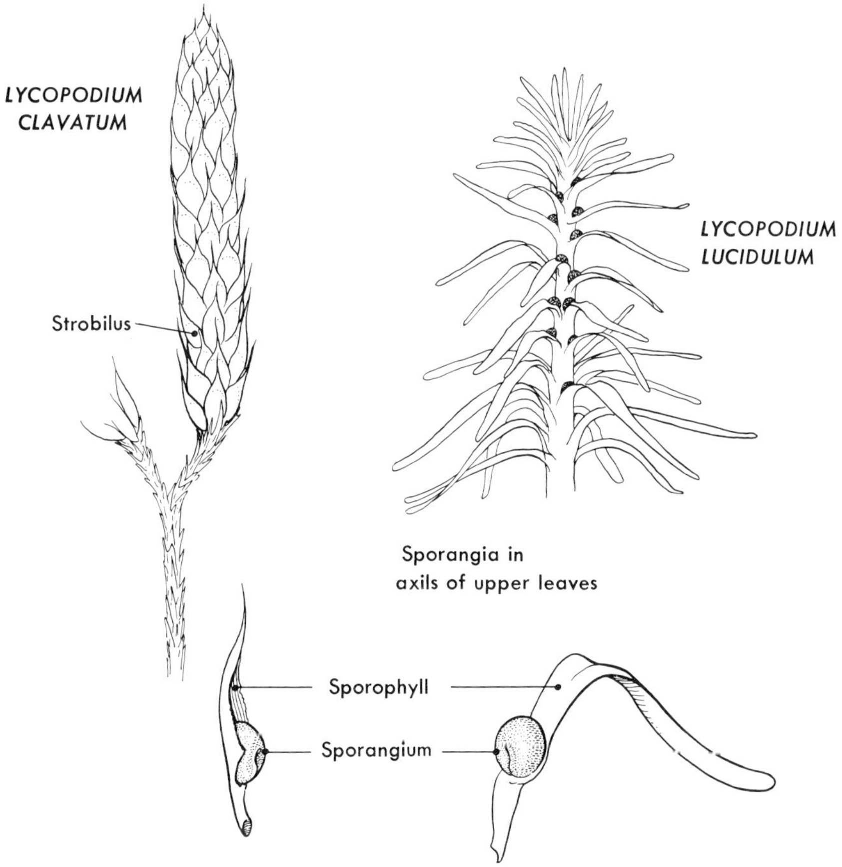

Fig. 23-2
Illustrations of two different species of *Lycopodium*, commonly called club mosses or ground pines.

of one of these. This plant gets its name from its ability to remain dormant, rolled into a loose ball and entirely free of attachment to the soil. When it rains or the plant is placed on moist soil, it opens very soon and turns green. Vigorous growth takes place, and spores are produced. As soon as the water supply is used up, it returns to its dormant condition.

Order Isoetales. Small, herbaceous, rushlike sporophytes, the quillworts, form the remaining extant order of this class. The plant (Fig. 23-3) in some ways resembles garlic because of the arrangement of the sporophylls. The root system is scanty, with dichotomous branching. The stem is short and thick, resembling a corm. All the leaves are sporophylls, the outer ones subtending the megasporangia and the inner ones, the microsporangia. Both kinds are located at the base of the leaves; therefore the plants lack strobili. In the gametophytes the sexes are separate. The male plant produces multiciliated sperm.

The 65 species grow mostly in wet places, completely or partly submerged in ponds and lakes; a few species are found in dry areas. They are found in the temperate regions of the world.

DIVISION 19. ARTHROPHYTA—HORSETAILS

The modern horsetails are the last of a great group of plants with treelike proportions that once inhabited the earth. Some of them grew to heights of 60 to 90 feet. The sporophyte is differentiated into roots, stems, and small scalelike leaves. The leaves are borne in transverse whorls upon the stem. The unique stems are jointed,

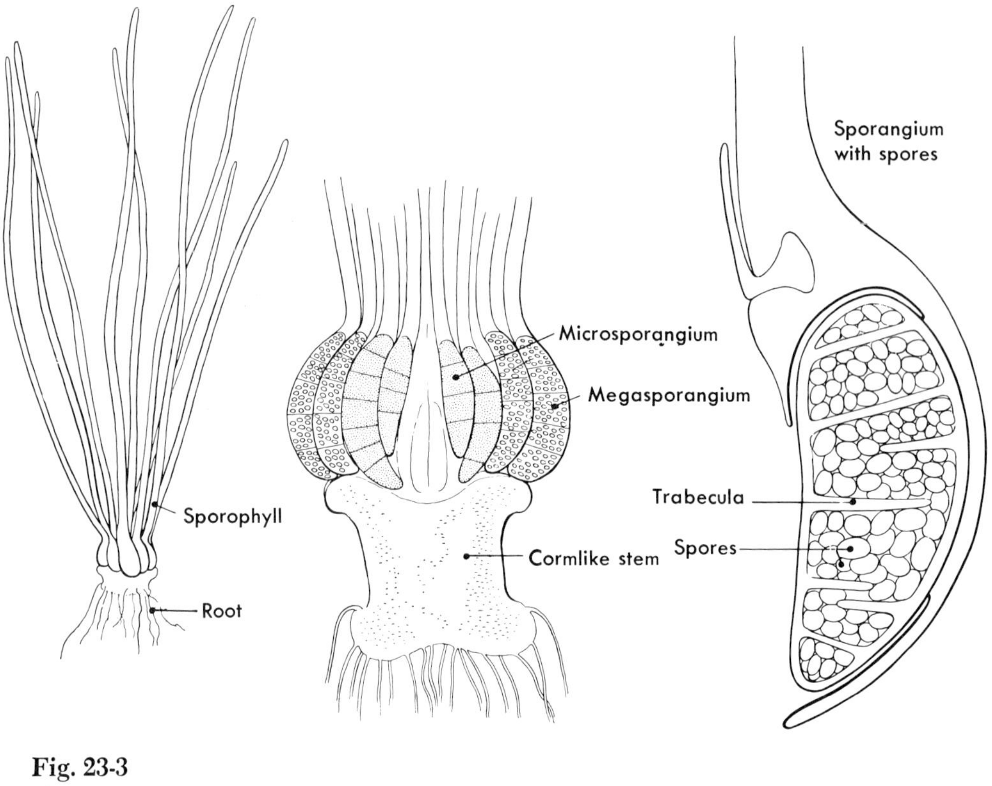

Fig. 23-3
The quillwort (*Isoetes* sp.) develops from a cormlike stem.

with distinct nodes, are hollow, and are longitudinally ribbed or furrowed. The cell walls of the epidermis contain silicon crystals, giving them a rough appearance. The sporangia are borne on special stems called sporangiophores, which are grouped into a compact strobilus. These branches are short lived and colorless. Other branches are green with longer aerial branches that give the vegetative part of the plant the appearance of asparagus. These may appear to be separate plants. However, they are connected by underground stems, the rhizomes.

The gametophyte develops separately as a small, green thallus. The life cycle is described in Chapter 14 (life cycle 13).

The horsetails are common in swamps, shallow ponds, meadows, damp shaded woods, and dry sandy embankments such as those found along railroad tracks. They grow from about 1 to 3 feet high. Only about 25 species have survived to modern times. For this reason they are sometimes called "living fossils."

The presence of silicon crystals in the stem has earned them the name "scouring rushes" because they were used by early settlers to scour pots and pans. The abrasive nature of the silicon is still used today by campers for the same purpose.

The fertile branches appear first in the spring with a strobilus. Spores have four spiral bands (elaters) that are coiled when moist but straighten out upon drying, aiding in catapulting the spores. The angiophores wilt and die after sporu but the vegetative branches remain thro out the growing season.

A spore germinates and develops into a small, green, somewhat branched prothallium, the gametophyte. This produces antheridia at the upper end and archegonia at the base of the short branches. The gametangia are produced at different times of the year, thus preventing self-fertilization. The sperm fertilize one of the eggs during wet weather, and a new sporophyte develops.

The division is divided into three orders, two of which are extinct. The other order, Equisetales, contains two families, one of which is extinct. All living species belong to the family Equisetaceae, the genus *Equisetum,* and inhabit all regions of the world except the Australian region.

QUESTIONS TO CONSIDER

1. What is meant by a "living fossil"?
2. List the new developments in the group of plants treated in this chapter.
3. What are the most likely ancestors of the vascular plants?
4. Referring back to the paleontological discussion of this group of primitive vascular plants, why are they no longer dominant?

DISCUSSION FOR LEARNING

1. Considering the answers to the above questions and the facts given in the paleontological chapter, discuss the reasons why the horsetails and lycopods were once a dominant group of plants.

Chapter 24

Ferns

Photograph by Don Normark, University of Washington Arboretum, Seattle, Wash.; from Russell, N. H. 1958. An introduction to the plant kingdom. The C. V. Mosby Co., St. Louis.

The division Pterophyta is characterized by the presence of true vascular tissue at the organ-grade level of organization. These plants are vascular plants one step more highly developed than those discussed in Chapter 23, but they are not yet seed plants; therefore they are often referred to as vascular cryptogams. They are typical, in our modern flora, of cool woodland glades such as the one in the photograph above, or, in tropical regions they may reach treelike proportions. They produce widely disseminated spores. In past geological ages the cryptogamic vascular plants were dominant, but now these have been replaced by the seed plants.

These cryptogamic vascular plants represent a separate division of extant plants, but they are the only remains of at least five ancient groups. The ancient origin of this group, known as far back as the Silurian period, makes it difficult to be sure whether they developed directly from the Psilophyta or whether they had common ancestry with the bryophytes. The former view is more generally accepted at the present time, but the lack of fossil evidence to bridge the gap between the two groups prevents any positive statements on the matter.

Ferns and their allies are generally distributed over all areas of the earth. The more primitive classes have more local distribution and now are of little importance.

A few species of ferns have some medicinal properties, but these are no longer

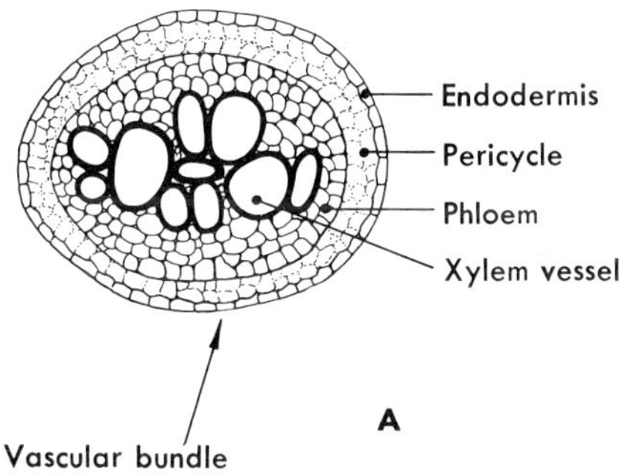

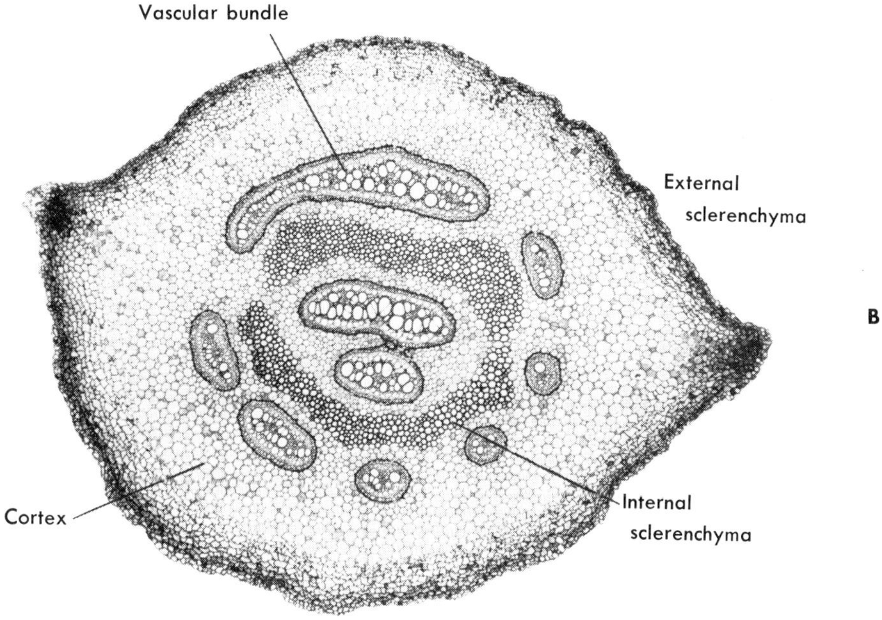

Fig. 24-1
Fern leaf and stem. **A**, Cross section of vascular bundle. **B**, Photograph of a cross section of a rhizome (×30). (**B** courtesy George H. Conant, Triarch Products, Ripon, Wis.)

used except locally by isolated groups. Several species are cultivated as ornamentals.

The development of true vascular tissue (Fig. 24-1) in the pteridophytes, as we have seen, made possible not only their greater size but also their firmer establishment of a terrestrial life. During the early geological periods when the ferns and related plants were dominant, conditions on the earth were more or less uniform, with an abundance of water on the land surface. One of the chief advantages of a vascular system is that the dead xylem gives rigidity to the plant as a skeleton does to an animal, and thus enables plants to grow to a larger size. This advantage aided them in their competition for space by exposing to the sun a greater area of leaf surface per individual. However, as time passed the land became dryer and the weather more severe. This resulted in a greater need for a well-developed root system to obtain water; otherwise, the plants would be forced back to swamps and shore areas. Only through this vascular system are they able to maintain their large size on land. The mosses and liverworts could not meet the competition on a drying earth.

Despite the addition of the vascular system, the pteridophytes are still dependent upon a fairly moist habitat because their sperm must swim to the egg for fertilization. Therefore the sexual part of their life cycle is restricted to the same habitat and condition as those of mosses. Thus in the life cycle there is a need for moist conditions during the gametophyte stage and dry conditions for spore dispersal during the fruiting stage of the sporophyte generation.

DIVISION 20. PTEROPHYTA—FERNS

The ferns have the highest development of alternate generations in which the two plants, the mature sporophyte and the gametophyte, are separate. Ferns are characterized by their well-developed sporophyte stage and prothallium. The sporophyte is differentiated into a rhizome, roots, and large leaves (Fig. 24-1), called fronds. A well-developed vascular system is present and consists of a vascular bundle with endomeris, pericycle, xylem, and phloem. Unlike other plants, the fronds of ferns exhibit circinate vernation (Fig. 24-2); that is, they develop by unrolling from the base to the apex. The sporangia of these plants are borne in clusters (sori) on the underside of the leaves (Fig. 24-3), on the leaf margin, or, rarely, in separate heads or in a sporocarp, a special leaf containing only the sori. Fronds bearing sporangia are called sporophylls (Fig. 24-3).

The prothallium is platelike, ribbonlike, or heart shaped and eventually becomes

Fig. 24-2
Young fern fronds unrolling as they grow in the spring, showing circinate vernation.

Fig. 24-3
Frond of a broad leaf fern in a rain forest in Ecuador. Note the rows of sori.

independent of the sporophyte. These prothallia are seldom observed but may be cultured from mature spores. When they can be observed, either or both kinds of sex organs are identifiable.

The ferns are the largest group of pteridophytes, having over 8,000 described species. The majority of the genera are known from tropical America and the Indo-Malayan–South Pacific area.

Classification. The two classes comprising the division Pterophyta are very distinctive. The first, the class Eusporangiopsida, develops massive sporangia partially or entirely embedded on the ventral surface of the leaf, with an indefinitely large number of spores surrounded by a many-layered sporangia wall. The other class, Leptosporangiopsida have simple layered walls and produce a definite number of spores, usually multiples of two, frequently forty-eight to sixty-four.

Only about 110 species occur in temperate North America, representing four orders and twelve families. The plants are widespread in their ecological distribution, but the majority of the species prefer damp shady places such as deep woods. A few species are cultivated in shady gardens. They also grow on rocks, cliffs, open fields, and especially land recently cut over, and a few species are aquatic. In tropical rain forests some species grow to a height of 50 feet, with fronds 12 to 15 feet long (Fig. 24-4). Many others grow as epiphytes in these forests and cover the trunks and branches of trees.

The ferns have a long fossil history, first appearing in the Devonian period. Of the nineteen or more known families, only five are extinct. Four families are unknown as fossils, and the remaining ten families have both fossil and modern representatives. Most of the families appear to have developed during the Mesozoic Era so that they are not a principal element in the formation of coal, as is sometimes thought. The group is probably as abundant today as in past ages. Their spore dissemination has enabled them to become widely distributed, although they probably do not occupy as much area now as they did when habitat conditions were more uniform.

Fig. 24-4
Giant tropical ferns with fronds over 10 feet high in a rain forest in Ecuador.

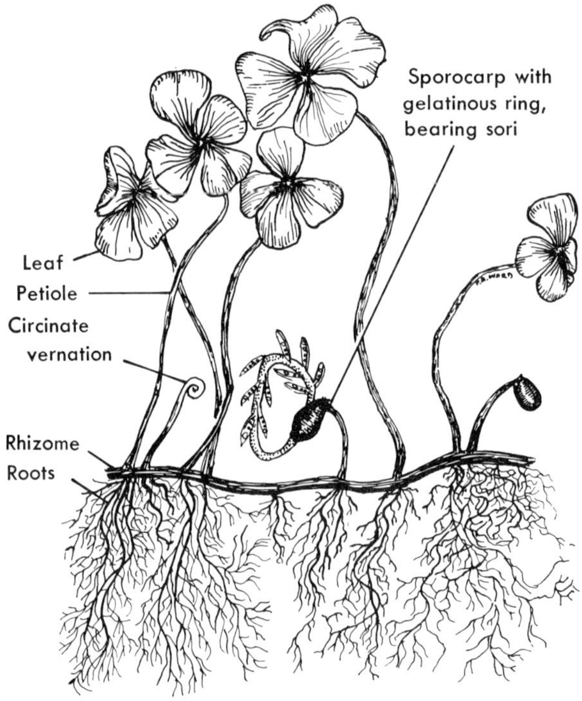

Fig. 24-5
The leaves of *Marsilea* sp. (heterosporous water fern) resemble those of clover.

Many botanists classify the ferns as closely related to the gymnosperms and angiosperms. The reason for this is that leaf gaps are present in the stele just above the departure of the branch veins. This morphological feature is shared with the conifers and flowering plants and is a fundamental indication of their close relationship to each other. Ferns are classified into family, genus, and species according to their frond shape and the position, shape, and arrangement of sori on the fronds.

One primitive family, the grape fern family, has a few solitary leaves. Part of the frond is nonfoliaceous and bears the sporangia, which are not grouped into sori. The blade of the frond may be simple, lobed, or variously compound.

The osmunda ferns, including the common cinnamon fern, have some fronds that are nonfoliaceous and are interrupted in the middle by the special pinnae bearing the sori.

The curly grass or climbing ferns may not be recognized as ferns because of their vinelike appearance and small, grasslike leaves or leaves resembling those of flowering plants. The sporangia are clustered into fertile fronds that resemble strobili, a feature of the several water fern families such as Marsileaceae (Fig. 24-5). The remaining several families have the more typical fern fronds.

The life cycle of ferns is given in detail in Chapter 14 (life cycle 14, *Polytrichium* sp.).

QUESTIONS TO CONSIDER

1. List the new developments found in the ferns.
2. What features lead to the subjection of these plants by the seed plants, that is, what did the fern fail to develop?

DISCUSSION FOR LEARNING

1. Since ferns are still a widely distributed group of hardy plants, discuss the possibility of this group again becoming dominant on the earth's surface.

Chapter 25

Primitive seed plants

The seed-bearing plants, sometimes referred to as phanerogams, are the peak of development of the plant kingdom. Spores provided the best means of dispersal during the ages of more or less uniform climate, but, as we have already demonstrated, seasonal changes brought about a different mode of life, that of producing seeds, a phenomenon shared by the last five divisions of plants. Each of these divisions has an ancient origin, and it is not certain whether they developed the seed-producing structures before they diversified, or separately after they branched off from their fernlike ancestors. It is certain, however, that the seed plants developed while the cryptogams were dominant and that they remained obscure, poorly adapted plants for many millions of years. One such plant, the ginkgo, pictured above, has survived to the present day and represents still another of our "living fossil" plants.

The seed plants are the most familiar and dominant forms of plant life on the earth, far outnumbering animal life in numbers of individuals. Everywhere one looks on land, except in the most barren deserts or arctic wastes, these plants are present.

They are characterized by dominant sporophytes, with the gametophytes reduced to small parasitic organisms enclosed in sporophyllic tissue, by the pollen, the microgametophyte, and by a small egg with simple accessory tissue, the megagametophyte. Typical seed plants have flowers or cones, both staminate and carpellate. However, not all seed plants are flowering plants or conifers, and, generally speaking, only the more primitive are woody. In many respects the life cycle of *Selaginella* sp. (life cycle 12) foreshadows the seed habit. In seed plants a seed is produced as the result of the union of a sperm nucleus with the egg nucleus in a very complicated manner, as explained in the section on pollination in Chapter 12, and in Chapter 14.

Seed plants have two principal means of seed development. The seeds of the primitive groups are exposed on the axis of a scale and are considered to have naked seeds, to which the term gymnosperm is applied. These seeds are produced in a strobilus, or in a cone (the conifers). The other group of seed plants, the flowering plants, have their seeds, as we have seen in previous discussions, enclosed by a covering produced by the sporophylls. The name, phanerogam means visible reproductive organs, which these plants have, as opposed to the spore-producing plants, the cryptogams, with their hidden reproductive organs.

Classification. Four of the five divisions of seed plants are shown in Table 25-1. The flowering plants, division Anthophyta, completes the list. These groups are the subject of the remainder of this chapter and of Chapters 26 and 27.

DIVISION 21. CYCADOPHYTA—CYCADS

The cycads or sago palms are a tropical or subtropical group of about 100 species. They are palmlike in appearance, with woody, thick, tuberous stems that are mostly subterranean. Some of them grow to heights of 50 feet, but they are very slow growing. Specimens are known to be more than 1,000 years old. Some produce very large cones, which may weigh nearly a hundred pounds, and others have individual ovules the size of large eggs. These plants are of interest to the botanist because of their retention of the primitive method of fertilization of the egg by motile sperm after they are carried to the female by windblown pollen.

Other than their cultivation as ornamentals, the species have little economic value. Some of the species produce edible seeds marketed only locally.

Table 25-1
Comparison of extant divisions of nonflowering seed plants

Division	Cycadophyta	Ginkgophyta	Coniferophyta	Gnetophyta
Habit	Palmlike	Tree with broad leaves	Trees or shrubs with needles	Prostrate forms with broad leaves or needles
Distribution	Tropical and subtropical only	Temperate regions under cultivation; native only in central Asia	Worldwide	Some desert regions of Old and New Worlds
No. of species	100±	1	520±	75±
Distinctive features	Sperm cell ciliate, motile, sporangia in strobili; leaves pinnately compound	Sperm cells ciliate; sporangia solitary or few, not in strobili; leaves simple	Sperm cells nonmotile; staminate strobili simple; stem with resin canals	Sperm cells nonmotile; staminate strobili compound; stems without resin canals

DIVISION 22. GINKGOPHYTA—GINKGOS

Only one extant species of this otherwise fossil group is known. This species has survived because it was kept in cultivation by Chinese monks. The group was known from fossils before it was discovered to have a living representative, the maidenhair tree, *Ginkgo biloba,* pictured at the beginning of this chapter. It was introduced into Europe in 1730 and into the United States in 1784. Since then it has been widely cultivated as a shade tree because it can tolerate smoke and a limited water supply. However, because the female ginkgo seed is foul smelling, only male trees are propagated, and these from cuttings, so that the undesirable trees may be avoided.

The tree grows to a height of 60 to 90 feet, with broad, fan-shaped, bilobed leaves quite unlike the more familiar representatives of the class (pine, etc.). The leaves are arranged alternately on the branches, which gives the tree a distinctive symmetry that can be identified at some distance. Fertilization by a small motile sperm attests to its ancient origin. Other members of this order were abundant during the Mesozoic period.

QUESTIONS TO CONSIDER

1. Define a "seed" plant.
2. What factors might have caused the extinction of the ginkgo tree if it had been left alone, and yet today, it is considered a hardy city tree?

DISCUSSION FOR LEARNING

1. Discuss the possibilities of discovering still more "living fossils," and tell which groups might have living remains.

Chapter 26

Cone-bearing seed plants and related plants

The conifers, or true cone-bearing plants, are widely distributed, forming a dominant group, generally cold adapted, that is, adapted for life in cool regions. They live in vast forests of single species in the northern hemisphere, such as the juniper forest shown in the photograph above. This single tree is one of a pure stand that covers the hills and mountains in the background. At higher altitudes in the mountains, pines, then spruce, take over and become dominant in place of the junipers.

Characteristics. The sporophyte is the dominant vegetative plant, producing two types of spores: the microspore, which develops into a pollen grain, and the megaspore, which develops into the female gametophyte. True seeds are produced. The vegetative plant has true roots, stems, and leaves. Most members of the class are woody plants, with an active cambium developing both secondary xylem (or wood) and phloem. All species are perennial. Most of these plants are evergreen; that is, they have green leaves throughout the year, with no particular time for leaf dropping. However, some are deciduous. Fertilization takes place by the transmission of airborne pollen in the same manner as spore dispersal in the cryptogams, thus eliminating the

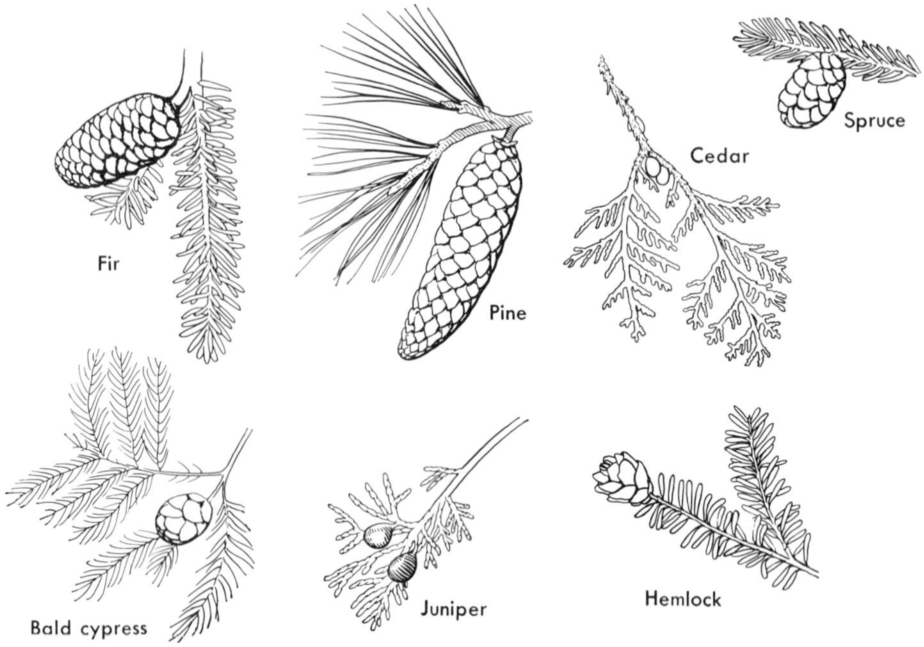

Fig. 26-1
Leaf structure and cones of typical conifers.

Fig. 26-2
Cypress and other conifers along the banks of a river in Louisiana.

necessity of water for fertilization, as in all other groups of plants. Like both the higher vertebrates, with their membrane-protected eggs, and the terrestrial arthropods, these plants are at last free of water as a necessity for reproduction (life cycle 15, *Pinus* sp.).

DIVISION 23. CONIFEROPHYTA—CONIFERS

This division, the largest of the non-flowering seed plants, includes the familiar pines, spruce, hemlock, juniper, cedar, yew, fir, redwood, and cypress, each with berries or distinctive cones (Fig. 26-1); all have representatives used extensively as lumber or grown for ornamentals. With few exceptions the leaves are needlelike and are borne on branchlets or in bundles. The 520 or so species are trees or woody shrubs that usually produce cones. A few, such as the juniper, produce a berrylike fruit. The sperm are nonmotile and are dispersed in pollen.

Seven families are known; of these four are native to the United States, and the others are grown as cultivated ornamentals.

Perhaps the group is best known to the public as Christmas trees, for which several species are used. On the other hand, the giant redwood tree ranks as the largest and oldest living member of the plant kingdom. These trees tower to a height of 360 feet, more than half the height of the Washington Monument, and some of them living today were seedlings when the Babylonian kings were ruling 4,000 years ago! Of equal but less spectacular interest are the knees of bald cypress trees growing with their roots submerged in water (Fig. 26-2).

Among the most valuable of the lumber trees are the western Douglas fir, the ponderosa pine, and the western white pine. In the east the southern white pine is used for wood pulp for paper and plastics, as well as for a source of turpentine and resins.

Only a few of the conifers are deciduous and are, therefore, not evergreen, such as the larch or tamarack and the bald cypress.

DIVISION 24. GNETOPHYTA—GNETES

This final division before the flowering plants includes a heterogeneous assemblage of about 75 species of plants that may well be representatives of more than one order. They are very different from the rest of the class. However, they all have naked seeds like the rest.

Included in this group is one of the most unusual plants in the world, *Welwitschia mirabilis* (Fig. 26-3) of the extremely dry deserts of southwest Africa. It consists of a huge tuberous root with a crown just above the surface. It produces two strap-shaped opposite leaves, the only ones produced during the hundred years or more of its life. These leaves continue to grow at the base and are beaten and frayed by the wind at the apex. A few members of this division also grow in the United States in the deserts of the southwest, for example, Mormon tea and joint fir. These are members of the genus *Ephedra* (Fig. 26-4), which produce cones and yield the medicinal alkaloid ephedrine formerly used to make nose drops. The synthetic product Neo-Synephrine has replaced the natural product. The plant remains an interesting member of a desert community where few persons would expect to see a gymnosperm. It may be seen at White Sands, New Mexico, as one of the few plants growing on the torrid, ever-shifting, pure gypsum mounds.

New features developed by the conifers and their relatives. The conifers introduced several features new to the kingdom Plantae. Undoubtedly this accounts for the great abundance of flowering plants that later developed by elaboration and refinement of these attributes. The production of two kinds of cones (Fig. 26-5) is a further development of the production of two types of spores, as seen in *Selaginella* sp. The permanent retention of the megaspore and the megagametophyte, with the megasporangium, and the development of a covering integument about each megasporangium permit seed development. The

Fig. 26-3
Unusual gymnosperm *Welwitschia mirabilis* found in a mist desert in western Africa. Photograph of a museum habitat group. (Courtesy Field Museum of Natural History, Chicago, Ill.; from Russell, N. H. 1958. An introduction to the plant kingdom. The C. V. Mosby Co., St. Louis.)

Fig. 26-4
The ephedra plant, common in the white sand desert of New Mexico and elsewhere in western United States, is a conifer.

A

B

Fig. 26-5
Typical pine cones (strobili). **A**, Staminate. **B**, Carpellate. (U.S. Forest Service.)

protection of the gametophyte by the sporophyte allows the production of fewer spores. The development of the new sporophyte within the old not only decreases the hazards of a double life cycle but also, by the addition of nutrient material and partial growth of the embryo, greatly increases the chances of survival once it has left the protection of the parent. Further simplification of the microgametophyte as it becomes a pollen grain allows it to function in place of the original spore. The process of pollination by the pollen grain and the formation of the pollen tube are permitted. These are necessary adjuncts to the new revised life cycle. The result is complete parasitism of both gametophytes upon the sporophyte and the establishment of the seed habit.

DISCUSSION FOR LEARNING

1. Carefully consider the statements just made in the "new features" section, and discuss.

Chapter 27

Flowering plants

The flowering plants are distinguished from the conifers fundamentally on the basis of seed formation. The presence of a complete flower is not an essential feature of the division. The seeds of the conifers, as described in Chapter 24, are naked, that is, not enclosed within a fruiting structure of ovarian origin because the ovules are borne directly on the sporophyll as a part of the sporangium. The ovules of the flowering plants are located within an ovary, which also takes part in fruit formation. The seeds, therefore, are never exposed during their formation.

The flowers, seeds, and general habitus of these plants, along with their many vegetative features are used for the classification of the orders, families, and species. The detailed morphology of these parts has been given in several previous chapters. This chapter will review the great diversity of form, such as we see in the beautiful *Datura* sp. pictured above.

The development of a flower instead of a cone is the most radical change from the conifer to the flowering plants. The advantages of this method of reproduction are primarily gained by the accessory structures, the flower parts, used to facilitate pollination and by the development of fruit parts to aid in seed dissemination. Additional protection is afforded the ovule by this kind of structure.

The economic importance of the group is of course obvious. The aesthetic value of flowers and other ornamentals is so much a part of our lives that it is taken for granted. The mind of man seems to be capable of understanding and of synthesis only through the organization of sensory data sorted to kinds, from which generalizations can be made. We take pleasure in this kind of exercise, and without entering further into more philosphical and psychological aspects it seems possible to state that any kind of organization will broaden both the pleasure and appreciation of any study. Therefore, since we know that these flowering plants are a part of our life, the following classification should help toward a fuller comprehension of this vast group of organisms.

We have not attempted a complete classification of the flowering plants, but rather, a classification of those plants most familiar to us, including all those used as examples in this text. The numbering system refers to the number of the order and the number of the family as listed and discussed in this chapter. Each common name and each scientific group name mentioned in the text is indexed by this number (the number preceding the family name is included in the index to plant names). In this manner one can easily locate in the classification any plant in the list. Even though no characteristics are given, by association of familiar plants, one will be able to note similarities and differences. We believe that a knowledge of the names of the flowering plants so often seen, used, and even eaten, offers recreation and relaxation to those who may find this of interest. By knowing the family grouping of common plants, one can further recognize plants new to them. With the aid of some of the many manuals now available, further identification of other plants can be made.

DIVISION 25. ANTHOPHYTA—FLOWERING PLANTS

The division Anthophyta contains one class, the Angiospermae, divided into two subclasses, the Dicotyledoneae and the Monocotyledoneae. The morphological characteristics of these two groups have been given in detail in previous chapters. The bean plant is often used as an example of the dicots, and the lily plant, while it has certain exceptional features, is representative of the monocots.

Several major classification systems have been proposed over the past two hundred years for the flowering plants, but a review of these would be beyond the scope of this text. No two books and no two monographs on these plants follow exactly the same system. There have been major revisions in the arrangement of plants in a classification system during the past few years. As taxonomists accumulate and assimilate more data, new evidence brings about changes in classification; therefore more changes can be expected. Modern classification merely reflects, by a linear listing, what is thought to be the phylogenetic or evolutionary development of plants. All such lists are limited by having two dimensions instead of three. Most of all, the student should not think of any list of extant plants as a list of ancestors and descendants: no modern plant taxon is the ancestor of another taxon.

One of the most definitive recent classifications and one that is readily available to the student is that outlined by Thorne (1968), which makes use of simple characteristics of the flower parts and gives a complete list and summary of the flowering plant orders and families of the world. His arrangement of families is followed in this text.

Thorne proposes that the magnolias are among the most primitive of extant species. The flowering plants have been analyzed using these as a starting point, and then a list of twenty-five superorders, fifty-four orders, and 312 families is given.

Certain features of the flowering plants are believed to be more primitive than others. Some of these trends are as follows:

1. Flower parts spirally arranged are primitive; those with cyclic or whorled parts are more advanced.

2. Flowers with many parts, the number of which may vary with the individual, are more primitive than those flowers with a few parts of a definite number, for example, five petals and five stamens.

3. Flowers with separate parts, for example, the individual petals free at the base, are more primitive than those with fused parts, for example, the petals fused to form a calyx cup.

4. Fusion of one type of flower part with another, for example, stamens fused to petals, is more advanced than those with free stamens.

5. Hypogynous flowers are more primitive than epigynous ones.

6. Plants with solitary flowers are more primitive than those with numerous flowers grouped into an inflorescence.

7. Monoecious plants are more primitive than dioecious ones.

8. Complete flowers are more primitive than incomplete ones.

9. Insect-pollinated flowers seem to be more primitive than wind-pollinated ones, although this is probably peculiar to certain groups and is not a generalization.

10. Woody plants in a given group are more primitive than the herbaceous members of that group; primitive plants tend to be perennials; biennials and annuals are more advanced.

These and a few other similar characteristics, along with a wide knowledge of the flowering plants, have been used to develop a classification. This classification is supplemented by detailed studies of all aspects of the species, such as their distribution, ecology, genetics, and morphology, all of which help to produce a sound and natural classification.

Subclass Dicotyledoneae

The subclass Dicotyledoneae is the larger of the two subclasses of the class Angiospermae and until recently was thought to be the more advanced. Now, however, because of the application of the foregoing phylogenetic trends, it can be seen that the monocots are the more advanced.

The subclass is divided into twenty superorders. Each of these groups is divided into several orders, some of which are further divided into suborders. Some orders contain only one family, but most have from two to as many as twenty-four families, with a world total of 269 extant families. Of these, about 160 occur in the United States and Canada, either as native plants, or as introduced plants that have become established under natural conditions. Many more are known as fossils. Some of the families are further divided into subfamilies and they in turn, into tribes. Every genus and species of plant is accounted for by being assigned to one of these families.

The following orders contain members that are familiar to all of us because either they are common members of our environment, are grown as ornamentals, or have some product of economic importance. There is hardly a family that does not have some member well known either to the field biologist or to the gardener, but all could not be included here. Students interested should refer to the references to this chapter at the end of this text.

Order 1. Annonales. This order is the largest in number of families (twenty-four) but has relatively few species. Among the familiar representatives of this order is the magnolia family, with the magnolia and tulip trees as members. These beautiful plants are among the relatively few trees in the United States with large showy flowers, in contrast to the many tropical trees that color the rain forest landscape during their flowering periods. The nutmeg family belongs in this group. Closely related are the laurels, the avocado with its edible fruit, the bay, the leaves of which are used for flavoring and contain cyanide, and the sassafras—all members of the laurel family. Other members of the family provide us with cinnamon and camphor. Finally, the members of the pepper family are plants that produce the seasoning black pepper.

1.4	Magnolia family—Magnoliaceae	
	Tulip tree	*Liriodendron tulipifera*
	Magnolia	*Magnolia soulangeana*
	Sweet bay	*Magnolia virginiana*
1.9	Nutmeg family—Myristicaceae	
	Nutmeg (=mace)	*Myristica fragans*
1.20	Laurel family—Lauraceae	
	Camphor	*Cinnamomum camphora*
	Cinnamon	*Cinnamomum zeylanicum*
	Mountain laurel	*Kalmia latifolia*
	Avocado	*Persea americana*
	Sassafras	*Sassafras albidum*
1.23	Pepper family—Piperaceae	
	Black pepper	*Piper nigrum*

Order 2. Berberidales. The largest family of this order is the crowfoot family, with its many species of buttercups and related plants, including the peony and hellebore, from which we formerly obtained an insecticide. The barberry family is infamous because the wild bush is one of the hosts of wheat rust. However, this family also includes the familiar Japanese barberry, an ornamental that provides a thorny fence around many dooryards, and the May apple, a common, eastern, spring wild flower. The poppy family is one of the best known of the six families in this order because of the species that yields the drugs opium and morphine. Plants of this family are grown widely as ornamentals also. Among the many wild species are bloodroot, a common, eastern, spring wild flower, and the prickly poppy of overgrazed, western arid lands. Bleeding heart, a delicate woods flower belongs to the family.

Fig. 27-1
Marsh marigold. (From Russell, N. H. 1958. An introduction to the plant kingdom. The C. V. Mosby Co., St. Louis.)

2.4	Crowfoot family—Ranunculaceae	
	Anemone	*Anemone coronaria*
	Columbine	*Aquilegia vulgaris*
	Cowslip (=marsh marigold)	*Caltha palustris* (Fig. 27-1)
	Larkspur	*Delphinium ajacis*
	Hellebore	*Helleborus niger*
	Peony	*Paeonia suffruticosa*
	Buttercup	*Ranunculus acris*
2.5	Barberry family—Berberidaceae	
	Japanese barberry	*Berberis thunbergii*
	Barberry	*Berberis vulgaris*
	May apple	*Podophyllum peltatum*
2.6	Poppy family—Papaveraceae	
	Prickly poppy	*Argemone platyceras* (Fig. 27-2)
	Bleeding heart	*Dicentra spectabilis*
	Poppy	*Papaver somniferum*
	Bloodroot	*Sanguinaria canadensis*

Order 3. Nymphaeales. The water-lily family, the most common of the two families of this small order, is widely distributed in the tropics and has long been cultivated in ornamental ponds.

3.1	Water-lily family—Nymphaeaceae	
	Common water lily	*Nymphaea odorata* (Fig. 27-3)

Order 4. Sarraceniales. Members of this order belong to one family only and are of interest because of their remarkable insect-capturing structures. Species of the pitcher-plant family are native to the bogs of the United States. The leaves of these species

Fig. 27-2
Prickly poppy.

Fig. 27-3
Large water lily. (From Russell, N. H. 1958. An introduction to the plant kingdom. The C. V. Mosby Co., St. Louis.)

are rolled into a fused tube partly filled with water. A nectarlike substance is secreted that attracts insects. They are trapped and die inside. Either bacterial action or enzymes cause their decay and digestion, after which they are absorbed by the plant to supplement the nitrogen-poor soil in which it grows.

4.1	Pitcher-plant family—Sarraceniaceae	
	Pitcher plant	*Sarracenia purpurea*

Order 5. Theales. Most members of the twenty-one families of this order are tropical, but two well-known plants of the tea family include the camellia flower and the tea plant, the leaves of which are used in making tea. The holly family and the Saint-John's-wort family are included here also.

5.5	Tea family—Theaceae	
	Camellia	*Camellia japonica*
	Tea	*Thea sinensis*
5.6	Holly family—Aquifoliaceae	
	American holly	*Ilex opaca*
5.19	Saint-John's-wort family—Hypericaceae	
	Mangosteen	*Garcinia mangostana*
	Saint-John's-wort	*Hypericum perforatum*

Order 7. Ericales. The heath family is the principal family of the three included in this order. It includes a great variety of plants, many of which are typical of the arctic and subarctic. Fruits such as huckleberries, cranberries, blueberries, and others are members of this family. Wintergreen used as a source of oil of wintergreen, is likewise a member, along with the remarkable Indian pipe, a flowering plant saprophytic on decaying vegetation such as a log or stump. Indian pipe is a plant that lacks chlorophyll and may be mistaken for a fungus unless it is examined closely.

Fig. 27-4
Indian pipe. (From Russell, N. H. 1958. An introduction to the plant kingdom. The C. V. Mosby Co., St. Louis.)

7.1	Heath family—Ericaceae	
	Wintergreen	*Gaultheria procumbens*
	Huckleberry	*Gaylussacia baccata*
	Indian pipe	*Monotropa uniflora* (Fig. 27-4)
	Azalea	*Rhododendron calawbiense*
	Rhodora	*Rhododendron canadense*
	Blueberry	*Vaccinium corymbosum*
	Cranberry	*Vaccinium macrocarpon*

Order 8. Ebenales. This small order with five families includes the famous ebony tree and the fruit persimmon.

8.1	Ebony family—Ebenaceae	
	Ebony	*Diospyros ebenum*
	Persimmon	*Diospyros virginiana*

Order 9. Primulales. This order contains the primrose family, several species of which are common garden flowers, and three other families.

9.3	Primrose family—Primulaceae	
	Fairy primrose	*Primula malacoides*

Order 12. Cistales. Sixteen families form this moderate size order. The begonia family contains several ornamentals. The violets are also familiar plants in this order, as well as the gourd family, an important family of vines, that includes squash, pumpkins, cucumbers, and gourds of all kinds as common members.

12.5	Violet family—Violaceae	
	Violet	*Viola odorata* (Fig. 27-5)
	Pansy	*Viola tricolor* var. *hortensis*

12.12	Rockrose family—Caricaceae	
	Papaya (=pawpaw)	*Carica papaya*
12.13	Gourd family—Cucurbitaceae	
	Watermelon	*Citrullus vulgaris*
	Cantaloupe	*Cucumis melo*
	Cucumber	*Cucumis sativus*
	Gourd	*Cucurbita argyrosperma*
	Squash	*Cucurbita maxima*
	Pumpkin	*Cucurbita pepo*
	Squirting cucumber	*Ecballium elaterium*
12.14	Begonia family—Begoniaceae	
	Tuberous begonia	*Begonia tuberhybrida*

Order 13. Salicales. All common members of this order are trees that bear their flowers in catkins, a pendulous string of flowers, the most familiar of which are the pussy willow, willow, poplar, aspen, and cottonwood.

13.1	Willow family—Salicaceae	
	Cottonwood	*Populus deltoides*
	Lombardy poplar	*Populus nigra* var. *italica*
	Quaking aspen	*Populus tremuloides* (Fig. 27-6)
	Pussy willow	*Salix discolor*

Order 15. Capparidales. A great many vegetables belong to the mustard family, one of the four belonging to this order. Among them are the radish, cabbage, turnip, kale, cauliflower, Brussels sprout, horseradish, broccoli, rutabaga, mustard, and kohlrabi. Many other species of the family are well-known weeds, some of which are mentioned in the Bible.

15.4	Mustard family—Brassicaceae	
	Golden tuft	*Alyssum saxatile*
	Horseradish	*Armoracia lapathifolia*
	Mustard	*Brassica campestris*
	Kohlrabi	*Brassica caulorapa*
	Rutabaga	*Brassica napobrassica*
	Kale	*Brassica oleracea* var. *acephala*
	Cauliflower	*Brassica o.* var. *botrytis*
	Cabbage	*Brassica o.* var. *capitata*
	Brussels sprout	*Brassica o.* var. *gemmifera*
	Broccoli	*Brassica o.* var. *italica*
	Turnip	*Brassica rapa*
	Radish	*Raphanus sativus*

Order 16. Malvales. The common hollyhock is an old-fashioned flower that used to adorn the backyards of many homes and is still a familiar sight, at least in rural areas. An examination of the "cheeses" formed by

Fig. 27-6
Quaking aspen.

the neat ring of tender seeds in the developing seed pod is a good mark of identification of this group, particularly the mallow family, to which the hollyhock, hibiscus, and cotton belong. A near relative, the linden family, includes a large number of tropical trees, one of which is native to North America, the important lumber tree, the basswood or linden tree. Four additional families are grouped here also.

16.1	Stercula family—Sterculiaceae	
	Cacao (=cocoa, chocolate)	*Theobroma cacao*
	Cola	*Cola acuminata*
16.4	Linden family—Tiliaceae	
	Jute	*Corchorus capsularis*
	Basswood (=linden)	*Tilia americana*
16.5	Bombax family—Bombacaceae	
	Kapok	*Ceiba pentandra*
16.6	Mallow family—Malvaceae	
	Hollyhock	*Althaea rosea*
	Cotton	*Gossypium arboreum*
	Hibiscus (=rose-of-Sharon)	*Hibiscus syriacus*
	Mallow	*Malva neglecta*

Order 17. Urticales. As the name implies, many species of this order are urticating; that is, they secrete substances that irritate the skin of mammals. However, the majestic elm, so symbolic of our peaceful village streets and now so threatened by the fungus causing Dutch elm disease that it may become extinct, belongs here also. In the same order are the mulberries, which furnish food for the silkworm moths as well as a delicious fruit, and the drugs hashish and marijuana, obtained from Indian hemp. The seeds are often included in canary bird food, and the stems also yield the hemp fiber of commerce. The nettle family contains a variety of these weedy plants. Only the three families listed here are included in the order.

17.1	Elm family—Ulmaceae	
	Elm	*Ulmus americana*

17.2	Mulberry family—Moraceae	
	Breadfruit	*Artocarpus altilis*
	Hemp (=marijuana)	*Cannabis sativa*
	Fig	*Ficus carica*
	Hop	*Humulus americanus*
	Osage orange	*Maclura pomifera*
	Mulberry	*Morus alba*
17.3	Nettle family—Urticaceae	
	Creeping charley	*Pilea nummularifolia*
	Nettles	*Urtica dioica*

Order 18. Rhamnales. This small order contains only two families, the buckthorn family with ceanothus, a mountain wild flower, and the oleaster family with the buffalo berry.

18.1	Buckthorn family—Rhamnaceae	
	New Jersey tea	*Ceanothus americanus*
	Cascara sagrada	*Rhamnus purshiana*
	Indian jujube (Chinese date)	*Zizyphus jujuba*
18.2	Oleaster family—Elaeagnaceae	
	Buffalo berry	*Shepherdia argentea*

Order 19. Euphorbiales. Seven families are assigned to this order, of which the Spurge family is one of the largest families of plants. Among the many species are such familiar plants as the poinsettia, the castor bean (which yields the dreaded and, incidently, very poisonous castor oil), the cassava (which yields tapioca), and the rubber tree of commerce. Many members of this family are plants that superficially resemble desert cacti and are often sold in drugstores and five-and-ten-cent stores as cactus. The candelilla is a leafless desert shrub (Fig. 27-7) that has been harvested for its high-quality wax used in polishing automobiles. The so-called Mexican jumping bean is a member of the family. Seeds of several species are infected by the larvae of a moth, and the jumping effect is produced by their movements. Persons who appreciate authenticity are shocked to find that even the jumping bean has been synthesized, since mechanical jumping beans are now sold in novelty shops in the United States.

Fig. 27-7
Candelilla, a plant growing in Big Bend National Park, Texas, produces a wax that has been much used for the production of candles and shoe and car polish.

19.1	Spurge family—Euphorbiaceae	
	Chicle	*Achras zapota*
	Candelilla	*Euphorbia antisyphilitica* (Fig. 27-7)
	Poinsettia	*Euphorbia pulcherrima*
	Rubber tree	*Hevea brasiliensis*
	Manchineel	*Hippomane mancinella*
	Cassava	*Manihot esculenta*
	Castor bean	*Ricinus communis*
	Mexican jumping bean*	*Sebastiania pavoniana,* also *Sapium biloculare*

*The bean "jumps" because it is infected by the larvae of a moth, *Carpocapsa saltitans,* family Tortricidae.

Order 20. Solanales. A host of familiar plants belong to this order, which contains only four families, all of which are listed here. The morning glory, phlox, and a variety of garden flowers are included. The important nightshade family, which provides us with a variety of plants such as

Fig. 27-8
Ocotillo.

the potato and tobacco mentioned previously, the eggplant, belladonna, from which atropine for dilation of the eyes for examination is obtained, the tomato, and the green or bell pepper. Two very curious plants, the ocotillo and the boogum tree are assigned to the ocotillo family. Both these plants are desert species. The former is distributed throughout the Sonoran desert region of southwestern United States and northern Mexico, and the latter, one of the world's most curious plants, is confined to a few square miles of Baja California, where it makes a weird forest of what appears to be giant carrots covered with spines and stuck into the ground upside down and twisted into grotesque shapes.

20.1	Nightshade family—Solanaceae	
	Belladonna	*Atropa belladonna*
	Pepper (bell or green)	*Capsicium frutescens* var. *grossum*
	Tomato	*Lycopersicon esculentum*
	Tobacco	*Nicotiana tabacum*
	Petunia	*Petunia hybrida*
	Chinese lantern plant	*Physalis alkekengi*
	Horse nettle	*Solanum carolinense*
	Eggplant	*Solanum melongena* var. *esculentum*
	Black nightshade	*Solanum nigrum*
	Potato	*Solanum tuberosum*

20.2	Morning-glory family—Convolvulaceae	
	Sweet potato	*Ipomoea batatas*
	Morning glory	*Ipomoea purpurea*
20.3	Phlox family—Polemoniaceae	
	Sweet William	*Phlox divaricata*
	Phlox	*Phlox drummondii*
20.4	Ocotillo family—Fouquieriaceae	
	Ocotillo	*Fouquieria splendens* (Fig. 27-8)
	Boogum tree	*Idria columnaris*

Order 21. Campanulales. The bellflower family contains a variety of small flowers, none of which have especially common species, except for the beautiful Canterbury bells. Only two other families are included in the order.

21.2	Bellflower family—Campanulaceae	
	Canterbury bells	*Campanula medium*
	Bluebell	*Campanula rotundifolia*

Order 22. Santalales. It is doubtful that many students are unaware of the tradition in connection with the members of one of the principal families of this order, the mistletoe family, the members of which are

parasitic on the branches of trees. Except for sandalwood most of the thirteen families of the order are obscure plants.

22.1	Staff-tree family—Celastraceae	
	Bittersweet	*Celastrus scandens*
22.8	Sandalwood family—Santalaceae	
	Sandalwood	*Santalum freycinetianum*
22.10	Mistletoe family—Loranthaceae	
	Mistletoe	*Phoradendron juniperinum* (Fig. 27-9)

Order 23. Oleales. The olive family belongs to this order and is one of the two member families. It may be somewhat of a surprise to learn that the familiar lilac and privet are near relatives of the olive and that the ash tree, which yields valuable wood for axe handles and similar wood products, also belongs here.

23.2	Olive family—Oleaceae	
	Forsythia	*Forsythia suspensa*
	Ash	*Fraxinus americana*
	Privet	*Ligustrum vulgare*
	Olive	*Olea europaea*
	Lilac	*Syringa vulgaris*

Order 24. Geraniales. Among the fourteen families included in this order can be found the flax plant, which yields by fermentation of the soft parts the fiber flax from which linen is manufactured. Other members include the familiar touch-me-not of eastern woodlands, the geranium, and the coca plant yielding cocaine.

24.1	Flax family—Linaceae	
	Flax (linen)	*Linum usitatissimum*
24.3	Coca family—Erythroxylaceae	
	Coca (cocaine)	*Erythroxylon coca*
24.4	Caltrop family—Zygophyllaceae	
	Creosote bush	*Larrea tridentata*
24.6	Geranium family—Geraniaceae	
	Lemon geranium	*Pelargonium crispum*
24.7	Balsam family—Balsaminaceae	
	Touch-me-not	*Impatiens biflora*

Order 25. Rutales. Most members of this order are trees or shrubs. Of the sixteen families belonging to this order, the citrus trees and the mahogany tree are the most valuable. Spices, including the myrrh mentioned in the Bible are obtained also from members of this order. The cashew family contains such poisonous plants as sumac. Maples, horse chestnut, and a variety of nuts such as walnuts, hickory, and pecan are members of the order.

Fig. 27-9
Mistletoe growing on juniper.

25.1	Rue family—Rutaceae	
	Lemon	*Citrus limon*
	Grapefruit	*Citrus paradisi*
	Orange	*Citrus sinensis*
25.2	Quassia family—Simaroubaceae	
	Tree of heaven	*Ailanthus altissima*
25.5	Mahogany family—Meliaceae	
	Mahogany	*Swietenia mahogani*
25.6	Bursera family—Burseraceae	
	Frankincense	*Boswellia carteri*
	Myrrh	*Commiphora myrrha*
25.7	Cashew family—Anacardiaceae	
	Cashew	*Anacardium occidentale*
	Mango	*Mangifera indica*
	Poison ivy	*Rhus radicans*
	Staghorn sumac	*Rhus typhina*

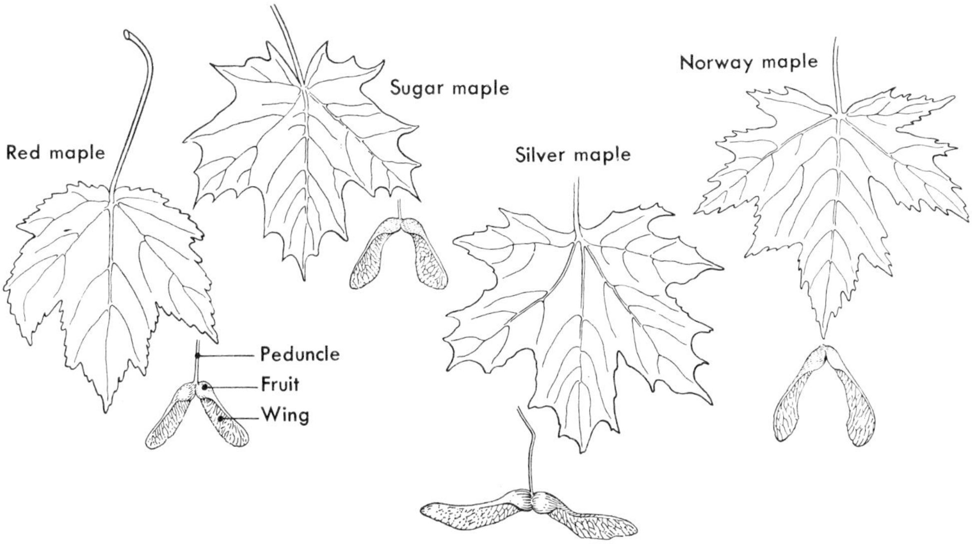

Fig. 27-10
Leaves and samaras of several species of maple.

25.12	Maple family—Aceraceae	
	Sugar maple	*Acer saccharinum* (Fig. 27-10)
25.13	Horse-chestnut family—Hippocastanacea	
	Horse chesnut (buckeye)	*Aesculus hippocastanum*
25.16	Walnut family—Juglandaceae	
	Swamp hickory	*Carya cordiformis*
	Pecan	*Carya illinoënsis*
	Walnut	*Juglans regia*

Order 26. Myricales. The interesting sweet-gale family, with strongly aromatic leaves, is the only family of this order.

26.1	Sweet-gale family—Myricaceae	
	Sweet fern	*Comptonia peregrina*
	Sweet gale (bog myrtle)	*Myrica gale*
	Bayberry	*Myrica pensylvanica* (sic)

Order 27. Leitneriales. The single family, the corkwood family, contains the corkwood, a shrub or small tree distributed throughout the southeastern states.

27.1	Corkwood family—Leitneriaceae	
	Corkwood	*Leitneria floridana*

Order 28. Chenopodiales. A variety of plants belonging to the thirteen families making up this order include such flowers as four-o'clocks, and such food plants as buckwheat, rhubarb, beets, chard, and spinach. Pigweeds, tumbleweed, a variety of other weeds, and the portulacas are included. The cactus family, in which the true cacti are found, show the diversity of this order. Although such plants as cholla, organ-pipe cactus, and saguaro are typical of the desert, the family is by no means confined to arid regions.

28.3	Four-o'clock family—Nyctaginaceae	
	Four-o'clock	*Mirabilis jalapa*
28.4	Mesembryanthemum family—Aizoaceae	
	Pigweed (in Arizona)	*Trianthema portulacastrum*
28.5	Cactus family—Cactaceae	
	Prickly-pear cactus	*Opuntia chlorotica* (Fig. 27-11)
	Cholla	*Opuntia versicolor*
	Organ-pipe cactus	*Cereus thurberi* (Fig. 27-12)
	Saguaro (giant cactus)	*Carnegiea gigantea* (= *Cereus giganteus*) (Fig. 27-13)
	Barbados gooseberry	*Peireskia aculeata*

Fig. 27-11
Prickly-pear cactus.

Fig. 27-12
Organ-pipe cactus of Mexico.

Fig. 27-13
Saguaro, giant cactus, with ocotillo in the right background and palo verde in the foreground, all characteristic plants of the Sonoran desert.

28.7	Purslane family—Portulacaceae	
	Bitterroot	*Lewisia rediviva*
	Portulaca (= Purslane)	*Portulaca oleracea* var. *sativa*
28.9	Goosefoot family—Chenopodiaceae	
	Saltbush	*Atriplex polycarpa*
	Beets	*Beta vulgaris*
	Chard (Swiss)	*Beta vulgaris* var. *cicla*
	Spinach	*Spinacia oleracea*
28.11	Amaranth family—Amaranthaceae	
	Tumbleweed	*Amaranthus albus*
	Pigweed (= lamb's quarters)	*Chenopodium album*
28.13	Buckwheat family—Polygonaceae	
	Buckwheat	*Fagopyrum esculentum*
	Bindweed	*Polygonum cilinode*
	Rhubarb	*Rheum rhaponticum*
	Dock	*Rumex crispis*

Order 30. Hamamelidales. The seven families comprising this order contain many trees and shrubs, including the witch-hazel and plane-tree families mentioned here.

30.6	Witch-hazel family—Hamamelidaceae	
	Witch hazel	*Hamamelis virginiana*
	Sweet gum	*Liquidambar styraciflua*
30.7	Plane-tree family—Platanaceae	
	Sycamore	*Platanus acerifolia*

Order 32. Fagales. The birch, alder, hazel, and hornbeam belong to the birch family. The only other family is the beech family, with the oak and beech of our forest lands as common and important trees. The ches-

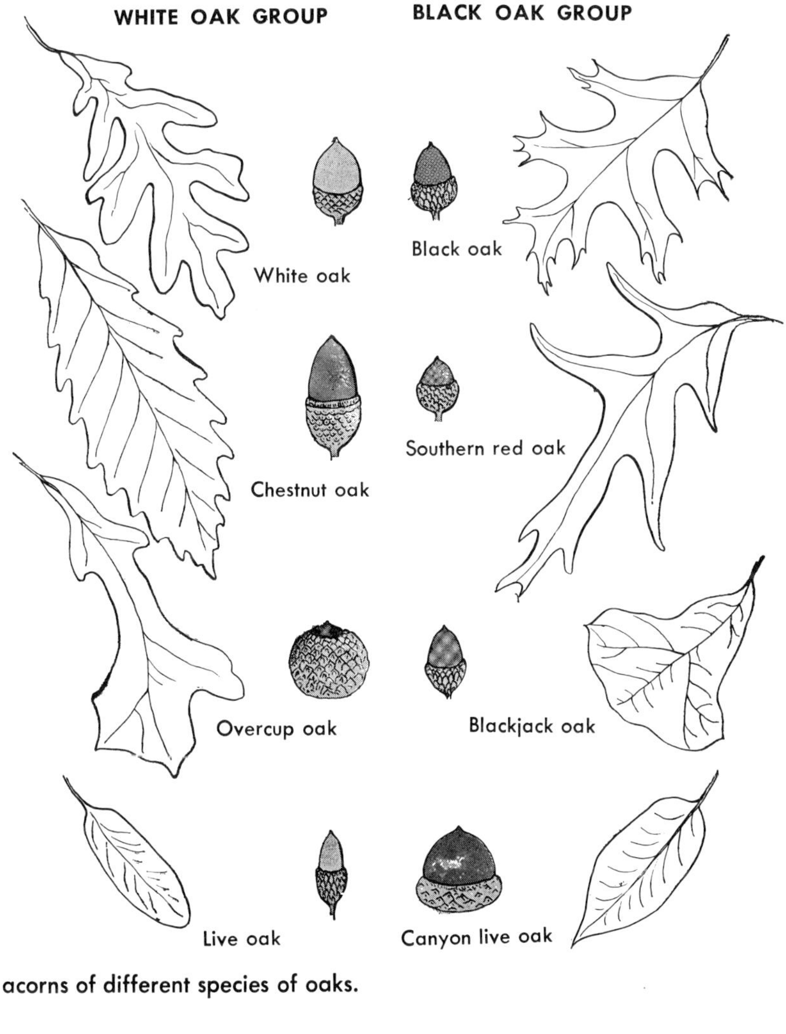

Fig. 27-14
Leaves and acorns of different species of oaks.

nut tree, formerly an important tree but now nearly extinct because of chestnut blight, a fungus disease, also belongs here.

32.1	Beech family—Fagaceae	
	White oak	*Quercus alba* (Fig. 27-14)
	Water oak	*Quercus nigra*
	Post oak	*Quercus stellata*
	Cork oak	*Quercus suber*
	Live oak	*Quercus virginiana*
	American chestnut	*Castanea dentata*
	Beech	*Fagus grandifolia*
32.2	Birch family—Betulaceae	
	Alder	*Alnus serrulata*
	Paper birch	*Betula papyrifera*
	Hazelnut	*Corylus americana*
	Filbert	*Corylus aveliana*
	Hornbeam (ironwood)	*Ostrya virginiana*

Order 34. Rosales. This is the second largest order of the flowering plants, with twenty families, and contains some of our most important food plants, as well as beautiful ornamentals. The largest family is the pea family, which contains not only the various peas, beans, and clovers, but also a variety of trees and shrubs, many of which have spines or thorns. One of the most interesting is the sensitive plant, which has the ability to fold its leaves and drop limply on the stem when it is touched. Plants of this kind are common in the tropics where they seem to go to sleep at night because they fold their leaves together and look very much denuded. Sometimes these plants form a major portion of a lawn in the tropics. Imagine the frustration of the operator of a lawn mower who sees a brown swath behind him because these plants have folded their leaves at the first touch of the mower, only to open them again after the machine is safely past.

The rose family includes the beautiful cultivated roses as typical members. However, equally as typical are the following fruits: blackberry, raspberry, prune, plum, cherry, apricot, peach, nectarine, loganberry, and many others listed below. The Venus' flytrap contains species that are capable of rapid closure of traplike leaves. An insect attracted to these leaves by a clear, sticky fluid secreted by gland cells springs the trap after contact with the surface of the leaf. This causes a mechanical stimulation that sets in action special cells, so that the leaves close by a rapid change in the turgor pressure of these cells. The fate awaiting the trapped insect is the same as when one is trapped in the pitcher plant.

A variety of wild flowers belong to the large saxifrage family, which includes not only the Venus' flytrap described above, but also gooseberries, currants, and similar plants.

34.1	Rose family—Rosaceae	
	Cockspur	*Crataegus crus-galli*
	Hawthorn	*Crataegus intricata* var. *straminea*
	Quince	*Cydonia oblonga*
	Strawberry	*Fragaria chiloënsis* var. *ananassa*
	Apple	*Malus sylvestris*
	Almond	*Prunus amygdalus*
	Apricot	*Prunus armeniaca*
	Sweet cherry	*Prunus avium*
	Plum (= prune)	*Prunus domestica*
	Peach	*Prunus persica*
	Nectarine	*Prunus persica* var. *nectarina*
	Pear	*Pyrus communis*
	Rose	*Rosa cathayensis*
	Blackberry (Blackcap)	*Rubus occidentalis*
	Raspberry	*Rubus strigosus*
	Loganberry	*Rubus ursinus* var. *loganobaccus*
	Bridal wreath	*Spiraea prunifolia*
34.4	Pea family—Fabaceae	
	Peanut	*Arachis hypogaea*
	Senna	*Cassia senna*
	Redbud	*Cercis canadensis*
	Beggarweed	*Desmodium tortuosum*
	Honey locust	*Gleditsia triacanthos*
	Soybean	*Glycine max*
	Kentucky coffee tree	*Gymnocladus dioica*
	Alfalfa (= lucerne)	*Medicago sativa* (Fig. 27-15)

Fig. 27-15
Red clover. (USDA photograph.)

	Sensitive plant	*Mimosa pudica*
	Locoweed	*Oxytropis splendens*
	Lima bean	*Phaseolus limensis*
	Kidney bean	*Phaseolus vulgaris*
	Pea	*Pisum sativum*
	Mesquite	*Prosopis juliflora*
	White clover	*Trifolium repens*
	Wisteria	*Wisteria floribunda*
34.7	Saxifrage family—Saxifragaceae	
	Hydrangea	*Hydrangea paniculata* var. *grandiflora*
	Gooseberry	*Ribes grossularia*
	Currant	*Ribes sativum*
	Venus' flytrap	*Dionaea muscipula*

Order 37. Myrtales. This order contains a variety of families, most members of which are shrubs or trees. The herbaceous members belong primarily to the myrtle family, which includes the common myrtle. The evening primrose belongs to a group of plants forming a separate family. A pest of waterways, the water chestnut, belongs here, as well as the tropical fruit, pomegranate.

37.2	Pomegranate family—Punicaceae	
	Pomegranate	*Punica granatum*
37.4	Water-chestnut family—Trapaceae	
	Water chestnut	*Trapa natans*
37.8	Myrtle family—Myrtaceae	
	Myrtle	*Myrtus communis*
	Allspice	*Pimenta dioica*
	Guava	*Psidium guajava*
	Clove	*Syzygium aromaticum*
37.10	Evening-primrose family—Onagraceae	
	Evening primrose	*Oenothera biennis*

Order 38. Gentianales. This order contains the gentian family, with many showy wild flowers, and the logania family with the genus *Strychnos,* which yields strychnine and curare, the latter a very poisonous substance used in the tropics to tip arrows and spears. It is used also in the physiology laboratory to demonstrate the physiological effect of such alkaloids on the muscles of frogs. Included here also is the dogbane family, which is a large family with few familiar species other than the periwinkle of gardens and lawns. The milkweed is included here too. The madder family with cinchona, the source of quinine formerly used in the treatment of malaria, completes the list of four of the five included families.

38.1	Logania family—Loganiaceae	
	Strychnine	*Strychnos nux-vomica*
	Curare	*Strychnos toxifera*
38.2	Madder family—Rubiaceae	
	Chinchona	*Cinchona ledgeriana*
	Coffee	*Coffea arabica*
38.3	Dogbane family—Apocynaceae	
	Dogbane	*Apocynum androsaemifolium*
	Indian hemp	*Apocynum cannabinum*
	Milkweed	*Asclepias syriaca* (Fig. 27-16)
	Periwinkle	*Vinca minor*
38.4	Gentian family—Gentianaceae	
	Fringed gentian	*Gentiana crinita*

Order 39. Bignoniales. This order contains a variety of ornamentals and weeds, included in ten families, with the snap-

Fig. 27-16
The winged seeds of the common milkweed emerging from the podlike fruits (follicles) in autumn. (From Russell, N. H. 1958. An introduction to the plant kingdom, The C. V. Mosby Co., St. Louis.)

dragon as one and the plantago family with the weed known as plantain as another.

39.5	Snapdragon family—Scrophulariaceae	
	Snapdragon	*Antirrhinum majus*
	Foxglove	*Digitalis purpurea*
	Mullein	*Verbascum thapsus*
39.6	Plantago family—Plantaginaceae	
	Plantain	*Plantago lanceolata*

Order 40. Cornales. The species of this group are mostly trees, shrubs, or woody vines, with fruits that are usually a berry. Some are biennial and produce roots and stems that provide us with food. The most important family is the grape family, one of the nine families included in the order. Closely related to this are the Boston ivy and Virginia creeper, both of which covered the walls of our academic institutions until the progressive engineers and economists decided that this touch of tradition was too costly to maintain. Mangrove trees, which make dense mangrove swamps in coastal areas, belong to the order. Two families, the dogwood family and the sourgum family, are other North American members of the group. Another of the principal families of this order is the carrot family, containing a number of spices such

Fig. 27-17
Flowering dogwood. Each "flower" is actually a cluster of small florets surrounded by four petaloid bracts. (From Russell, N. H. 1958. An introduction to the plant kingdom. The C. V. Mosby Co., St. Louis.)

Fig. 27-18
Poison hemlock flowers. (From Russell, N. H. 1958. An introduction to the plant kingdom. The C. V. Mosby Co., St. Louis.)

as caraway, celery seed, dill, anise, and others. Many vegetables such as celery, parsnips, carrots, and others and a number of common weeds are included.

40.1	Mangrove family—Rhizophoraceae	
	Mangrove	*Avicennia nitida*
40.2	Grape family—Vitaceae	
	Virginia creeper	*Parthenocissus quinquefolia*
	Boston ivy	*Parthenocissus tricuspidata*
	Grape	*Vitis vinifera*
40.3	Sour-gum family—Nyssaceae	
	Swamp tupelo	*Nyssa aquatica*
	Sour gum	*Nyssa sylvatica*
40.4	Dogwood family—Cornaceae	
	Dogwood	*Cornus florida* (Fig. 27-17)
40.9	Carrot family—Araliaceae	
	Dill	*Anethum graveolens*
	Celery	*Apium graveolens* var. *dulce*
	Caraway	*Carum carvi*
	Water hemlock	*Cicuta douglasii*
	Poison hemlock	*Conium maculatum* (Fig. 27-18)
	Carrot	*Daucus carota*
	English ivy	*Hedera helix*
	Parsnip	*Pastinaca sativa*
	Anise	*Pimpinella anisum*

Order 41. Dipsacales. Five families have been grouped into this order, with honeysuckle and teasel as familiar representatives.

41.1	Honeysuckle family—Caprifoliaceae	
	Honeysuckle	*Lonicera sempervirens*
41.4	Teasel family—Dipsacaceae	
	Teasel	*Dipsacus sylvestris*

Order 42. Lamiales. This large order contains seven families, with many wild flowers of such families as the verbena family and the mint family. The borage family has as some of its species the hound's-tongue, forget-me-not, and others.

42.2	Borage family—Boraginaceae	
	Hound's-tongue	*Cynoglossum officinale*
	Forget-me-not	*Myosotis scorpioides*
42.5	Verbena family—Verbenaceae	
	Verbena	*Verbena hybrida*
42.7	Mint family—Lamiaceae (= Labiatae)	
	Hyssop	*Hyssopus officinalis*
	Lavender	*Lavandula officinalis*
	Peppermint	*Mentha piperita*
	Spearmint	*Mentha spicata*
	Bells of Ireland	*Molucella laevis*
	Basil	*Ocimum basilicum*
	Marjoram	*Origanum vulgare*
	Rosemary	*Rosmarinus officinalis*
	Sage	*Salvia officinalis*
	Savory	*Satureja hortensis*
	Thyme	*Thymus vulgaris*
	Vetch	*Vicia sativa*

Order 43. Asterales. The largest of all plant families, the aster or sunflower family with about 20,000 species of rather showy flowers, many of them under cultivation, is the only member of this order. These plants may be readily recognized by their compound flower heads surrounded by a single outer row of petals, the ray. They are chiefly herbs. Among the many familiar examples are the aster, sunflower, ragweed, marigold, thistle, zinnia, chicory, lettuce, artichoke, and dandelion.

43.1	Aster family—Asteraceae (= Compositae)	
	Ragweed	*Ambrosia trifida*
	Burdock	*Arctium minus*
	Aster	*Aster laevis*
	Sticktight	*Bidens tripartita*
	Chrysanthemum	*Chrysanthemum morrifolium*
	Endive	*Cichorium endivia*
	Chicory	*Cichorium intybus*
	Thistle	*Cirsium arvense*
	Yellow cosmos	*Cosmos sulphureus*
	Artichoke	*Cynara scolymus*
	Dahlia	*Dahlia pinnata*
	Sunflower	*Helianthus annuus*
	Lettuce	*Lactuca sativa*
	Goldenrod	*Solidago speciosa* (and many others)
	Marigold	*Tagetes lucida*
	Dandelion	*Taraxacum officinale* (Fig. 27-19)
	Salsify (= vegetable oyster)	*Tragopogon porrifolius*
	Cocklebur	*Xanthium spinosum*
	Zinnia	*Zinnia elegans*

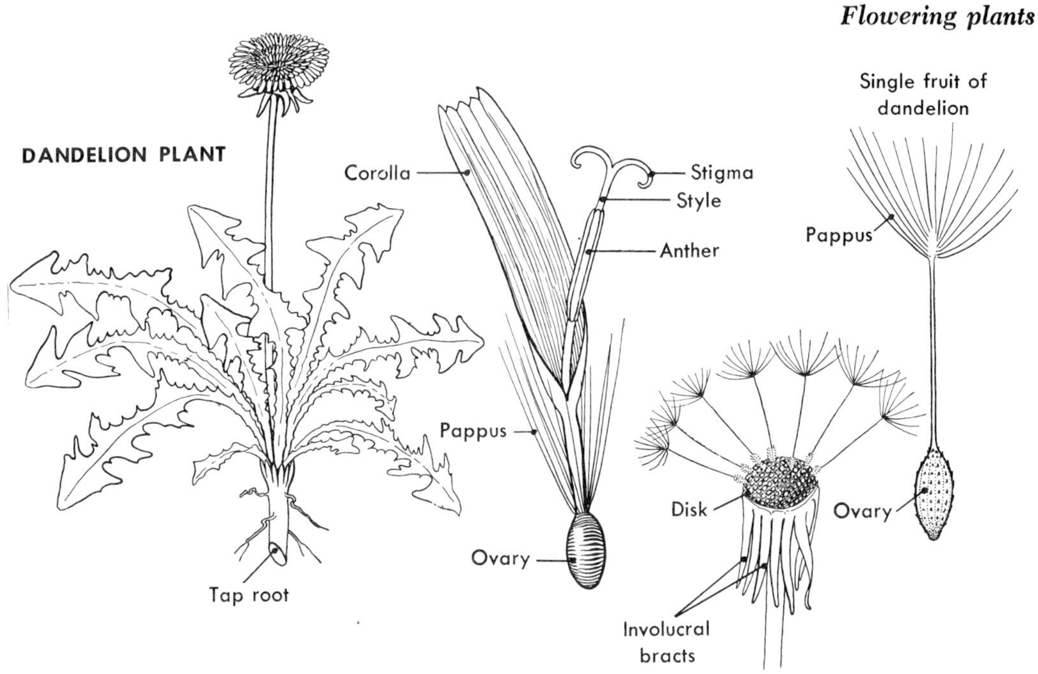

Fig. 27-19
Dandelion plant, a common weed, and a good example of a plant producing achenes.

Subclass Monocotyledoneae

The peak of plant development is found in the monocots if the current theories of the phylogeny of plants are correct. Most of the species of this group are herbaceous. A few, such as the palm and bamboo are woody. Their wood is much different from dicot wood and is not usuable as lumber except for special ornate construction or, in the case of bamboo, for lightly constructed buildings in tropical and subtropical areas. In general the seed embryo produces a single cotyledon instead of two, as in the dicots. The leaves most frequently are lanceolate structures with parallel veins, although in the palms they may be compound structures.

The subclass is divided into eleven orders and forty-five families. From the standpoint of the economy of mankind the grasses far exceed in value all other plants. Grasses are one of the primary sources of food. Bread, as is obvious to everyone, comes from flour, which is made from the seed of a few species of grasses. Meat comes from livestock that feed entirely on plants, mainly many species of grasses.

Order 44. Alismales. The arrowhead plants of ponds and marshes are among the relatively few common species of this order. The order, with three families, is confined to aquatic or marshy areas.

44.2	Water-plantain family—Alismaceae	
	Arrowhead	*Sagittaria graminea*

Order 45. Zosterales. The six families included here are a group of water plants, called pondweeds, common in freshwater ponds and streams.

45.2	Arrow-grass family—Juncaginaceae	
	Arrow grass	*Triglochin palustre*
45.3	Pondweed family—Potamogetonaceae	
	Pondweed	*Potamogenton natans*

Order 46. Najadales. Among the tropical species is a plant important for its role in the transmission of disease. The najas plants make large floating mats of vegetation in

rivers and lakes in tropical America, especially in the Panama Canal. These plants provide hundreds of square miles of breeding ground for mosquitoes, especially for certain species of *Anopheles* mosquitoes, which act as vectors of the disease malaria. Unfortunately, it has been impossible to eliminate these plants, so that malaria remains a threat to the Panama Canal area. It is the only family included in the order.

46.1	Najas family—Najadaceae	
	Najas	*Najas marina*

Order 48. Liliales. The lily family contains many beautiful flowers, most of which are readily identifiable on sight as members of this family. Even when the flowers are not nearby, reminders of the family are present in the odoriferous seasonings, onion, garlic, and leek. Also included in the order are a variety of aquatic plants making up several families, as well as the yam family and others such as the iris. The most beautiful members of the order are the orchids, which comprise a large family of mainly tropical plants. There are a total of eight families included in the order.

48.1	Lily family—Liliaceae	
	Sisal hemp	*Agave sisalana*
	Onion	*Allium cepa*
	Leek	*Allium porrum*
	Wild garlic	*Allium sativum*
	Asparagus	*Asparagus officinalis* var. *altilis*
	Queen-cup	*Clintonia uniflora*
	Autumn crocus	*Colchicum autumnale*
	Lily of the valley	*Convallaria majalis*
	Adder's-tongue (= dogtooth violet)	*Erythronium americanum*
	Lily	*Lilium canadense*
	Narcissus	*Narcissus tazetta*
	Greenbriar	*Smilax herbacea*
	Tulip	*Tulipa gesneriana*
	Century plant	*Agave americana* (Fig. 27-20)
48.3	Yam family—Dioscoreaceae	
	Yam	*Dioscorea opposita*

Fig. 27-20
Century plant in blossom.

48.6	Iris family—Iridaceae	
	Crocus	*Crocus susianus*
	Gladiolus	*Gladiolus gandavensis* (hybrids)
	Iris	*Iris versicolor*
48.8	Orchid family—Orchidaceae	
	Orchid	*Cattleya labiata*
	Lady's slipper	*Cypripedium acaule* (Fig. 27-21)

Order 49. Arales. Two interesting families of the four included in this order occur in the United States. The jack-in-the-pulpit, a common, spring woods flower, and the skunk cabbage belongs to the arum family. The duckweed family would hardly be suspected of being even a flowering plant. The

Fig. 27-21
Yellow lady's-slipper. (From Russell, N. H. 1958. An introduction to the plant kingdom. The C. V. Mosby Co., St. Louis.)

common species of *Lemna* are small floating plants consisting of a leaflike thallus and a few roots. Most reproduction is vegetative, and the simple little reproductive pouches that comprise their flowers are seldom seen.

49.1	Arum family—Araceae	
	Jack-in-the-pulpit	*Arisaema triphyllum*
	Skunk cabbage	*Symplocarpus foetidus*
49.2	Duckweed family—Lemnaceae	
	Duckweed	*Spirodela polyrrhiza*
	Wolffia	*Wolffia punctata*
49.4	Cattail family—Typhaceae	
	Cattail	*Typha latifolia*

Order 50. Arecales. This order contains one family, the palm family, composed of hundreds of species, all of which are tropical or subtropical. Dates and coconuts are among the many products these important woody plants produce.

50.1	Palm family—Arecaceae (= Palmae)	
	Betel nut	*Areca cathecu*
	Coconut	*Cocos nucifera*
	Date palm	*Phoenix dactylifera*
	Saw palmetto	*Serenoa repens*

Order 53. Commelinales. The importance of this order with its fourteen families cannot be overestimated, since it contains the grasses. The sedges are typical of bogs. The grasses (Fig. 27-4) probably constitute the best known family of plants because of their great importance. Bamboo also belongs to the family. Many botanists are specialists in this group alone, as systematisists, ecologists, and managers of range lands. The pineapple family also includes many epiphytes, so typical of tropical rain forests, and includes the Spanish moss of southeastern United States.

53.1	Pineapple family—Bromeliaceae	
	Pineapple	*Ananas comosus*
	Spanish moss	*Tillandsia usneoides*
53.4	Pickerelweed family—Pontederiaceae	
	Water hyacinth	*Eichhornia crassipes*
53.7	Sedge family—Cyperaceae	
	Sedge	*Carex morrowii*
	Papyrus	*Cyperus papyrus*

Fig. 27-22
Bamboo, a variety used for construction. (USDA photograph.)

53.8	Spiderwort family–Commelinaceae	
	Spiderwort	*Tradescantia virginiana*
53.14	Grass family–Poaceae (=Gramineae)	
	Wild oat	*Avena fatua*
	Oat	*Avena sativa*
	Bamboo	*Bambusa bambos* (Fig. 27-22)
	Bermuda grass	*Cynodon dactylon*
	Crabgrass	*Digitaria sanguinalis*
	Holly grass	*Hierochloë alpina*
	Barley (malt)	*Hordeum vulgare*
	Millet	*Pennisetum glaucum*
	Rice	*Oryza sativa*
	Sugar cane	*Saccharum officinarum*
	Rye	*Secale cereale*
	Sorghum	*Sorghum vulgare*
	Wheat	*Triticum aestivum*
	Corn	*Zea mays*

Fig. 27-23
Banana plants in Ecuador.

Fig. 27-24
Traveler's palm, so-called because the cuplike leaf bases hold water, which travelers are said to drink. It is a close relative of the banana, and not a palm.

Order 54. Zingiberales. Banana, plantain, and traveler's palm are examples of the five families included in this last order.

54.1	Banana family—Musaceae	
	Plantain	*Musa paradisiaca*
	Banana	*Musa paradisiaca* var. *sapientum* (Fig. 27-23)
	Manila hemp	*Musa textilis*
	Traveler's palm	*Ravenala madagascariensis* (Fig. 27-24)

Chapter 28

Fungi and lichens

Fungi

The fungi, long thought to be a specialized offshoot of algaelike plants, now are placed in a separate kingdom. They have distinctive features that place them neither in the plant kingdom, as usually conceived, nor, obviously, in the animal kingdom. The common puffball pictured above shows only the developing spore-producing body. Much of the plant is hidden as hyphae in the deadwood below. It is their saprophytic means of food getting that makes these organisms an important part of the biota.

The fungi most familiar to everyone are those that cause food to mold. This fascinating group of organisms plays an important but subtle role in the biological economy of this planet. The fungi are so universally distributed, yet so perfectly integrated into the maze of biological interrelationships, that a nonbiologist rarely senses their presence. Farmers, physicians, and ecologists know the importance of the group because of the plant and human diseases they cause, as well as their role in decay, soil formation, and food production.

Some fungi are saprophytic or parasitic plants unable to manufacture their own food because they lack photosynthetic pigments; others are autotrophic like the chemosynthetic bacteria. The production of chlorophyll-bearing cells and specialization for direct absorption of nutrients permits this radically different group of plants to maintain themselves in all aquatic and terrestrial habitats. As a group they are specialized and not really an advance over the algae because they have no structural complexities that cannot also be found in at least some of the algae.

Fungi may be defined as a kingdom having nucleated cells surrounded by cell walls, having tissue differentiation, but lacking organ systems. The cells lack photosynthetic pigments, making it necessary for them to directly absorb nutrients either as saprophytes or parasites. Their body is usually composed of branched filaments; these are sometimes interwoven to form a complex body of large size, for example, the mushroom. A few species are unicellular.

In the fungi the branching, filamentous thallus is known as a mycelium, a single branch of which is a **hypha.** Hyphae in the substrate upon which the fungi are feeding are called rhizoids, whereas those that connect one set of rhizoids with another are called stolons. The mycelium may be a multinucleated coenocyte with no transverse septa; when transverse walls are present, it is called **septate.**

At one time the bacteria were considered to be fungi, but they are so different in their lack of differentiated nuclei and their unicellular structure (rarely multicellular) that they are now treated as a separate group. It is now believed that the two groups originated independently.

Fungi have some characteristics resembling animals. Some of the true fungi have cell walls composed of a chitinlike material called "fungi cellulose," which is similar to the material in the body wall of arthropods. They also store their food in the form of glycogen, or "animal starch," a carbohydrate typical of animals rather than plants.

Fungi are rather difficult and challenging. It is necessary to grow the fungi so that all stages are available for study. Often critical microscopic examination is necessary before the species can be identified, which partly accounts for the fact that the group is poorly known, except for the economic species.

Classification. The kingdom is composed of over 80,000 described species, grouped into eight rather poorly defined divisions. The origin of the group is still a matter of speculation. Few are known definitely as fossils, and their rather radically different mode of life obscures the evidence that shows relationships. The unique relationship peculiar to the lichens, that of two taxonomically distinct organisms, representing two different kingdoms, arranged into an obligate association with distinctive body form, makes their assignment in a classification system very difficult. However, they may be treated with the fungi, provided it is realized that in each case the species of organisms involved in the association are at the same time classified in their respective groups. The deuteromycetes or imperfect fungi are also an unnatural group. All the species are temporarily assigned to a single group pending the discovery of further knowledge of their reproductive processes.

Reproduction. The great variety of reproductive mechanisms in the fungi allows more than a few general statements. Asexual reproduction is by means of spores, either produced from unspecialized hyphae or in specialized structures, the sporangia. All but the deuteromycetes reproduce sexually by the production of and union of gametes. The gametes may be motile; however, more often, the sex organs themselves are contiguous, and copulation of the gametes is effected by the growth of hyphae. In some cases the gametes depend upon insects or water for fertilization. Because of the many structural differences of the sex organs and the complications in the life cycle, many special terms are used in describing the life cycles of fungi. These are related to the standardized system used throughout this book in the following sections and in Chapter 14.

Biological significance. The fungi as a group contribute little new to the overall general phenomena manifested by organisms because they represent a closed sideline. However, their complicated life cycles and involvement with other organisms as saprophytes and parasites demonstrate how complicated protoplasm can be even when appearing simply organized. The inherent properties of the nucleus and the successive morphogenesis of a generation of fungi under control of these nuclei amply exemplify the potential of complex, highly organized proteins and of the enzymes they secrete. The mechanisms involved are poorly understood and embody the very nature of life itself. How the fungi came to have such a complicated life cycle as that of wheat rust, for instance, will not be understood for some time. It does show the paths organisms may take as changing conditions drive them to further and further specialization in their struggle against extinction. It is well known that spore-bearing plants are poorly adapted for present conditions on the earth. Once, when the seasons were uniform the world over, the plants reproducing entirely by spores were dominant, and perhaps the fungi were free-living green plants—terrestrial representatives of their algae counterparts. Like many groups that struggle to survive after changed conditions render them no longer fit as dominant organisms, they have reverted to a saprophytic or parasitic life, making use of more advanced organisms to aid them in their propagation. By so doing, they have regained, at least in part, their former position, but as a result, they are so changed that their existence is entirely dependent upon the success of their secondary role as subordinate organisms in the great cyclic changes typical of all life processes, among the most important of which is the decay cycle discussed in Chapter 17.

DIVISIONS 26, 27, AND 28— SLIME MOLDS

Slime molds are funguslike plants that pass through a vegetative stage called a plasmodium, a multinucleated mass of protoplasm having an indefinite, ameboid form instead of the thallus typical of fungi. Cell walls are completely lacking in this stage. However, reproductive stages are very different. Fruiting bodies and spores are produced like those of the true fungi. The definite cell walls covering the spores are the only convincing evidence that slime molds are plants. Even so, they are sometimes classified as animals.

Classification of slime molds. Four groups of organisms are generally known as slime molds. One has been discussed already in Chapter 18, the division Plasmodiophoromycota, the endoparasitic slime molds. The other three form the subkingdom Gymnomycota as follows:

Division 26. Myxomycota	Plasmodial slime molds
Division 27. Acrasiomycota	Cellular slime molds
Division 28. Labyrinthulomycota	Cell-net slime molds

Most of the 450 species live as saprophytes on moist, rotting wood or on moist organic soil. A few species parasitize seed plants and cause plant diseases. The group is undoubtedly very ancient in origin and is now widespread throughout the world. Many are brightly colored, the fruiting bodies particularly, but none have photosynthetic pigments.

Plasmodium. The plasmodium is a flat, slimy body capable of slow locomotion, with a thick, veinlike arrangement that at first glance may be mistaken for branching filaments. The edges of the vegetative body produce pseudopods like those of amebas, and nutrition is accomplished in the same manner. The organic material is surrounded by the pseudopods and incorporated into the cytoplasm, where digestion takes place. All waste products are left behind as the plasmodium moves on.

Reproduction. Under suitable conditions, usually at the end of a rainy period, the entire nature of the plasmodium changes. When food supply or moisture is lessened, the plasmodium becomes round, thick, and less transparent, and the color is intensified. If the plant has been living inside a log, it creeps out to an exposed surface, after which the entire body then rounds up into a shape characteristic for the species. Division of the plasmodium into distinct cells, each with a single nucleus, takes place. When this transformation has been completed, the plasmodium changes to one or several sporangia or fruiting bodies, which may be stalked or sessile. The capsule portion is a network of filaments, the **capillitium** (Fig. 28-1). The shape of the entire sporangium is characteristic of the species, in contrast to the uniform, nonspecific nature of the plasmodium. Fruiting bodies may be variously colored, and the surface of the capsule as well as the capillitium may be sculptured with a pattern peculiar to the species.

Spores are produced by the capillitium and are released as soon as they are mature. They pass through the network of capillitia and are dependent upon the wind for distribution. Dry conditions facilitate this dispersal, and their microscopic size permits them to reach even the upper atmosphere so that they are blown for great distances, all of which accounts for the widespread distribution of the species.

The spores may remain in a dormant condition for either a short or long period of time until they reach a suitable place for germination. On germinating, they give rise to naked swarm spores, which move about by means of two flagella, or in some species spores develop into small nonflagellated ameboid cells. Both types can swim about in the water of the soil or on moist wood. They behave as gametes, two of opposite strains fusing to form a zygote, which usually develops directly into a plasmodium, with only nuclear division instead of cell division. The developing zygotes may annex others to form a single, large plasmodium composed of several individuals. The plasmodium presumably is diploid; therefore, meiosis probably takes place at the time of spore formation.

The species that parasitize seed plants do not produce fruiting bodies but instead

Fig. 28-1
The capsule portion of the slime mold is a network of filaments known as a capillitium. (Courtesy General Biological Supply House, Inc., Chicago, Ill.)

produce a mass of spores within the host. These in turn form zoospores, some developing directly into a new plasmodium and others, acting like gametes, fusing to form a zygote and then a new plasmodium.

DIVISION 29. OOMYCOTA—OOSPHERE FUNGI

This is a group of water molds, white rusts, and downy mildews, some of which are serious plant diseases. They are characterized by a biflagellate zoospore, with one flagella directed forward and one backward, borne in sporangia.

The fungus body may be a unicellular thallus or a filament, some of which are profusely branched. Asexual reproduction consists of the formation of zoospores, which are, in the higher genera, retained within the sporangium, and it, in turn, functions as a spore and germinates to form the mycelium.

Sexual reproduction is usually by means of gametes of two types produced in two different kinds of gametangia. The formation of oospores gives the group its name. These originate in an oogonium. The sexual cycles of many species remain unknown, but in one, the *Saprolegnia* sp. or water mold, it has been followed through (life cycle 17). The oospheres of this group are fertilized by nuclei of the antheridium formed from the same somatic hyphae. The zygote thus formed is termed an oospore. These eventually germinate to produce new mycelia.

Economic importance. Several species are of great economic importance. Their role in preventing crop yields is of great concern and has resulted in costly control. The potato famine of Ireland in 1845 was the direct result of crop failure caused by a member of this class, *Phytophthora infectious.* Another species, *Plasmopara viticola,* threatened the wine industry of France by causing downy mildew of grapes. There are many other examples. The recent development of organic fungicides has resulted in more effective control measures for these plant diseases.

DIVISION 30. CHYTRIDIOMYCOTA—TRUE CHYTRIDS AND RELATED FUNGI

This group is characterized by the production of motile cells with a single posterior, whiplash flagellum. They are found generally in water, soil, rarely as parasites on algae and some economic plants, or as saprophytes on dead organic material. Their life cycle is similar to that of the true molds.

DIVISION 31. ZYGOMYCOTA—CONJUGATION FUNGI

Most members of the division, the most common member of which is bread mold, have coenocytic hyphae that are filamentous, branched, and multinucleated. The sporangia produce a large, indefinite number of spores. The sporangia may be borne on special reproductive hyphae, where they produce either motile or nonmotile spores.

Reproduction. Two generations occur in these plants: the asexual generation, which produces spores, and the sexual generation, which produces gametes. One more or less typical life cycle is described in Chapter 14 (life cycle 18, *Rhizopus stolonifer*). The spores are produced on special hyphae, the sporangiophores, as a branch of the mycelium or thallus of the sporophyte. Two types of spores are produced by members of this class: zoospores, which are uniflagellated or biflagellated, and **aplanospores,** which are nonmotile. Both are discharged into either the water or the air. All aquatic species and many terresterial species have flagellated spores. Therefore these terrestrial species depend upon rain for their dispersal. Spores germinate under suitable conditions and produce new sporophytes.

Sexual reproduction takes place by the production of unicellular gametangia (Fig. 28-2) characteristic of all algae and fungi. Some species discharge free-swimming gametes that unite in water to form motile

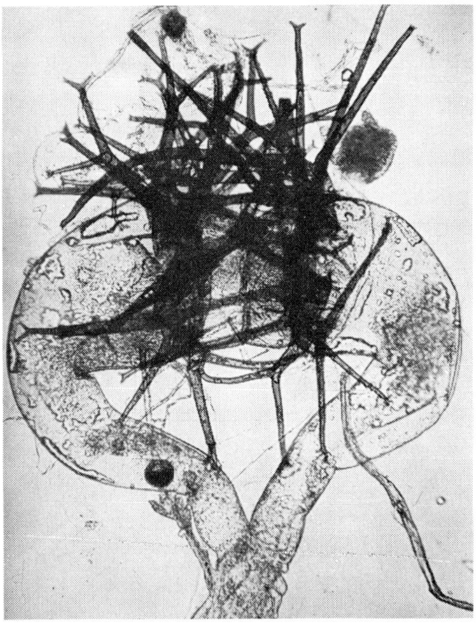

Fig. 28-2
Enlarged view of an immature zygospore of *Phycomyces* sp. showing the suspensions with branched outgrowths and fused gametangia. (Courtesy Carolina Biological Supply Co., Burlington, N.C.)

zygotes. In others the gametes are retained in the gametangium. Fertilization takes place by union of the two gametes of opposite sex. True sexual differentiation does not occur in the group. The differences are physiological and are distinguished as plus and minus strains. The zygote remains dormant for some time. If it is uninucleated, it is then called a resting spore; if it is multinucleated, it is a resting sporangium. Meiosis takes place within the developing zygote, so that the vegetative plant is haploid.

Biology. The primitive members of the division are saprophytes living on dung and decaying organic matter. Bread mold is among this group. Others are parasites on stored fruits and potatoes, some on plants and animals. The fly fungi are parasites of insects, and one of this same group is the cause of a serious disease of humans. At least one, *Mucor* sp., causes asthma and rhinitis, as previously mentioned. Finally, still others parasitize protozoans and nematodes.

DIVISION 32. ASCOMYCOTA—SAC FUNGI

The ascomycetes are both saprophytic and parasitic. All species produce spores in sacs called asci, which will contain two, four, or eight ascospores. It is usually eight. Although a true mycelium is absent in yeasts and similar forms, when present, the mycelium is divided into sections by walls or septa. Motile cells are not produced by any of the species.

Among the familiar species in this class are the yeasts used in breadmaking, the black and green molds, the powdery mildews, the cup fungi, and the edible morels and truffles that are a real delicacy. Many species are of great economic importance, causing a number of plant diseases, whereas others are of great benefit in curing human disease, for example, the green mold, *Penicillium* sp., which is responsible for the flavors of Roquefort and Camembert cheeses as well.

Classification. The ascomycetes comprise one of the largest groups of fungi, with about 30,000 species. The division is divided into two major groups: those that form ascogenous hyphae and those that lack these structures. The latter group is smaller, with only about 250 species, and includes the yeasts. The other subdivision has many orders.

Vegetative structure. The yeast and related groups form round or oblong nucleated cell masses that may be held together in poorly organized chains. The rest of the sac fungi form mycelia such as those of the phycomycetes and others. The mycelia are **septate;** that is multinucleated portions are divided by crosswalls, which give the mycelium the appearance of being cellular. However, there is a pore in each of the septa large enough to permit a steady flow of cytoplasm throughout the entire plant.

Reproductive structures. The reproductive organs of the ascomycetes are varied and complicated. Special terminology is applied to the parts because they deviate from the normal for other plants. Asexual reproduction is by both simple and complex processes. Some yeasts reproduce asexually by a process of fission or equal division of the protoplasm. A septum is formed between the two more or less equal parts. Budding, also characteristic of yeast, is the constriction of a small portion of protoplasm from the mother protoplasm. This part may stay attached to form a small chain, but usually the buds become detached. Fragmentation may occur either naturally or artificially and results in the formation of new individual plants. When fragmentation is natural, **oidia** are produced. This is the separation of a hypha into individual, multinucleated cells. If the cells develop a thick wall before they separate, they are termed **chlamydospores.** However, the production of conidia is the principal method of spore formation. Conidia are produced in chainlike fashion at the end of special branching hyphae, the conidiophores.

Sexual reproduction is by the development of asci on the same mycelium that produces the conidia and also involves the association of two compatible nuclei brought together in the same cell by one of several methods. However, they do not fuse but instead remain in close association to undergo several divisions forming **dicaryotic** cells, with two closely associated cells that do not fuse. Nuclear fusion eventually takes place before the production of ascospores in the ascus. The process of asci development is very similar to that of sporangia or in this case conidiophores, except that there is also the close association of two strains of plants. The process involves bringing the antheridial nuclei to the ascogonial nuclei. The ascogonium produces a trichogyne through which the antheridial nuclei pass to pair with the ascogonial nuclei. Various modifications of these structures aid copulation in the different species. After the two nuclei are brought together, ascogenous hyphae grow out of the ascogonium. A pair of associated nuclei enter one of the hyphae, nuclear division by mitosis takes place, and an ascus mother cell is formed by the growth of a crosswall that separates the apical portion of the hypha from the ascogonium. Fusion of the two nuclei takes place to form a zygote. This is followed by meiosis, so that eventually eight haploid ascospores are formed. Meanwhile the adjacent sterile haploid hyphae have been stimulated to develop a mass of sterile hyphae that enclose the ascogenous hyphae and their asci. These fruiting bodies or **ascocarps** are of three kinds: the cleistothecium, a closed type, the perithecium, an ascocarp with an ostiolate opening, and the **apothecium,** an open, saucer-shaped ascocarp. All these vary with the species and are used in the description and classification of sac fungi.

Life cycle. The mycelium of a sac fungus develops from either an ascospore or a conidium. Only the latter are multinucleated. The mycelium develops conidiophores that produce numerous conidia (Fig. 28-3) throughout the summer months, resulting in large numbers of asexual generations (life cycles 19, *Schizosaccharomyces* sp., and 20, *Microsphaera* sp.)

Later in the growing season the same mycelia that produce conidia will develop ascogonia. Nearby, an antheridium develops on the end of a special hypha, either on the same mycelium or on a separate one. Compatible nuclei are brought to the ascogonia, the formation of asci takes place, and ascogenous hyphae develop. The zygote is formed by the eventual union of the nuclei in the ascus mother cell. Meiosis follows to result in

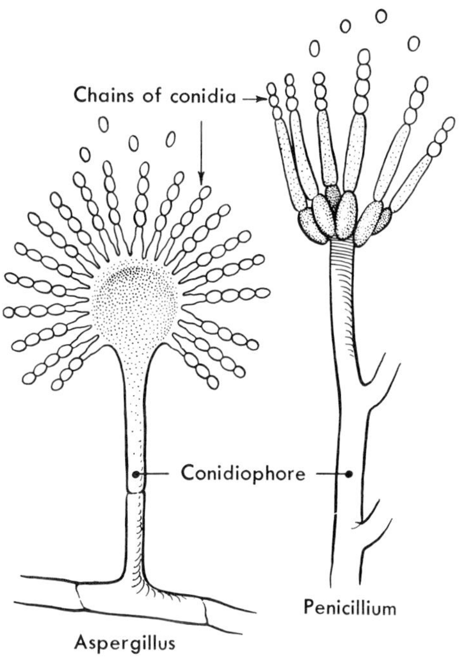

Fig. 28-3
Blue and green molds. Both *Aspergillus* sp. and *Penicillium* sp. produce conidia (spores) on upright structures called conidiophores.

four haploid nuclei, each of which divides by mitosis to form the eight nuclei that eventually become ascospores. The ascospores are formed by the development of a wall surrounding a portion of the cytoplasm near each of the nuclei.

Economic importance. The fermentation of sugar to produce alcohol is dependent upon yeast. The entire wine and liquor business is built around this simple plant process. Baking makes use of yeast for the generation of the carbon dioxide that causes bread and cake to rise. It is probable that the sac fungi have been domesticated along with the grains and grapes needed for these foods and beverages. The economic importance of these relationships is obvious. Other yeasts occur on plants, in the soil, and in the body cavities of some insects, where they enter into a symbiotic relationship. A few are pathogenic, causing diseases of the skin and lungs.

The blue and green molds are cosmopolitan in their distribution. A few are parasitic and cause plant disease, but most of them are saprophytic, and several are very beneficial in the production of food and in medicine. As already mentioned, *Penicillium* sp., by its enzymatic hydrolysis of dairy products, is used in the production of cheese and for the production of high-grade citric acid. Members of this same genus cause storage rot in some fruits. The antibiotic action of penicillin, now well known, was accidentally discovered by Fleming in 1928, but not until 1940 was it used as a medicine. The production of enzymes that inhibit the growth of bacteria and prevent the cell from forming a wall accounts for this remarkable action. At the same time penicillin is nontoxic to human beings. Production of penicillin and related drugs has become a large industry in the United States and other countries. A closely related fungus, *Aspergillus* sp., also has antibiotic properties, but the enzymes it produces are pathogenic to man. Several are the causative agents of diseases of the ear and lung.

Powdery mildews such as *Erysiphe graminis* affect various grains, and *Erysiphe cichoricearum* affects some fruits and vegetables to cause great loss to farmers by damaging these products and bringing about low production. The ergot fungus *Claviceps purpurea* and others are parasitic on cereals, causing crop loss. In times past, grains used from infected plants caused **ergotism** in man and livestock because of the presence of alkaloids. Holy fire, epidemics of which are described in ancient literature, was presumably a disease caused by these fungi. The drug ergot, obtained from infected plants, is sometimes used in obstetrics to produce contractions of the uterus.

The genus *Neurospora*, which belongs to an order related to the powdery mildews, has become well known as an experimental organism in genetic research (Chapter 10). Actually it is a bakery mold known as red bread mold, which infests bakeries and causes considerable damage.

DIVISION 33. BASIDIOMYCOTA—CLUB FUNGI

The spores of club fungi are produced on the outside of special spore-producing bodies, the **basidia,** and are called **basidiospores** (Fig. 28-4). They are uninucleated and haploid and are definite in number, usually four per basidium. The club fungi are macroscopic, forming a mycelial mass having a characteristic shape and growth on plants.

The club fungi are commonly called smuts (Fig. 28-5), rusts, mushrooms, deadly amanita, bracket fungi (Fig. 28-6), puffballs, jelly fungi, and stinkhorns. This group, like the previous one, is of economic importance because of its beneficial nature, as well as its harmful nature. Mushrooms (Fig. 28-7) form a part of our diet; growers produce millions of dollars, worth each year. Beneficial as these fungi may be, others, such as the smuts and rusts, cause crop loss valued at millions of dollars annually, and still others attack wood and wood products, especially in tropical and subtropical areas, necessitating a substantial annual outlay for preservatives to prevent this damage. Needless to say, the group is generally distributed throughout the world.

Classification. Club fungi are divided into two groups, one typified by the rusts

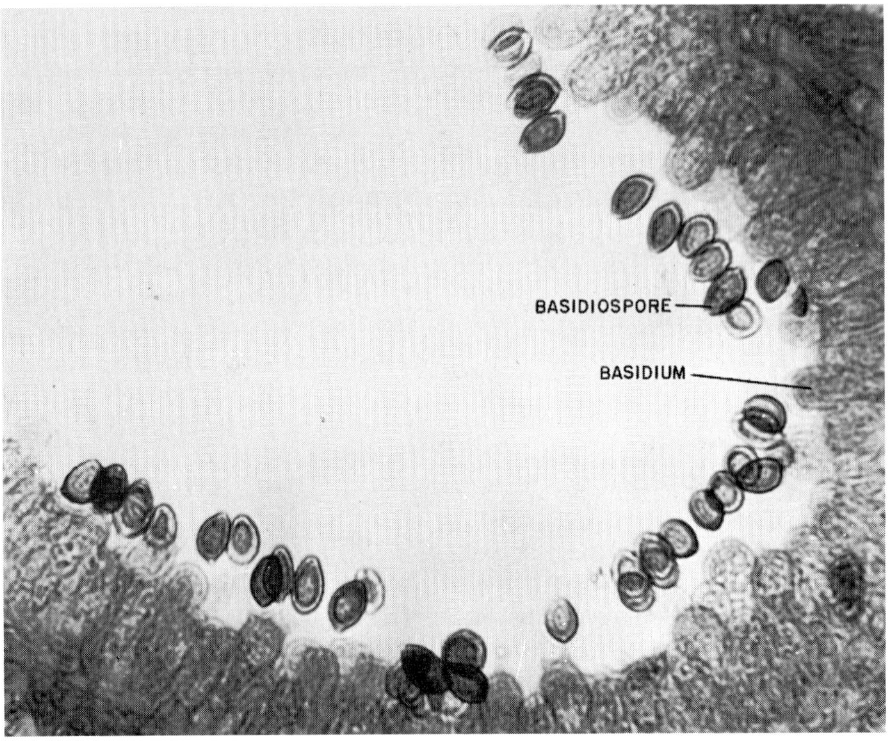

Fig. 28-4
Cross section of the fruiting body of a club fungus showing the basidia with sterigmata and basidiospores. (×530.) (Courtesy George H. Conant, Triarch Products, Ripon, Wis.)

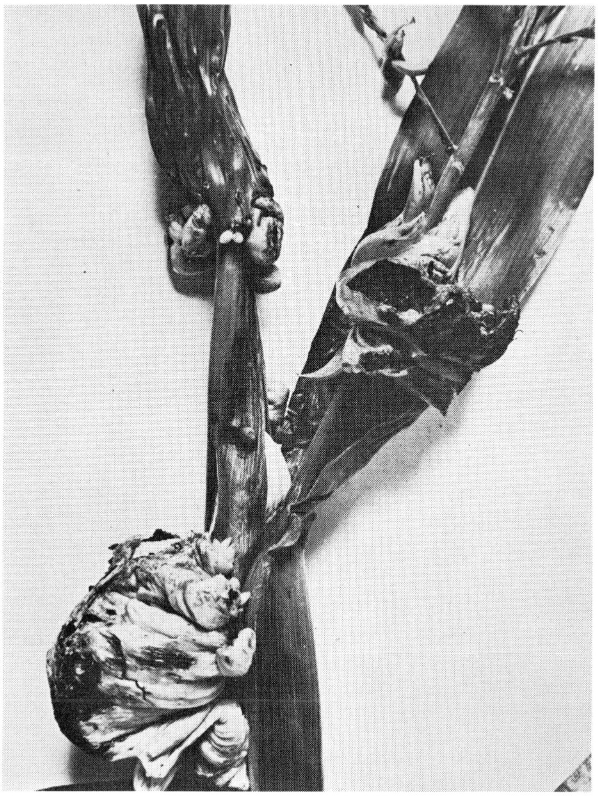

Fig. 28-5
Common smut on stalks, sheaths, and leaf of corn. (USDA photograph.)

Fig. 28-6
Bracket fungi.

Fig. 28-7
Fruiting bodies and stalks of the common edible cultivated mushroom. (USDA photograph.)

and jelly fungi and the other by the bracket fungi, mushrooms, puffballs, truffles, and similar forms. Each group is divided into several orders and many families. Representatives of each subclass are described in Chapter 14. There are about 25,000 species.

Vegetative structures. The mycelium of these plants is septate; that is, complete crosswalls are present, separating, in mature growth, binucleated and uninucleated cells to form the plant. The rhizoids penetrate the substratum to absorb nourishment and form masses in decaying, moist wood, on leaves, or in rich soil, where they spread out in fan-shaped fashion and are noticeable because of their white, yellow, or orange color. Development from the basidiospore is complicated and goes through three stages, with characteristic features for each stage.

The sporophyte produces fruiting hyphae or basidiophores, with apical basidia. By a complicated process the basidium produces basidiospores as a part of the asexual cycle. The sexual cycle involves the conjugation of hyphae of different strains, as in the other classes of fungi. The higher members of the class produce the basidia in a large, complex fruiting body, the **basidiocarp.** In these plants the main body goes unnoticed, whereas the basidiocarp has the characteristic shape of the plant.

Reproductive structures. Sexual reproduction is so reduced that most species have no sex organs. Regular somatic hyphae, uninucleated basidiospores, and oidia serve in sexual reproduction, which usually involves the conjugation of separate strains. Formation of the sex cells is complicated, however, by the genetic factors determining sexual compatibility. The sexual cycle involves the formation of binucleated cells by the union of the cytoplasm of two uninucleated cells of the primary mycelia or by the formation of binucleated sperm from oidia. The latter unite with somatic hyphae. The union of the two nuclei forms a zygote-like structure that is comparable to the diploid zygote found in other fungi. It develops into a basidium, by means of which meiosis again produces basidiospores. The sexual phase of most basidiospores is unknown, but it is presumed to be the same in all species.

Life cycle. The vegetative plant begins by one of three methods: (1) a haploid basidiospore (the spore) germinates into a primary uninucleated mycelium, one of the cells of which fuses with another uninucleated spore, (2) the hyphae of different strains produced by two basidiospores conjugate, or (3) two uninucleated basidiospores of different strains fuse. This in effect is the sexual cycle. The union of two uninucleated cells by one of these processes produces binucleated hyphae. These grow into a larger vegetative plant that develops basidia and eventually basidiospores, thus completing the cycle. The production of basidiospores involves the fusion of nuclei and meiosis in the following manner. The basidium may be septate or nonseptate, depending upon the subclass to which the species belongs. The septate basidium as found in smuts and rusts is formed from the germination of a binucleated basidiospore. On germination the two nuclei fuse, after which meiosis follows and four haploid nuclei are formed. They migrate into a short club-shaped hypha that grows from the spore. Each of the nuclei are then delimited by a septum that cuts them off from the remainder of the hypha. They often transform into conidia. The nonseptate basidium characteristic of the gill fungi are binucleated, with nuclei of opposite strains. As the basidium develops, the two nuclei fuse and then divide meiotically to form four haploid nuclei. These migrate into four little projections at the tip of the basidium where they develop into basidiospores to be shed. Thus it is apparent that there is little similarity between this life cycle and the relatively simple alternation of generations found in other plants. Yet the basic elements of that fundamental plan are there. Fig. 28-8 will serve to illustrate this. This life cycle is hypothetical and rather simplified. The club fungi discussed in Chapter 14 illustrates actual cycles (life cycles 21, *Puccinia graminis*, and 22, *Agaricus campestris*). The many variations of life cycles found in other species will be omitted from this introductory discussion. It should be noted that the binucleated condition in these plants is the result of the fusion of two nucleated cells without the fusion of their nuclei, and that this condi-

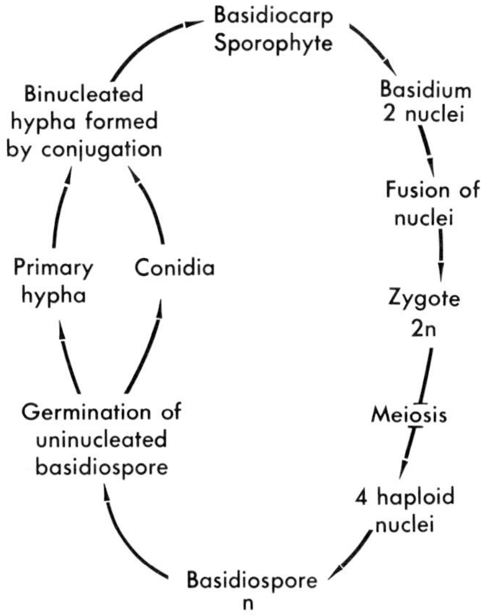

Fig. 28-8
Life cycle of a hypothetical basidiomycetes.

tion is retained for some time during the growth of the plant. Thus we may find in a given plant two types of hyphae, those with uninucleated cells and those with binucleated cells. An interesting mechanism has developed, called **clamp connections,** which ensures that the sister nuclei arising from the division of the nuclei of binucleated cells become separated in the two daughter cells during cell division. The clamp connection is a special structure formed during nuclear division as a short branch arising between the two nuclei, x and y, as shown in Fig. 28-9. Nucleus y migrates into the clamp connection and divides at the same time x divides; x remains in place, whereas x' migrates toward the former location of y. The clamp connection grows toward the position of x and y' moves back to its original position. A crosswall is formed in the clamp connection between y and y'. Then, after x' has moved into position, a cell wall appears between the two new cells; y enters the cell with x through a new opening in the clamp connection, and the two daughters cells have nuclei of opposite strains preserved.

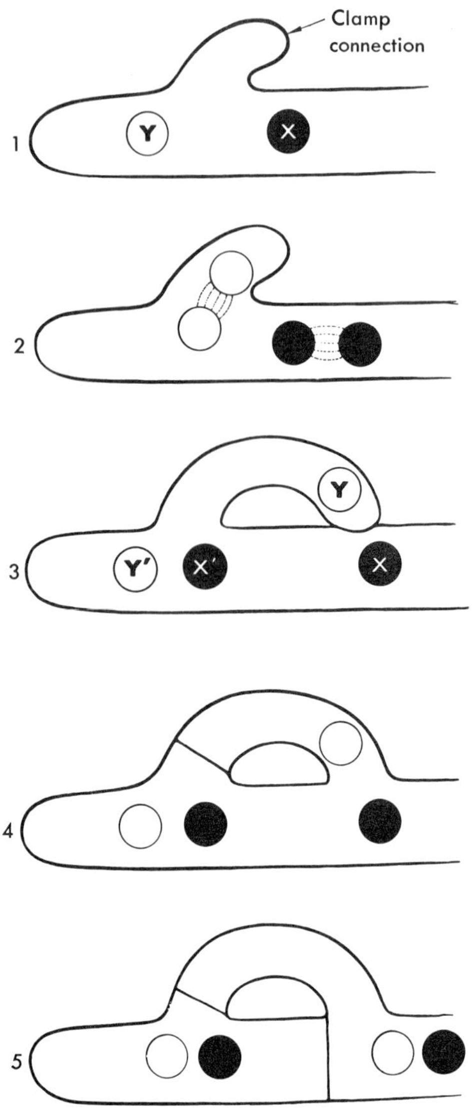

Fig. 28-9
Clamp connectors in the basidiomycetes.

Some important club fungi. Among the many kinds of club fungi, the following deserve special mention either because of their economic importance or because of their academic interest. No attempt will be made to give complete life cycles of these species, but some of the interesting features of their life processes will be mentioned.

The jelly fungi are so named because their fruiting bodies have the consistency of jelly. Many of the species are white, bright yellow, or orange and live on the branches and trunks of living or dead trees. They are most abundant in the fall, winter, or spring. Some species are extensively eaten as food in the Orient. Other species in the group are parasitic on flowers and some on scale insects. An interesting host relationship is noted in the latter species. The fungus builds a hyphal mat over the scale colony and penetrates the insect's body. They do not kill the insects, but render them sterile.

The rusts, order Uredinales, form a large group, causing great loss to cultivated plants such as pine and many cereals. The life cycle of wheat rust is discussed in Chapter 14 and is typical of the group. Two hosts, barberry (Fig. 28-10) and wheat (Fig. 28-11), are involved, through either of which the fungus can be controlled biologically. The elimination of the noncultivated, but equally essential, host plant will prevent the infection of the cultivated plant. A related species causes the cedar apples or galls of cedar trees, and another causes white-pine blister rust, resulting in the destruction of important lumber trees. The primary host of this fungus is found on the leaves of currants and gooseberries. However, since these plants are of less value than lumber, it is they that are eradicated for effective and permanent control.

The smuts are related to the rusts. They get their name from the black, sooty spores they produce. Corn smut, wheat smut, loose smut of oats, and onion smut are a few examples of economically important species. Unlike the rusts, these species are not obligate parasites.

The large basidiomycetes include a wide variety of plants; many are well known and several are edible. Some cause plant disease and others produce galls. Many are brightly colored and are often pictured in full color in natural history books.

Fig. 28-10
Aecidium on the lower surface of a barberry leaf, the rust spots. (USDA photograph.)

Fig. 28-11
Spores of wheat rust being collected from a stem. (USDA photograph.)

The bracket fungi are distinguished by their large basidiocarps growing on the sides of trees. When they are young, they are soft, but as they mature, they become hard and woody in nature. It is this group of plants that causes wood rot and necessitates large expenditures for wood preservation.

Agaricaceae is a large family including the mushrooms and toadstools. A few are edible, but many are poisonous. There is no positive way to distinguish the edible ones from the poisonous ones, except with the help of an experienced mycologist. Therefore it is best to exercise extreme caution in picking wild mushrooms to eat. Many species may be eaten, however, and a serious student can learn to identify those that can be used. In this same family is the famous and deadly amanita. Toadstools and other Agaricaceae often grow in such a way as to form what is termed a "fairy ring." This is a ring of darker green grass caused by the increase of available nitrogen in the soil due to the growth and death of fungal mycelia. The succeeding seasonal growth of the fungus is approximately equal and radiates out from the original starting site. This eventually forms a ring. The fruiting bodies appear around this ring in the fall (Fig. 28-12).

The puffballs (see photograph at the beginning of this chapter) form an interesting group, and as far as is known their round, fruiting bodies are all edible; no poisonous species are known. Basidiocarps sometimes grow to basket-sized proportions. Unfortunately, they are difficult to market; otherwise, their steaklike flavor would soon make them a popular delicacy.

Fig. 28-12
"Fairy ring" of toadstools.

By contrast there are the Phallales or stinkhorns, a near relative. These plants emit a fetid odor that attracts flies. Their mouth parts in turn pick up sticky spores that are disseminated after passing unharmed through the alimentary tract.

The bird's-nest fungi have a most remarkable mechanism for the dissemination of **peridioles** containing the basidiospores. The fruiting body of this fungus is cup-shaped and resembles a bird's nest. The cup is called a splash cup—when it rains, drops of water strike the cup and splash out the peridioles. These complicated structures have a cord and an adhesive **hapteron** at the end of the cord. When the raindrops splash out the peridioles, the cord and sticky hapteron are extended. The peridiole flies through the air, strikes a leaf or stem, adheres to it by the hapteron, and ties itself fast, the momentum of the peridiole winding the cord around the stem.

DEUTEROMYCETES—IMPERFECT FUNGI

The deuteromycetes include those fungi known to be or thought to be the monidial stages of Ascomyceteae or, rarely, of Basidiomyceteae. They are called imperfect fungi because no sexual stage is known. These stages are perhaps undiscovered or may no longer exist. At any rate, the group is believed to be an unnatural one, merely representing a dumping ground for those fungi that lack the necessary stages for placement in the proper classes of the division. As many as 10,000 species remain in this group. Many more were once placed here, but they have since been removed upon discovery of the so-called perfect or sexual cycle. The only known means of reproduction in the imperfect fungi is by conidia. It is probable, therefore, that most species are really ascomycetes.

Despite the artificial nature of the group, many of the species are of importance to man because of their disease-producing effects. *Streptomyces* sp., *Alternaria* sp., and *Hormodendrum* sp. have been mentioned as causing asthma and rhinitis. Ringworm infections and athlete's foot, as well as other types of skin infections, are caused by these plants. Others cause low-grade infection of the heart, lungs, brain, meninges, and bone. A few of these diseases are serious in nature, and all are difficult to treat.

Morphology. Conidia are usually developed on conidiophores generally distributed on the hyphae, or some may be grouped into small fruiting bodies. A variety of arrangements occur in the many species of the group. Pathogenic species produce other types of spores to which several terms have been applied, depending upon their method of formation. The hyphae of deuteromycetes are septate. The mycelium may be extensive in such subcutaneous forms as those causing ringworm. Some species are known only by their sterile hyphae; no form of reproduction has ever been observed.

Classification. There are four orders. Sphaerossidales produce conidia in pycnidia, globose or flask-shaped bodies on short conidiophores. The Melanconiales produce conidia in acervuli, flat open heads of generally short conidiophores growing side by side and arising from a mass of hyphae. Moniliales reproduce conidia by different means from those found in the other orders, or they reproduce by oidia or by budding. The order Mycelia-Sterilia has no known means of reproduction.

Lichens

The lichens are not taxonomically a separate group. They represent symbiotic associations of two plants, an alga and a fungus, both of which are otherwise classified in their respective groups. They are, therefore, combination plants and can only be classified artificially. However, because of their importance in ecological associations, soil building, and as food for certain animals, it is convenient to group them together for general discussion.

Component plants. Species of two groups of algae, the Chlorophyta and the Cyanophyta, form the algal component. The fungal elements are mostly ascomycetes, with a few species of basidiomycetes represented in one group. It is by the fungal species that the lichens are artificially classified. All species are named according to the fungus elements in the association.

Nature of association. Considerable controversy still exists among botanists about the nature of the association of these plants, which indicates that further study of the group is necessary. It is generally taught that the fungus provides moisture retention and protection for the alga and, in turn, receives nourishment from the chlorophyllic alga in a perfect symbiotic arrangement. However, there is some evidence that the alga is held captive by the fungus, which may well be cultivating the former for its own purpose, so that the relationship is merely one of host and parasite.

Reproduction. Vegetative multiplication may take place by any portion of the body developing into a new lichen or by granules, called soredia, composed of one or more algal cells surrounded by fungal hyphae breaking off and forming a new plant.

Spore formation also takes place (Fig. 28-13). The fungus element forms spores in a special cup-shaped structure. In ascomycetes these spores are in sacs or asci. The sacs rupture and spores are discharged. If the spore settles down in contact with the proper alga, it develops into a lichen; if not, it dies. The algal element reproduces independently and is spread by rain.

Importance. The lichens are important elements in the formation of soil, dissolving and disintegrating rocks. Lichens form an important food source for caribou, musk ox, and reindeer. Lichens on barren plains and mountains of western Asia and northern Africa formed the manna of the Israelites and are still known as bread of heaven. Some of the lichens have medicinal properties; others are a natural source of many dyes such as litmus; still another, the oak moss, is used in southern Europe as a fixative for perfumes.

Classification by types. There are three types of lichens: (1) crustose lichens, which form a crust on trees, rocks, and soil, (2) foliose lichens, with leaflike thalli, the

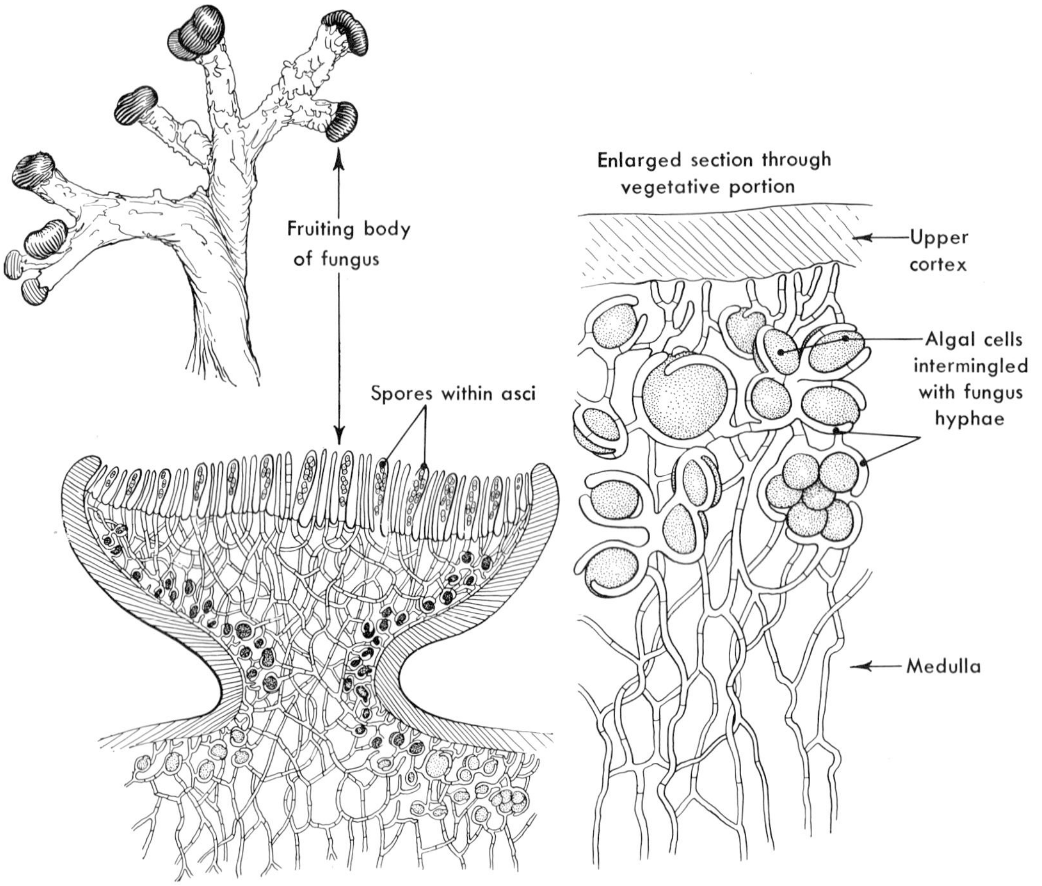

Fig. 28-13
Details of a lichen showing the relationship of the fungus and alga.

upper and lower surfaces of which are different, and (3) fruticose lichens, with the major portion pendant or erect.

In the crustose lichens the upper hard layer is formed of fungal hyphae. The lower layer is composed of algal cells intermixed with hyphae. Finally, there is a loosely woven mass of hyphae forming the substrate. A familiar example is the common *Graphica scripta* on tree bark or on stones.

The foliose lichens (Fig. 28-14) form flat lobes, some of which adhere to the substrate by strands of hyphae. Beneath are pores that allow free passage of air into the algal layer. These are common forms on stones in exposed places.

The fruticose lichens (Fig. 28-15) develop from a primary thallus, which may be crustose or foliose. They form branches of woven hyphae, with algae within the branches. Among the familiar examples are old-man's-beard and reindeer moss.

SPECIAL TERMS USED IN CHAPTER 28

aplanospore	chlamydospore	hypha
apothecium	clamp connections	oidium
ascocarp	dicaryotic	peridiole
basidiocarp	ergotism	septate
capillitium	hapteron	

QUESTIONS TO CONSIDER

1. List three reasons why the fungi are considered now to be a separate kingdom.

Fig. 28-14
Foliose lichen.

Fig. 28-15
Fruticose lichen. (Courtesy Carolina Biological Supply Co., Burlington, N. C.)

2. Why are the bacteria not included in this kingdom?
3. Describe the food-getting process as found in the fungi.
4. Why is the fungus body often compared to an iceberg?

DISCUSSION FOR LEARNING

1. Discuss the reasons why some of the fungi are beneficial, but some are the cause of very serious diseases.
2. Discuss a way you think the lichens may have evolved.
3. Discuss the problem the fungi have with a binucleated cell. How might have this come about? Do you think it indicates a "wrong way" of development or evolution?

References

In the bibliography that follows, selected references for the subjects covered in the text are given. Many of these contain more complete bibliographies that should be consulted for further references. For the most recent information on any biological subject, consult *Biological Abstracts*.

Chapter 1

Altman, P. L., and D. S. Dittmer [ed.]. 1964. Biology data book. Federation of American Societies for Experimental Biology, Washington, D. C.

Flanagan, Dennis [ed.]. 1966. Information. Scient. Amer. **215**:64-260 (Sept.).

Holman, H. H. 1962. Biological research method. Hafner Publishing Co., Inc., New York.

Nagel, E. 1961. The structure of science. Harcourt, Brace & World, Inc., New York.

van Norman, R. W. 1963. Experimental biology. Prentice-Hall, Inc., Englewood Cliffs, N. J.

Chapter 2

Arber, A. 1938. Herbals. Cambridge University Press, New York.

Gabriel, M. L., and S. Fogel [ed.]. 1958. Great experiments in biology. Prentice-Hall, Inc., Englewood Cliffs, N. J.

Grieve, M. 1959. A modern herbal. Stechert-Hafner, Inc., New York.

Hanson, N. R. 1958. Patterns of discovery. Cambridge University Press, New York.

Howell, F. C., and the editors of Life. 1965. Early man. Time Inc., Chicago.

Journal of the history of biology. 1968. Vol. 1, *et seq*. Belknap Press, Harvard University, Cambridge, Mass.

Linton, R. 1957. The tree of culture. Alfred A. Knopf, Inc., New York.

Locy, W. A. 1925. The story of biology. Garden City Publishing Co., Garden City, N. Y.

McKenzie, A. E. E. 1960. The major achievements of science. Cambridge University Press, New York. 2 vol.

Mendel, G. 1941. Experiments in plant hybridization (Translation of a paper published in 1865). Harvard University Press, Cambridge, Mass.

Peattie, D. C. 1936. Green laurels, the lives and achievements of the great naturalists. Simon & Schuster, Inc., New York.

Reed, H. S. 1942. A short history of the plant sciences. Chronica Botanica Co., Waltham, Mass.

Rodgers, A. D. III. 1949. Liberty Hyde Bailey, a story of American plant sciences. Princeton University Press, Princeton, N. J.

Singer, C. 1959. A history of biology. Rev. 3rd ed. Abelard-Schuman, Ltd., New York.

Watson, J. D. 1968. The double helix. Atheneum Publishers, New York.

Williams, B., and S. Epstein. 1961. Plant explorer David Fairchild. Julian Messner, Inc., Publishers, New York.

Chapter 3

Anderson, E. 1952. Plants, man and life. Little, Brown & Co., Boston.

Lanjouw, J. [ed.]. 1966. International code of botanical nomenclature. International Bureau for Plant Taxonomy and Nomenclature of the International Association for Plant Taxonomy, The Hague, Holland.

Savory, T. 1962. Naming the living world. John Wiley & Sons, Inc., New York.

Chapter 4

Bailey, L. H. 1949. Manual of cultivated plants. Rev. ed. The Macmillan Co., Publishers, New York.

Bailey, W. R., and E. G. Scott. 1966. Diagnostic

microbiology. 2nd ed. The C. V. Mosby Co., St. Louis.

Kingsbury, J. M. 1964. Poisonous plants of the United States and Canada. Prentice-Hall, Inc., Englewood Cliffs, N. J.

Milner, H. W. 1953. Algae as food. Scient. Amer. **189**:31-35 (Oct.).

Montgomery, F. H. 1964. Weeds of the northern United States and Canada. Frederick-Warne, New York.

Schery, R. W. 1952. Plants for man. Prentice-Hall, Inc., Englewood Cliffs, N. J.

Schwanitz, F. 1966. The origin of cultivated plants. Harvard University Press, Cambridge, Mass.

Smith, F., G. Walton, and H. Chapin. 1954. The sun, the sea, and tomorrow. Charles Scribner's Sons, New York.

Weiss, F. J. 1952. The useful algae. Scient. Amer. **187**:15-17 (Dec.).

Chapter 5

Baldwin, E. 1964. An introduction to comparative biochemistry. Cambridge University Press, Inc., New York.

Elsasser, W. M. 1966. Atom and organism. Princeton University Press, Princeton, N. J.

Epstein, H. T. 1963. Elementary biophysics selected topics. Addison-Wesley Publishing Co., Inc., New York.

Jellinck, P. H. 1963. Biochemistry, an introduction. Holt, Rinehart & Winston, Inc., New York.

Lehninger, A. L. 1965. Bioenergetics. W. A. Benjamin, Inc., New York.

Mahler, H. R., and E. H. Cordes. 1966. Biological chemistry. Harper & Row, Publishers, New York.

Pauling, L., R. B. Corey, and R. Hayward. 1954. The structure of protein molecules. Scient. Amer. **191**:51-59 (July).

Speakman, J. C. 1966. Molecules. McGraw-Hill Book Co., New York.

White, E. H. 1964. Chemical background for the biological sciences. Prentice-Hall, Inc., Englewood Cliffs, N. J.

Young, L. B. [ed.]. 1965. The mystery of matter. Oxford University Press, Inc., New York.

Chapter 6

DeRobertis, E. D. P., W. W. Nowinski, and F. A. Saez. 1965. Cell biology. 4th ed. W. B. Saunders Co., Philadelphia.

Freeman, J. A. 1964. Cellular fine structure. McGraw-Hill Book Co., New York.

Giese, A. C. 1968. Cell physiology. W. B. Saunders Co., Philadelphia.

Kennedy, D. [ed.]. 1965. The living cell. W. H. Freeman & Co., Publishers, San Francisco.

Kennedy, D. [ed.]. 1967. From cell to organism. W. H. Freeman & Co., Publishers, San Francisco.

McElroy, W. D. 1964. Cell physiology and biochemistry. 2nd ed. Prentice-Hall, Inc., Englewood Cliffs, N. J.

Mitchell, J. S. 1960. The cell nucleus. Academic Press, Inc., New York.

Roodyn, D. B., and D. Wilkie. 1968. The biogenesis of mitochondria. Methuen & Co., Ltd., London.

Swanson, C. P. 1964. The cell. 2nd ed. Prentice-Hall, Inc., Englewood Cliffs, N. J.

Watson, J. D. 1965. Molecular biology of the gene. W. A. Benjamin, Inc., New York.

Watson, J. D., and F. H. C. Crick. 1953. A structure for deoxyribose nucleic acid. Nature **171**: 737-738.

Chapter 7

Ackley, M. E., and P. B. Whitford. 1965. The chemistry of photosynthesis. Appleton-Century-Crofts, New York.

Baron, W. M. M. 1963. Organization in plants. Edward Arnold (Publishers) Ltd., London.

Gates, D. M. 1962. Energy exchange in the biosphere. Harper & Row, Publishers, New York.

Greulach, V. A. 1955. Plant movements. Scient. Amer. **192**:100-106 (Feb.).

Levitt, J. 1969. Introduction to plant physiology. The C. V. Mosby Co., St. Louis.

McElroy, W. D., and B. Glass [ed.]. 1961. A symposium of light and life. Johns Hopkins Press, Baltimore.

Meyer, B. S., D. B. Anderson, and R. H. Bohning. 1960. Introduction to plant physiology. D. van Nostrand Co., Inc., Princeton, N. J.

Slatyer, E. O. 1967. Plant-water relationships. Academic Press, Inc., New York.

Steward, F. C. 1964. Plants at work. Addison-Wesley Publishing Co., Inc., New York.

Chapter 8

Black, C. A. 1957. Soil-plant relationships. John Wiley & Sons, Inc., New York.

Daubenmire, R. F. 1959. Plants and environment. John Wiley & Sons, Inc., New York.

Odum, E. P. 1959. Fundamentals of ecology. 2nd ed. W. B. Saunders Co., Philadelphia.

Odum, E. P. 1963. Ecology. Holt, Rinehart & Winston, Inc., New York.

Platt, R. B., and J. Griffiths. 1964. Environmental measurement and interpretation. Reinhold Publishing Corp., New York.

(See also references for Chapter 15)

References

Chapter 9

Bold, H. C. 1957. Morphology of plants. Harper & Brothers, New York.

Hawkes, J. G. [ed.]. 1966. Reproductive biology and taxonomy of vascular plants. Pergamon Press, Long Island City, N. Y.

(See also references for Chapter 11)

Chapter 10

Baker, H. G., and G. L. Stebbins [ed.] 1965. The genetics of colonizing species. Academic Press, Inc., New York.

Beadle, G. W. 1963. Genetics and modern biology. Amer. Philosophical Soc., Philadelphia.

Bonner, D. M. 1961. Heredity. Prentice-Hall, Inc., Englewood Cliffs, N. J.

Darlington, C. D. 1958. The evolution of genetic systems. 2nd ed. Basic Books, Inc., Publishers, New York.

Dobzhansky, T. 1951. Genetics and the origin of species. 3rd ed. Columbia University Press, New York.

Ehrlich, P. R., and R. W. Holm. 1963. The process of evolution. McGraw-Hill Book Co., New York.

Ford, E. B. 1960. Mendelism and evolution. 7th ed. John Wiley & Sons, Inc., New York.

Gardner, E. J. 1968. Principles of genetics. 3rd ed. John Wiley & Sons, Inc., New York.

King, R. C. 1965. Genetics. 2nd ed. Oxford University Press, Inc., New York.

Levine, L. 1969. Biology of the gene. The C. V. Mosby Co., St. Louis.

Merrell, D. J. 1962. Evolution and genetics. Holt, Rinehart & Winston, Inc., New York.

Moore, R. 1968. Evolution. Rev. ed. Life nature library. Time Inc., Chicago.

Ravin, A. W. 1965. The evolution of genetics. Academic Press, Inc., New York.

Stebbins, G. L. 1950. Variation and evolution in plants. Columbia University Press, New York.

Chapter 11

Barth, L. J. 1964. Development: selected topics. Addison-Wesley Publishing Co., Inc., New York.

Berrill, N. J. 1961. Growth, development, and pattern. W. H. Freeman & Co., Publishers, San Francisco.

Bold, H. C. 1957. Morphology of plants. Harper & Brothers, New York.

Bonner, J. 1965. The molecular biology of development. Oxford University Press, Inc., New York.

Ebert, J. D. 1965. Interacting systems in development. Holt, Rinehart & Winston, Inc., New York.

Eames, A. J. 1961. Morphology of the angiosperms. McGraw-Hill Book Co., New York.

Esau, K. 1960. Anatomy of seed plants. John Wiley & Sons, Inc., New York.

Foster, A. S., and E. M. Gifford, Jr. 1959. Comparative morphology of vascular plants. W. H. Freeman & Co., Publishers, San Francisco.

Lee, A. E., and C. Heimsch. 1962. Development and structure of plants. Holt, Rinehart & Winston, Inc., New York.

Sinnott, E. W. 1960. Plant morphogenesis. McGraw-Hill Book Co., New York.

Sussman, M. 1964. Growth and development. 2nd ed. Prentice-Hall, Inc., Englewood Cliffs, N. J.

Thimann, K. V. 1950. Autumn colors. Scient. Amer. **183**:40-43 (Oct.).

Chapter 12

Naylor, A. W. 1952. The control of flowering. Scient. Amer. **186**:49-56 (May).

Salisbury, F. B. 1957. Plant growth substances. Scient. Amer. **196**:125-134 (April).

(See also references for Chapter 11)

Chapter 13

Biale, J. B. 1954. The ripening of fruit. Scient. Amer. **190**:40-44 (May).

Grant, V. 1963. The origin of adaptations. Columbia University Press, New York.

Jacobs, W. P. 1955. What makes leaves fall. Scient. Amer. **193**:82-89 (Oct.).

van der Pijl, L. 1969. Principles of dispersal in higher plants. Springer-Verlag, Inc., New York.

Chapter 14

Bonner, J. T. 1965. Size and cycle, an essay on the structure of biology. Princeton University Press, Princeton, N. J.

Chapter 15

Darlington, P. J., Jr. 1965. Biogeography of the southern end of the world. Harvard University Press, Cambridge, Mass.

Daubenmire, R. 1968. Plant communities. Harper & Row, Publishers, New York.

Gleason, H. A., and A. Cronquist. 1964. The natural geography of plants. Columbia University Press, New York.

Good, R. 1953. The geography of the flowering plants. 2nd ed. Longmans, Green & Co., Ltd., London.

Hanson, H. C., and E. D. Churchill. 1961. The plant community. Reinhold Publishing Corp., New York.

Küchler, A. W. 1964. Potential natural vegetation of the conterminous United States. Amer. Geographical Soc. Sp. publ. no. 36. (Map and book), New York.

Oosting, H. J. 1956. The study of plant communities. 2nd ed. W. H. Freeman & Co., Publishers, San Francisco.

Richards, P. W. 1952. The tropical rain forest. Cambridge University Press, New York.

Shelford, V. E. 1963. The ecology of North America. University of Illinois Press, Urbana.

Went, F. W. 1955. The ecology of desert plants. Scient. Amer. **192**:68-75 (April).

Chapter 16

Abelson, P. H. 1956. Paleobiochemistry. Scient. Amer. **195**:83-92 (July).

Andrews, H. M., Jr. 1947. Ancient plants and the world they lived in. Cornell University Press, Ithaca, N. Y.

Derek, V. 1963. Principles of paleoecology. McGraw-Hill Book Co., New York.

Knopoff, L. 1969. The upper mantle of the earth. Science **163**:1277-1287.

Oparin, A. I. 1957. The origin of life on the earth. 2nd ed. Academic Press, Inc., New York.

Rutten, M. G. 1962. The geological aspects of the origin of life on earth. Elsevier Publishing Co., Inc., New York.

Walton, J. 1953. An introduction to the study of fossil plants. Adams & Black, London.

Chapter 17

Whittaker, R. H. 1969. New concepts of kingdoms of organisms. Science **163**:150-160.

Chapter 18

Breed, R. S., E. G. D. Murray, and N. R. Smith. 1957. Bergey's manual of determinative bacteriology. 7th ed. The Williams & Wilkins Co., Baltimore.

Clayton, R. K., and M. Delbruck. 1951. Purple bacteria. Scient. Amer. **185**:68-72 (Nov.).

Salle, A. J. 1961. Fundamental principles of bacteriology. McGraw-Hill Book Co., New York.

Thimann, K. V. 1963. The life of bacteria. The Macmillan Co., Publishers, New York.

Wollman, E. L. 1956. Sexuality in bacteria. Scient. Amer. **195**:109-118 (June).

Chapter 19

Goodheart, C. R. 1969. An introduction to virology. W. B. Saunders Co., Philadelphia.

Chapter 20

Leedale, G. F. 1967. Euglenoid flagellates. Prentice-Hall, Inc., Englewood Cliffs, N. J.

Prescott, G. W. 1964. How to know the freshwater algae. William C. Brown Co., Publishers, Dubuque, Iowa.

Chapter 21

Smith, G. M. 1933. The fresh-water algae of the United States. McGraw-Hill Book Co., New York.

Smith, G. M. 1955. Cryptogamic botany. Vol. 1. Algae and fungi. 2nd ed. McGraw-Hill Book Co., New York.

Taylor, W. R. 1957. Marine algae of the northeastern coast of North America. 2nd ed. University of Michigan Press, Ann Arbor.

Tiffany, L. H. 1958. Algae: the grass of many waters. Charles C Thomas, Publisher, Springfield, Ill.

Chapter 22

Cobb, B. 1956. A field guide to the ferns. Houghton Mifflin Co., Boston.

Conrad, H. S. 1956. How to know the mosses and liverworts. William C. Brown Co., Publishers, Dubuque, Iowa.

Smith, G. M. 1955. Cryptogamic botany. Vol. 2. Bryophytes and pteridophytes. 2nd ed. McGraw-Hill Book Co., Inc., New York.

Chapter 23

Gleason, H. A., and A. Cronquist. 1963. Manual of vascular plants of northeastern United States and adjacent Canada. D. Van Nostrand Co., Inc., Princeton, N. J.

Gleason, H. A. 1963. The new Britton and Brown illustrated flora of the northeastern United States and adjacent Canada. Hafner Publishing Co., Inc., New York. 3 vol.

(See also references for Chapter 27)

Chapter 24

(See references for Chapter 23)

Chapter 25

(See references for Chapter 23)

Chapter 26

(See references for Chapter 23)

Chapter 27

Abrams, L. 1923-1940. An illustrated flora of the Pacific States, Washington, Oregon, and California. Stanford University Press, Palo Alto.

Baerg, H. J. 1955. How to know the western trees. William C. Brown Co., Publishers, Dubuque, Iowa.

Benson, L. 1957. Plant classification. D. C. Heath & Co., Boston.

Fassett, N. 1957. A manual of aquatic plants. University of Wisconsin Press, Madison.

Fernald, M. L. 1950. Gray's manual of botany. 8th ed. American Book Co., Boston.

Hutchinson, J. 1959. The families of flowering plants. Oxford University Press, New York. 2 vol.
Kearney, T. H., and R. H. Peebles. 1960. Arizona flora. University of California Press, Berkeley.
Lawrence, G. H. M. 1951. Taxonomy of vascular plants. The Macmillan Co., New York.
Lawrence, G. H. M. 1955. An introduction to plant taxonomy. The Macmillan Co., New York.
Lloyd, F. E. 1942. The carnivorous plants. Chronica Botanica Co., Waltham, Mass.
Muenscher, W. C. 1944. Aquatic plants of the United States. Cornell University Press, Ithaca, N. Y.
Petrides, G. A. 1958. A field guide to trees and shrubs. Houghton Mifflin Co., Boston.
Porter, C. L. 1967. Taxonomy of flowering plants. 2nd ed. W. H. Freeman & Co., Publishers, San Francisco.
Small, J. K. 1933. Manual of the southeastern flora Hafner Publishing Co., Inc., New York.
Thorne, R. F. 1968. Synopsis of a putatively phylogenetic classification of the flowering plants. Aliso **6**:57-66.

Chapter 28

Alexopoulos, C. J. 1962. Introductory mycology. 2nd ed. John Wiley & Sons, Inc., New York.
Bessey, E. A. 1950. Morphology and taxonomy of fungi. Hafner Publishing Co., Inc., New York.
Christensen, C. M. 1951. The molds and man. University of Minnesota Press, St. Paul.
Clements, F. E., and C. L. Shear. 1931. The genera of fungi. Hafner Publishing Co., Inc. New York.
Gray, W. D. P. 1959. The relation of fungi to human affairs. Henry Holt & Co., New York.
Ramsbottom, J. 1953. Mushrooms and toadstools, a study of the activities of fungi. The Macmillan Co., New York.
Smith, A. H. 1963. The mushroom hunter's field guide. Rev. ed. University of Michigan Press, Ann Arbor.

Glossary

The technical names in boldface type in the text are defined in glossary that follows. The term is indicated in this manner only once in the text, usually the first time it is used. The list of terms at the end of each chapter is, in most cases, therefore, a list of new terms for study.

Å (see Ångström unit).

abscission zone the separation layer of cells formed at the base of some leaves and other organs that permits these parts to be shed.

accessory fruit a fruit composed of parts of the flower in addition to the ovary.

achene a small dry, one-seeded, indehiscent fruit with the seed coat and fruit wall unfused.

activator a substance that activates a chemical reaction.

active absorption the taking of ions into a cell against diffusion pressure as the result of the expenditure of energy.

active secretion the exchange of ions when the principal ions in the cell are at a higher concentration than the solutions outside.

adenosine diphosphate (ADP) a chemical substance present in active cells during metabolic activity and involved in the metabolic economy of the cell; the end product along with inorganic phosphate that results in energy release from adenosine triphosphate (ATP; cf.).

adenosine triphosphate (ATP) an energy-yielding compound with high-energy bonds involved in the metabolic release of energy within the cell.

adnate two or more series of parts united in a flower.

ADP (*see* adenosine diphosphate).

adventitious bud a bud that arises on a part of the stem other than its axil or apex.

adventitious root a root that develops from parts other than the primary root or one of its branches.

aeciospore a spore formed within an aecium; a type of spore produced by the rusts.

aecium a group of binucleated hyphal cells at one stage in the life cycle of a rust that produce spore chains.

aerial stem a stem that grows above the ground.

aerobe an organism that uses free oxygen during respiration.

aggregate fruit a fruit composed of a cluster of ripened ovaries from a single flower.

akinete vegetative cells of a filament or trichome that are capable of generating new filaments.

allele one gene of a pair of alternate genes found in the same position on homologous chromosomes.

allopatric populations of two species occupying separate and distinct ranges.

allopolyploid polyploid organism with more than two sets of homologous chromosomes that arise through hybridization of different species.

alternate leaves leaves occurring at different levels successively on opposite sides of the stem.

alternation of generations alternation of a spore-producing stage with a gamete-producing stage in the life cycle of a plant.

amino acid organic compounds with the general formula NH_2-R-COOH, which combine to form proteins.

amylum stars star-shaped aggregates of cells filled with starch found in certain bryophytes; these cells are capable of asexual reproduction.

anabolism constructuve stages in metabolisms, e.g., assimilation; opposed to catabolism.

anaerobic organism an organism that does not use free oxygen in its respiratory process.

anaerobic respiration respiration without the use of free oxygen; fermentation.

anaphase stage of mitosis in which the chromo-

some halves separate and begin to move to opposite poles.
androecium the stamens, collectively; male reproductive organs of a plant.
Ångström unit (Å) a unit of length, equal to one tenth of a millimicron, or one ten millionth of a millimeter.
annual a plant that grows from seed and produces seed in one growing season.
annual ring the layer of wood or xylem formed during one year's growth of cambium.
annulus a row of specialized cells in the wall of a sporangium that aid in the release of spores.
anther the pollen-bearing part of a stamen; a pollen sac or microsporangium.
antheridiophore a disk-headed stalk that bears antheridia.
antheridium a male gametangium in which sperm are produced.
antherozoid motile male sexual cells in antheridia.
anthocyanin one of several substances of a blue or violet pigment found in flowers, leaves, and stems.
antiauxin any of several organic substances that inhibits or controls the growth stimulation of the auxins.
anitbody a globulin type of protein produced within an organism that opposed the action of an antigen.
anticodon the amino acids that seek out the proper codon during protein assembly.
antigen a foreign substance, usually a protein, that causes the production of specific antibodies within an organism.
antipodal one of the three nuclei opposite the micropyle of the embryo sac.
antitoxin a type of antibody that neutralizes a toxin.
apetalous without petals on the flower.
apical meristem the embryonic tissue at the tip of a stem (or root); also called promeristem.
aplanospore a nonmotile spore of algnae; an encysted spore of fungi.
apomixis asexual reproduction, of which there are many types, that substitutes for or replaces sexual reproduction.
apothecium a saucer-shaped or cup-shaped ascocarp that bears asci on its extended or exposed surface; characteristic of the lichens and certain fungi (discomycetes).
archegoniophore the portion of the thallus of some cryptogams that develops the archegonium.
archegonium a multicellular female gametangium of higher plants.
ascocarp sporocarp of ascomycetes, the sac fungi.
ascogonium the female gametangium of ascomycetes, the sac fungi.
ascospore a spore produced by the ascomycetes within an ascus.
ascus a saclike structure that contains ascospores.
ascus mother cell a cell that produces an ascus.
asepalous without sepals.
asexual not involving the fusion of sex cells or the fusion of their nuclei; without sex organs.
atom the smallest particle of an element that can exist alone or in combination with other atoms and elements.
ATP (*see* adenosine triphosphate).
autecology the study of an individual organism or individual species in relation to the environment.
autogamy the asexual fusion of nuclei.
autopolyploids polyploid organisms with more than two sets of homologous chromosomes.
auxin growth-regulating substance found in plants; several types.
auxin precursors substances that can be converted by a plant into one of the auxins.
axil the angle formed by the axis and any organ arising from it.
axile placentation the placenta on a central axis in the ovary.
axillary bud the bud located in the axil of the leaf; also called lateral bud.

bacillus a rod-shaped bacterium.
backcross crossing a hybrid with one of its parents or parental forms.
bacteriophage a virus parasitic on bacteria.
bark the tissue that lies outside the vascular cambium or the periderm and tissues external to it in woody stems or roots.
basidiocarp a sporocarp of the club fungi.
basidiospore a spore produced by a basidium, by the basidiomycete fungi.
basidium the spore mother cell of the basidiomycete fungi.
berry a fleshy fruit with the entire pericarp fleshy.
biennial a plant that grows a taproot and a rosette of leaves during the first growing season and produces flowers and seed during the second growing season.
binominal system a system of nomenclature in which the name of a species is composed of two words, the generic name and the specific name.
biochemistry the study of the chemicals and chemical processes found in living organisms.
bioluminescence emission of light by living organisms as the result of internal oxidation of cellular substances.
biome a large ecological habitat, such as tundra, desert, or evergreen forest; a major life zone of plants and animals.

biota all living organisms, as individuals, living at one time.

bivalent chromosome homologous chromosomes when the two are associated during synapsis in meiosis.

blade the thin, expanded portion of a leaf.

botany the study of some or of all the organisms belonging to the kingdoms Monera (all), Protista (part), Fungi (all), and Plantae (all).

bract small, modified leaves or scales subtending a flower or its pedicel.

Brownian movement the random movement or agitation of small particles in suspension.

bud an embryonic or young shoot that has the capacity to produce leaves, stems, flowers, or roots.

bud scales protective leaflike coverings of buds.

bud scar the scar left after a bud has fallen off.

bulb an underground storage leaf bud composed of enlarged and fleshy scalelike leaves.

bulbil a thickened axillary bud that may be detached and produce a new plant; also an aerial bulb.

caline a root-, stem-, or leaf-forming growth regulator.

calyx collective term for the sepals of a flower.

cambium layer of meristem that is responsible for the lateral growth of a stem or root.

capillarity the effect of a thin tube on fluids, causing the surface to be elevated or depressed.

capillitium a network of threads in the capsule of the fruiting body, especially in the slime molds.

capsid the protein coat of the head of a virus.

capsule the sporangium of a bryophyte; the dry dehiscent fruit of an angiosperm; the slimy covering around certain bacterial cells.

carbohydrates organic compounds composed of carbon, hydrogen, and oxygen, with the general formula, $C_nH_{2n}O_n$.

carbon dioxide cycle the circulation of carbon dioxide during food synthesis and respiration.

carotene reddish orange pigment found in many plant cells.

carotenoid yellow, orange, red, and purple pigments present in plant fats.

carpel female part or megasporophyll of an angiosperm flower.

carpellate cone megasporophylls of conifers.

carpogonium part of the procarp; cells resulting in the sporocarp after fertilization.

carpospores the spores produced by the carpogonium after fertilization.

carposporophyte a small parasitic plant located on the gametophyte of certain algae.

catabolism destructive stages in metabolism, e.g., respiration.

catalyst a substance that alters the rate of a chemical reaction without permanent change during the reaction.

category a hierarchical arrangement of levels of taxa, each level a category; e.g., the taxon Rosaceae belongs to the category family.

caulocaline a growth-regulating substance synthesized in roots but necessary for stem growth.

cell small mass of protoplasm, usually containing a nucleus or nuclear material; the fundamental unit of structure in plants.

cell division (*see* mitosis).

cell membrane the outer living and limiting boundary of protoplasm in a cell.

cell plate the early stage of the formation of the new cell during mitosis.

cell wall a cell secretion covering plant cells, usually composed of the carbohydrate cellulose and certain minerals.

cellulose the carbohydrate forming the walls of plant cells.

central placenta a stalk that extends from the base of the ovary into the locule.

centromere the constricted part of the chromosome to which the spindle fiber is attached.

chalazal the end of the megaspore mother cell opposite the micropyle.

chemosynthesis the formation of a more complex compound from simpler compounds by chemical action; the formation of organic compounds from inorganic compounds by some plant cells.

chiasma the point at which two chromosomes cross and fuse during meiosis.

chlamydospore a thick-walled resting spore.

chlorenchyma a tissue composed of parenchyma cells that contain chloroplasts.

chlorophyll the green pigment in plants necessary for the process of photosynthesis; a methylphytol ester of the chlorophyllins.

chloroplast a cell organoid that contains chlorophyll.

chlorosis yellowing of a plant due to a mineral deficiency.

chromatid one of the two parallel halves of a fully contracted chromosome as it appears during the late prophase of mitosis.

chromatin the nucleic acid of a nucleus as it shows up as threads or rods after staining.

chromatography a process used to separate and show the component amino acids of a protein.

chromoplast a body (plastid) in the cell that contains a pigment.

chromosome characteristic and deeply staining body in the nucleus composed primarily of DNA; bears the genes or determiners of heredity.

or other component part of a complex organic molecule.

circinate rolled on its axis with the apex as its center; coiled inwardly.

clamp connections a special structure found in some fungi during cell division to ensure the proper distribution of the set of two nuclei for each cell.

cleistothecium an ascocarp in which the asci and ascospores are developed within a closed envelope.

climax a steady community reached at the peak of development for a particular area.

clone a group of plants developed asexually from a single plant.

closed system a process maintained and concluded without input or output after its original construction.

coalescent fused parts, particularly in reference to flower parts.

coccus a spherically shaped bacterium.

codon the nonnucleotide residues placed in proper sequence used to code the sequence for protein formation.

coenocyte a group of protoplasmic units, each with a nucleus, enclosed in a common cell wall.

coenzyme a nonprotein substance that forms the active portion of an enzyme system.

cohesion the molecular attraction of water molecules causing sap flow.

coleoptile a protective growing sheath covering the plumule of the seeds of grasses and similar plants during germination.

collenchyma a supporting tissue made up of somewhat elongated cells, the walls of which are thickened with cellulose, particularly at the corners; like parenchyma cells, they are living.

colloid the state of a substance composed of particles that are intermediate in size and that will neither settle out nor form a true solution; opposed to crystalloid.

colony a group of contiguous cells held together by cell cement, usually derived from a single ancestor, growing on a solid surface.

columella the axis or central part within a sporangium, capsule, or carpel.

community all the living and nonliving components of the environment of a particular area, usually as a part or division of an ecosystem.

companion cells elongated parenchyma cells adjoining sieve tubes in the phloem tissue.

complete flower a flower that bears sepals, petals, stamens, and pistils.

complex permanent tissue xylem and phloem.

complex tissue tissue consisting of two or more kinds of cells that perform the same function.

compound leaf a leaf in which the blade is divided into a number of leaflets.

compound pistil a pistil formed by the fusion of two or more carpels.

computer an electronic apparatus used to carry out repetitious and highly complex operations at high speed.

conceptacle a hollow structure containing sex organs.

cone a specialized branch that bears sporophylls, each subtended by a bract.

conidiophore the organ that bears conidia.

conidium a propagative cell, asexually produced and separated from the parent.

conjugation a process involving the fusion of isogametes to form a zygote.

continental drift a theory which proposes that there is and has been movement of the continents by actually drifting apart.

converter gene a gene capable of changing a color-determining gene permanently to one determining another color.

core the central part of certain fruits, as in the apple; the true fruit, exclusive of the accessory tissue.

cork a tissue composed of cells whose walls are impregnated with a fatty substance called suberin.

cork cambium the layer of meristematic cells that produces cork; also called phellogen.

corm a short, stout, erect underground stem in which food is stored.

corolla collective term for the petals of a flower.

cortex the tissue lying between the epidermis and the endodermis of a stem or root.

cosmopolitan generally distributed; ubiquitous.

cotyledons seed leaves or food storage parts of a seed.

cristae baffles in mitichondria.

crossing-over an exchange of chromatin material between paternal and maternal chromatids during pairing of the chromosomes at meiosis.

cross-pollination transfer of pollen from the anther of a flower of one individual to the stigma of the flower of another individual.

cryptogams a term sometimes applied to plants reproducing by spores instead of seeds.

crystalloid crystallike; opposite of colloid.

cupule a reproductive structure producing gemmae; the cup of such fruits as the acorn.

cuticle the outermost skin or external pellicle containing the epidermal layer of cells.

cutin a waxy substance secreted by epidermal cells.

cystocarp the portion of the gametophyte of certain algae that bears the carposporophyte.

cytokinins plant growth inhibiting substances.

cytology the study of cells.
cytoplasm the protoplasm of the cell other than the nucleus.
cytoplasmic inheritance inheritance by means of cytoplasmic influence as opposed to inheritance by chromosome influence only.

data facts or statistics, either historical, or derived from calculations or experimentation; data may be processed in a computer and retrieved in a form that differs from that stored.
data processing the storage and retrieval of data.
deciduous falling at the end of a season of growth, as leaves and flower petals; usually refers to trees other than evergreen trees.
defective virus a virus that cannot reproduce except in combination with a helper virus.
dehiscent fruit a dry fruit that splits open on maturity for release of seeds.
deoxyibonucleic acid (DNA) one of two nucleic acids composed of nucleotides formed from a nitrogenous base, the sugar 2-deoxyribose, and phosphoric acid.
determinate flower an inflorescence in which the first flowers to open are at the apex or the center of the cluster.
diakinesis the last stage of the prophase of meiosis in which the definitive chromosome has been formed and the nuclear membrane is still intact.
diatomaceous earth earth composed of shells of dead diatoms.
dicaryotic containing two separate and genetically different nuclei within the same cell.
dictyosome small storage vesicles associated with the golgi complex of the cell.
differentiation the development of a cell or an organism into a mature or permanent cell or mature organism.
diffusion the process by which molecules or other particles intermingle as a result of their thermal motion.
diffusion pressure the force or pressure developed when particles or ions move from a region of greater concentration to one of lesser concentration.
digestion the preparation of raw food material for absorption by the cell.
dihybrid a hybrid involving two characteristics or two pairs of alleles.
dioecious presence of staminate and pistillate flowers on different plants.
diplanetism the phenomenon involving the production of zoospores that swarm, then encyst, followed by the production of a second type of zoospore that again swarms before germination.
diploid with two sets of homologous chromosomes; the 2n number.
diplotene the stage during prophase of meiosis when the chromosomes separate.
disaccharide a carbohydrate, $C_{12}H_{22}O_{11}$, in the form of a crystalline solid, soluble in water, sweet to the taste; a sugar.
DNA (*see* deoxyribonucleic acid).
dominant refers to a characteristic dominant in inheritance, one which overshadows some contrasting characteristic, the recessive.
double fertilization union of one sperm nucleus with the egg nucleus and another sperm nucleus with the fused polar nuclei.
drupe a fruit with a thin exocarp, a thick fleshy mesocarp, and a stony endocarp.
dry fruit a fruit with a leathery, papery, or woody pericarp.
dynamic equilibrium a system in which the interior is in balance and kept in balance, by physiological processes, with the exterior.

ecology the study of the relationship of a plant to its environment.
ecosystem all the living and nonliving components of the environment of a particular kind of habitat as a pond, desert, or similar system.
ecotone the interface between biomes.
egg female gamete or sex cell.
egg nucleus one of the nuclei of the embryo sac.
elaiosome an oil-secreting body on the outer surface of seeds; the oil is an attractant to ants.
elater an elongated structure produced in the sporophyte of the liverwort and on the spores of horsetails that aids in the dispersal of spores.
embryo a plant produced as the result of gamete union and development but before rapid growth at time of germination.
embryo sac the cells in the ovule in which the embryo is formed.
embryonic region a region of actively dividing cells.
emulsion the suspension of droplets of a liquid within another liquid.
endocarp the inner layer of a pericarp.
endodermis the layer of cells just inside the cortex, particularly in a root.
endosperm the food storage portion of many seeds; a part of the female gametophyte.
environment the nonliving and living surroundings of an organism.
enzyme an organic catalyst secreted by a cell and used in the cell metabolism.
epicotyl that part of the embryo in the seed that produces the shoot.
epidermis the tissue that covers and external surface such as that in leaves, young roots, and stems; usually one layer thick and covered with cutin; primarily a protective tissue.

epigynous a flower with the ovary beneath the flower parts.
epiphyte plant that lives on the surface of other plants.
ergotism a diseased condition of animals or man produced from eating grasses or grains infected with an ergot fungus.
eucaryotic cell cell with definite nucleus surrounded by a nuclear membrane.
excretion the elimination of the waste products of metabolism.
exine the outer coat of the pollen grain.
exocarp the outer thin layer of a fruit; the "skin."
experimental design the plan for the procedure to be used to investigate a scientific problem.
experimentation to test by comparing through measurement or controlled processes to discover information or gather data; valid experiments must be repeatable.
extracellular outside the cell.
fascicular cambium the vascular cambium within the vascular bundle.
fat organic compound composed of carbon, hydrogen, and oxygen in which the hydrogen and oxygen are not in a ratio of two to one as in carbohydrates; 3-carbon compound, called glycerin, combined with fatty acids; solid at ordinary temperature; one of the lipids.
fatty acid even-numbered carbon chains, either saturated or unsaturated.
feedback the return of some of the energy output of a system reintroduced as input to effect a correction or control.
fermentation a transformation occurring in organic substances, such as a carbohydrate, caused by the action of an enzyme.
fibrous root root composed of many slender roots, most of which are the same size; typical of monocots.
filament a series of cells linearly arranged in a continuous series, e.g., spirogyra plants; the stalk of a stamen.
flagellum a whiplike thread attached to some cells (spores, sperms, and bacteria) that makes motility possible.
fleshy fruit the fruit in which all or most of the pericarp is soft and juicy.
flow chart a diagram showing the steps in data and information analysis and processing; also refers to a diagram of any system of input and output with details of the process.
flower a group of modified leaves specialized for reproduction and borne on the end of a short stem in most cases.
flowering hormones the plant hormones that initiate flower primordia and their formation and development.
flowering regulators the plant regulators that affect flowering.
fossils the remains or impressions of plants and animals preserved in the various strata that compose the earth's crust; usually extinct species are represented.
frond the leaf of a fern or similar large leaves.
fruit a fully developed and ripened ovary or ovary cluster.
Fungi multinucleate (except various slime-molds) organisms; nuclei (with nuclear membrane) within walled and often septate mycelial syncytium.
funiculus attachment of the ovule to the placenta.

gametangium a special structure producing gametes.
gamete a sex cell; reproductive cell.
gametophore the stalk of the gametophyte in mosses on which the antheridia and archegonia are formed.
gametophyte a sexual generation; gamete-producing generation; haploid (n) generation.
gamopetalous a flower with a fused or continuous corolla.
gamosepalous a flower with a fused or continuous calyx.
gaps the unfilled variation possibilities between species; the differences between species leave gaps between one series of individuals belonging to a species and a second series.
gel the solid state of a colloid.
gemma an asexual or vegetative growth from a parent body (usually liverworts) that is capable of growing into a new plant.
gene the bearer of the hereditary traits.
gene pool the total available genes in a breeding population.
generative nucleus one of the nuclei of the pollen grain concerned with fertilization.
generic name the name of a genus to which a species is assigned.
genetic drift changes in the gene frequency due to the survival of small samples of genes from year to year.
genetics the study of the mechanism of heredity.
genotype the genetic make-up; the gene composition of an individual.
gens (pl., gentes) a concept of species that includes time, space, and form.
geotropism growth movement stimulated by gravity.
germination growth or development of a plant from a seed or spore.
gibberellin growth regulator inducing increased stem elongation, particularly in dwarf mutants.
gill flat, spore-producing plate on the underside

of the pileus of agarics, e.g., the gill of a mushroom.

globule the sperm-producing structure in chara and similar plants.

glycerin part of the composition of fat; 3-carbon chain compound.

Golgi complex membranes of the vacuolar system lacking ribosomes.

gonidia special cells, asexually produced and nonmotile, that separate from the parent gametophyte to produce daughter plants.

grana small granules embedded within the matrix of the chloroplast of a cell.

ground meristem a tissue that will develop into the pith and cortex of stems.

growth an increase in mass either through cell enlargement or through increase in the number of cells by mitosis.

guard cells the cells that surround the stomata of leaves.

guttation the accumulation of droplets of water at the hydrathodes; formation of water droplets on leaves due to root pressure.

gynoecium the pistils collectively.

haploid having one set of chromosomes; the *n* number characteristic of the gametophyte.

hapteron a discoid outgrowth, highly adhesive, by which a peridiole is fixed to some substratum.

haustorium a specialized hypha of the fungi that is absorptive; a modified, suckerlike, adventitious root found in some higher plants.

helper virus a virus that combines with a defective virus for reproduction.

herbaceous stem a stem that is composed of soft tissue, usually green and not woody.

herbarium a collection of preserved, mounted, identified, and systematically arranged plants.

heredity the study of the inheritance of variations.

heterocyst a large, apparently dead cell that causes the separation of sections of an algal filament; responsible for dividing algal filaments into hormogonia.

heterogamous producing two kinds of gametes on the same vegetative plant.

heteroploidy having different chromosomes of the set present in different numbers; unbalanced polyploidy.

heterosis a characteristic that is the result of a cross between two varieties to produce a new variety that exceeds both parents in desirable qualities; hybrid vigor.

heterothallic with two different kinds of filaments or thalli.

Hill reaction the chloroplast reaction named after Robert Hill, who first discovered it; the production of oxygen by isolated chloroplasts or by preparation from them.

hilum seed scar; point of attachment of the seed to the ovary wall.

histology the study of tissues.

holdfast cell the anchoring cell of an alga, usually without chlorophyll.

homologues the equivalent of certain parts in different organisms thought to have the same evolutionary origin, although they may have the same or different function.

hormogone a cell of a hormogonium.

hormogonium small segments of cells broken from certin filamentous algae capable of growing into new filaments.

hormone a growth substance.

hybrid the result of a cross between two individuals that differ in one or more characteristics.

hybrid vigor (*see* heterosis).

hybridization the result of a crossing of two organisms that differ in one of more characteristics.

hydathodes specialized cells in leaves that are responsible for the formation of droplets of water or other fluid not associated with stomata.

hydroponics the art of growing plants in nutrient solutions without soil.

hydrotropism growth movement in response to water.

hypertonic solution a solution with a greater osmotic pressure than an isotonic solution.

hypha the filaments of the fungi.

hypocotyl embryonic stem portion of an embryo of a seed just below the cotyledon attachment.

hypogynous a flower with the calyx, corolla, and stamens arising beneath the ovary.

hypothesis the organization of available data and information into a system for testing and verification.

hypotonic solution a solution in which the osmotic pressure is less than an isotonic solution.

imbibition the process by which solids (especially colloids) adsorb water and swell.

imperfect flower a flower that lacks either stamens or pistil.

incomplete dominance the dominance of one gene over the other not complete so that the phenotype shows features of them both.

incomplete flower a flower that lacks either sepals, petals, stamens, or pistil.

indehiscent fruit a fruit that does not open or split upon maturity.

indeterminate flower an inflorescence in which the tip continues to grow so that the first flowers to open are at the base or outside the cluster.

indusium the membranous covering of a fern sorus.

inflorescence a cluster of flowers on the same peduncle.
information documents communicated or received concerning a particular fact or circumstance; information is stored and retrieved without change.
information retrieval system (IRS) a process involving computer machinery used to code and file documents and references to documents so that they may be rapidly retrieved for use.
inorganic compound a chemical compound usually not containing carbon in the form found in living organisms or their products.
input the gathering and processing of documents.
integument the covering of the ovule, usually develops into the seed coat.
interaction of genes two or more pairs of genes producing a specific effect in combination only.
interfascicular cambium the vascular cambium that is found between the vascular bundles.
internode the area of the stem between two successive nodes.
interphase the resting and growth stage of the cell between mitoses.
intracellular within the cell.
introgressive hybridization incorporation of genetic material from one population into that of another by repeated backcrossing of two distinct species.
inversion a reversal in the gene arrangement.
irregular flower a flower the floral parts of which are not symmetrical, e.g. the sweet pea.
IRS (*see* information retrieval system).
isogamete a gamete that can fuse with another of the same size, structure, and behavior to form a zygote.
isolating mechanism any of several factors that inhibit breeding or the circulation of genes in the gene pool.
isomerism the relation of two or more chemical compounds that contain the same number of atoms of the same element but differ in structural arrangement and properties.
isotonic solution a solution in which the osmotic pressure is the same on either side of the membrane.
isotope any of the atoms of an element with a different atomic mass than those normally found in a molecule.
isotope tracer an element of radioactive properties used to tag a molecule so that its metabolic pathway may be traced.

kelp a large brown marine alga, e.g., fucus.
Krebs cycle a sequence of aerobic, enzymatic reactions that effects the breakdown of the acetyl group (derived from pyruvic acid or some other source) to carbon dioxide and water with a release of energy that is used in the formation of the high-energy phosphate bonds of ATP; also called the citric acid or tricarboxylic acid cycle.

lacustrine a type of soil characterized by a gray, silty sediment found on the bottom of lakes.
land bridge a connection of land or isthmus of former existence over which animals and plants migrated to new areas.
lateral conjugation conjugation or union between two adjacent cells of the same filament.
lateral root (*see* secondary root).
layering the rooting of stems by bending them down and covering them with soil.
leaf axil angle between the petiole of the leaf and the stem.
leaf mosaic pattern formed by the leaves of a plant.
leaf scar the scar left by a leaf after it has fallen off.
lenticel a pore in the bark or epidermis of the stem through which air enters the living tissues beneath.
leptotene stage in the prophase of meiosis in which the chromosomes are threadlike.
lethal gene a gene that causes death in the homozygous state or frequently an abnormal effect in the heterozygous state.
leukoplast colorless plastid.
liana a vine or climbing stem.
lignin an organic material associated with cellulose in cell walls of plants, especially sclerenchyma cells.
ligule tonguelike appendage at the top of a leaf sheath in grasses.
***linkage* group** a series of genes on the same chromosome inherited as a unit.
lip cell thickened cells along the edge of the annulus of the fern that are the first to separate on dehiscence.
lipid one of many organic compounds that have the common property of being variably soluble in organic solvents and only slightly so in water (*see* fats).
literature information in printed form and distributed to libraries and individuals.
locules cavities in the ovary in which ovules are found.
locus a place or position; the place on the chromosome at which a gene occurs.
lysogenic bacterium a bacterium harboring certain latent viruses.
lysosome small storage vessels in some cells.

macronutrients the ten elements found to be essential for the nutritional needs of plants and required in large quantities compared to micronutrients (cf.).

macrosporangium a cell producing macrospores found in the conceptacles of some algae.

matrix the basic molecular fabric of the cell.

medullary ray vascular ray.

megaspore a spore that develops within a megasporangium and produces a megagametophyte.

megasporophyll a leaflike organ that bears one or more megasporangia.

meiosis reduction division in the germ cell chromosomes at maturation; reduction from the diploid (2n) to the haploid (n) number of chromosomes.

meristem embryonic tissue of plants undergoing cell division and differentiation giving rise to additional tissue.

mesocarp the middle layer of a fruit wall.

mesophyll the several layers of cells containing chloroplasts that are found between the two epidermal layers in the leaves.

metabolism both the constructive and destructive chemical processes (anabolism and catabolism) that occur in a living organism.

metaphase stage in mitosis in which the chromosomes become arranged in the center of the cell (on the equatorial plate).

metaxylem xylem that matures last from procambium.

microgametophyte the plant that develops from a microspore.

micronutrients elements essential for plant nutrition but only in minute quantities; trace elements.

micropyle opening in the ovule through which the pollen tube enters.

microsome small granules or particles of cell substrate composed of lipids and enzymes.

microsporangium a sporangium that produces microspores.

microspore a spore that develops into a male gametophyte; a young pollen grain.

microsporophyll a leaf or similar structure that bears a microsporangium.

microtechnique the methods of preparing material for microscope study.

midrib the largest vein in the net-veined leaf.

mitochondria minute granules, filamentous or round bodies present in the cytoplasm of the cell, the site of protein synthesis.

mitosis a type of cell division in which the daughter cells have the same number of chromosomes as the parent cell; usually differentiated into stages called prophase, metaphase, anaphase, and telophase.

modifying gene a gene having only a slight effect by itself but which modifies the expression of other genes.

mole the gram molecular weight of a molecule.

molecular biology the study of the molecular architecture of the cell, and cellular processes at the molecular level.

molecule two or more atoms of one or more elements combined to form the smallest single particle of a chemical substance.

Monera a kingdom of organisms characterized by being unicellular or a mycelium, the cells lacking nuclear membranes, mitochondria, and advanced cilia.

monoecious possessing both male and female reproductive organs on the same plant but in separate flowers.

monohybrid offspring resulting from a cross of two individuals involving only one pair of genes.

monosaccharide a simple sugar, soluble in water, e.g., glucose, fructose.

monosomic loss of one chromosome from a set because of nondisjunction during meiosis.

morphogenesis the process that involves the origin, development, and differentiation of plant tissues and organs.

morphology the study of form and structure.

multicellular the entire organism composed of a few to many cells, usually more than twelve.

multiple fruit cluster of several ripened ovaries produced by many flowers on the same inflorescence.

multiple gene gene that acts with another gene to produce a single characteristic.

mutagenic agent a substance that causes mutations.

mutation a characteristic different from any in the ancestry of the individual possessing it and capable of being transmitted to offspring.

mycelium a mass of hyphae forming a network that produces the fungus body.

n ***number*** an indication of the chromosome number without specifying the exact number of chromosomes; indicates the haploid number.

naked flower a flower with neither calyx nor corolla.

neck canal the narrow canal that extends the length of the neck of the archegonium.

nectar sugary secretion produced by the nectaries of the petals.

nectaries glandular hairs at the base of petals that secrete nectar.

nitrogen cycle the series of steps in the fixing and release of nitrogen in the environment.

nitrogen-fixing bacteria the bacteria (*Rhizobium* spp.) living symbiotically with certain legumes or freely in the soil and able to use free nitrogen gas from the air in their metabolism.

node the area of the stem at which the leaves or buds are attached.

nucellus the megasporangium of an ovule located inside the integument and enclosing the female gametophyte.

nuclear sap the watery fluid within the nuclear membrane suspending the nucleus and the nucleolus.

nucleolus a spherical or ovoid body in the nucleus that disappears during mitosis; usually associated with specific chromosomes.

nucleoprotein basic protein of the histone and protamine type and complex residue proteins found in the nucleus primarily but also in the cytoplasm.

nucleus a differentiated portion of the protoplasm within a cell that generally controls the metabolic and reproductive activities of the cell.

nucule the egg-producing structure in the Charophyceae.

nullisomic loss of both chromosome pairs of the set.

nutrition all the processes involved in bringing food to the individual cells of an organism.

oidium a thin-walled hyphal cell formed from the fragmentation of a hypha or from an oidiophore.

oogonium a unicellular female gametangium in which one or more eggs are produced.

oosphere an unusually large, nonmotile female sex cell.

oospore a thick-walled spore that develops from an oosphere usually after fertilization.

open system a system depending on input and an energy source.

operculum the lid that covers the moss capsule.

opposite leaves two leaves (or buds) present at the same node on opposite sides of the stem.

organ a group of tissues combined to perform a specific function.

organ system a group of organs interconnected structurally and functionally, e.g., the flower of a plant.

organic compound substance that contains carbon, except carbon dioxide and the carbonates that have a mineral origin.

organism any individual plant or animal.

osmosis the diffusion of water from a region of greater water concentration to one of lesser water concentration through a selective membrane.

osmotic pressure the pressure resulting from the resistance offered by a cell to the diffusion of water during osmosis.

output the results of information and data processing.

ovary the enlarged base of a pistil that bears the ovules.

ovule portion of the ovary that develops into a seed.

oxidation the chemical union of oxygen with some other substance with the release of energy.

oxygen cycle the circulation of oxygen molecules through plants during photosynthesis and respiration.

pachytene the stage during prophase of meiosis when crossing-over takes place.

paleomagnetism the direction of the magnetic poles of the earth shift from time to time as shown by the record in ancient rocks by the direction of their magnetized particles.

palisade layer one or two, sometimes more, rows of mesophyll cells found just beneath the epidermis.

palmate palm or hand-shaped, with a number of veins radiating from the petiole of the leaf.

palynology the study of pollen, particularly ancient pollen and its distribution.

parenchyma a simple tissue composed of large thin-walled cells undifferentiated in shape and containing conspicuous intercellular spaces.

parietal on the wall of the ovary.

parthenospore an unfertilized oosphere that develops a new organism.

passive absorption absorption of any substance by simple diffusion.

pathogen an organism capable of producing a disease in another organism.

pedicel the stalk of an individual flower of an inflorescence.

peduncle the stalk that bears a single flower or the main stalk of an inflorescence.

peptide an amide derived from an amino acid.

perennial a plant that continues to grow for several to many years, producing flowers and seed after the first few years; a woody plant.

perfect flower bearing both stamens and pistil in the same flower.

perianth the calyx and corolla, collectively; covering of an archegonium.

pericarp the wall of a mature ovary.

pericycle a layer of parenchyma cells found between the phloem and the endodermis of roots; the outermost boundary of the stele.

periderm a term that includes cork, cork cambium, and phelloderm.

peridiole a free chamber containing basidiospores, as in the fungi.

perigynous a flower in which the sepals, petals, and stamens encircle the ovary but are detached from it by a cup-shaped receptacle that rises above the base of the ovary.

perithecium a spherical or flask-shaped ascocarp with an apical pore.

permeation the passage of a substance through the cell membrane.

petal one of the floral parts just inside the calyx, usually white or colored.

petiole a distinct stalk supporting the blade of a leaf.

phage another name for a bacteriophage.

phanerogams a name sometimes applied to the seed-producing plants.

phelloderm the secondary tissue formed by the cork cambium at the inner surface.

phellogen the cork cambium.

phenotype visible hereditary characteristics; the appearance of the individual.

phloem a complex vascular tissue of a plant; the product of the vascular cambium; chief function is to conduct food down the stem.

photoperiodism growth and development responses of plants to different periods of exposure to light.

photosynthesis the manufacture of carbohydrates from carbon dioxide and water in the presence of chlorophyll, utilizing light as the energy source.

photosynthetic capable of carrying on photosynthesis; manufactures food in the presence of light.

phototropism growth movement induced by light stimulus.

phycocyanin a blue pigment present in some of the algae.

phycoerythrin a red pigment found in some of the algae.

phyllocaline a complex substance, probably a purine, necessary for the growth of the mesophyll of leaves.

physiology the study of the life processes and functions of the entire plant, as well as of the individual organs and tissues.

phytohormone any plant hormone.

phytol a long-chain alcohol derived from hydrogenation of vitamin A.

phytophagous organisms that feed on plant parts.

pileus the umbrella-like cap of the mushrooms and allies.

pinna a primary division of a pinnate leaf; a leaflet.

pinnate divided or arranged like a feather.

pinnately bicompound the primary leaflets are compound.

pinnately compound a series of leaflets arranged on opposite sides of the leaf rachis.

pistil ovule-producing part of a flower consisting of a single carpel (simple pistil); composed of a stigma, style, and ovary.

pistillate flower having pistils but no stamens.

pith parenchyma tissue located in the center of a stem surrounded by xylem.

pith ray the same as a medullary ray.

placenta the organ of the ovary that bears ovules.

plankton free-floating aquatic organisms.

plant hormone any hormone found in plants.

Plantae multicellular or organisms with walled and frequently vacuolate nucleate cells and almost always with photosynthetic pigment in plastids.

plaques open spaces in a bacteria colony where lysis of bacteria has taken place through viral action.

plasma membrane the interfacial membrane of a cell.

plasmodium a multinucleated mass of naked protoplasm; the vegetative stage in the life cycle of the slime mold.

plasmolysis shrinking of the protoplasm from the cell walls as a result of loss of water through osmosis.

plastid organoid in the cytoplasm of the cells of most plants (Plantae) and many Protista, usually containing chlorophyll; the area of metabolic activity concerned with the synthesis of carbohydrates.

pleiotrophic effect a single gene controlling both morphogenesis and physiology as a single character, although these may be unrelated; e.g., rate of growth and leaf shape.

pleuropneumonia-like organisms (PPLO) a small viruslike organism, actually a small bacterium.

plumule the bud of the embryo, composed of apical meristem, that will develop into the shoot.

plurilocular with more than one locule.

polar nucleus one of the nuclei of the embryo sac that unites with the other polar nucleus to form the primary endosperm nucleus.

poles the opposite ends of a cell during mitosis between which the spindle appears.

pollen a microspore and young male gametophyte of seed plants.

pollen grain pollen.

pollen sac microsporangium, containing pollen, located within the anther.

pollen tube tube that contains the sperm nucleus and extends down through the style.

pollination transfer of pollen from stamen to pistil.

polymorphic more than one form or habitus in a single species.

polypeptide a compound that is composed of amino acids but has a lower molecular weight than a protein.

polyploidy some multiple of the haploid number of chromosomes other than the diploid.

polysaccharide the combination of several monosaccharide units, e.g., cellulose, starch.

polysomic a diploid individual with one chromosome represented three or more times.

pome fruit in which the seed cavities are surrounded by the fleshy floral tube and receptacle.

population a breeding population is a unit of reproduction held together by bonds of fertilization and hereditary descent with a continuity in space and time; the individuals are variable within the population; populations differ one from another and may or may not be connected in space.

PPLO (*see* pleuropneumonia-like organism).

prickles sharp, pointed, weak outgrowths of the stem not associated with vascular tissue.

primary permanent tissue permanent tissue formed by the apical meristem.

primary root the main root.

procarp a stage in the development of the carposporophyte of certain algae.

procaryotic a condition of a cell in which the nucleus is not differentiated.

proembryo immature embryo.

promeristem a region in which growth takes place by increase in number of cells rather than by increased cell size.

prophase the stage in mitosis in which the chromosomes become distinct and organized.

protein organic compound composed of carbon, hydrogen, oxygen, nitrogen, and usually sulfur; mainly composed of amino acid chains.

prothallium (*see* prothallus).

prothallus the gametophyte stage of ferns and other plants resulting from the germination of the spore; bearing sex organs; a prothallium.

Protista a kingdom of living organisms characterized by usually being unicellular or colonial-unicellular with true nuclei and chromosomes.

protonema the threadlike growth resembling an alga from which the moss gametophyte develops.

protoplasm a term applied to the entire contents of a living cell.

protostele a stele, in which pith is absent, formed by xylem surrounded by phloem.

protoxylem primary xylem or the first xylem to mature at any level of the root or stem.

pycnium in rusts, a pocket at the surface of the host in which pycniospores are produced.

rachis a part of the leaf or flower to which the leaflets are attached; main axis of a compound leaf; petiole of a fern frond.

radiation of species the adaptation of many closely related species to a variety of habitats so that they have become reproductively isolated. e.g., the species of *Iris*.

radicle embryonic root.

raphe a ridge on a seed leading away from the hilum or scar of the funiculus.

realm a geographical division of the earth based on the faunal components.

receptacle the expanded end of a stem to which the floral parts are attached.

recessive one of contrasting characters not apparent in the presence of a dominant gene.

record any durable packet of information used in communication of information.

refugia isolated areas harboring plants and animals not exterminated by a previous glaciation or simular factor.

region of elongation that portion of a growing root or stem in which cells elongate.

region of maturation in a root, the same as root hair region.

reproduction the process of duplication of individuals.

residue the individual amino acids of peptides.

respiration the production of chemical energy from the organic components of living protoplasm within organoids.

rhizocaline a growth regulator that is synthesized in leaves and necessary for root formation.

rhizome a horizontal underground stem.

rhizomorphs a mass of twisted hyphae behaving as an organized unit such as is found in the mushroom.

ribonucleic acid (RNA) one of the two nucleic composed of nucleotides formed of a nitrogenous base, the sugar ribose, and phosphoric acid.

ribosome small bodies composed of ribonucleic acid and protein, the site of protein synthesis.

RNA (*see* ribonucleic acid).

root a primary organ of vascular plants the functions of which are anchorage and absorption.

root cap the group of protective epidermal cells surrounding the root tip.

root hair slender outgrowth of the epidermal cells in the zone of differentiation of roots.

root hair region the zone of differentiation in the root tip where the epidermal cells become specialized as absorbing cells; area in which root hairs develop.

root pressure the pressure caused by the movement of water into a root.

root tip the meristematic apex of a root in which mitosis is occurring.

runner a long, slender, nearly leafless horizontal stem.

sap water and minerals or water and sugar flowing in the vascular tissue of a plant.

saprophyte an organism that obtains its food from nonliving organic matter.

scalariform conjugation conjugation between cells of two different filaments giving the appearance of a ladder.

scales small appressed leaves or bracts, usually vestigial.

science knowledge that has been accumulated by the use of the scientific method and organized under the direction of a scientist.

scientific method the system used to investigate a problem or answer a question by gathering data and information, coding this for processing by forming a hypothesis, repeating to gather more data or confirm the information, and presenting the results for use in other investigations.

scientific name (*see* specific name).

scion the shoot or stem that is detached and then attached to the stock in grafting.

sclerenchyma supporting tissue of dead cells, the walls of which are uniformly thickened with lignin.

secondary growth an increase in the amount of vascular tissue in a stem or root.

secondary permanent tissue permanent tissue formed by the cambium and other meristematic tissue.

secondary root a root lateral to the primary root.

secretion release, by a cell or multicellular gland or a cell part, of a manufactured substance, including also its production and passage.

seed embryonic sporophyte in the dormant condition, the endosperm or food storage in the embryo, and the seed coat.

seed coat the modified integument of the seed, usually composed of the integument and the nucellus.

segregation the term applied to the separation and distribution of homologous chromosomes among the different gametes during meiosis.

selectively permeable the ability of a membrane to allow the passage of some substance but not others.

self-pollination transfer of pollen from the anther to the stigma of the same flower or another flower on the same plant.

semipermeable permeable to some substances but not to others.

sepal the outermost floral part; a division of the calyx.

septate divided into cells; having crosswalls in reference to hyphae.

sere a stage in succession before reaching a climax.

sexual a type of reproduction involving the fusion of gametes or sex cells and their nuclei.

shoot the stem and the leaves of a plant, especially those newly developed.

shrub woody plant that usually branches profusely at or near the ground level.

sieve tube a series of phloem cells with perforated end walls arranged end to end to form a tube.

simple fruit the result of the maturation of the ovary of a single pistil.

simple pistil a pistil with only one carpel.

simple tissue a tissue in which the cells are essentially alike in structure and function.

sol the fluid state of a colloid.

sorus a cluster of sporangia on a fern frond.

source document any representation of information that becomes input to a data processing operation.

species (pl. species) kind of organisms composed of interbreeding individuals with more characteristics in common than with any other organism; genetically related organisms.

specific name the name of an organism in Latin composed of the generic and the species names; the name of the species.

sperm male gamete or sex cell.

spermatium a nonflagellated male gamete or gametangia in the red algae and some fungi.

spindle threadlike condensation of the cytoplasm formed during mitosis and to which the chromosomes become attached.

spirillum corkscrew-shaped bacterium.

spongy layer the layer of cells of the mesophyll of a leaf that contains large air spaces.

sporangiophore a structure that bears one or more sporangia.

sporangium a capsule or structure in which spores are produced.

spore an asexual reproductive structure, usually unicellular, that is capable of giving rise to a new plant.

sporogenesis the formation of spores (*see* sporulation).

sporophyll a spore-bearing leaf.

sporophyte an asexual generation; spore-producing generation; diploid (2n) generation.

sporulation the formation of spores.

stamen pollen-producing organ of a flower; usually composed of an anther and a filament.

staminate cone microsporophylls of conifers.

staminate flower flower with stamens but no pistil.

starch sheath the inner layer of cortex when it contains large amounts of starch.

statospores resting spores characteristic of the diatoms.

stele the conducting region that occupies the center of the root, stem, and leaves, and composed of xylem, phloem, pericycle, endodermis, and pith.

sterigma the short stalk that subtends the basidiospore, the conidium, or the sporangium.

sterol complex ring compound that is chemically related to fats but resist saponification, e.g., cholesterol.

stigma the part of the pistil that receives the pollen, usually located at its apex.

stipe a stalk such as that found in the mushroom or algae.

stipule small leafy structure at the base of the petiole of a leaf.

stock in grafting, the basal portion of the stem to which another piece, the scion, is attached; in genetics, a race or group of genetically similar organisms.

stolon a horizontal rooting shoot in higher plants (also called a runner); in fungi, a horizontal hypha.

stoma opening in the epidermis surrounded by two guard cells.

stone cell sclerenchyma cell that is similar to a fiber but much shorter; form the outer stony parts of the pits of cherries and the gritty portion of pear fruit.

strobilus a conelike collection of sporophylls borne on a stem or axis.

style the stalklike portion of the pistil that bears the stigma.

suberin waxlike waterproof substance present in the walls of cork cells.

subfractionations the particles of a cell as separated by an ultracentrifuge.

substrate the substance acted upon.

succession the passing through seres toward a climax.

swarm spores zoospores freely swimming after release from the cell chamber.

symbiosis two kinds of organisms living together with some mutual benefit.

sympatric population of two species with overlapping ranges.

synapsis the pairing of homologous chromosomes preparatory to meiosis.

syncytium a tissue or structure containing many nuclei but not divided into separate cells.

synecology the study of groups of organisms that are associated together as a unit.

synergid one of the two functionless nuclei of the embryo sac.

syngamy the fusion of the sperm and egg nuclei in the cytoplasm in sexual fertilization.

systematic botany a study of classification, identification, evolution, genetics, ecology, and distribution of plants.

taproot root in which the primary root is enlarged and the secondary roots are few, as in many vegetables, i.e., biennials.

taxon a group of organisms named and assigned to a category, e.g., Rosaceae is a taxon of plants of the category family.

taxonomy the classification and identification of plants; the cataloging of the species.

teliospore a thick-walled resting spore found in the rusts and smuts in which nuclear fusion occurs.

telium the group of binucleated cells that produce teliospores.

telophase stage in mitosis in which the chromosomes are regrouped in the daughter nuclei and the nuclear membrane reappears.

tendrils may be either modified stems, leaves, or leaflets, or stipules that are capable of twining around some object.

terminal bud the bud located at the very tip or apex of a stem.

tetrad a group of four cells, usually spores, produced by two divisions of a spore mother cell; also a group of four chromosomes formed at meiosis.

tetraspore one of the four spores in a tetrasporangium that will develop into the gametophyte.

tetrasporophyte a plant that resembles the gametophyte in the red algae and produces tetraspores in the tetrasporangium.

thallus a simple plant body lacking true roots, stems, and leaves; characteristic of algae, fungi, liverworts, and so on.

theory the result of the testing and verification of hypotheses by the addition of more data and information; a concept.

thorn a modified stem forming a sharp-pointed woody structure of various sizes and shapes.

tissue a group of similar cells having the same function.

toxin poisonous secretion of a living system.

trace elements elements necessary for plant growth but used in extremely small amounts; micronutrients.

tracheid the fundamental type of xylem cell that is elongate with tapering ends and, when mature, is nonliving.

translocation the movement of food from the point of manufacture to the site of its use within the plant.

transpiration the loss of water vapor from a leaf by evaporation and diffusion.

tree a woody plant whose main stem rises above the ground before it branches.

trichogyne the distal receptive end of the ascogonium.

trichome a hair or bristle; a filament of globose cells.
trifoliate three leaflets attached to the petiole of the leaf.
trihybrid a hybrid for three different characteristics.
trumpet hyphae filaments with large cells inflated at ends that serve in a vascular capacity in some brown algae.
tube nucleus one of the nuclei of the pollen grains not involved in sexual fusion.
tuber a thick, fleshy stem modified for food storage.
turgid a cell is turgid when the pressure resulting from osmosis is great enough to exert a force against the inner surface of the cell membrane, as in a hypotonic solution; distention as a result of turgor pressure.
turgor pressure the pressure developed as a result of diffusion of water into a living cell; pressure exerted against the cell wall.

ultrastructure cell components smaller than the usual range of the light microscope, or from about 2,000 Å to 2 Å.
unicellular the entire organism composed of one cell.
unilocular with a single chamber.
uredium a short filament of binucleated cells producing uredospores.
uredospore a binucleated spore produced by the uredium.

vacuolar system organoids of the cell that serve as storage and communication channels.
valence the degree of combining power of an element or a radicle.
vascular bundle a group of conducting tissues (xylem and phloem) found in a plant.
vascular cambium the meristematic tissue that produces phloem toward the outside and xylem nearer the inside; responsible for the lateral growth of stems and roots.
vascular ray thin radial layers of xylem and phloem extending from the pith to the cortex.
vascular tissue conducting tissue of stems, roots, and leaves; xylem and phloem.
vasculum a metal can in the shape of a flattened cylinder with a side door and a shoulder strap, used to store plants in the field that are gathered for pressing later in the laboratory.
vegetative body the part of the plant concerned with all functions except reproduction.
veins vascular bundles in the leaf that can be seen as ridges on the surface.
venter the enlarged end of an archegonium within which the egg develops.
vessel a conducting tube in the xylem.
vine a climbing stem.
viron the mature virus outside a cell.
virus a noncellular organism-like unit composed of a nucleic acid core and a protein shell functioning only within a host cell.

weed an undesirable plant, no particular kind.
whorls three or more leaves (or buds) located at the same node.
woody stem a stem that is usually thick and composed chiefly of xylem or wood.

xanthophyll yellow pigments present in plant cells and sap.
xylem complex vascular tissue of a plant; the product of the vascular cambium; wood; chief function is to conduct water up the stem.

zone of differentiation the area of the stem or root just behind the zone of elongation in which the cells are developing into the primary permanent tissues.
zone of elongation the area of the stem or root just behind the apical meristem in which the cells are undergoing some modification, particularly increase in size.
zoology the study of some or of all the organisms belonging to the kingdoms Protista (part), and Animalia (all).
zoospore a motile asexual spore.
zygospore a spore formed by the fusion of two gametangia that enters a resting stage.
zygote a fertilized gamete.
zygotene the stage during prophase of meiosis when synapsis takes place.

Subject index

NOTE: The index has been expanded, made more detailed, and generally cross-referenced to aid in the rapid location of information. To facilitate efficient use of the indexes, the user first should find the noun concerned, for example, "cell," and then look for pertinent modifiers or restrictors, e.g., "membrane," or "wall." Page numbers in boldface are major topics; those followed by *f* indicate a figure, or *t*, a table. Books and journal articles are listed by reference to author or editor only, except for a few journals, which, like names of organizations, are alphabetized under the first main word in the title. Organisms have been indexed separately (see p. 468).

Index to organisms

All organisms mentioned in the text, both the Latin name and the common name, are listed here. In the case of the flowering plants, in addition to the page number, the order and family number is given (see p. 390). A page number followed by *f* indicates a figure.